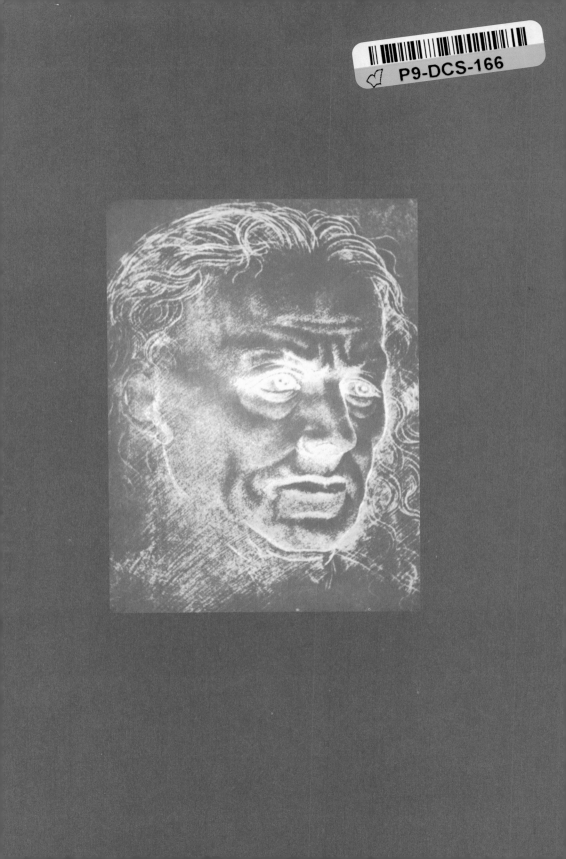

THE BODY HAS A HEAD

A _Cass Canfield_ BOOK

# THE BODY
# HAS
# A HEAD

Gustav Eckstein

1817

HARPER & ROW, PUBLISHERS

NEW YORK, EVANSTON,
AND LONDON

FIRST EDITION

LIBRARY OF CONGRESS CATALOG CARD NUMBER: 62–14529

*For Martha Keegan who has helped me to write*

# CONTENTS

## II. LIFE

### Sum of the Forces That Resist Death

## III. GROWTH

### 1 Becomes 1,000,000,000,000,000

## IV. ENVIRONMENT

### Up from the Sea

## V. MUSCLE

### Ivory Until It Moves

## VI. HEART

### Regular Day-and-Night Deliveries

## VII. BLOOD

### Five Quarts

## VIII. LUNG

### Rock of Earth in Raiment of Silk

## IX. DIGESTION

### It All Works Like a Hungry Dog

## X. METABOLISM

### Tallow to Flame

## XI.  KIDNEY

### Shrewd Housewife

## XII.  ENDOCRINE

### "La Comédie Humaine"

## XIII.  DEFENSE

### The Investment

## XIV.  NERVE

### Carrying the Torch

## XV.  BRAIN

### Cold Earth of Life

## XVI. SENSE

### Message from the World

## XVII. VOLUNTARY

### Threading a Needle or Felling an Oak

## XVIII.  INVOLUNTARY

### Night Watch

## XIX. THE HEAD

### The Rest Exists for This

# INTRODUCTION

The intent of this book is to make the human body more familiar to anyone who owns one.

During a summer vacation years ago I read back into history to find where the idea of the body as a machine began. That idea—our body a machine—has been a mainstitch in the fabric of thought for centuries. To contemporary man his body *is* a machine. He accepts this early in his life, takes it literally, not as analogy, and feels no irony. I did not read that summer as the scholar reads but as I liked, much or little, without a scholar's caution, twenty open journals or volumes lying around. Then, one day, I found myself drawn to the side, recognized that I was passing milestones on a road that led to our present-day understanding of the body, our experimenting with it, our resolve to wrench from it its secrets. The last we do sometimes with a passion that could make us wonder whether it might not be to compensate for our disappointment-to-rage at having been so coldly dropped into the unspeakable dimensions of the galaxy of the Milky Way, as if, since the century has spent so much of itself to prove itself doomed, we singly wanted the sadist satisfaction of having faced the entire doom, the entire mechanistic truth, singly.

That was how this book got started. The first sixteen short chapters, the History, are flashbacks, scant biographical sketches, each of a person who more than others was responsible for placing one milestone on the long road. Abruptly, that road looked as if it were ending in the first half of the seventeenth century, actually went on, may even have gone on more precipitately, but I could stop. The short chapters turned out not history, not biography, not pursuit of the idea of the body a machine, yet something of all three, and introductory for this book as a whole.

After that the book keeps to a line, *Life, Growth, Environment,* a special environment inside us and curious, then the organs, *Muscle, Heart, Lung,* roughly each, roughly how it works, roughly how the parts of the total work together, how the body is provided with a shrewd *Defense,* so on, short chapters, conceptions as much as facts, pictures as much as dynamics, the hope being that there will rise before the reader a better understanding of himself.

I have taught for more than a generation in a medical school, a city hospital and its clinics across the street, my subject human physiology, the everyday operations of the body. By the curriculum this was crammed into four months of one semester. Too brief, though the brevity added tension, helped at least the teacher to realize again that the little that one knows is always making one wish one knew more, making one realize how fast the detail of knowledge changes, how what one studied yesterday could seem wrong today, and how it escapes one, needs to be relearned, to escape again. The earnest student might for these reasons say earnestly that a book like this should never have been written, that science is too large, too specialized, too often maltreated by the unspecialized, should be reported in technical language by expert to expert. A commendable view. No sane author would think he could be expert or tenth-expert, or want to be, over such a sweep. Notwithstanding, were a reader to come away from any book with an overall impression of his own body, that book would require, besides sweep, selection, and an overall impression would be worth coming away with. Another student will be yet more meticulous, say that even in the time a long manuscript passes through the press science has moved elsewhere, ought in a wildly productive age like ours not to be reported in books but in abstracts with a computer to retrieve the essence. Again a commendable view. It would be more applicable to some chapters than others, and in spite of the speed of change the bastions of science do tend to stand.

A literary tone fell of itself over the writing, was allowed to, because the wonder of the body, met at every turning, inspires moods of wonder, of philosophy, of far distances, thoughts of universe or universes, of the majestic vague, frequently of our smallness in immensity, but also the reverse, of our private immensity in the presence of those particles of which there are always more and more, and of which finally we are constructed. Our age is so engrossed with those particles.

The human mind is the body's master and the book's destination. It is a mind that sometimes seems half-ashamed of its extraordinariness, forgets its unparalleled story that began in dimness so-and-so-many millions of years ago to arrive finally at the power of a single mind to write *Hamlet*. There are persons who experience no difficulty telescoping mind into body, others do. The mind-body debate in this century has not the bitterness nor the brilliance it has had, is in the opinion of many an empty subject. Yet each person in every hour lives with his mind, and an impression of the mind-body question including some of the modern evidences would be worth coming away with also.

In any case, I have written a book about his body and his head for a mid-twentieth-century reader and a book needs an introduction, and this is it.

# I
# HISTORY

Better Than Any Machine
of Human Invention

# HOMER

## World Before Naturalists

If Homer was one man, which I believe, he lived in the ninth century B.C., or the eighth, or the seventh. Herodotus thought eighth. It took a long time to produce a Homer. The mind bogs down when it tries to imagine the travail to advance from a first animal brain to a human brain, from primitive sounds and signs to a language, to an alphabet, to this poet who in words expressed a total world. Homer seems yesterday when one thinks that way.

To Homer all natural forces, the flaming sun, the shipwrecking wine-red sea, were either gods or the contrivance of gods, and so too was what went on in the houses and the streets, so too the hewing of the timbers for the building of the ships, the blowing of the bellows for the forging of the shields, the maidens down at the river doing their washing, so too chicanery, treachery, badinage, passion, victory in battle, or defeat. Homer's unequaled vigor of statement does probably express the unequaled vigor of that time. The Scamander River protested to Achilles that the corpses of his victims were causing the river to overflow, and when Achilles did not listen the river called a god to help, who flooded the land, and Achilles was saved from drowning only because he in turn called a god. Cause and effect were personal. Under their heroic masks those gods were human beings, but they did not die, and like human beings were by turns mysterious and not-at-all-mysterious. Bedroom and drawing room and courtroom proceeded on Olympus (where there was no snow, no rain, no humid hot weather) much as they proceeded here below, and since things up there did mirror things down here man took on an added stature. He was not merely of earth but of sky. "On earth as it is in heaven" would come later, but the recognition was there then.

The human body to Homer stood forth now in striding beauty and now in a heap of hacked-off limbs. Guts spilled onto the plain. A spear was thrust at one side of the skull of a hero, cracked it open, came out the other. Eyes were popped from sockets. A warrior was cut through at the neck from behind, the blade reappearing at his chin, and his body wilted. It was known to Homer that a blow on the temple or on the orbits of the eyes might wipe out life. Much like that was known, doubtless more than is said or suggested in *Iliad* or *Odyssey*. In them the record is one of gore and death and monstrous ferocity done upon the body, but the reason for its skills, its intricacies, had not yet begun to excite man's analyzing imagination. Homer's panorama was one of poetic grace and rough force, neither accounted for physiologically.

# THALES

## *Who Measures Is a Scientist*

Two hundred years passed by, or three hundred. Now there were naturalists in our sense. Trying with slaughter of sheep and heifers to flatter the gods while they were making administrative decisions on Mount Olympus was giving way to watching the storm advance in the valley. Thunder and lightning were no longer only expressions of the anger of Zeus, but thunder was wind ripping a cloud, and through the rip flashed the blue sky, which was the lightning. Fancy? No. At least, fancy was trying to come to grip with fact. The gods did not settle all things; some went on rules of their own. Among the early naturalists the name Thales stands high. Thales thought the magnet had a soul, whence its power to move iron— soul and movement, cause and effect. Thales was a Greek. He traveled to Egypt. There the Nile regularly rose over its banks, receded again, and the Egyptian knew how to measure and re-establish rapidly the boundaries of each man's land. Thales was interested in the measuring but also in the principles of measuring. That no doubt long had been part of the human mind, was more so now, would become more and more so right down into our time. Measuring could give answers. Numbers had naturalistic besides philosophic meaning. Up to then no one in the Greek world had been too much concerned with that. Thales is credited with bisecting a circle and proving the halves equal. Is credited with calculating the height of the Pyramid of Cheops from the length of its shadow, did it by a geometry of triangles simple today, not then. Calculated the distance of ships at sea as a modern surveyor might. Was a clever man. Accurately predicted an eclipse. Which suggests he was also a lucky man. Hundreds of years after his death people would say that he led thinking from theology to physiology, and though physiology could not have meant what it means to us, the reasons for the skill of the body had begun to excite man's analyzing imagination. A direction had begun to show in the history. Possibly even the gross consequences of perpetual measuring upon the measurer had begun to show; peering back from our computer age we are tempted to believe that this must have been the case, that there was loss where there was gain.

# DEMOCRITUS

## Atom

Two hundred years again passed by. The Greek was restless. The Greek was wanting to find the bottom of things, the bottom of the bottom of the bottom, wanted the ultimate, the unit of the universe, the final finality, wanted that illusion, without the worry of suspecting it was illusion. Here are some words of Democritus: "By convention sweet is sweet, by convention bitter is bitter, by convention hot is hot, by convention cold is cold, by convention color is color. But in truth there are atoms and the void." *Atomos* meant uncut. Of the atomists the most distinguished was Democritus. To him atoms were small, in rapid motion, met at random, differed in shape, and their shape determined their action. (Still believed with some changes to fit the centuries.) To Democritus soured wines and acids that pricked the tongue were composed of atoms with sharp points. On the contrary, if atoms slipped easily in and out and nudged a passage between other atoms, they must be round. Flame had round atoms and so did the soul. It followed that reality was atoms and space and atoms and space and atoms and space, and was discontinuous. (Still believed.) The search for the ultimate, the irreducible, never again would cease. It would go where Democritus' wildest imagining could not have imagined. He lived four centuries, probably, after Homer, fifth century B.C. By the twentieth A.D. the ultimate would have become a relation of energy and mass definable to the physicist, no doubt lucid to him, always somewhat murky to most of us. An electron-microscope would attempt to photograph the atom and display it before us, not succeed, quite. The unseeable nevertheless would be split. The splitting would result in a bombed Hiroshima, result in nation after nation adding itself to the proud list of those where science is paramount and bombs are built, with the probability of more Hiroshimas yet to come, and end-of-war, if achieved, would not be because thinking man had delivered himself beyond bombs but because he did not want to die. His instinct to escape destruction would have maneuvered him into a bargain with others who had the same instinct. Trapped man would be, body and mind, by atoms. Trapped so that even a solitary madman who wished to witness a bombed earth, and bombed it, when he had done so might not dare look back on what he had done, if he were there to look. Trapped man would have been by this reduction of a universe to a unit that was almost infinitely small and approached the infinitely powerful. Cheated by atoms. Cheated of his chance to win victory with his love of life by his repugnance to death. His familiarity with pent-up forces would have saved him, he not saved himself.

To Homer the pent-up had been the contrivance of gods. Poseidon lifted mountain-high the waves of the sea. To our century the pent-up was the contrivance of no one. It just was.

However, the scrutiny of the atom would have had other results, practical results, a new astronomy, a new physics, a new chemistry, a new physiology, a new medicine. Body and head, body and brain, body and mind would under that new scrutiny reveal a detail within detail undreamed of heretofore.


# SOCRATES

## *Seminar*

Socrates asked. Socrates answered. Usually he did both, asked, did not wait for an answer, answered. He was a great intellectual, who talked. The manner of the talk would be remembered through millennia, the artificiality, the annoying precision, the flashes of wisdom, the genuine feeling. Socrates was serious. Socrates was moral. Socrates was not easily turned aside, was both patiently and impatiently insistent. It was always said that his wife Xanthippe nagged, but he also nagged, and we do not know too much about her and we do know about him. Indeed, he bludgeoned men to that precision, to stating exactly what they meant. They must define each term. They must listen to each question. His own questions he slanted so that the questioned would look into himself and discover the truth that had been there all of the time. This was the Socratic method. It would survive. The relentlessness, the coolness, the vague or less vague fretfulness, would like a ghost from Attic streets be standing somewhere behind each chair at each seminar in the mid-twentieth century, at each panel of arguing experts, a poor pallid replica of the old ghost, of course.

His method was his milestone.

The bright Athenian youth flocked to him, came from all parts of the city, might come before dawn, stay through the night, worshiped him, the foolish youth, forsook felicity to learn, found felicity in this seminar of the streets. It was a situation sure to enrage someone. Furthermore, Socrates was fearless as to what he asked or whom it galled, turned on respected citizens in public places and asked them questions they could not answer, which was bad. Might be icily critical, which was bad. Might be successfully sarcastic, which was bad. He joined no party, which was bad. After a while he managed to offend all parties. The outcome? He was accused of impiety, brought to trial, found guilty. What surprises us, and surprised him, was that the vote of the jurors should have been close. He spoke of this: "Had thirty votes gone over to the other side, I would have been acquitted." Instead of regarding his trial as a blemish on Athens we should probably regard it as a tribute, consider that no later people might so far have suppressed their distaste toward an unpleasant person in favor of justice as did those Greeks. Even during his trial and after his conviction he remained his usual gadfly self, inquiring, pronouncing, persecuting words. He conducted this seminar like the others, unruffled, lucid, sometimes as a man not concerned, as if the issues debated actually were more important than whether he lived or died, as if death actually were a life-to-afterlife passageway. Even on the long day of his execution he asked and he answered. The waiting on that execution has remained one of the breathless events of history, like the waiting of Joan of Arc, of Servetus who discovered the circulation of the blood through the lungs and was burned to death for heresy. Socrates preparing to die has thrown color on all he was, and on his method that might otherwise have been too rational, too arguing-for-the-sake-of-arguing to be that thoroughly remembered. Is it necessary to die to be remembered? In those closing hours he was so human, as one says. He was gentle with Xanthippe, who was crying and beating herself, did not want her on the scene, and not his infant son. "Crito, let someone take her home." He bathed his body. This continued awhile. Later the infant was brought back and two young sons, and he talked to them and to the women of his family, dismissed them. Not much was said now, but over every phrase lay a twilight that even at our distance we could see, that pained us, that added to the meaning of the world, that modified our manner of thinking. He drank the hemlock and died.

# PLATO

## *The University*

Plato described that death. He gave us the Socrates we know best. He kept himself ostensibly out of his immortal *Dialogues* but must have realized, being the writer he was, that this would only weave him in so that even time could not rip him out. Plato reported those disputatious seminars of the Athenian streets, later brought seminars within walls, brought learning and philosophy within the confines of the Academy, that has correctly been described as the first university in the world. Universities would expand, cover the earth, become a power, sometimes good, sometimes not so good, in the spiritual then more and more in the social affairs of the millennia, and, this was sure, would increasingly nurture the sciences of living body and head.

Plato revered Socrates, did what he could to keep bright the faces of the man, but seemed himself to have been different in genius and in tact, seemed not one to have gotten himself into such a quandary that it would be necessary to test his readiness to die for his thoughts, or, having gotten himself in, would have discovered a way out. This would not mean that he might not risk what was dangerous. He might and did. But he was highly equilibrated, lived with success until he was eighty.

Plato theorized about the earthly—how it reached toward the unearthly. Theorized about government—how it could be made to bring to the greatest number of citizens the fullest realization not of their rights but of their capacities. (Worth consideration in the twentieth century A.D.) There had been theory enough before Plato—the straight thinking with the unembarrassed wrangling of the Greek and other worlds—and upon this he built. But to theorize with talent in any age one must be born with the temperament, and Plato was. He had one of the strongest and yet light-spirited intellects of all time. This has never been doubted. University dialogue would because of him, though diluted by the centuries, have more give and take, more ostensible tolerance, less humor, more urbanity, more of the conscious, less of the unconscious.

Plato theorized that beyond the apparent was the ideal, the apparent a materialization of the ideal. To say that differently, the apparent had its own inner unapparent truth. Earth, air, water, fire, each of those four of Empedocles and everything else was referable to its own perfection, the universal. Reality merely approximated the universal, mostly distorted the universal. A soul crammed into a body became merely a mind; at least, one has the faraway impression that that might have been his way of regarding mind. Often one has been satisfied that one was keeping up with his argument but not satisfied that one was keeping up with him. At least, though, he would not have wasted his time or a pupil's on, say, a particular triangle. He might use a geometric form in illustration, had curiosity about geometric form, but what interest could there be in a particular triangle? What pleasure? Euclid a century or two later might amuse himself with that. Why dwell on a single isosceles triangle when there was existent the lovely principle, triangle? Why a single puny example when there was the universal? Similarly with good and evil: not a good man, but good; not an evil man, but evil. And behind good and evil? Virtue. The sophists argued that there were many virtues. Plato and Socrates argued one. That one was the intimate of knowledge, and knowledge could be learned, and taught. Plato was a teacher and a writer who in disputatious Athens defined and redefined the universal. Whether that was for the eventual well-being of Athens has been doubted. Whether it was for the eventual growth of science has not been doubted. It was. Various scholars enjoy explaining how Plato spanned from ancient science to modern science. He had Democritus's atoms, too, but precision in their shape. He brought number to shape.

When he theorized about the human body he also brought fantasy (possibly it was not fantasy to him, but conviction) to the concrete. He would say: "The gods made what is called the lower belly to be a receptacle for the superfluous meat and drink, and formed that convolution of the bowel, so that the food might be prevented from passing quickly through, and compelling the body to require more food, thus producing insatiable gluttony, and making the entire race an enemy to philosophy and music, and rebellious against the divinest element within us." He would say, would theorize, that the flesh of the neck served as an isthmus to separate the mortal below from the divine above. He would theorize that the lungs served as cushions to prevent the undisciplined heart beating itself to shreds during bouts of passion.

Twenty-three hundred years afterward all of this might seem quaint as statement and quaint as physiology, but the direction in the history that had begun to show in Thales was plainly pushing forward.

# SOPHOCLES

## The First Freudian

Athens was a place the size of Waukesha, Wisconsin. Twenty-one thousand citizens. There were also ten thousand foreigners and forty thousand slaves. In that city in its great period were living the following: Thucydides, the almost unparalleled historian. Herodotus, historian also, his mind filled with strange biological observings and stranger biological interpretings. (Biology had begun pushing forward.) Thucydides died at seventy-one. Phidias, builder of the Parthenon, died at sixty-eight. Pindar, poet, died at seventy-four. Pericles, statesman, orator, writer, died at sixty-six. Socrates drank the hemlock at seventy. Then there was Plato, first unofficial president of the first unofficial university. And Democritus, atomist, in his imagination was seeing the atom, with it reaching into an unbeknown future. And Aristophanes, via a pricking irony letting the air out of various windbags, now and then letting blood out, as when he helped hound Socrates to his death by jabbing at him from the stage, Socrates sitting there in the audience. Aristophanes could do this and still be a poet. That was Athens. Aeschylus, the first of the three mighty writers of tragedy, lived until 456 B.C., well into the great period. Sophocles was second. Euripides third. These all were in that city. A busy place. An angry place. So distinguished. In the theatre you might turn your head and, two rows back, that gentleman there was Praxiteles, the sculptor.

Fate was substance and shadow in that theatre. Fate to us seems much the externalization of what is in our own hidden depths.

Would that Athenian audience have understood this? We do not know. Depths were in those Greeks, this we do know, and the deepest depths, the most illumined by the poetry, were in the dramatists. Revenge, jealousy, madness, war, patricide, incest, all the sweet themes of the theatre had fate beneath to start a blaze, then for dramatic contrast the blaze damped, let to smolder while the chorus lamented and retold and anticipated, the air heavy with suggestivenesses and meanings.

Sophocles was the most polished of the three. The most limpid. The most detached. The most probing. In his quiet there was always an urgency. He never let go his story. He began with what seemed tragedy's highest feelings, yet the feelings mounted, as an agonizing thirst mounts, pulled from within to the point of pain, each emotion transferred to each spectator as he sat waiting for thirst and pain.

Sophocles was an actor in his own plays. He was born at Colonus. It was said he was a handsome man. A general in the army. A rounded, a perfect Athenian. Lived to be ninety. Lived between the battles of Salamis and Marathon, therefore not only within the great period but at its peak. How we cling to every dim datum about him. How almost bland his life and still what comprehension of psychological forces. (Psychology was pushing forward.) Year after year, almost, he won first prize for tragedy. He wrote one hundred and twenty plays. Seven remain. Upon seven we base our convictions. *Oedipus Rex* was his greatest. The whole afterworld agreed but not the Athenian critics. *Oedipus Rex* did not win first prize.

Its argument ran so. Oedipus, the King, in an altercation at a crossroad slew his father, not knowing who he was, then wedded and bedded his mother, not knowing who she was. Patricide and incest. This had been foretold. Many years elapsed. Pestilence fell on Thebes. The citizens cried out to their King. An aged priest spoke for them. An oracle brought down from the Pythian temple pronounced pollution was within the city, and it must be swept out. The King vowed it should be. But what was this pollution? Where was it? A soothsayer, unwilling to tell what he knew, refused, was bitten to it by the King, who in a rage accused the soothsayer, who in turn accused the King. The King was the pollution! For this seeming falsehood the King might have struck the soothsayer dead, instead dallied with the accusation, was lured by the dalliance, fol-

lowed threads, sharpened his questions, could not halt his self-pur-
suit, by a marching compulsion uncovered fact on fact, until nothing could save him from the recognition that it was he who did the
double deed. His guilt stood forth as clear as sunlight—one thought
one saw the literal sunlight falling on that crowded theatre of
Dionysus—so clear that he ripped from his wife, who had hanged
herself, the brooches of gold that fastened her robe, lifted them
high, brought them down, gouged out his two eyes, blotted out
earth and sky, presented to his audience his white face with its two
blood-streaming sockets; theatre and tragedy.

How much of this took place at the behest of the hidden? How
much was chance? How much that appeared was appearance only?
What were these cool gods? What was this relentlessness? What
events were symbol? How much in gouging out his eyes was he
gouging out some other organ? A long and penetrating probing of
the normal and the abnormal, the direct and devious, the bizarre
ways of our human mind must have preceded the writing of such a
play. The play would haunt the future. It would be interpreted and
reinterpreted. It would rise in the psychiatry of the twentieth century, sit there in the King's place, lash on the human mind to continue its perpetual harassed self-analysis.

# HIPPOCRATES

## *Learning from the Sick*

He was Dr. Hippocrates, of Cos, Greece. (A contemporary, somewhat later, of Socrates and Sophocles.) "Redness of the face, eyes
fixed, hands distended, grinding of the teeth. . . ." That was his
economical description of a man who had had a stroke. Often of an
evening this physician walked by the harbor—the evidence lets us
reasonably say that. He was serious. He also was bearded. (Beards
since then have come and gone.) He reduced a fisherman's dislocated collarbone, diagnosed an intermittent fever. Watched sickness. Watched himself. Watched men. He said so. What unexpect-
edness he saw in a human body, heard in it, felt with his fingers, he

wrote down. He had stubborn conviction as to what the practice of medicine might teach the practitioner. "I think that one cannot know anything certain respecting nature from any other quarter than from medicine." Those words too are from somewhere handed down. They are astonishing words. He claimed no great cures. To his fellow physicians he was a fellow. To the Coan school of Medicine, that was in competition with the Cnidian school, he added instruction and fame, both schools spreading their learning and their often opposite views through the ancient world, and then through the rest of the human world.

A marble bust of him is in the British Museum, a style more than a likeness, and better so, because what do we know of him as a person? Another marble poses him with his right shoulder and his chest bare and a gown thrown over his left arm. Then, two thousand years after his death, the Spanish realist, Velázquez, painted his portrait. Hippocrates may have looked like that portrait one decides as one keeps gazing at it.

From fragments such as these one attempts to piece together his daily life.

From a vast mass of medical writings gathered under his name, the Hippocratic writings, that could not have been the writings of a single man, one pieces together how this physician probably thought.

There is an anecdote that when he was young he was an arsonist, had to flee from his town because he set fire to the library—and why he did it was quaint—so that no one might know what he knew. He got to be quite old. The years 460 B.C. to 390 B.C. are accepted. Granting that he recognized he was unusual, which is probable, he could hardly have anticipated that men at this far point of time would still be saying that he was the Father of Medicine.

Born in Cos, studied in Cos, practiced in Cos, practiced in Athens, died in Larissa. That was the itinerary.

The Hippocratic oath—repeated by medical students about to become physicians—he wrote, or part wrote, or possibly did not write, but it believably attaches to him. It bespeaks the rules of an honorable profession, and it bespeaks some of the mundane realities of the medicine of that time. Hippocrates had done what he could to rip that medicine from the supernatural. It should not lean on the supernatural. He objected to anyone saying disease was a punishment from the gods. It had natural causes. The physician must pay attention "to the patient's habits, regimen, and pursuits; to his conversation, manners, taciturnity, thoughts, sleep, or absence of sleep, and sometimes his dreams, what and when they occur; to his picking and scratching, to his tears; to the alvine discharges, urine, sputa, vomitings; to the changes of diseases from one to the other; to the deposits, whether of a deadly or a critical nature; to the sweat, coldness, rigor, cough, hiccough, respiration, eructation, flatulence. Whether passed silently or with noise. . . ." All of that might be written this noon were there a pen to write it, but there is none.

Epilepsy, he insisted, ought not to be called divine. There was no divinity in it. It was a disturbed brain. He noted how the epileptic wanted to be alone, and this had nothing to do with any fear of the divine but was the epileptic's shame at having his fit in public. Different if the epileptic was a child. "Little children at first fall down wherever they may happen to be, from inexperience, but when they have been often seized, and feel the approach beforehand, they flee to their mothers, or to any other person they are acquainted with, from terror and dread of the affection, because being still infants they do not yet know what it is to be ashamed." When he spoke like that he must have seemed a father, any loving father. He learned from all the ages of the human being. He drew truth from the disease itself, from the physician himself, from the man in the sick man.

Something like that was Hippocrates' milestone.


# ARISTOTLE

## *Animals Added*

Here is what Aristotle said: "If there is anyone who holds that the study of the animal is an unworthy pursuit, he ought to go farther and hold the same opinion about the study of himself." With his clear mind (conceivably too clear) in that later still golden age of Greece he went from animal to man, man to animal. Sharks. Rays.

Sponges. Octopi. Cuttlefishes. Sea urchins. Deer. Weasels. The energy of that biological search (not necessarily biological thought) has never been equaled. Aristotle examined all. Dragged the Mediterranean for specimens. Raked the land. The comparing of animal with animal—comparative biology—begins with him. After him the comparative would increasingly include man. So much begins with Aristotle. So much of earlier Greek searching, adventuring, finds its focus in him, and from him reflects toward what will be the decline of Greece. He was keen. He was critical. He dared to be enthusiastic, conceivably too enthusiastic, because while some Greeks were undoubtedly reasoning themselves out of all common sense, he may have worked too hard, wandered too far and wide. But, to read him is like a fresh cup of morning coffee. "The elephant's nose is unique owing to its enormous size and its extraordinary character. By means of his nose, as if it were a hand, the elephant carries his food, both solid and fluid, to his mouth; by means of it he tears up trees, by winding it around them. In fact, he uses it for all purposes as if it were a hand. . . . It is the opinion of Anaxagoras that the possession of hands is the cause of man being of all animals the most intelligent." So, already in the fourth century B.C. someone was not only taking for granted that man was an animal but was speaking freely of it, and someone was struck by the intimacy of hand and mind. "But, it is more rational to suppose that this possession of hands is the consequence not the cause. . . . The invariable plan of nature in distributing the organs is to give each to such an animal as can make use of it. . . . For, it is a better plan to take a person who is already a flute-player and give him a flute, than to take one who possesses a flute and teach him the art of flute-playing." Though Aristotle may be wrong about hand and mind, Anaxagoras right, that is not important. What is important—about Aristotle's milestone on the road running through the centuries—is how in his reasoning he went from animal to man, man to animal. "Quadrupeds find it no trouble to remain standing, and do not get tired if they remain continually on their feet—the time is as well spent as lying down, because they have four supports underneath them. But the human being cannot. . . . His body needs rest; it must be seated. That, then, is why man has buttocks and fleshy legs, and for the same cause has no tail; the nourishment gets used up for the benefit of the buttocks and legs before it can get as far as the place for the tail." Though his reasoning here may be upside-down, though we know that in obvious respects it was wrong, though we always have realized that we might not be reading exactly what he wrote, none of that is important. What is important is how in his observation and his speculation he went from animal to man, man to animal. In the sciences of the human body this comparative approach was henceforth a part of all thinking of any size. "The human tongue is the freest, the softest, and the broadest of all. . . . On the one side, it has to perceive the various tastes. . . . It has, also, to articulate the various sounds and to produce speech, and for this a tongue which is soft and broad is admirably suited, because it can roll back and dart forward in all directions. . . . That is why the clearest talkers, even among the birds, are those that have the broadest tongues." He concerned himself with everything, birds, talk, poetry. There is his realistic theory of tragedy, with which all subsequent writing on tragedy was apt to begin, even late psycho-analytic writing, the theory so trenchant in Greek ideas of art that one has needed sometimes to remind oneself that Aristotle did not write the great Greek tragedies, only confidently explained them, left us his theory of the tragic.

The gods on Olympus, throughout, were like shadows fading before the accumulating light. They were stepping down not without dignity but stepping down. The ruthless myths were losing their ruthlessness and other ruthlessness was coming in. Aristotle, with his force, his intellect, his statement, was supporting the new facts, or the new illusions that would grow in power and produce eventually naturalistic science and mechanistic logic. He was a naturalist who embraced not a fraction of nature but the whole sweep. He was a logician who did himself formulate the famous first principles of logic. Around him was still the youth, the gaiety, the lightness of Athens, but these were lessening, and there might in the future be left only the logic, and logic without youth, gaiety, lightness, and creation becomes mathematics, analysis, presumption, and peremptoriness.

Aristotle had unique talent, was highly intelligent, possibly was never quite mature, and he was fearless. A fearless boy can be rasping. His reasoning about tongue and talker may be upside-down. We know it is in part wrong. But that is not important. What is important is how in his reasoning he went from animal to man, man to animal.

# LUKE

## *Made in His Image*

Greece declined. The myth and the reality were in the glory of the sunset, where they would stay. The Hebraic world had begun earlier and was where it was. Rome had grown to  power. Julius Caesar was born and died. Jesus Christ was born and died. That last gave a twist to history. Christianity's strengths, its weaknesses, its sentiments, its sentimentality, its tendernesses, but especially its way of regarding the sick—not the newness in that but the stress put upon suffering and dying—would affect all later thinking about body and mind.

Luke, the beloved physician, what do we know of him? Turn to the Bible. It will surprise you. You will be sure there was more and that you have forgotten. Humility he had, we may say this confidently because of the Christmas story, but if Bible fact is looked for, there is not much. Nevertheless, we were not wrong in what we thought we remembered, because in Luke, the physician, we saw Christ, the physician. Luke had taken from Christ the compassion that we thought was Luke's alone, and of Christ we knew much, howsoever we interpreted what we knew. Another meaning had come to the physician. Mortification of the flesh. The halt. The lame. The blind. The healing of the leper. Made whole by faith. The Samaritan. The resurrection. There were a hundred new phrases, with many implications. The idea and the ideal of the human body were changing, changing. It never again would be what it had been to the Greeks. That shine was off it. It had become a garment. It was dust. It was a thing to throw away. A mystery touched it, gave it importance, reminded it of its brevity, then left it. There had been similar thinking in different corners of the world, but this now was sewn with such insinuation into the cloth that it would have taken a fervor equal to that which sewed it in to pluck it out, and there were no Greeks to do it and they alone would have had the intensity and the cunning; probably not even they. A Greek might have spoken of pallor in Christ's face as Luke spoke of it. A Roman might have. But what they saw there and what they thought of it would not have been what Luke saw and thought.

# GALEN

## *The Clinician*

Rome in turn declined. The Western Empire—that we mostly think Rome—passed its peak.

Dr. Claudius Galen, of Rome.

For four years Galen had been surgeon to the gladiators in Pergamum, where he was born, must often have looked across the sea to Rome. "That is the size of city for me." Six years, then, this Greek physician practiced in Rome with a wild vitality. After that, we do not know much of his whereabouts but what he was at we know. He was writing the most voluminous physiological and clinical work of all antiquity, volumes of blunt fact and free interpretation; with this was welding an authority that would reach through twelve centuries. "Galen and Aristotle," men said. "In the name of Galen." "Galen did." "Galen spoke." "Galen wrote." The writing reveals his range.

*Animal Experimentation.* He probed the alive animal—watched its organs while they were in action. Was saturated with the temperament of the vivisector. Cut into swine and oxen and the Barbary monkey without benefit of anesthesia. Cut open the alive skull, looked in, saw the pulsations. Cut the nerves to the larynx, paralyzed the voice, concluded that the control of the voice must be from somewhere above where those nerves came out of the brain. Associated what he saw in the vivisected with what he saw in the sick human being. Dr. Claudius Galen, of Rome. Before death a patient had had difficulty breathing and at post-mortem, Galen recollected, a tumor had been found in the brain, reported the facts with cool clinical-pathological-conference acumen. Later, the part of the brain where the tumor was pressing or gouging would be known as the respiratory center. A slave, Galen recollected, had committed suicide by holding his breath.

*Anatomy.* Probed the dead human body—was saturated with the temperament of the dissector. Of the twelve nerves that go from brain to head, trunk, chest, abdomen, he knew seven. (This was the second century A.D.) Left rules for dissecting a brain. Gave permanent names to parts of it. Knew two of the three membranes that wrap it.

*Physiology.* Probed the alive human body—was saturated with the temperament of the medical investigator. Probed with imagination. Was interested always in the work of the kidney. "Now, the amount of urine passed every day shows clearly that it is the whole of the liquid drunk which becomes urine, except for that which passes away with defecation or passes off as sweat. This is most easily recognized in winter in those who are doing no work but are carousing, especially if the wine be thin and diffusible. These people rapidly pass almost the same quantity as they drink." An informal medical observation but important. Later, physiologists would be measuring the quantities exactly in hundreds of circumstances, would be continuing a line that can be said to join the measuring of Thales to the measuring of Galen to the measuring of any modern anyone cares to mention.

*Obstetrics.* Probed the parturient—was saturated with the temperament of the obstetrician. "The midwife does not make the parturient woman get up at once and sit down on the chair, but the midwife begins by palpating the os as it gradually dilates, and the first thing she says is that it has dilated enough to admit the little finger." Obstetricians would be saying that eighteen hundred years afterward, tonight.

The sick are always with us, but the talent of a Galen occurs only sporadically through centuries.

Galen's activities in that Roman world went all directions. His mind went all directions. His rationalized explanations went all directions. He was vivisector, dissector, physiologist, obstetrician, neurologist, psychiatrist, every species of clinician, businessman, and writer. The writing was savagery sometimes. Apparently he had a bad disposition, thought like Rome, not like Greece. He detested, harangued. One fancies one hears him. One sees the man in his literary style. Meanwhile, love thy neighbor was whispered in the streets, and religion also added its tone to this complicated personality that would dominate twelve centuries—dominate them the more because of the platform from which he spoke, a Greek clinician speaking from a Roman world into the millennium that followed the decline of the Western Empire, the speech rumbling and thundering and instructing.

His milestone was conspicuous. He sank it with assurance. He seemed by intellect and temperament to anticipate the modern multiple approach to the understanding of body and head.

# DE CHAULIAC

## Plague

A man lies sick. A doctor has been sent for. This now is a thousand years after the fall of Rome. The doctor arrives on horse. He does not dismount. He is as groomed as the horse. A member of the family comes to the door. From the horse the doctor makes some formal inquiry, hardly deigns to look at the member of the family, nods to his assistant, who is not on horse and stands at a respectful distance. The assistant acknowledges the nod, disappears into the house, takes the sick man's pulse, asks him to stick out his tongue, does this (we of our age immediately suspect) for the effect on the sick man's mind, reappears at the door with a sample of the sick man's dejecta. He on horse bestows a glance on the dejecta. Knows the date of the sick man's birth, the hour, the astrological conjunctions, the magical and alchemical possibilities, if the man was rich has already sent his ring for blessing to the King, by that combination of actions is expecting to change the humours, finally goes into a séance with himself, comes out with a prescription for rhubarb. There were exceptions, but the common run of doctors conversed of Galen and Aristotle, the realist Galen, that different realist Aristotle, and they knew less and less of both. As for the living body, they rarely laid hand on it, and as for the dead, the dissection of two corpses in a lifetime might give a man eminence, at least add eminence to one who already had it. This was true for Guy de Chauliac, who practiced a brilliant surgery in Avignon, France, between 1342 and 1368. It is necessary to view the body for its wonder, as the Greeks did, or think it the be-all and end-all, as our century mostly does, in order that medicine may reap its greatest successes. One cannot think it gross and still concentratedly study it. This would not mean that the Middle Ages, the Dark Ages, were not finding other facts, especially anatomic facts about the body, and facts about man's mind, the potential in his reasoning, the potential in his character. Discipline was exercised for the sake of discipline. Logic was practiced for the sake of logic.

Also, the Dark Ages were not dark, not literally. The sun did rise every morning. Birds sang. There was a hot bun at supper. Babies laughed, fell asleep, the moon came out, its magic drove lovers into each other's arms. And doctors continued to converse of Galen and Aristotle, and to dispute of the status and the rights of barbers versus physicians.

300 A.D. 400 A.D. 1100 A.D. 1300 A.D. Then—plague struck Europe. Bubonic plague. Black Death. The Great Dying. It struck again and again. In the fourteenth century. In the fifteenth. In the sixteenth. The first definite European date was 1347. Earlier it had ravaged Asia and Africa. Apparently it was brought to Europe by ship late that year. The next three years, 1348, 1349, 1350, were the frightful years. A quarter of the population of Europe was wiped out. And plague kept striking. It burnt out in Sicily. It struck in Italy. In Spain. In Oxford, England. It kept striking. Eighty thousand dead in Danzig. Forty thousand dead in Paris in the one year 1466. Before the entire toll had been taken sixty million were dead. We of this century have not yet been able to match this even with a war. The hearts and the eyes were on the ground. "The Lord gave, and the Lord hath taken away; blessed be the name of the Lord." Stark death was on every side, in the streets processions of flagellants whipping each other, strange dancings to exhaustion, roving mystical cults. Read Defoe's *Journal of the Plague Year*. Recall Boccaccio's *Decameron*. See again Raphael's agonized painting *La Peste*. The sick were deserted and their houses burnt. A starving wretch

touched the stone of a house where plague broke out in the night and he was hanged next day for witchcraft. Another was hanged for pollution of the wells. Elizabeth I shut herself up in Windsor Castle, a gallows outside for anyone who dared come there from stricken London. Whole towns were evacuated. Transportation stopped. Speech all but stopped. The doctor, poor pompous ass with an engraved voice, stood revealed. The bankruptcy of his calling stood revealed. He saw nothing, knew nothing, was nothing. There were powerful exceptions. Guy de Chauliac, the surgeon, was one, and rare. The human body had become a shroud. What the doctor believed the unchallengeable truths of Galen and Aristotle was a heap of broken type. This fellow could not perform the simplest palliative acts. He could not ever have thought his way (even if the antecedent facts had existed in his mind) from the disappearance of plague each cold winter and its reappearance each warm spring, to fleas on a black rat. He died and was himself buried with the priest. The medicine of the Middle Ages was largely buried with him. The disillusion of human beings in their doctor swept medicine toward the modern era where all the sciences one after another and sooner or later would come in or be drawn in to the study of the living machine. Not plague alone wrought this change, because the revival of learning touched medicine as all else, but plague put the granite hand into the doctor's silk glove, one must believe.

Guy de Chauliac, the surgeon, did not shut himself into a castle, did not run away from Avignon, as did others of his profession when plague struck their cities. He stayed in Avignon and worked. People called him Guy.

# PARACELSUS

## *Metals in Medicine*

Dr. Philippus Aureolus Theophrastus, of Basel, Switzerland.

This one was crazy, a little. And so sane. We are in the second quarter of the sixteenth century. Milestones fly—in the pages of history. It was St. John's Day. At the celebration of the students at the medical school, by way of launching his first course of lectures, Theophrastus Bombastus von Hohenheim (Paracelsus) threw into the fire the books of Galen. In the announcement of his lectures he stated that they would be not from Galen but from Paracelsus, would be delivered not in Latin but in the vernacular, German.

This was the new world. This was the end of the millennium. This was air let into the palaces, the houses, the churches, the polluted wells, the cellars.

The birth of Paracelsus occurred in Einsiedeln, a small place. It was the year after Columbus discovered America. Paracelsus' father was the village doctor. Later the family moved to a mining town where there was a mining school, and Paracelsus learned chemistry, especially of metals, worked in the mines and in the mine laboratory. Then began his *Wanderjahre.* All his life he enjoyed travel, and traveled. "Those who sit behind the stove eat partridges, and those that follow after knowledge eat milk broth." His contemporaries thought him braggart and violent, and he never said or did anything to give them reason to change their minds. "After me and not I after you—ye of Paris, ye of Montpellier, ye of Swabia, ye of Meissen, ye of Cologne, ye of Vienna, and those that are on the Danube and the Rhine, ye islands of the sea, thou Italia, thou Dalmatia, thou Athens, thou Greece, thou Arabia, thou Israelite, after me and not I after you—there will be none of you remain in the farthest corners on whom the dogs will not . . ." Here was a man, who was the doctor, in the presence of another man, who was the patient, and the doctor would not have denied, indeed would have proclaimed, that his personality was part of his cure, often all of the cure. But, the question was: "Did he have his medical diploma?" That worried respectable people at the time, has ever since. According to Paracelsus he did have. According to him he studied far and wide. "I therefore attended the universities for many years, in Lisbon, through Spain, through England, through the Mark, through Prussia, Lithuania, Poland, Hungary, Wallachia, Transylvania, Croatia, and the Wendian Mark, as also other countries not necessary to enumerate. And in all corners and places I industriously and diligently questioned and sought for the true and experienced arts of medicine. And not only with the doctors; but also with the barbers, surgeons, learned physicians, women, magicians who practiced that art; with alchemists; in the cloisters; with the noble and the common, with the wise and the simple." He ruled out neither magic nor astrology as possibly intruding in human affairs, but practiced as if nature and natural law alone were to be met head-on, or their aid solicited.

It had got to be the year 1526. Paracelsus was preparing to settle down as a doctor in Strassburg, but was invited to the medical school in Basel, accepted, and that was when he threw the books into the fire. One sees him there, with his Henry VIII grandeur and thrust of body and dress, and one understands that such a one might not wish to rise from the ash of dead books. The faculty at Basel did not enjoy him. The great humanist Erasmus did. The faculty wanted him dropped. He petitioned city council. It would not have looked well for city council to have supported the faculty immediately. An event mediated. A burgher of Basel had been ill and no doctor had helped him, so he promised a reward to any doctor who would, one hundred guldens. Paracelsus cured him. The burgher, cured, paid ten guldens. Paracelsus took the matter to court. The judges decided in favor of the burgher. Paracelsus told the judges what he thought of their justice, therefore must steal out of Basel under the cover of night. He had been in that city, usual city, two years, and faculty and council were rid of him without the slightest self-sullying act.

In spite of all the energy he flung around him Paracelsus had energy left for experimental investigation. He was a doctor of medicine who was a chemist in an era when it was rare for a doctor of medicine to be a chemist, turned to profit what he had learned in the mines, started chemical pharmacology. He introduced metals into the pharmacopoeia. No one yet knew that a tiny amount of some metal might be essential for life, the body that lacked it not surviving, but we cannot be sure of what Paracelsus did not know, or guess. Galen had prescribed mainly herbs; Paracelsus prescribed antimony. Paracelsus was the first to employ the word *zinc*. He argued for the healing value of salts, sulphur, mercury. His arguments do not sound like our arguments, though to hear any similarity one might need to hear both arguments four hundred years hence. What is certain is the heritage that this heretic handed down to us to this day, and one knows that if one had been ill in Basel and

free to call whomever one liked it would have been Dr. Theophrastus Bombastus; in Rome, Dr. Galen; in Athens, Dr. Hippocrates.

Professional modern opinion passing on Paracelsus' skeleton has
held that he did not die of a blow on the head. (The blow was
purported to have been received in a tavern brawl.) We are not
sure, but in any case the hour came that always seems to him to
whom it comes to come too soon. Throughout his life he had been a
loyal Catholic though he had burnt a papal bull before Luther—not
inconsistent for Paracelsus. Throughout his life he had been no
atheist though he had slashed savagely at all the churches—not
inconsistent for Paracelsus. His death took place at Salzburg. His
last will and testament he had dictated to a notary three days
earlier. The words on his tomb are: "Here is buried Philippus
Theophrastus, distinguished Doctor of Medicine, who with wonderful art cured dire wounds, leprosy, gout, dropsy, and other con-
tagious diseases of the body, and who gave to the poor the goods
which he obtained and accumulated. In the year of our Lord 1541,
the 24th day of September, he exchanged life for death."

Plague still was abroad that September. Epidemics still would
come and go, though never again with the rapid rush over the world
as in the fourteenth century. The fresh intelligence of  Paracelsus
would have done more than who knows how many hundreds or
thousands of physicians to light the spirit that would draw medicine
and physiology and the causes of disease out of corners.


# LEONARDO

## *The Human Body in the New World*

The Renaissance arrived early in Italy. It crept in. From imper-
ceptible sources suddenly dawn was everywhere. In that soft light
all human beings must have been observing the human body for the
artists to have observed it with such fervor. Art had become again a
force in the advancing history. How quiet Leonardo is. His very
name suggests quiet, Leonardo da Vinci. The *Mona Lisa* took four
years to paint. We know without Vasari telling us that the model

was a lovely woman and that her portrait resembled her. Vasari said of the portrait: "To look closely at her throat you might imagine that the pulse was beating." This would not have been said two centuries earlier, the human eye not having then that quickly shifting vision, for art, for science. And true, it could occur to one that a surgeon cutting into that throat would find close by the vagus nerve and the jugular vein the pulsing carotid artery where that body landmark ought to be. But, he would not find her secret. The *Mona Lisa* differs from her model in that she will not die. (Sometimes genius is both so out-of-this-world and so enduring that one could fancy it to offer evidence for immortality.) When Leonardo was a boy he studied painting with Verrocchio. In the course of the years of his life he designed mills, invented instruments, drew plans for a war tank, a steam gun. He thought like an artist, and thought like a scientist, employed both capacities as they could be in the Italian Renaissance, had violence, had intelligence, had peace of mind. He constructed clay models, sketched with color on cambric and linen, talked, talked, talked. How convincing that talk must have been. He persuaded a man that a certain temple might be lifted from its foundations and moved to another place, and the man believed it, did not realize how grossly impossible it was until Leonardo and his voice had vanished. He slid open the door of a vendor's bird cage, let the birds escape, paid the vendor, vanished. He had mind and time for play. It would have served him right if Fate that sixteenth-century morning had released into the sun a silvery airplane, let that cross the Florentine sky, the phenomenon to Leonardo's eyes an impossibility, but Leonardo's imagination for the mechanical at once setting to work to make the impossibility possible. He himself drew plans, every schoolboy knows, for a flying machine. The psychic laziness of that physician on horse was gone from the world for a while, as was the aggression of Paracelsus that seemed always needing to prove itself, and there was left this controlled intensity of Leonardo. De Chauliac had dissected two human bodies. Leonardo dissected thirty. He brought the knowledge learned from the dead to the living, his paintings and the drawings of the living having always under them the firm structural framework that he had found in the dead. He has been blamed—and when he was coming down to his own end it seems he blamed himself—that he started so much and left so much unfinished. Some persons blame him still. However, had he finished what he started, and one doubts he ever meant to, the loss would have been ours, because we would have been trading for completed works that consumed his strength the wonder of beholding the human body in the many guises he beheld it in, at the peak of inspiration, unbelabored, unchilled. The shapes that the mouth takes in battle. The unseeing eye of a thoughtful child. Heads rotated in passion. Torsos twisted. Arms forcibly flexed. Abdomens bent. Figures kneeling in adoration. These are as perfect often as those of the Greeks, never as young, more actual, more touched by something that we do not miss in the Greeks, a darker psychology, and this not in the face alone but falling like a shadow over the whole. Christ as he sits at the middle of the table in *The Last Supper* definitely has a body. Above it is a human head, a human mind, a human sadness, and this does not subtract but adds to that body's divinity.

# VESALIUS

## Dissection in the Morning

It would have been one of the sights of this world—Andreas Vesalius, the Belgian, dissecting. A legend (or is it history?) has it that a robber was hoisted by a chain and slowly burned to death, a roast for wild birds that picked and picked till finally there was left the skeleton. Vesalius never could resist a skeleton. At night he let himself get locked outside the city walls, tore off single bones, brought these one after another into the city as there was opportunity, assembled them, remounted them, then, on being questioned by the authorities, presented them with the skeleton. So goes a legend (or is it history?). It was sin besides outrage to hash the image of God. True, the Church had become more liberal, this now well into the sixteenth century, but there was always the question of what a corpse would do at the Resurrection if its parts got scattered.

Andreas Vesalius dissecting—in Paris. In Paris labored the famous anatomist Sylvius. Sylvius read aloud in the gradually dis-

credited texts of Galen while a demonstrator cut at a corpse to
demonstrate what Sylvius read. If a part mentioned in the texts
could not be found in the corpse, and if Sylvius was forced by the
headstrong Vesalius to admit an error in Galen, he commented that
man's body had changed since Galen, and not for the better. Sometimes Galen had dissected a dog body, and the demonstrator was
dissecting a human body.

Andreas Vesalius dissecting—in Padua. He had received an appointment at the University of Padua. He impressed everyone.
There could be no two opinions, Italy was the center of the world in
anatomy, Padua the center of Italy, Vesalius the center of Padua.
He was twenty-three. Corpses, corpses, corpses, he wanted corpses.
A legend (or is it history?) has it that he stole them when he could
get them in no other way, waited for the living to die if they looked
enticing, cut them up, explored the anatomy of the human being as
it never had been, threw in a force that made the dead all but
literally come alive. The errors in Galen mounted. There were
hundreds. By 1543 Vesalius's immortal work, his atlas of anatomy,
*De Humani Corporis Fabrica,* was published. His fame reached
over the earth. He was twenty-nine.

Back in Paris, Sylvius was continuing to read Galen and continuing to denounce Vesalius. Sylvius hated Vesalius. Many anatomists
may have hated him—and why not? He withdrew to Spain, to
Madrid, had been summoned there as physician to the court, had
become less the anatomist and more the great and, possibly, the
fashionable doctor. He lamented that in Madrid he could not "get
hold of so much as a dried skull, let alone the chance of making a
dissection." His unprecedented talent was no longer exercised, a fact
that might have turned anyone's wisdom to peculiar uses. Back in
Belgium, Maximilian d'Egmont was ill. Vesalius not only predicted
that Maximilian would die but when, the day, the hour. "The dread
anticipation occupied the Count's mind. He called his relatives and
friends to a feast, distributed gifts, declared his last wishes, took
formal leave, waited with suppressed emotion, and at the hour predicted, died." All young doctors ought to be told this, and some old
doctors too, reminded of the power in a prognosis.

Throughout these latter years a pupil of Vesalius, named Fallopius, quietly dissected. Fallopius published his own atlas of anat-

omy. Fallopius was not as great an anatomist, not by far, but he had some greatness, and he was working. Vesalius began to hate Fallopius. A final legend (or is it history?) has it that Vesalius got permission to dissect a corpse in order to discover what had been the nature of the malady, and when the corpse was opened the bystanders saw the heart beat. Unlikely, but Vesalius had been an envied man and the luster was off him, and the youth. The Church felt defiled. The family of the deceased was not satisfied with a prosecution for murder and appealed to the Inquisition. The Emperor intervened in Vesalius's behalf, ordered him to make a pilgrimage to the Holy Land, and it was on his return from there that either he was stricken with fever or he was shipwrecked, swam ashore, died of exhaustion, this on the Mediterranean island of Zante, and a ship-companion, a stranger, a foreigner, placed a stone by the dissector's body. *Requiescat in pace.* The accuracy or inaccuracy of the legends touching this great man seem of no importance now, though they add their quality to a biographical sketch, and help the figure of the genius to stand forth from this special background. His was a milestone so clear that it would compel the future to be clearer still, and anatomy often to speak directly of the body's action, the machine's design reveal what the machine does or can do.

# HARVEY

## *Quantitative Experiment*

It would have been one of the sights of this world—William Harvey vivisecting. Frogs. Pigeons. Eels. Crabs. Oxen. Sheep. Shrimps. Serpents. Slugs. Wasps. Hornets. Flies. Chicks. Snails. Mice. All of them have blood in them and William Harvey's sharp dark eyes looked into them all. The variety of the animals turns one's thought back to Aristotle—nearly two thousand years before. Harvey was using all of them to quench his curiosity, that seems quenchless, to find every proof he could that the flow of the blood was continuous, from the heart back to the heart, an irregular circling but a circling.

This was the seventeenth century—Shakespeare's great decade near the beginning of it, Newton's near the end.

Harvey spoke of the difficulty his eyes had to follow the beat of an animal's heart. Today, any freshman medical student follows with ease, sees—because he knows beforehand what he ought to see. To see first, first in the world, anything, is psychologically a completely other process than to see second. Also, to see with those eyes that always measured was a completely other process than to see with a seventeenth-century pedagogue mind, or any century's pedagogue mind.

Add a pointed dome of head, a lean face, a clean figure, short, a man who likes to carry a dagger and in the Stuarts' and Cromwell's London must.

Harvey measured in a human corpse the quantity of blood the heart can hold, then in a sheep the quantity its heart can hold. The two quantities were comparable. Next he bled a sheep. The sheep's entire blood was not more than four pounds. At each beat the heart forced out some. The total forced out in half an hour must be greater than all there was in the sheep or—Harvey stated this in different ways—must be greater than the sheep's body needed for nutriment, greater than could be produced by the nutriment eaten by the sheep, greater than the weight of the sheep. Therefore, the blood pumped out by the heart could do nothing else than return to the heart. Harvey calculated this carefully. He made it inescapable that that blood must return. It was ejected, went in a circle, re-turned. It circulated. The blood did circulate. Harvey was not the first to use that word *circulate,* and he was not the first to conceive the blood as departing from a point and again arriving at that point, but he was the first to prove it. With many reasonings and testings he established the fact of the circulation of the blood, and he established, we rightly say, the quantitative experiment. This was Harvey's milestone. With the quantitative he carried the history plungingly forward. He brought number to the sciences of the body. Hereafter the living would be investigated with mathematics. Mighty discoveries would accrue from that, though mathematics at a later date would sometimes become dogma and men not see what was before their noses because their yardsticks got in the way.

Add to this brief portrait the detail that completes all our human brevities, the signs and symptoms of an apoplectic stroke. He discovered that he had a "palsey," "and he saw what was to become of him, he knew there was then no hopes of his recovery. . . ." The apothecary was sent for "to lett him blood in the tongue, which did little or no good; and so ended his dayes." The hour that seems to him to whom it comes always to come too soon, had come to Harvey. It was the year 1657. Harvey was born in 1578. So, he had reached seventy-nine, a good age even in our day. Shakespeare was fourteen the year Harvey was born. One wonders what the boys and girls in their respective English streets thought of those two, William Shakespeare, William Harvey. Perhaps not much. Newton would live till 1727, be eighty-five when he died. Shakespeare was dead at fifty-two.

# DESCARTES

## *"Give Me Matter and Motion"*

I said in the Introduction that I had read back into history to find where the idea of the body as a machine began. I had thought it must have aggregated during centuries. But no. At least, a first statement came full-blown from the mouth of one man. "The body is a machine made by the hand of God, infinitely better than any machine of human invention." Those were the words of René Descartes. They would be repeated, paraphrased, stand somewhere at the back of all later thinking and experimentation on the animal body, including man's. Henceforth the body—viewed from so many sides through so many centuries—was a machine. The idea was a part of the modern mind. It would also help in its way to steer that mind toward its twentieth-century dogma of meaninglessness. Descartes would have regretted this.

A geometrician, Descartes invented analytical geometry. A physiologist, he wrote the original description of a reflex. A philosopher, he arrived at the conclusion *Cogito, ergo sum,* three Latin words known to many who know no Latin. Contrive another three, just three, three words, put them in as happy an order, and men will be speaking your name A.D. 2500, if that happens to interest you. "I think, therefore I am." For a time Descartes had been a soldier. Then he seemed to experience the need to learn everything afresh and from the foundations, a need some men experience, and in order to accomplish it settled away from his countrymen, far away, out of Paris, out of France, in Holland. His mechanistic thoughts he expressed neatly. "Give me matter and motion and I will make a world." Expressed them too neatly, possibly, but that was his temperament, Gallic, thrifty, terse.

Those years in Holland were happy years, a village, a house hidden by sand dunes, the sound of the breathing sea playing an accompaniment to his mathematics. Small warm wonderfulnesses are in every one of our lives, along with the rest.

The seat of the soul to René Descartes was in the brain. He knew where. Was geographic about it. The pineal gland. The pineal was not paired as were other parts of the brain, was located in the center, and he mistakenly believed it to dip into the fluid-filled cavities that Galen had described. To a mathematical mind like Descartes's the fact of this one structure in the midst of the twos of the brain seemed proof. Descartes's reasoning went as follows. The soul was in the pineal. He did not say the soul was the pineal; it resided in the pineal, temporarily; he was a Catholic and his faith was firm. The soul stirred. The pineal moved. It acted like a valve, guided the flow of the vital spirits through those cavities, the ventricles. Everybody in that day believed in vital spirits. These traveled out the hollow tubes of the nerves—everybody believed in hollow tubes—and at the other end of the tubes the muscles contracted. As simple as that. As machinelike. The human machine to Descartes was a mechanical machine, whereas to us it is in addition chemical, thermal, electro-magnetic. The dog to Descartes was a machine, too, but no soul. In the pineal of the human brain, then, there was a device that enabled the body to make material contact with the immaterial. Descartes compared other devices of the body to the devices in the Royal Gardens: if a visitor stepped on a particular tile a Diana sprang automatically up in front of him, and if the visitor rushed after the Diana and stepped on another tile, a Neptune intervened with his trident to block the way. The fountains round about were playing.

Descartes lay on his couch. He did not doze as he lay. He kept thinking. He wrote. During those twenty years in Holland he went back to Paris only three times.

In body-height he was average. His dress was fashionable. He had an impressive shape of head. His eyes were the eyes of meditation, the eyelids prominent.

Letter writing was the chief means of spreading knowledge in that busy century. Anyone wrote freely to anyone. Many all over Europe corresponded with Descartes. Christina, Queen of Sweden, was wanting to know how to live happily and still not annoy God. Who has not wanted to know that? The French Ambassador advised her to write to Descartes. She did. It flattered him very likely, a queen, and soon Christina was inviting him to Sweden. It proved a long journey, and when he arrived in the autumn of the year he regretted he had come, found Sweden cold, especially the hearts. A Frenchman, and fifty-three, old for that day, he was accustomed to rising late and having breakfast in bed. Christina, a Swede, and twenty-three, preferred her lessons by candlelight before sun-up. At that wretched icy hour Descartes had to walk to the palace, or ride in a sleigh, caught a cold which became pneumonia. The Queen--we are all so foolish—sent to this Frenchman a Dutch doctor. That was nearly to send a German. The doctor concluded he must bleed the Frenchman. Moribund, Descartes rose from his bed. "You shall not sacrifice one drop of French blood." He said that over and over, lingered between delirium and lucidity, commended his soul to God, and died. This was 1650. The Queen—we are all so foolish— wanted to bury this Catholic in a Lutheran graveyard. To that the French Ambassador forcibly objected and the corpse went to the graveyard for those children who had not been baptized, the found- lings, the little bastards. Years later, Descartes's greatness evident in the world and in France, his body was exhumed, taken to Paris, buried again, and as if that were not enough, exhumed again, buried again, in St. Germain des Prés, lies there now, what is left of it. Impossible to escape the irony that this corpse, which during its lifetime had been the shining new philosopher who proclaimed across the centuries and down into our day that the body is a machine, should have been thus bumped from junkyard to junkyard.

# II

# LIFE

---

Sum of the Forces
That Resist Death

# ASTRONOMER

## *"In the Beginning"*

To the astronomer—what is life?

Before World Wars I and II had shrunk our earth to the United States and the Soviet Union, before astronauts and cosmonauts and intimacy with the moon, and radio astronomers with big-dish telescopes or horn antennae reaching over billions of years, and X-ray detectors perched on rockets and shot into space and sending back their news, before astronomy had given up enlightened star-gazing and become the present-day fighting science, a man in a parlor car reading the five-cent *Saturday Evening Post* might start a conversation that today would make a high-school student shudder.

"Where do you think life began?"

"You mean—just life?"

"Life."

"Read once germs floated in from out there—you believe it?"

"That would only shift the scene."

"You mean, how'd it begin out there?"

"Right."

It is likely that at the top of our atmosphere particles are escaping, others arriving. Some of the arriving might be spores (seeds) and they can be tough, might not be killed by the cold of space, might slip alive through those belts supposed lethal that wrap the earth. Life, by that notion, might have arrived ready-made. Newswriters report scientists as worrying that when the spacemen arrived from the moon and other places they might carry along infections worse than the plague, also contaminate the places they came from. Earth contaminate the solar system! Following that arrival of the seeds, allow the right number of years, millions or billions, and there would appear on this promontory a geranium, a fish, Julius Caesar, Bob Hope. The whole likelihood is not likely at all.

Many in that parlor-car era believed something else, that the earth and every sparrow on it was created at a stroke, as were night and day, the date April 1, 6000 B.C. Some scientists, later, also believed in creation at a stroke, but an earlier date, April 1, 4,600,000-000 B.C. Others believed that too recent, put it 10,000,000,000 B.C. To the nonprofessional it often seemed the astronomers were just picking dates, giving one at the meeting of the American Association for the Advancement of Science in December, another at the meeting of the American Astronomical Society in the spring. Previous to any date there would have been nothing, and out of the nothing a universe, with stars, satellites, quasars, galaxies, meteors, comets, pulsars, collidings everywhere, grand commotion.

Some believed that there was an original lump as large as the orbit of Jupiter, highly compressed, a density of billions of tons per cubic inch, matter and anti-matter squeezed tight, tighter, tightest, then an explosion, called big bang, energy enough released to produce the universe we think we live in, radiations traveling and traveling and stimulating the eyes and thoughts of twentieth-century astronomers. After the explosion there would have been an expansion, temperatures rising, in five minutes all common elements, the molecules of earth and air and sea with their atoms and the contents of their atoms, the entire marvel blasted from that lump, that vagrant. If 'twere so, 'twere fast. If 'twere true, 'twere odd. Others believed, instead of an original lump, an original gas and dust, a grand nebula, that shrank, and out of the shrinking came the astronomer's universe, small nebulae, suns, our sun, life. Others, not necessarily thinking of the Bible, mumbling over their coffee, repeated that life always had been, always would be, no beginning, no end. However generated, life would have gone on generating, and the fact that that primordial life was not found anywhere was because the earth, sterile once, had become overrun by bacteria that ate the primordial as fast as made.

A famous scheme was by a Russian. He began with an earth. It had a temperature not too low in winter, not too high in summer, not too long a day, not too long a night, steam steaming off oceans that covered much of its surface. The sun, younger then, its heat kindled by bursting bombs as one gas changed to another, neutrinos burning protons at the furnace's center, the earth's position relative to the sun producing tropics and arctics, and the moon "with white fire laden" ripped from earth and leaving a hole that is the Pacific, or hijacked from space, or some other explanation, shedding its beams and drawing up tides and letting them slide back, winds tearing, sea currents driving, volcanoes erupting, lightning crashing, radiation from the lightning, radiation from the cosmos, radiation, radiation, radiation, and there came to pass somehow somewhere sometime, first hesitantly, then boldly, that which is only less aggravatingly strange than mind, life.

Who has not asked himself whether some star out there, one of our galaxy of the Milky Way, or some cluster of galaxies, or some of the billions within reach of our telescopes, or the unreachable that may be billions and billions of light-years away, did not on this or that of its unseen planets have the conditions that wombed life here, so wombed it there, a different life but life, different living bodies but bodies, different minds but minds, then highly technological single minds, they not understanding us, we not them, this sequence occurring in who knows how many corners of the universe, and everywhere planets exploding and throwing out meteors with the possibilities of life, hence life and life and life, it getting so cheap.

A distinguished American astronomer thought life less probable among the stars we see but probable among those our telescopes do not see. There would have been mother stars and baby planets, and on one or many of them, life—subhuman, human, superhuman, extrahuman—and again after a time mind, sane mind, insane mind, genius, a Galileo, many Galileos annoying their fellow-citizens with arguments as to the character of the heavens from their positions and signalings. Evidence of interstellar signalings of natural origin, as astronomers say, or from intelligent beings, as a few risk to say, has excited investigation with the largest telescopes, computers thrown in, making it understandable that a Harvard astronomer would reasonably deduce even an academic life, comfortable chairs, some lofty creature sitting out yonder signaling, some lofty creature at this end listening with college-professor aplomb for a college-president's "I say there!" It would be like the first time a telephone rang in our house.

There remains the person convinced that our earth is the one-and-only, segregated, hoarding, nothing like us among masers and galaxies, and that this is the way things ought to be, man the center of earth, it the center of the stars, the dome of the sky once every twenty-four hours turned by a hand on a crank. That egotistic narcissistic autistic paranoiac, the 1969 psychoanalyst assures us, dwells in the unconscious of every one of us. However that, the life of earth began on earth, and a long time before René Descartes's machine better than any of human invention.


# PHYSICIST

## *"Wherein There Is Life"*

To the physicist—what is life?

He may be the lean Princetonian. He may be the MIT engineer. He gazes not so much outward into incredible space as inward into the incredible atom.

Erwin Schrödinger was a physicist, died not long ago, born Austrian, small-bodied, had a roaming imagination, highly cultured, could reflect over the whole area that a good mind reflects over, in mid-life thought once he would change his profession from physics to philosophy, teach physics and think philosophy. He was thirty-nine years old, to be exact. He stumbled, one feels, onto his equation for wave mechanics, became one of the world's greatest physicists, won a Nobel Prize, for nearly twenty years remained an émigré in Ireland, returned to Vienna to live out what was left, not much. From Dublin where he was teaching at the university he released the manuscript of a short book, *What Is Life?*, that quickly became a minor classic. It did not require the reader to be a physicist, and he would learn what Schrödinger thought the physicist might contribute to the question.

Electricity, magnetism, gravitation, heat were chapters in our old physics textbooks. Today's textbooks take up the same themes in the light of later developments, also take regular side-trips into the atom and over the chalkboards is written piously Einstein's $E = mc^2$.

None of this seems promising as contributing to the question. Schrödinger too thought that the old physics we studied in high school and the Soviets study in kindergarten would not help much. The old was based on statistics. He described situations where the disorder of random movements is given statistical order. Life is not like that. Life draws order from order. One need not be a physicist to see orderliness persisting in families over centuries, as the Hapsburg lip, *Hapsburger Lippe,* which fascinated Schrödinger as it might anyone who in some museum has stood before the portraits of Velázquez, fearlessly painting those Hapsburgers who march on parade before him. But, right in our own neighborhood, despite crossbreeding, reckless diluting of wives with husbands, there will bob up in a thirty-second cousin an unmistakable pair of blue eyes.

Schrödinger considered the unit of that old physics to have a pattern that kept repeating, like wallpaper. The unit of the new had no such repeating pattern. It was like a tapestry of Raphael. The chromosome belonged to the new.

Each chromosome—and this now was unexpected—Schrödinger suggested, was an aperiodic crystal. The physicist studies crystals. The aperiodic would be those where the facets do not repeat. (Since the publication of Schrödinger's book the study with the X ray of crystallized organic molecules has made possible the most important genetic discovery of our time.) Inside that aperiodic crystal would be the genes, that make us what we are, characteristic for characteristic, and by ceaseless duplication keep us what we are, you and I and the kangaroo. Those cocky gene-molecules bequeath our past to our future, predict and produce each member of the President's cabinet, each player on the Detroit baseball team, each pilgrim in the *Canterbury Tales.* Each gene-molecule to the physicist is as clear as the tip of Manhattan, though that tip is large and the atom small. Schrödinger asked: Why is an atom so small? He answered: It is because we measure everything in relation to our own body, which is large. He repeated the story of the English King who stretched out his arm and established the yard as the distance between his chest and his fingertips. Nowadays we accept that a grouping of atoms joined with other groupings of atoms can produce in nine months a summing-up like Johnny Appleseed.

Schrödinger touched then on evolution. He called ours the age of the evolutionary idea, his carefulness placing evolution not in the realm of reality, where of course he thought it was, but in the realm of ideas. The new species resulted not as a selection from among the continuous slight variations always occurring in the living, as Darwin believed, because those slight ones are not inherited, result from quantum jumps. Nature for some still unexplained reason operates in jumps. If quantum does apply to evolution, and it does, that again justifies the physicist trying to help us comprehend the life that is in us, though our passion to comprehend more than we already do, in a bad hour strikes us as pathetic—large comprehending brains but a span of life too short.

Schrödinger does with his vivid writing get into us a picture of a physicist peering at life.

He concludes with entropy. He defines it. He clarifies it. He employs some mathematics, but he never allows it to become too difficult. By the dictionary entropy is the number that expresses the unavailable energy in a thermodynamic system, not an exhilarating definition. What Schrödinger does is emphasize the tendency of all physical systems to go into disorder. Order goes toward disorder. It goes toward what he calls positive entropy. The living is not like that. In the living, order goes to order. Life has absolute order. Life draws order from order. The living creature, or that molecule, that aperiodic crystal, is the most orderly thing on earth. How does the living keep up its order? By feeding on negative entropy, Schrödinger says. And so long as it does it avoids maximum positive entropy, which is death. Sounds like the reasoning of a medieval monk in a black habit and a black hood. The regularly-added can be tea with a slice of cake. The winner of the Nobel Prize suggests just that: the positive entropy that is the cost of living is canceled out by the negative entropy brought in on a dinner plate by the white-coated waiter. Nothing topsy-turvy in preventing death by adding life, though it surprises us that it should be woven into a physicist's *What Is Life?* It surprises us also that this is not a romantic science-writer writing the book but a neat-minded mathematically-minded physicist. Schrödinger was above all a poet, able to feel wonder, state wonder. Already in the eighteenth century Marie François Xavier Bichat, another poet, came to a similar conclusion, defined life as the sum of the forces that resist death.

# CHEMIST

## *"Whose Seed Is in Itself"*

To the chemist—what is life?

The chemist gives us our clothes of rayon or dacron, our artificial arteries of the same materials. Cleanses our skin with detergent. Manufactures foods without soil, rain, sun. Accounts for our immunity—part way. Accounts for our heredity—part way. Sometimes his success leads him into extravagance, like the nightmare thought that he will alter our chromosomes, manipulate them as old Gregor Mendel his peas, pick out a bad gene, replace it with a good one, then gets alarmed at a Meet-the-Press; if he has provided us a new society will our moral stature be equal to it? On the occasion of a late chemical synthesis the *New York Times* noted that experts "feel it conceivable that man will be able to make an exact duplicate of a genius, such as an Einstein, with DNA." No doubt, keep sperm and egg cells on ice until the propitious moment for bringing them together. One shrugs. One shudders. Desoxyribonucleic acid (DNA) is the molecule. It is the double helix. It has us jailed. We are forced to stay in it while it methodically goes on methodically making exact copies of itself, in you, in me, keeps you you, keeps me me.

Long back the chemist plunged into the fight against disease. Viruses were one cause. He studied them. He came to understand more and more about them. He discovered how sexual those trifles are. He went back into history and calculated how ancient they are. The virus of the common fever blister was annoying the two Misses Monkey, who were wanting to look their best at the coming-out party, millennia before anything like us had put in our appearance. We appeared. We became infected. Proven virus diseases ran to hundreds: yellow fever, smallpox, poliomyelitis, measles, German measles, mumps, influenza, herpes, respiratory-tract diseases, digestive-tract diseases, then animal diseases, foot-and-mouth disease, cattle plague, hog cholera, and insect diseases, and plant diseases, among the last a kind called mosaic disease because there grew on the tobacco leaf a mosaic that spoiled it for smoking.

Tobacco mosaic virus was investigated early by a winner of a Nobel Prize. He purified it. He crystallized it. A ton of sick tobacco plants yielded a teaspoonful of crystals. Everybody at the time was repeating that. Those crystals might have been put into a bottle, the bottle put into a tomb of an Egyptian Ptolemy, who put strange things into their tombs, and after twenty-two hundred years, a lonely crystal taken from the bottle, encouraged to be intimate with an innocent tobacco leaf, conceive on it, and the dance go on, a dance on the coffin lids of leaves.

By usual criteria a virus does seem alive. It can exist like a parasite (live on life), multiply inside of life (its seed is in itself), infect (show aggression), mutate (change to another species). Accept life as host—breed inside the host—evolve a new kind inside the host—murder the host—go for a vacation to the bottle. All of this with the years was worked out in sometimes strange detail. A raw virus penetrates part way into a bacterium, leaves its protein coat outside, and inside the bacterium breeds, and soon there are many, and every one of the many has a coat. One is ashamed that one gets used to the strange.

Looked at from the other side, a virus is an exact molecule, a matter-of-fact chemical. It is nucleic acid plus a protein. The Nobel-Prize-winning chemist back there—that was 1935—already considered the virus to be the missing link between the living and nonliving, which was exciting in 1935, almost dull today. How to define life nevertheless still gives trouble. In 1935 we were taken by the hand, told of virus molecules larger than some bacteria, others smaller than some protein molecules, and that life was a region along a scale of sizes, low on the scale the lifeless, high the living. Too simple. Not true. But as correct as it could be in 1935, or as mistaken, and no one in an hour of scientific enthusiasm should allow himself to forget that everything is correct or incorrect or more correct or less correct always in relation to its time and place in history.

It is years since a chemist first startled us by bringing together chemicals abundant eons ago (methane, hydrogen, ammonia, water), flashing them with an electric arc, laboratory lightning, and producing what and how, as he fantasied, had been produced at the start, life. Molecules had cooked themselves in the direction of life. (No chemist would say cooked.) Eons earlier the inorganic presumably was cooked in the direction of the organic. And eons earlier still something else presumably was cooked in the direction of the inorganic. Earliest—hydrogen. A universe all hydrogen. Before hydrogen was the vague. The vague → hydrogen → other molecules → the inorganic → life → mind.

Chemists speak of two billion, six billion years for these accomplishments, molecules, then more molecules, then giant molecules from pigmy molecules, then life from giant molecules, hence you, hence me, hence mad dancing Vaslav Nijinsky. Two billion would strike one as an adequate span for the juggling among pre-molecule and post-molecule possibilities if the First Book of Moses called Genesis is to have scientific credibility. Jehovah, recall, did the whole of it in six days.

Virus. Gene. Cancer. Life.

Those four, sometimes friends, sometimes foes, that chemist winner of the Nobel Prize predicted would prove interchangeable, and it should please him today for seeming to have predicted so well. Notwithstanding, when chemist sits down with chemist to lunch, what they piece together, can that be it? Has not something critical got lost? Are those chemists themselves convinced? Will they be satisfied tomorrow morning? Would some chemist entertain us, please, drop something into a pan, stew it awhile, something else with Top Hat, White Tie and Tails leap from the pan and sing and strut for us, abruptly in its strutting stop, wane into sadness, wax into gladness. Whatever contempt a chemist must feel for a non-chemist who expects him to produce in a pan what took an astronomy and a physics and a chemistry incalculable ages, it would be logical, and fashionable, in 1969 to stand to one side and admire a virus as a chemical, then go to the other side and admire it as life. A modest chemist somewhere is dreaming he will fabricate the Top Hat too. Perhaps he will. My God, perhaps he has!

# CYTOLOGIST
## *"And He Rested"*

At last, at last—but that was long ago—Nature achieved it. After thousands of millions of years of selection, or thousands of billions, if that sounds bigger, she earned her Seventh Day. She built a spot of life, and it had a wall. There it is—under the microscope.

Before the gossipy seventeenth century no eye ever had seen through such a wall into such a house. Nothing to see with! Microscopes did not arrive at a fell swoop, yet by the seventeenth century Leeuwenhoek the Dutchman owned 247, had himself ground 419 lenses, and Jonathan Swift was versifying:

> So naturalists observe, a flea
> Hath smaller fleas that on him prey;
> And these have smaller still to bite 'em;
> And so proceed ad infinitum.

Swift was right. Swift anticipated how the human eye and mind would travel inward, inward, inward, and Leeuwenhoek had begun it. Leeuwenhoek had magnified the spot. He had made the invisible visible—with the light-microscope. Later the spot would be magnified thousands of times, and one day a hundred thousand and even hundreds of thousands—with the electron-microscope. In a drop of pond water there were more lives than even the accountant mind of the seventeenth century would have had the patience to count. A life might be complete within one wall, or might require 1,000,000,-000,000,000 walls, as ours, or ten times that number if the turnover from our birth to our death is reckoned in.

Robert Hooke of that century did the christening. *Cell.* Robert Hooke thought of it as a small box. The richness of that jewel-casket even a mind like his could not have guessed. After generations of study came the definition that most of us have heard: a cell is the unit of structure and function. A liver-cell does in the small what an entire liver does in the large. A liver-cell has a wall, has a nucleus (a smaller spot inside the bigger spot), and nucleoli (still smaller spots inside the smaller spot), and other shapes, granules. And every microscopic and in our day submicroscopic item invites experiments to clarify the nature of the cell—dedicated to staying alive as long as it can and to starting a family tree.

In consequence there emerged a new professional, the student of the cell, the cytologist.

How does it look when alone? How when among its neighbors? A piece of flesh is cut out, hardened, sliced into thin sections, a section stained, that single cell observed, sketched, photographed, the whole neighborhood observed, sketched, photographed. Stains reveal the architecture, sometimes reveal also what that architecture accomplishes. The light from the microscope's mirror reflects up through the stained specimen, the human eye peers down, the wandering human fancy sees a church window, blues and reds and darks and lights. Tissue after tissue is studied—spleen, kidney, brain—as it has been, and the years of an era have gone by, a happy often sleepy era. How many dear people have fallen asleep over their microscopes and stayed asleep.

Robert Hooke's cell was the large one of the outer bark of the cork tree. Robert Hooke was the original cytologist. He was a remarkable man, had an imagination that reached outward to the astronomical, inward to the microscopic. The talkative John Aubrey recognized how remarkable. "As he is a prodigious inventive head, so is he a person of great vertue and goodness. Now when I have sayd his Inventive faculty is so great, you cannot imagine his Memory to be excellent, for they are like two Bucketts, as the one goes up, the other goes downe. He is certainly the greatest Mechanick this day in the world." Aubrey said that Newton's idea was Hooke's, that Hooke wrote the idea to Newton in a letter, and who knows? Among high and low such transactions do occur. Hooke came of an interesting family. "His father died by suspending him selfe."

The cytologist trespassed inward. He dissected the cell. A special dissecting-microscope was invented and special slides. One fine glass needle, rigged to the microscope, held the cell tight while another needle ripped it apart, the eye looking down on the scene through the eyepiece. A fine end of a glass pipette was plunged into a cell, it given a shot of something, and a fact previously incomprehensible was comprehensible. The inquiry increasingly moved toward meanings. Why that director-nucleus? Why those nucleoli? That centrosome? Mitochondrion? Ribosome? Many ribosomes? All were clear-cut parts, were describable, were separable, could be subjected to experimentation, and be better and better understood. Sometimes these activities were gay, sometimes not. Sometimes a discovery burst forth in half a minute, most times crept forth in half a decade. Tissue after tissue was studied—spleen, kidney, brain— and the years went by, the years of the prime of the dissecting-microscope. Today it is the electron-microscope.

The cytologist trespassed inward. An organ was scissored from a guinea pig's abdomen, fed into choppers, grinders, macerators, produced a ruthless-looking concoction. Cunning spinning devices centrifuged the concoction into layers, and the layers were studied. Or the investigator picked out a single cell, literally a single cell, worked with that. Items were absorbed, adsorbed, dissolved, solvents jiggled, temperatures jiggled, reactions jiggled. Cells were artificially grown. Three centuries after Robert Hooke had named the cell the microscopic content was all but forsaken for the submicroscopic. The electron-microscope had followed the dissecting-microscope that had followed the ordinary microscope that had followed the unaided human eye, the whole harvest dropped willy-nilly into the lap of the chemist.

The human mind had reached to the molecules, especially the large molecules, the macromolecules, the joinings of two or more into aggregates, and then had reached to the still finer, the atoms, their arrangements. Billions of cells were probed. Billions and billions of molecules were probed. Billions and billions and billions of atoms. Billions of dollars were spent. Whence the impression has gone abroad that for the next millennia man's eye and mind will be on these small matters. What else is there? Atoms and ions and molecules? "Atoms and the void," said old Democritus twenty-four hundred years ago.

In the course of this the cell, the boxed-in spot of life, has become a thing almost gross. Its reality is almost vulgar. It is so big. It occupies an unmysterious place in the mind of a man. "And he rested on the seventh day from all his work which he had made."

# III

# GROWTH

====

## 1 Becomes
## 1,000,000,000,000,000

# FEMALE

## *Lamps Trimmed and Burning*

She has plenty of them. They have been counted. They may not fall short of half a million. Only one needs to go forth each lunar month, so a simple arithmetic makes clear that the total used in a female life is only one tenth of one percent of her inheritance. In a modern marriage two may be expected to get to college. Experts formerly believed her allotment was in her when she was born but later evidence suggests she goes on making them. Eggs. Like enough to a hen's, containing yolk enough to keep a new life from starving during its first inexperienced days. Only a boiled egg is unmysterious. For greater dignity our species calls ours the human ovum—largest cell of the human body.

An ovary is a factory for eggs. Even a fashionable lady manufactures eggs in her two ovaries. An ovary is also a storehouse for eggs. Each is hung to its side of the pelvis, is more than an inch long, half that thick, has the shape of a shelled almond, they say. Sigmund Freud could have told them why they say "shelled almond," but he is dead.

A female whale has an ovary weighs three pounds. Odd to think of an ovum in an ovary in a whale who is keeping her lamps trimmed and burning.

A tube—Fallopian—points in the direction of the ovary, is wide-open there, like a funnel, its fringed border in ceaseless motion. Funnel and tube are the open roadway when that egg sallies forth. Careful nature is almost careless here, because an egg if it happened not to slip into the funnel might into the abdomen, get lost, does occasionally, is fertilized there, rarely. For any routine lunar month funnel and tube are sufficiently near. Inside the tube are microscopic lashes, have a rhythmic beat, create a current, paddle along that small important object, paddle it fast, its escape impossible once it is caught in the millstream. At the other end of the tube is the brood-chamber—the egg's destination. If fertile, it will dwell there a time. That brood-chamber, the uterus, has the shape of a pear (shelled almond . . . pear . . . ), its cavity in the intervals between pregnancies a mere chink, tremendous at term, its muscle-cells grown seventeen to forty times their interim size, and their myriad number myriadly increased. The pregnant uterus is one single-minded muscle when it needs to be.

At the tip of the pear an opening leads into still another passage, three inches long, joins the brood-chamber to the outside world. At its entrance the caller is received.

Everything except the ovaries are the female's accessory reproductive organs. Everything is enclosed within the bony framework of the pelvis. Everything is ready. Nature has no diverting thoughts. Nature is seldom forgetful. Later, at that entrance a small blurry transient will be ushered out.

Far and wide over the female body is a feeling of welcome. Far and wide the lamps are kept trimmed and burning. By those lamps one sees in a special way, separates the sexes, distinguishes the female abdomen, chest, waist, hip, small of the back, hair on the head meeting the brow in a bowlike curve. Large breasts. High-pitched voice. A temperament that varies female to female but is female, a quality of mind, a manner of thinking, of sighing, complaining, understanding, misunderstanding, talents all its own, old often though the body still is young, and when modified by life modifies in its female direction. That temperament especially keeps the lamps trimmed and burning, waits.

# MALE

## *Into the Town*

Into the town comes the swashbuckler, has something to sell. That is his role, or his illusion. Assault is his physiology.

He is the owner of two testicles. Carries them outside in front of him, a unique idea that even the male mind never quite gets used to. They are in a hanging pouch, skin mostly, no fat, sweat glands to help cool them. The reason for the air-conditioning could be that the male sex cell, the spermatozoon, manufactured by the testicles, dies if subjected to the high temperatures inside. A fever kills spermatozoa in catastrophic numbers.

Loops of tubes and loops of tubes and loops of tubes, hundreds, closely packed, lead to a delta of channels draining into a dozen large tubes—that is the testicle. The dozen drain then into a single tube, twenty feet long, *feet,* microscopically thin, coiled, also closely packed—that is the epididymis. A small organ, the epididymis, tacked to the top and back of the testicle, the total of its channels, its corridors, leading in one direction, out. That seems forever the spermatozoon's ambition, out. Either it is on the road on its travels, or it is gathering itself together back at the start, where in layers of cells the production of the spermatozoon proceeds stepwise from the most external layer to the most internal, and there the finished marvel emerges, a no-nonsense construction, spare, needlelike, nervous, dancing, essentially a nucleus with a tail, a head with a tail, funny head, deformed often, a Mardi Gras affair, and whereas the ovum is the largest cell of the human body the spermatozoon is the smallest. As slim as an iron filing and lured by a magnet, small spermatozoon, sad in its self-important way.

A duct continues the travels, runs under the skin of the groin, up over the top of the pubic bone, down beneath the bladder, approaches its companion of the other side, and, as if this were not too much already, receives there its own offshoot that delivers a glistening fluid. Everything suggests a rich concoction brewed in stages. The final lap is via a tunnel, the urethra, that traverses eventually the frank instrument, a common passageway for urine and spermatozoa, insulting, as if Nature even at this late date did not take seriously our place in her plan. All the accessory male parts are supplying ingredients to the fluid that is the liquid world in which the spermatozoon first floats as a physical object, then races on its own propulsive power. Once it has struck its stride it may attain a speed of more than two inches an hour.

The swashbuckler when he entered the town carried with him his secondary sex characteristics—angular look, visible muscularity, low voice, a mind that touches with maleness everything that body touches, and a confidence born of the knowledge that he is possessed of the talent to keep life alive by fertilizing an egg.

# MATING

## *The Presentation*

They meet. They partly know, partly blunder, partly learn. It states a fundamental fact about us—some learning attaches even to the most instinctual performance. In the scale of animals the higher the creature the greater is the capacity to learn, and the more manipulated the instinct. A young male chimpanzee does not just know how to mate, but fumbles, a distinguished primate-student describes to us, and a female chimpanzee, this is more amazing, has to learn how to take care of her infant, is clumsy with her first-born.

The male's duty is to deliver the spermatozoa deep inside the female. That duty—he forgets how long before his day it was assigned—he takes seriously even when for some reason he wants to laugh. Nature has great interest that this delivery should be guaranteed to the full, 60,000,000 to 200,000,000 spermatozoa in a cubic centimeter, and two to four of these at each encounter. If fewer than 60,000,000 the male is apt to be sterile, though he may be sterile even if he has the 60,000,000, and fertile sometimes when he has not.

Methinks I see those 60,000,000 needlelike ambitions.

A combination of senses, sight, touch, smell, stimulate the male, the act advancing until at a moment with a burst those spermatozoa are released, after which the entire male body relaxes. In the female the mating is apt to be quieter. In her as in him there are serial contractions, their mechanical purpose, they say, being to help the male complete the release, and to assure the spermatozoa arriving duly where they should, in the uterus. At the end of the act her body relaxes. That mating is complete.

# CONCEPTION

## *Toast to the Future*

In the warm wet dark an egg slips out of an ovary, does once in twenty-eight days, but only on the average, as every woman knows. A finger, to phrase this elegantly, points to one egg in an ovary packed with eggs, and that egg, which is nestling in its follicle, sidles over to the wall of the ovary. The wall thins at that point. A blister forms. The blister bursts. The egg is free. It enters the tube, starts toward the brood-chamber. If the blessed event is to occur this month it must soon, and within the tube, because the journey down the tube, four inches, takes three days, and an egg is dead in hours. That journey is rapid at the start, rapid at the finish, slow in the middle, where the egg appears to loiter expectantly. Meanwhile the appointed spermatozoon has been coming the other way, if the mating has been properly timed, or is lucky. Or unlucky, should this not be what the lady and gentleman meant. Spermatozoon and siblings enter the necklike entrance to the uterus, hustle up, literally head up, thence on out into the tube. Millions are hustling. Why millions? Some believe this is not merely nature's usual making sure, but the spermatozoon besides being a living cell is a chemical able to dissolve the jelly that embeds the egg. Millions do the work of dissolving, one small spermatozoon sneaks in, and that woman has conceived. Here is the moment to toast the future. The Japanese reasonably consider us a year old when we are born but even a Japanese waits till the actual birth to go to the front of his house and hang verses on the trees.

Human beings are always at one game or another with Nature,

want a conception when they want it and do not want it when they
do not want it, and in order intelligently to use their wills in this
matter would like to know the time of ovulation (discharge of the
egg from the ovary). By various lines of evidence that is close to the
middle of the twenty-eight days. Professionals in the field have
placed the period of no-conception between the twentieth day of
one cycle and the eighth day of the next, which does not leave much
time, yet the situation might be satisfactory if the twenty-eight days
were dependable, but they are not. Nature does not like being
cheated at her game of chess, though she always has been considerably, and more so of late, but still not enough according to the
population explosion statisticians, and the newspapers and the journal *Science* and the church debating the consequences.


# ZYGOTE

## *Genius or Blockhead or Plain Citizen*

Every cell in the human body has forty-six chromosomes, as thirty
days hath November. For years it was thought forty-eight, also forty-
seven, but now forty-six. The chromosomes have our genes, our
genes have our heredity, and of the study of genes there is no end,
as of the squabbles called seminars where DNA and the brethren
RNA are batted back and forth, the chemist resolved to dissolve the
magic, all magic. One does hope he may not in the process make our
dull minds duller.

Only the sex-cells have half as many chromosomes, the reduction
to half, to twenty-three, occurring in the female and male primary
organs.

At conception a twenty-three-male goes up to a twenty-three-
female, tips his hat, and without further fuss they are united in a
forty-six-zygote. That cell—zygote—is the first cell of the impend-
ing new human life. It is the fertilized ovum. It has in it the contribution of both parents, and their parents, and the parents back to
Jehovah. It possesses the entire stupendousness. Were we to begin
soberly to contemplate it, would that send some men down onto

their knees in prayer, others up onto their feet cursing this Trojan gift of still another life?

The zygote's sex also has already been decided. There are two kinds of male cells and which kind won that illustrious race up the uterus and out the tube decided. Ova have sex chromosomes of one kind, X. Spermatozoa have two kinds, X and Y, the Y smaller, a runt, but important. No geneticist would call it a runt. He something like woos his chromosomes. He maps them, for horse, for zebra, for man.

If in the tube Y met X, boy. If X met X, girl. Ergo, the zygote was not only the first cell of the new life but had upon it the vestment of man or of woman. What a moment! The clock stops in Tristram Shandy's kitchen, or it should, except that nothing stops in this forward-moving world.

Such fascination was in what so far has taken place, and is taking place, and will take place, one might forget the something along-side, underneath, behind, above, wherever one is inclined to place it: namely, this zygote had besides the potentiality of a human body, the potentiality of a human mind different from every other. As for bodies, nature mills them out like table legs, though these table legs have differences from other table legs: namely, each one is in itself different, some, from every other. It is conventional nowadays to emphasize the influences from without, physical and chemical, school and society, death and life, that play upon the zygote. Many evidently do, and upon what ensues, but as evidently, and this fact is disturbing sometimes and satisfying sometimes, that zygote was from the start touched with destiny. Say genetic destiny, if that eases you, or merely old-fashioned destiny, if you have left any talent for superstition. Either way, where it concerns a genius one feels knocked down. Edmund Baehr complained one day to God, that He should sit up there in the sky with His pailful of the water of genius beside Him, dip in His finger, splash a drop on this one, on that, on a third, get bored, pick up the pail, pour the whole of it over the zygote of Beethoven.

## PREGNANCY
### *Ten Lunar or Nine Calendar Months*

The zygote divides. Two daughter cells result. These divide. Those divide. The chromosomes are heralding the dividing by their own dividing. This multiplication beggars description, truly, can be observed directly only in a mouse or other nonhuman species, but we may safely assume that in those earliest hours a family man and his house-mouse are brothers under the skin. The dividing goes on. Goes on. If the ovum met no one in the tube it would simply have been one more ovum that died and was sloughed off as in most lunar months. The dividing goes on. Soon there is a lump of cells, called mulberry mass, that hollows, fills with liquid, and to one side of the hollow is growing the future President of the United States, or someone greater, or someone less, but whoever he is he has henceforth the rights and privileges of a still more amazing title, Embryo.

About one week after conception the entire product, hollow and that growing to the side, Pullman and passenger, attaches itself to the inner lining of the uterus, digs in, implants, and here is the beginning of that famous connection between mother and offspring, the placenta.

The dividing goes on. Goes on. Besides increase in number of cells, a separation into kinds is occurring. How? No one yet understands. Why is it studied? Why has it this sincerity, this life-drive behind it? One suspects the answer is only that if we did understand it some of us might live longer, and what is more important than that? The freshman college student learns and in a way knows the three kinds. Each kind contributes to the building of special tissues and organs of the growing body. Try to keep always in mind that that small zygote had in it these unnumbered dynamic possibilities. Embryologists have pieced together the steps that simultaneously are taking place in different directions at different rates, have killed countless animal mothers, robbed them of their embryos, studied these, studied also the human embryo when occasionally they could get hold of an early one, filled boxes with microscopic slides, sketched, photographed, named. And no one understands, either, the hell-and-heaven meaning of this, if there is meaning, if it is not just psychically blind mathematics; and one does have minutes of illusion that one knows, knows something, when the embryologist shows one under the microscope how this follows that, this becomes that, that follows that, that becomes that.

While you have been reading the embryo has been bulging farther and farther into what was originally that chink of cavity in the uterus. Embryo is growing. Uterus is growing. Placenta is growing. Umbilical cord, that makes the deliveries between mother and fetus, is growing. Nature spares (or is it deprives?) the mother from most of this. She is just another pregnant woman. On her side of the placenta blood is moving slowly in what has become a lake or bayou into which the ends of the blood-vessels of the embryo dip. The two bloods, mother's and embryo's, do not mix but nourishment goes across from mother to embryo; wastes go the other way, the mother adding these to her own and disposing of both through routine channels. The great source feeds and cleanses the small spring. In the meantime the developing parasite lolls there quite lordly, hands in his pockets.

That woman carries about inside her an expanding universe. At a moment she begins to sense it—it is so gross—and having begun to sense it, senses it more. Whatever way one looks back over the course—ovum, spermatozoon, zygote, embryo, perfect small foot of a perfect small fetus—the word *marvelous*, the word *miracle*, is not enough. Her disposition too, the mother's, has been changing. She is annoyed at one special kick from that small foot directed straight at her stomach. Once the young lord stands on her backbone and delivers an oration. Her appearance is changing, oh yes. She may be annoyed at that too. How can she be? How can she be fussing with her gaudy tent of dress designed to hide the most astonishing truth in the whole of our astonishing world? She may be more beautiful in her face now. That's an old wives' tale, she says. But if that is not an old wives' tale—if she does have it—she will lose it again, she will lose this particular beauty. It is not hers to keep, just loaned, and the loan passed from woman to woman to woman.

# FETAL BRAIN
## *Rake's Progress*

The developing parasite has a head. In the head is a brain. Associated with the brain is a mind—or is it only, will be associated? Pity him who lacks the curiosity to speculate on how the brain got started, what that rake's progress is.

When the zygote became the mulberry mass, along its back was a line of cells more active than others. These were the groundbreaking. Upon these would rise this most special part of this new human being, the base for his thinking, dreaming, gaiety, disillusion, swindling, lechery, tenderness. Four weeks after conception the head-end of the line had become so sculptured that any tyro could see the dawn of the later divisions of the adult brain. Everywhere the fetus was changing. (For the first three months in the womb a human creature is an embryo, after that a fetus.) Six weeks after conception the dawn-brain had become the gross daylight-brain. Three months, the same tyro could see that this was not only brain but undeniably human brain, shape of a lima bean. Four months, shape of a globe, unnaturally smooth on the surface. Five months, back of the globe rounded out. One thinks of the many brains of the many fetuses that might have been statesmen or TV commentators but instead added each its bit to our knowledge by dying soon enough. Seven months, all the main grooves on the outside had appeared but the surface still too smooth, still suggesting brain in the butcher's pan, edible, animal. Eight months, one no longer found oneself shrinking with some unformulated feeling from an object that should be our kind and was not quite. Nine months, all as it should be, what one was familiar with from drawings and photographs, a grooved baby brain waiting in a womb. The reason for the grooving, we are told, is that the outermost surface has been enlarging more rapidly than the underneath, and it could not do that and still keep itself within the skull unless it folded. So it folded. So, more surface. At maturity only a third of that outermost surface would touch the skull, the rest be down in the grooves. Because of that more-surface, the mind associated with this brain could go on conquering chimpanzee and gorilla and dolphin and everything except death. If one day it should have conquered death, how dull life would be.


# BIRTH

## *"Into This World and Why Not Knowing"*

Tristram Shandy's father in the kitchen was worried about Tristram, who was to arrive on earth that night. Nothing would be surer cure for the neurosis of expectant fathers than to read *Tristram Shandy* in their own kitchens. They would learn, for instance, that there had been fathers before them. They would learn that they were not required to be as calm as the obstetrician. Of course, if they fully pictured to themselves what was going on, what macroscopic and microscopic turmoil, what pushings, what twistings, they might never in their lives, not the fathers and not the obstetrician either, have a completely calm hour again.

> Into this world and why not knowing,
> And out of it like water willy-nilly flowing.

Why this birth occurs when it does, why not earlier, why not later, what starts it, experts give reasons, nobody clearly knows. It occurs if it occurs on schedule ten cycles after conception, ten moons. The uterus at that time gets irritable, but why? Of course, the fetus is stretching, poking the uterus, and would it not be amusing if the whole cause were one final savage satisfying stretch? The placenta at the ten-cycle period is having its blood source squeezed upon, some of it cut off by the pressure, spots of tissues dying from lack of oxygen, and the irritation of the dead tissue might be the cause. But, the cause could be something simple like dice having been thrown long back, and "ten" having come up, and "ten" having been written into the lunar calendar of the species. Whatever the cause it can act later, act earlier, even three months earlier, "preemies" being common and most of them surviving and growing to sturdy boys and girls and grandmothers and grandfathers and obstetricians and midwives and Macbeth and Macduff; in fact, why should the kitchen clock be such a law?

Childbirth—called also parturition, labor, travail, other Biblical or just excellent terms—is divided into three stages. The three is mainly that everyone may understand what everyone else is talking about. All through the pregnancy the uterus indulged in mild contractions, training conceivably, and toward the end these were more frequent. One result was the steady angling of the package into a position mechanically right for delivery. In the first stage, small contractions became large contractions, each beginning feebly, rising to a crest, staying a while, fading out, attended by pangs, the episodes five to ten minutes apart. Pressure was thus built up inside the brood-chamber. If the fetus in spite of Nature's long experience was still not in a right position, the obstetrician is sure to be a fellow with clever fingers able to turn a passenger around without too much disturbing his composure. However, even an obstetrician can have trouble. Had Tristram's—a family doctor no doubt—been equipped with one of those gigantic obstetrician forceps, and had he deemed it necessary to reach into that half-dark and wrench Tristram into place, how the author of *Tristram Shandy* would have made that violence vibrate like grand opera to the rafters, made us witness the shambles in and around the bed, even see the drop of blood over there on the floor.

Usually it is the head that is most in a hurry, anxious to be out and get things started, but the obstetrician shoves it back, the head shoves the other way, a fight between obstetrician and head, the total creature meanwhile moving downward, the uterus behind gathering up the slack, the membranes bulging, those membranes that through ten lunar months were the thin inner wall of the fluid-filled Pullman that provided this passenger with the finest of possible accommodations in this finest of possible worlds. At last the membranes burst and from then on the passenger is constrained to take the bumps of our world direct. Body follows head. A small red or a small pale guest steps out, or is dragged out. Instead of a head, it may have been only the back of a head, or one shoulder may have got stuck, or a face may have appeared as if it wanted to see how things are, or a breech have presented itself, or one arm, and a spectacle that is, an arm reaching through; it adds to the difficulties.

During most of this the woman will have given voluntary help, as Galen assured the Romans she would.

A half hour or so later, the contractions that had stopped begin again, have still a job, to expel placenta and membranes, with which expulsion this female body is freed of its months of responsibility and has nothing to do but vote and cook supper. Soon after the guest's arrival the obstetrician tied the umbilical cord and cut it with scissors, with confidence. A cat does the same, with confidence, without scissors. The tying and cutting were delayed until the cord no longer pulsed, the mother's heart no longer able to pump blood through the narrowing narrows of that cord. Finally, the obstetrician touched the raw flesh with an antiseptic and fastened a sterile pad over the top. Maybe that tying and cutting should have been the occasion for giving this fetus his ticket of admission to infancy. He was a member of the human race henceforth. Maybe, also, some small formal mention should have been made of the famous first cry, the new citizen's first proclamation, excited by a slap on the backside in the barbarous days, his first swallow of free air on this not very free planet. Tired obstetrician is permitted now to go home to his own family.

Every human being every so often in his bathtub looks down with thoughtfulness at his navel, certificate that he likewise crossed the Rubicon, though he does not believe it really. We believe in our birth no more than in our death.

# GROWTH-CURVE

## An Uneven Slope

Ovum, spermatozoon, zygote, embryo, fetus, infant, child, bride, middle-aged matron, old lacy lady, growth-curves could be plotted for each. Each is growing in its or his or her way. Each complicated growth has a complicated rate. Each could be plotted: increase or decrease of size or weight against hours, days, weeks, months, years. Even the duplication of the single cells, and even the duplication of the molecules that build the cells, may hasten, may slow, at a

moment almost to zero. Growth-curves have been plotted for the
single organs, each with its rate while maintaining a harmony with
all the other rates. He who plots, puts marks on graph paper and
connects the marks with a line, is a scientist, and curves are a language for moderns. Curves could be plotted for girls, boys, the
normal, the abnormal, for birds, beasts, swimming fishes, curves,
curves, curves, and from the lot of them there might be written the
definitive *Natural History of Growth;* then, as happens, without a
single curve some student from whom nobody expected anything
might make the one important discovery about growth made in half
a century.

During the first months in the womb there was the growth from
mulberry mass to creature complete though small, not yet able to
survive outside his Pullman. After those first months his chances of
survival were better and better, and the nearer to birth the more the
Pullman could be regarded as a luxury, he in there rolling along, not
worrying about who was to meet him at the station, able to jump
out practically when he liked, a type of luxury reserved for the
human species and only a few others, most dumped out earlier. That
might be plotted, too, place of creature in the hierarchy of creature
against the time the mother dumped it, rid herself of it, as she
appears to do.

For three months at the start of the pregnancy the gain in weight
was slow (a slowly rising curve), sped during the last three months
(a steeply rising curve), the peak reached at birth or thereabouts,
with, immediately after birth, a brief loss in weight because the
newborn had not begun to eat (a briefly falling curve), in ten days
this loss made good (the curve back where it was), then for months
and years, in general, a gain (the curve rising). Preschool. School.
Postschool. In inches, the fetus made its greatest advance during the
middle three months of the pregnancy, then a decline. The mechanics of curves could grind depressingly into each of us the
inescapable in what he likes to think is in his case just a trifle
escapable, a trifle unmechanical: namely, his getting big enough
and then too big for his breeches.

If the weight was eight pounds at birth it would be roughly
twenty at one year, twenty-five at two, a gradual increase from then
on with a tendency to loss at four or five, followed by a methodic

rise to puberty. From about eleven the girl would run ahead, weight and height, but the boy overtake her, pass her, might pass her also in some powers of mind. After full height the young man would continue to accumulate weight, till he was thirty. This is all average. Exceptions are on every side. In the elderly there is loss both in weight and height, but "elderly" comes at different times in different persons, fast in any case, many at sixty already earnestly sunning themselves on the sands of Miami, apparently not having read in the newspapers that a human life might today reach one hundred and ten. Every item of this could be converted to curves. Loss in the elderly might be a wasting in each cell, or a reduction in the number of cells, or both. As to organs some begin aging sooner as some began their rapid growth sooner. Why a body should age, when cells grown artificially in a test tube outside the body may not, is a problem herewith bequeathed to the younger generation, that probably already has solved it.

# POSTNATAL BRAIN
## *Rake's Further Progress*

What, meanwhile, has happened to that brain that was to conquer gorilla and chimpanzee and everything else? At birth, fetal brain became postnatal brain. This grew rapidly. Enormous is the topmost part of the nervous-system at birth, enormous all that has to do with seeing and hearing. Big head, big eyes, small nose, small everything else, is what impresses anyone when first he sees an infant daughter lying unworried in the bend of her father's elbow. Once out of the womb this daughter every minute is getting into trouble, is in danger of her life, requires that big head, requires the defending nervous-system, though at birth this topmost part is still not ready to do much defending. It is about twelve percent of our body then, two percent when we are full-grown. One-fiftieth of us is the scene of our torments and delights, one five-hundredth of the lion, wherefore the lion must have its great muscular strength and we can get along with our comparative weakness. The weight of an average human brain at birth—the brain lifted out of the skull of a stillborn and put onto a scale—is 350 grams. At one month, 420 grams. At one year, half the adult weight. At seven years, nine-tenths. That last fact—his daughter at seven years having nine-tenths of all the brain she ever will have—could possibly help the mystified father somewhat to understand why she, now in her second year at school, is already as smart as he, lacks only experience and a deposit of memories. While the growing brain has been brought almost to a standstill the rest of the body is steadily gaining. As for the skull, it also has been brought almost to a standstill, the stimulus for the growing skull having long been thought to be the growing brain, and if a brain does not press outward—that is the idea—the skull will not properly develop and any irregularity up at the top is an unhappy circumstance for a human being. From the beginning nerve-cells have been dying, and since about as many as there ever will be are present at birth, that infant daughter in this respect is already on the decline.

> And so, from hour to hour, we ripe and ripe,
> And then, from hour to hour, we rot and rot.

# GROWTH THEMES

## *How Far by How Soon*

The phenomenon growth could be discussed from a dozen points of view—geometry, heredity, health, dietetics, others. Each would bring in its knowledge, its sophistication, its dogmas.

Geometry could be the theme—a body to become a body must fall in with the forces that give it its lines and angles. The mathematics of the direction of the outward push from inside the cell, with the wall of the cell resisting, and the pushed cell pushing the cells around it, they resisting and pushing, finally in terms of geometry the cell's size and shape explained. There would be a geometry for the "living" wall and a geometry finally for every "living" molecule. As with the single cell, so with the single organ, the pushes, the resistances, the organ's size and shape thus eventually explained. And as with the single cell and organ so with the entire body, it scooping its spot from time and space, the sun and moon scooping their spots, and at the end of a life or a light-year the spots shriveling again. Little moon. Little sun. Little Jason. Geometry could discuss growth, and has, if one chooses to call all of that growth.

Heredity could be the theme—bring in its knowledge, its sophistication, its dogmas. Dumpy mothers have dumpy daughters often, small mothers small daughters. True, a ten-pound four-ounce baby boy did literally leap from the womb of his seventy-eight-pound mother into the amazed arms of a green medical intern, but it is too soon to write that baby's story, and he may yet when his growing is finished be a father as small as was his mother, have a body like hers. Japanese are short and have shapes like Japanese. Single features may be race begotten, also family begotten, like that jaw of the Hapsburgs. Heredity is stern. Doctors used to warn the expectant mother to curb her appetite lest her baby be too large at birth but obstetrical experience teaches that, though she ought for excellent reasons not overeat, if she nevertheless does the baby will probably be born no bigger. The reverse is true also, except in grimmest famine; in ordinary starvation the fetus quietly consumes its starving mother. To be sure, the fetus might be deprived of a single item which its mother because she was starving was unable to supply, and that can be bad for a fetus. Most of the time the body attains the size and shape that was its fate, that was its heredity, that was written into the gene-script.

Both excitants of growth and checks on growth might take orders from the script. A check might operate indirectly. The growth of the heart is limited by the amount of blood circulating through its substance, and the growth of the body by the amount of blood the heart can deliver, and so would an indirect check have prevented a human body growing to an inhuman one. The kidney is big, not bigger. An eyelash is long, not longer. That gene-script is employing its chemical agents, of course, chemical drudges that do blind labor for it, supervise the rates of growth, or they do blind labor for the ancestors, or for the devil, or, if anyone can still afford to see life in pink chiffon, for the guardian angel.

Health could be the theme—a body to grow must have health and that might be defined, freely, as all the multitudinous molecules operating harmoniously, proper order in time, proper order in space. Verily, that health may be a minimum. Should a children's hospital summon all its medical acumen to keep one of the inharmonious little ones alive, nature might not insist that this one be too happy while it lived, might for instance give it a useless body and a sharp mind, let it be wheeled through the corridors with its thin legs dangling while it intelligently watched the other children at their play.

Food could be the theme—a body to grow must have food. Growth of ovum and spermatozoon was owing to molecules that were eaten by the mother; growth of zygote, to molecules eaten by the mother; growth of embryo, fetus, nursing infant, still to molecules eaten by the mother. Each fed on her; the nursing infant merely fed more conventionally. Later the young master did no longer require or desire that authorized cannibalism, nursing; rudely shoved his mother aside, drew his chair to the supper table, brought his own spoon into the foray.

Food-factor could be the theme—each body, and indeed each cell of each body, takes from the food that comes to it what it requires to manufacture itself. This has been studied in the bacterial cell; that is comparatively simple. A bacterium may get all the carbon it needs from a sugar. If that bacterium is changed, made to mutate by, say, X ray, it and its descendants require something besides that sugar to get their carbon. That something the geneticist calls a food-factor.

# MISGROWTH

## *Giant and Dwarf*

Giants do not grow to specification. Dwarfs do not. Many human beings do not, in some detail. Many animals. Giants exceed specification, may keep growing, upwardly, until they are more than eight feet tall, long arms, long legs, long hands, long feet, bones of the torso huge. Dwarfs stop short. Though not growing to size a dwarf may grow to fame, be right pleasant to look on, become the playmate of the Infanta, have his portrait painted by Velázquez, a well-proportioned small man with an immense head who doubtless experiences the satisfaction of having sons and daughters of socially acceptable proportions. The Infanta sometimes considered her dwarf not a playmate but a plaything and that was surely nice of her. Giants of the fairytales and of the irony of Jonathan Swift have giant natures, are offspring of a fictitious race, as also the giants of Richard Wagner, but the giants of the circus are not fiction, are offspring of chemical abnormality, molecule abnormality, have genes with an atom displaced, bumped a submicroscopic bump. They are real, we say. Now and then a prize fighter has been such, a big fellow, a heavyweight who through all the months of training before the moment he stepped into the ring was frightened. But sometimes a giant will be muscularly feeble, sexually feeble, speak with a high-pitched voice, and that giant also experiences the satisfaction of living in a society where like prefers like.

The bones of our arms and legs have long shafts with marrow in them, and expanded ends. For a time after birth the ends are not united with the shaft. If there is one kind of chemical abnormality, the union is delayed, the shaft grows excessively, the result is a giant, not a distasteful type. If the abnormality occurs after the union, the shaft does not grow excessively, but other parts of the skeleton do, also the soft tissues, the result is not a giant. This body is not tall. This misgrowth is of an abominable character. The person may for a time not know he is changing, seeing himself as he does every day in his mirror, but one morning he must admit, with perplexity, then with horror, then with disgust and defeat, that last winter's gloves are too small because his hands are too large, his shoes pinch because of his feet, his total body lacks symmetry in a manner befitting the damned. He stares at his jaw. The ugly thing pokes out. His teeth stand apart. His chewing is so difficult he could prefer to starve. His cheekbones jut. The arches above his eyes project till they shame the architect. Nose and ears bulge. The vertebrae have gotten porous, the spine has given way, collapsed, poor hunchback, his hands reach his knees. Acromegalic he has been classified. He is careless of his clothes because he is careless of his life. Some say Punch was one of them. Some say the village blacksmith.

Then, there is that dwarf unpleasant to look on, that other type of dwarf, limbs too short, belly protruding, navel protruding, preternaturally old, skin that hangs and is cold and with a peculiar soft thick pad under it, bones of the face insufficient, especially of the nose and the orbit of the eyes, bones of the cranium oversufficient, a swollen tongue between thick lips that do not close, and, to pretty the picture, hair in patches, thinned eyebrows, half-calcified decayed teeth, stupid to idiocy in appearance and in fact. He sleeps. Dickens knew him. In him too a chemistry has gone wrong.

Other chemistries can go wrong, genetic wrong, a gene dropped in or out or substituted. The result may be trifling. May be merely an emphasis. Some such emphasis is in each of us, more than 1969 can explain, though it explains so much, a private chemical change bequeathed to us from somewhere in the past. May also be miserable chemical change. Warped human chromosomes, warped human bodies, warped human brains, warped human minds are all over the world. (This same fact of mutation, but advantageous mutation, has somewhere in the past made our biological evolution possible.) In his eagerness the chemist-geneticist may sometimes claim to understand what he cannot yet, may claim that in the future all abnormality not only of body but of mind will prove chemical abnormality, as it appears. What he can safely claim is that the future human family, if its cleverness has not wiped it out, ought because of accumulating chemical knowledge have fewer bitter experiences with its problems of growth than do the unknowing rat and hog.

We are born what we are, and if that was not lucky, we can make it worse by adding to it our thoughts, but who can escape his thoughts? How the giant or the dwarf thinks of what he sees in the shopwindows when he is pretending to study the new spring styles is a force in his life, also what others let him know they think. A human being is not simply cells. There is a mind attached. This may wish often it had been born a tree.

# OBESITY

## *Size*

Obesity is not growth. It is increase. There can be a tenfold difference in the amount of fat in two human bodies. Fat is a poor conductor of heat, so the fat man sighs in summer and gasps and has
the air conditioner going full blast, nevertheless fans himself (why
with that ridiculously small fan?), but he is mighty comfortable in
winter. Undoubtedly a body can have too much fat, past all common sense in some, no chance to enjoy it, spend it, just possess it,
cart it around day and night. Those persons are the obese. To give
them a name takes a kind of care of them, gets them out of our
mind, puts them in a class, provides them a nationality, a communal
defense so they do not need to fight it through alone. It might be
more accurate simply to write them off as sick. The greatest quantity of their fat, half of it, is under their skin, where fat is the worry
of half the world. Their weight increases, their size increases, but
growth is always either increase in the number of normal cells or in
the mass of each single cell, and not fat cells only.

Why ever is there excess fat? Answer us that, O Lord, for we
fumble and do not understand. We could of course, should perhaps,
blame the dark lady, heredity, who got absent-minded, preoccupied
at the wheels, shifted something too far or not far enough, wrecked
a fat control. Might instead have tinkered with the temperament,
splashed around in it like an artist in an expansive mood, produced
what inspired a man (who was fat already when he was a bed-
wetting small boy) always toward the couch, toward the elevator,
toward the slow walk, second plateful, long hours on his backside
looking at TV or the administrator's budget sheet, or a fishpole, or a
blonde who could not bring herself to look at him but had brought
herself to marry him because he was the son of so-and-so. By such
indirection heredity might manage it. A fat disposition foreordained
in the zygote and come to fruition in a fat carcass runs through
families, through generations, unhappy wretches haunted to consume more than is required to heat them and to get them their

exercise. "For God's sake, I tell you I am hungry!" He is. Stop nagging at him who has been nagged at enough. He has given up. He has taken a job as a cook and does not think cooking respectable. He just eats, adds too much and subtracts too little, tastes, as a cook must, finds it good, eats, and on his summer holiday consumes doughnuts with a self-conscious nonfattening drink while resting in a hammock. Has waddled his way into an imbalance between the consuming and the exercise, which two must balance if fat is not to settle in grotesque amounts in grotesque places.

When the dark lady wrecks a different control she leaves a body bloated, and that is not growth either, a dragged-out-of-the-river look, this one suffering not from excess fat, obesity, and not from excess water, edema, but he is the victim of another misgrowth, a dilute glue everywhere under his skin. When she wrecks a still different control the muscles themselves get big; not powerful, big.

Human beings laugh in this world when they should not. We laugh at fat. The wine god was fat. Bacchus laughs, so as to be laughing before he is laughed at, gets us into a rocking humor so we may not see what is before our noses: namely, him. Something queer is in this. He flies into a rage, knows how silly that looks, dissolves it in a laugh. Why does a naked fat woman in a painting of Rubens make human beings laugh? They ought to scream with pain. People go to the theatre to see Falstaff—that is Shakespeare dramatizing the evidence in his century. Fat men like to play Santa Claus—that is our century. Do they play Santa Claus to conceal their bulk in a costume? Do they wear that fake red cotton and fake white so as to be able to spill out old fat jokes and no one able to object because it would spoil the fun for the children?

The yearning to eat excessively goes by an excessive name, hyper-phagia. *Phagia* means hunger and *hyper* means hyper, and there is an experimental procedure that induces such successful hyper-phagia in a rat that the experimenter stands in revulsion before his success, a rat that eats, eats, eats, gets bigger, bigger, bigger, poor fat rat. (The experimenter restrains him and he, poor fat rat, gets what is called a stress stomach ulcer.) Does a rat laugh in a manner that the human eye misses? Another experimental procedure keeps a rat lean. Human beings too. She drinks coffee at breakfast, skips lunch, and is nervous all through dinner, keeps lean.

The top-end of the obese as of the skin-and-bones is the head. The trouble, the explanation, is in his head, or her head, we say lightly, make the head responsible for the whole fattish situation even while we are unsure whether the head is responsible for itself. In a man's or a woman's life or her thinking about her life there can be that which lets her friends literally see her enlarge. One young woman fell out with her mother, ate determinedly for a week, gained thirty pounds. Disappointment, frustration, chagrin, those words from the lexicon of the damned stand for drives that are weighable in pounds. The bored gain. They gain in spite of dreaming of a svelte social career. Is that heredity? Is the boredom? The dreaming? The wife who was snared? The husband who was the son of so-and-so and looked such a bargain? The brat who was unwanted? The other who was wanted too much and was overfed with coddling and food? Is nothing just chance? Is all causality? Is all written with a fine pen by the bookkeeper? Is it all in that original molecule of DNA? A new original molecule for each new individual—how fantastic. Is there no escaping that molecule, that cursed double spiral? When one stares at it a while—at the old model of it modified for each textbook so that that author may feel original too—one can bear to stare no longer. Double spiral, bah! We say of a fat man that he is doing this thing to himself. We say he is eating to avoid facing his misery. He eats at each crisis, at each disappointment, like any addict. He tries to forget in mashed potatoes. Apparently that is easy to do. Tries to forget her who has forgotten him. Is being a fool DNA? Anyway, she eats and drinks and fattens and he eats and drinks and fattens and both do it to take attention from the gnawing at their hearts. Dogs do that. The treatment for all, dogs, woman, man, short fat boy jeered at in the schoolyard, is a treatment of their lives. We say that lightly too. The newspapers make life stories out of it. But she was born fat! Yes, she was born fat. Poor woman, she whines that she is eating no more than the happier members of her family and maybe is telling the truth. She says her sister eats the same food she eats but she gains while her sister loses, and even this might begin to be understood if the minds, the top-ends, could have been eliminated before one started on the biochemical calculation. But how to eliminate the top-ends? And if they were born with their top-ends?

# OLD AGE
## *The Downslope*

This gentleman who comes to mind should by the rules have been
dead years before, at least should have been on the downslope, but
never a sign. Then one day there was a sign. He did not tie his tie.
He put on a clean white shirt but did not tie his tie. A bow tie. His
ties always were crisp. Also, instead of taking his walk around the
block and then around the shopping center, he took it in the grass
around the house. His body rolled a bit. Last week he let an appointment wait, five minutes, and that was unprecedented, would
have wounded his pride three years ago. One year ago in Mariemont
Square a suburbanite failed to recognize him, and, as always happens, a second suburbanite on the same morning did the same, but
the second hurried back, burst out over-enthusiastically that he had
not changed in thirty years.

Did this make him glum? Why should it? The downslope is fact,
is natural, is a stage of growth. The growth-curve rises, plateaus,
falls.  His was falling rather rapidly of late, he pointed out. It
amused him. His body had shrunk, an inch. He made the measurement himself. It is possible to shrink an inch and be not in the least
bent. He also weighed less. In ten years he had gone from 146 to
132. Always had weighed himself in the bathroom and measured his
height, in the morning before breakfast, was regular about everything. He had quietly and accurately and all-around looked into the
face of approaching old age, turned that corner as he had the others,
with interest, not too much interest, he said.

He was born lucky to begin with, insisted it would have been
luckier not to have been born at all, been one of his mother's three
miscarriages. Aggressive by the end of the first five minutes when
the midwife (this was passed down through the family) caught the
athlete by one foot and just missed letting him land on his head, he
fought so. "It's a boy!" She said that. The mother did not say much.
The growth-curve fell briefly after birth, as it should, then rose, then
went its inexorable course. A boy, therefore long legs that got
longer. The body as a whole was on the short side but robust. With

birth began the methodic tearing off of the pages of the calendar, but who in those middle European hills would have dreamed that in this instance the tearing off would not cease for more than ninety-three years?

Five years old. Seven years old. Eleven years old. At eleven he emigrated from the old country. That second cutting of the umbilical cord he did himself, no more hurt than the first. Never had seen the ocean. Crossed it in winter. The crowd in the steerage was seasick, not he. A new country. Rain did not give him a snotty nose, snow freeze up his joints, fears keep him locked in his boarding-house.

Marriage. Conception. Four conceptions. Started four growth-processes around him while his own continued. Those four always knew from the time they knew anything that they had a father with a healthy body and a healthy mind. The latter kept growing luxuriantly.

Prime. A satisfying intelligence. A gay cynicism. An amused irreligion. Possibly the irreligion was over-vocal. His advancing years could have provided a life-insurance company with data for supreme old age. In this living machine of René Descartes as once there had been normal acceleration of parts now there was normal deceleration. He would have agreed that he was a Descartes machine. It would have satisfied the intelligence and the cynicism and the irreligion—the human body a trim machine. His was. It made use of all the advantages of the planet, green things, flesh broiled, fried, stewed, and at the other extreme the advantages of human history, human learning. Then there came that morning in Mariemont Square when with a sardonic chuckle he took official note of the downslope. He had too much health, some thought, also too much cleanliness. That of the body remained as on that first day the midwife gave him the first bath. Recently the baths may have required a few minutes longer but he came out scrubbed. (Socrates before he drank the hemlock scrubbed his body too.) The fiftieth birthday was behind him. The sixtieth. The seventieth. The eightieth. A noisy glittering party at the ninetieth. The girls of twenty and forty and sixty kissed him and he enjoyed it and slept well that night. He slept well all nights, always had, slightly fitfully of late.

It was the greatness and decline of one man's empire.

What happens in the healthiest and often even in the young is that a single system of the body is not as brisk as the others, and in old age definitely is not. His skin was soft, pink, a shade more transparent, some flakiness here and there, more dead cells rubbing off. The single cell was probably able to do less work. The single organ was able to do less. The stream from the bladder was lazier even though the drainpipe was not pressed on. At the ninetieth birthday party the guests were saying that his back was as straight as ever. "How he marches, that philosopher there," remarked a stranger who from a distance observed his confident locomotion. Possibly his digestion was less exuberant. "How that man eats fruit," remarked his daughter, "and candy!" A touch of old-age diabetes? Without a doubt he spoke more deliberately, but also uniquely lucidly, had his private sense of the funny, said the downslope ends in the horizontal, where one can at last lie flat on one's back until the trumpet calls.

No doctor ever had the honor of taking his pulse. Took his own. Had for years. It may have become an odd pulse even to him. Always had been slow, was faster now, not much. His brain according to growth-curve statistics would have grown smaller by so-and-so-many million cells. Then, one morning, unannounced he departed this felicity a while. Someone noted that he did that as all else, chose his time. He himself had said that the best time was early in the morning. He had got bored rising every day, eating three meals, going to bed, ninety-three years of that. People were surprised he should look so spruce in the coffin. People say just about what they say at other times when they gossip next a coffin.


# CANCER

## *Out of Bounds*

Decorum was in all so far. Decorum is a fact of normal growth. Throughout the long history of the ninety-three-year-old gentleman decorum was everywhere. Control. In the giant and the dwarf the processes of growth had a different inherited base, but upon that base the processes maintained decorum. Control. An obese Negress will move oh so decorously in Harlem at midnight, on the boardwalk in Atlantic City, in Paris as she steps from restaurant to restaurant to seat herself majestically on two chairs at the round center table in Le Beau Chat, and no Frenchman says of her, "badly-put-together American." A big man waltzes smoothly around the great Christmas tree in Genesee, none of his parts show disproportion, never did day to day while he was getting that big. Normal growth, the blessed normal, has a middle-class regularity, the correct amount deposited in the Building Association every Thursday night, despite that some types of cell are dividing frequently, as on the outside surface of the body, and some types infrequently, but both types maintaining the regularity.

In the artificial growth of cells in a test tube the growth of each cell is correct and decorous. Control. It is built into the processes of that growth. Millions of dollars in grants to dozens or hundreds of laboratories have been appropriated for the study of these cells that grow older as cell-cultures but remain young as individuals and accomplish nothing spectacular nor disastrous, even the disastrous types, so long as they stay there in that tube. Study in a test tube would seem ideal, except it makes some not immediately calculable difference whether a cell does or does not have the influence of the cells of a living body around it.

During the development of the human embryo where growth performs such extravagant feats there is perfect regard for time and space, the cells multiplying, changing type, differentiating so as to become the different parts of the body. Control. Then comes birth, and after birth we speak mostly in sociological terms. "Harry has gained five pounds again, dear boy." That remark takes the last frolic out of growth. The cool precisions are going on inside Harry, though one asks why they bother when one looks into Harry's silly well-slept face. In the growth-process first to last there is never a feeling of too fast, too much, no time excess, no space excess. Each cell seems individually to know when to stop dividing. Order. Order. Order.

Then one happy summer evening something changes. Growth becomes an ogre. Growth becomes a process with a goal that no respectable citizen can respect. The clock has been tampered with. No dependable growth rate enables a dependable scientist to construct a dependable curve. Mathematics has deserted the universe. No tidy mind is keeping watch from the top balcony. Our beneficent God is off playing poker.

The causes of cancer?

Under the microscope the sight suggests wild young cells run amuck among their elders, undisciplined cells gone like vandals among the disciplined. Each cell no longer knows when to stop dividing. Control is gone. Order is gone. This is a different kind of cell. These are different chromosomes, it may be in number, it may be in structure. The DNA is different, the enzymes different, the overall chemistry different. That chromosomal difference can be seen under the microscope. That chemical difference can be tested by the regular methods of the chemist. Which might say not too much about what started that gross speed that again and again catches even the experienced physician unprepared. Was it something inside the cell? At the surface of the cell? Inside the nucleus? Inside the membrane around the nucleus? Inside the chromosome? Inside a gene? An upset pattern in those nucleic acids? Did the upset occur before birth? Was it after conception? Was it before conception? Was there an inherited factor? Must there be an inherited factor even if it does not usually show itself? Or did the upset occur after birth? Or both after and before? Did something after act on something before? Or was it nothing like that? Most patterns of atoms in a cancer cell might be the same as in a normal cell, a few different, but that few might be enough. The few had the destructive drive. Some experts think this. Or the few might have had lifted off them the normal control on the normal cell-division. Whatever it was, or wherever it was, the chemistry of the good clean cells of God's clean earth was now driving toward an altogether different kind of growth, or freeing something to allow an altogether different kind, and there has resulted a growth that is mad. These seemed normal cells, and now they are cancer cells.

Causes of cancer have been claimed, denied, hailed, scorned, a slippery minority established, meanwhile tremendous step-on-step world-wide work. Even that minority may be only exciting causes that give a determined nudge to one universal cause. Viruses could give the nudge. Viruses could be the cause. Viruses could play the role at one time and not at another, could enter us frequently but be effective only in one condition of the chemistry. Viruses could operate on the first day they entered, or the sixteenth, or the two hundred and sixteenth, or any day in between. A rainy day. A sunny day. Even the hour of the day—the place that the hand of the clock stood. There are pathologists who believe viruses responsible for all cancers. The well-known chicken cancer (sarcoma) is definitely caused by a virus, proven more than fifty years ago, and in a gloomy moment one could think that in spite of all the grants nothing as crucial has been discovered since, and undoubtedly that discovery has been underneath mountains of speculation and experimentation. If one virus does excite cancer, then another virus might conceivably prevent or cure, the right virus at the right instant for the right cancer. This would not give the person dying this month much hope, but would give hope. How the world of men would honor the man who established as fact any similar surmise, any similar hint come to him at the bedside or in the laboratory, any clue that led to the chill plague's beginning unmistakably to withdraw, leave our little lives a little less threatened. Meantime, no cancers in man have been proven caused by a virus, though there is suggestive evidence, and for leukemia, a cancer of the blood-forming tissues, considerable evidence.

But, the cause might not be a virus at all, or a virus rarely, might be some altogether different dynamics, a mutation, the production "spontaneously" of a different type of molecule, this the result of any of many factors that became a chemical factor, any chemical factor that became a physical factor. Interrelations in science have been so multiplied that any explanation suggests another explanation, which could be the truth about the nature of cancer, as it is about our daily lives, as it is about the era in which we live. We are in an unpredictable world.

The popular explanation for cancer has often been chronic or acute irritation, or inflammation. "He had an odd bruise last year." That bruise became something else. Tolstoy motivated his gnawingly brilliant novel *The Death of Ivan Ilyich* on such an "odd bruise." He also throughout made the reader feel that that might not be the truth though Ivan was convinced of it.

Carcinogen is a summarizing term. It applies to anything that is cancer-causing. Carcinogens at least for the present have range. Radiations. Poisons. The friendly cigarette that kicks us out of life in the course of giving us relief from life. The exhaust of automobiles. The temperature of our body. Something easily turned topsy-turvy in the growth of aging cells whether these be in a young person or an old. Normal growth might have in it an agent or agents that halt, that inhibit what would otherwise run to the abnormal, some inhibiting chemical among chemicals, some inhibiting molecule among molecules, and the cause of cancer then the failure of that chemical, that molecule.

Pathologists and all the others who come into the field speak of resistance in the animal body and keep trying to find what this means in terms of molecules. Pathologists speak of the slow preclinical stages of cancer, of an initial hesitant alteration overtaken by a rapid alteration, in gut, thyroid, bladder, bile duct, breast and uterus of the woman, prostate and stomach of the man, and they look for a chemical meaning for the alteration, a precancerous chemical. Pathologists may in a particular tumor be uncertain whether this one is malignant or benign, killing or kind, but that is rare. Pathologists do usually speak of cancer in the plural, cancers, many cancers, while laymen speak in the singular, cancer, one disease. If all cancers were one cancer, one deviltry in the midst of the ostensible varieties of the phantasmagorias, the deviltry keeps itself hidden. How the world of men would honor the man. . . . It could be many men, or appear many men, then the history when finally written reveal that the essential was one man, as so often in the story of achievement, one mind, or one mind clinching the work of many minds.

The net result of the carcinogen?

A great rapid growth in the neck of a five-day-old-infant (a rotten joke), or a leisurely fatal wart on the leg of an eighty-five-year-old woman (fair enough), in either case a procreative cockeyed increase of cells, a piling on, cell on top cell, more headlong, less headlong; when more headlong, the whole chaotic process from the first acceleration to that final deceleration, death, being in a bad case a matter of months, even weeks, even one week; or such has been the appearance. Nothing in normal growth resembles this except some special tissue for some brief period in the embryo. On the contrary, a cancer of the skin, like the eighty-five-year-old woman's wart, may have grown for a decade and one scarcely saw an enlargement. A cancer may lazy along in the cervix of the uterus for nine years, fourteen years, abruptly break through, spread and scatter, invade and metastasize, that is, transfer cells to an unrelated part of the body.

Cancer cells are in some sense tougher than normal ones, abler to survive under adverse conditions, to live anywhere, need not recognize their own kind, and be associated only with their own kind, or so one might think of it, but can thrive among other kinds, make room for themselves in blood, brain, kidney, or cleave a path like a drunken loon who shoots his way as he advances. Surely cancer cells do themselves look as if they had lost their normal control, or even look as if they had lost their self-control. May on the contrary show a kind of bashfulness, wait ten years after a breast operation, then quietly flower in the line of the surgeon's old incision. One ought not to use the word *flower* even though seeds do seem to have lain in the ground and waked. What waked them? What disturbed chemistry waked them? (Surely the chemists are at it.) What desecrated this garden? A fresh young garden it can be, a twenty-year-old girl with a rapidly growing lethal tumor in her bladder. "Like a rose." A surgeon said that when he peered into her bladder.

A surgeon, or a scientist whose daily eight-hours it is to study this reckless multiplication, may speak thus irreverently, at least speak composedly; the rest of us cannot. We feel revulsion besides fright —revulsion also where a nodule of cancer has been experimentally planted in the pouch of a hamster's mouth and has grown as big as the hamster's head, or bigger than the whole hamster. Even dim revulsion at those cancer cells that grow as a culture in a test tube, produce descendants, a progeny that goes on and on and on. We cannot watch calmly when nature is in this abandoned humor. We are outraged at some suspicion of personal. If only it were. If only a tongue were stuck out from a gargoyle's face that we could spit into before it ducked. We would enjoy that. We would relish an embodied adversary that was strikable, a definable someone who could be held for this that has devastated our neighbor, laid him so white and frightened and weary and gasping in his bed, used his blood-vessels as containers filled with blood, a warm rich nutriment for growing things, then choked the vessels. We loathe the bully. Loathe the sneaking habit.

Cancer cells are usually well on the march before the patient feels anything, before the physician finds a lump, a tumor. They have stolen an unsportsmanlike advantage before there are enough of them to crowd an organ or jam a tissue and produce symptoms. In some cancers abnormal cells do normal but excessive work, as a cancerous ovary secreting excessive ovarian extract, a cancerous thyroid secreting the hormone thyroxin, the results of the secretion on some activity of the body catching the patient's or the physician's attention. An old idea placed the origin of the anarchy in a single awry cell. Possibly it is. But not possible to test the possibility, and no pathologist would hang a diagnosis, as is said, on the look of a single cell. He wants numbers. A reasonable number will convince him. Parenthetically, artificial growth outside the body from a single cancer cell is feasible, and from a gob of cells is accomplished at will.

The appearance of the dividing cancer cell?

As in the early stage of any cell-division the chromosomes in the nucleus split and arrange themselves in figures, mitotic figures. When the physician or surgeon snips out a bit of questionable tissue, the biopsy specimen, the pathologist examines it under the microscope and if it is cancer he can diagnose it by those figures. In a modern hospital the specimen may be sent over an automatic delivery system from the operating room to the pathologist and the report come back over the operating-room TV, an irony in the promptness of reporting an insidiousness about which there may be nothing to do but wait for the dismal end. A rapidly growing cancer shows mitotic figures everywhere. The cells look like haste. The chromosomes are crooked or fractured. The nucleus is geographically misplaced. The design in that spot of walled life has lost that smug decorum. The cell's parentage may be unrecognizable, and the bastard multiplies. The progeny of cancer cells are cancer cells. The massing of such cells may so squeeze the neighborhood, so distort the picture, that even the specialist does not know whether this is kidney, brain, liver, lung. Blood-vessels streak through at unexpected points. The scene could be thought aboriginal, undifferentiated, wrecked, as if life had not yet found its assignment, or had found it but had lost its pride and whatever intention it ever had of becoming the crisp nonagenarian. The price tags have been removed. Growth is splitting fees with the gravedigger.

Why does the stricken human being die?

The products of the cancer cells kill, in some. A small tumor may kill quite mysteriously, no explanation. The outlet to a necessary gland gets plugged by cells, the body deprived of a necessary secretion, in some. Blood-vessels get plugged and blood does not flow where it must, in some. The mass of the tumor may squeeze and thus from the outside block the flow of secretion or blood. Streaking blood-vessels may be eroded by the cancer, with fatal hemorrhage, in some. A vital control line may be cut across, in some. The malignant cells may and usually do sooner or later, and sooner in one type of cancer than another, free themselves from the original lump and travel and procreate in soil elsewhere, consume the good earth, luxuriate while the normal cells suffocate and starve. A surgeon's knife can extirpate part or all of the growth. Radiation can dissolve it. In rare instances the cancer arrests spontaneously, not meaning that the arrest did not have a cause but the cause unknown. Often the cancer eludes knife and spontaneous arrest and cobalt and X ray, to disport at leisure. The invader has lodged upon the peaceful countryside, an unashamed illegitimate relative determined to stay, and this land will not be fallow again. To the human mind this is discouraging in 1969, but might stand as only a low point in history in 1999, for which possibility there are signs. All is not completely discouraging. Furthermore, physicians are thinking. Surgeons are thinking. Chemists are thinking. Physicists are thinking. Geneticists are thinking—thinking among molecules. Virologists are thinking. We know more about normal growth, all of us. It interests us. It should.

# IV
# ENVIRONMENT

===

## Up from the Sea

# CONSTANCY

## A Fresh Idea

To René Descartes the body was a machine, a machine made by the hand of God and better than any of human invention. Descartes was religious. He was French. Lucky for him were both those facts. His idea had no irony. It did have greatness.

In the nineteenth century there was another fresh idea, Claude Bernard's. He also was French. To him also the body was a machine but in it he conceived what he called an *internal environment.*

By environment we mean usually something outside, the slum, Hollywood, Thoreau's pond at Walden, the beasts, the fishes in the green sea, the people, all the people, the underprivileged, the overprivileged, whole nations, especially nations that abut on our territory, abut, that is, on our minds; and then there is the moon. Everything outside. All of that a man considers his environment, but it is his *external environment,* and Claude Bernard's was internal.

In the two centuries between Descartes and Bernard the living machine was more and more examined for the chemical, had become a chemical machine. Bernard recognized this. He was a genius, a physiologist, had lofty imagination, was limited by the techniques of his day, as we by the techniques of ours, but he could see that a human body, anything living, lived by a multitude of balances. He knew that body and brain were built of cells, and he emphasized, postulated, because he could not yet prove, that around the cells there was always fluid. In his time the fluid would have been mostly but not altogether the blood. Cells + fluid-around = a living body. It would make a body a vast interconnected catacomb walled by cells, the corridors filled with fluid. A Japanese undersea spear fisherman and a high-and-dry investment broker on the fifty-seventh floor of the RCA building were equally 1,000,000,-000,000,000 cell-islands plus the fluid-around.

Bernard being French his phrase for the fluid-around was *milieu intérieur*. We today frequently use the French phrase.

The fluid-around bathed the cells, watched over them, prevented them getting too hot, too cold, kept them physically constant, chemically constant, therefore body and head constant. Creatures required this constancy if they were to remain healthy or, as said, alive. Bernard's words were, "All the vital mechanisms however varied they may be have only one object, that of preserving constant the conditions of life in the internal environment." He compared the body to a hothouse. The 1,000,000,000,000,000 cells were the flowers. Outside a hothouse the weather could be fickle, scorching summer, January cold, but for the flowers the environment did not change. They had the constancy. The higher in the scale of creatures the more exacting was this requirement. The low could endure larger ups and downs. The frog could. We could not.

Bernard argued that experimental medicine—he said experimental medicine, not experimental physiology, suggesting that, like Pasteur, the health of human beings was at the back of his experimental thinking—ought to have its special instruments as other sciences theirs, the meteorologist the barometer, and ought to use those special instruments to study the internal environment. The lengths to which this would be carried, the shell-within-shell-within-shell revelation of our body, this way of looking at it, and at the dog's body, and the cat's, all parts of all bodies, no one of his day could have foreseen, not he either.

# WATER

## *Essential to It*

Bernard's *milieu* is mostly water. Inside and outside of the cells there is water. Human body, dog body, their guts, their eyes, and cockroach body, also the trim daisy and the portly turnip, every-

where are run through with water. Water flows. Water drops from
the sky.

To Bernard, water was the ordinary water, out of the hydrant, out
of the cistern, not the stuff of intricate internal structure that the
physical chemist today describes.

Bernard's constancy requires water—or his constancies, because
there are many. Blood-cells contain it. Blood-cells travel in a blood-
stream that is largely water. So, water travels in water. And the
stream streaks through flesh that is watery cells plus the watery
fluid-around. If the creature is a swimming fish it has around the
whole of it also the water of the sea, and the human fetus, that fish,
has the water of the mother's sac; and there is the watery mother.
Much water is in each of us: by later studies, 50 percent in our cells,
15 percent around them, 5 percent in our blood. A quarter of a
century ago a Canadian ventured the notion that the fluid-around
was essentially the ancient sea. We came up from the sea, an earlier
model, an earlier earth, were complicated then and became more so.
For ages we went back and forth between sea and land, finally
stayed on the land, on one or another of the juttings hoisted by
frightening force out of the sea, and we brought along inside us
some of that sea. But our private sea on analysis proved not to have
the same salt concentration as the modern Atlantic and Pacific, was
more dilute—then the Canadian came with evidence that the
ancient Atlantic and Pacific were more dilute also.

Increasingly, with increasing learning, we seem on all sides to be
finding that we are simply the old chemicals of earth in new
combinations.

Water is scant in the universe, none on the hot stars that are half
the universe, none on our sun. A frequently discussed question is
how much on the other planets, trifling in any case, Mars given less
than it was, Venus more. The astronauts should tell us precisely.
They have told us reasonably of the moon.

Substances that combine with water are fabulously altered. The
water molecules themselves have a latticed structure. In ice they are
rigidly bonded together, and still considerably bonded when the
water is liquid, kicked about helter-skelter when steam. Water has
such intelligent physical properties, boils at a heat suitable to
coagulate our breakfast egg, freezes suitable for sherbet, dissolves

sugar, dissolves table salt, allows the salt molecules to slip in between and make connections with the water molecules, or the other way. Some of the water of the body is intracellular (inside cells) and some extracellular (outside cells but inside us), this said in consequence of Bernard's thinking but not said in his day, and in our day said with mathematical descriptions but always with some new qualification. We suspect we do not even now know all the pushes and pulls put upon water, and not all the pushes and pulls exerted by water. Water fraternizes freely with those giant molecules that begat life. Lucky for life that it arrived on the earth and discovered the place such a spa. Had the earth not been a grand resort with geysers and mineral springs, life as we know it could not have arisen from the eternally agitated but lifeless molecules at our corner of the Milky Way.

Who cannot recall the shock when first he heard that an average-sized man has a hundred pounds of water in him, the half-worry that he himself might leak. However, a creature of water soon convinces itself that water is the best material to be made of. Bags we are, water bags primordially, bags perpetually balanced in the universe's perpetual balancings, bags that keep their shape, move in a water-laden atmosphere over the watery outside of a sphere that pirouettes with tide-producing moon. The moon is dry. To dry a corpse, to desiccate that—it has been done—is not a courteous way to measure water. Other ways are neater. By all ways we are two-thirds water, 73 percent of our fat-free weight. A new corpse continues to give off insensible water for a time, then that ceases, a balance reached. If identifiable water molecules are injected into the blood of one not a corpse they distribute themselves equitably throughout Bernard's fluid-around within two minutes, as if the body were impatient to establish the new balance.

It goes this way: injection, imbalance, two minutes, new balance, all molecules at relative peace once more.

On a bad midnight when our bowels come to our rescue, flush us out, wash us clean of every chunk of rotten roast beef eaten in the honky-tonk, employ that vulgarest of our outlets, we might hear again in the air the lucid nightmarish Samuel Taylor Coleridge's "Water, water, every where."

# ACID-ALKALI

## *Small Swings*

Acid-alkali can speak for the other constancies.

Our body must not be too acid, not too alkali, only small shifts permissible. The pH says where the body stands. That hieroglyphic, bandied nowadays by high-school chemistry students, is not uncomplicated Egyptian picture writing, but not complicated either. It came after Claude Bernard. It tells, in numbers, where the internal environment stands as to acid-alkali.

"The pH of a man's blood or his urine or the fluid-around in the corridors between his cells is the negative value of the logarithm to the base 10 of his hydrogen ion concentration."

With that now thoroughly understood, we continue.

A living cell produces, finally, carbon dioxide, and since water with carbon dioxide produces carbonic acid, acid is added to the body. Whenever wherever in the water-drenched flesh a water molecule consorts with a carbon-dioxide molecule, the body receives a vanishingly small push to the acid side, the pH falls. A working muscle produces lactic acid—the body to the acid side. Rest after work removes lactic acid—body to the alkali side. The stomach manufactures hydrochloric acid that mixes with the digesting food, virtually leaves the body—to the alkali side. The pancreas manufactures alkaline juice, also mixes with the digesting food—acid side. We eat acid in pickles. Eat alkali in baking soda. In an illness like diabetes the body handles fat abnormally—acid side.

On, on, on, known and unknown, calculable and incalculable pushes, but despite all pushes the acid-alkali constancy remains unbelievably constant. By human yardsticks only the smallest back-and-forth is permissible. Descartes's machine operates best when Bernard's internal environment is somewhat on the alkali side, around pH 7.4.

In short, each of us is an exact cell-dotted slightly alkaline marsh.

Consider the solution of an acid. Acid molecules disport among water molecules and electrified particles, hydrogen ions, appear. The solution tastes sour because of them, stings, burns, damages because of them. Democritus two thousand and more years ago said: "Atoms of sour wine have sharp points that prick the tongue." Those were hydrogen ions. Weak acids have few, strong many. Water itself splits, vanishingly triflingly, and in pure water there are only one-ten-millionth as many hydrogen ions as water molecules.

That last, one-ten-millionth, written in decimals would be 0.0000001. See that on the chalkboard. See a printed list of a chemist's solutions: alongside each its hydrogen-ion concentration. See a gastro-enterologist's recordings of an afternoon of human stomachs. See a researcher's data-sheet: bloods of three guinea pigs, seven tadpoles, one camel. The decimals hypnotize. In water, acid ions multiplied by alkali ions would be one-ten-millionth multiplied by one-ten-millionth and that would come to 0.00000000000001.

Logarithms rescue us from that, make a man feel less inept, allow him to write the hydrogen-ion concentration of water as simply pH 7, neutral.

Pure stomach juice is pH 1, far to the acid side, pancreatic juice, pH 8, far to the alkali side. Always difficult to believe that two quarts of concentrated hydrochloric acid have been produced by our stomach since this same hour yesterday, two pounds of carbonic acid in body and head. The mind feels scalded but the flesh is not. The shuttlecock internal environment keeps bobbing composedly around pH 7.4.

How this spinning constancy of the living machine maintains itself every medical student understands, every science student, every student of Sanskrit on this bespectacled planet. When necessary for the body to lose acid it is breathed out from the lungs, urinated from the kidneys, or fails to be cast out, stays, but stays harmlessly because it has disguised itself in a weak acid plus the weak acid's salt, an ingenuity called buffering, and chemists and Nature employ buffering with the greatest of ease.

Thus could one say that the acid-alkali constancy, that one illustrative constancy of the body, has been accounted for, roughly. It moves between pH 7.35 and 7.45, *in extremis* between 7.00 and 7.70, between death by acidosis, with coma, and death by alkalosis, with convulsion.

As to how little this represents in actual hydrogen, the gas that started the universe, its weight was calculated a quarter of a century ago by Joseph Barcroft, for his amusement. Barcroft never lost his sense of humor when handed a sackful of numbers. He said that the weight of this hydrogen varied from one to five grams, one-thirtieth to five-thirtieths of one ounce, this distributed in all the blood of all the citizens of Great Britain. All the citizens, so we can imagine how trifling the amount in one citizen.

It would have pleased Bernard in Paris to have learned of this calculation in Cambridge, a gentleman on the other side of the Channel so vividly verifying the constancy of his constancy. Now, if the calculation had been wrong! Neither gentleman would have raised an eyebrow. Barcroft died, casually, on the way from his laboratory in a bus.


# CLAUDE BERNARD

## *He Who Conceived It*

What kind of man?

The earliest fact is registered in the village registry. He was born at a usual convenient hour between midnight and morning, July 12, 1813. Later he was baptized. The middle name was his paternal grandfather's, Claude, the full name, Pierre Claude Bernard. That ceremony took place at Saint-Julien, canton Villefranche. Bernard *père* was a failure, a wine-grower, and the mind of the son was from the beginning fed through eyes and ears and nose from the earth of a French countryside. He never cut his connection with that countryside, on his vacations must get his hands into anything that had to do with the making of the wine, an excellent Beaujolais. One supposes it ought to be possible for someone with an instinct for biography to move from such a seed in the youth to the flowering in the man, to the labor of a life, to the discovery of an all-pervading principle. The man, the physiologist, years afterward would reflect on the relative value of facts, how the right ones lead to right hypotheses, the hypotheses to further facts, and so on. Could a profile be written on those lines? Facts there were. He was a reserved boy, aloof really. One felt this as much as found it in the record. His teacher said Claude was only average. A priest was struck by him, the early severity possibly, good stuff for the church. After a time he was apprenticed to a druggist—folded square papers for powders, for eighteen months. At night he might slip away, not frequently of course, this being stern old Europe, to the nearby city of Lyon, to the theatre, wrote a skit for the theatre, *La Rose du Rhône,* which was produced and he earned sundry francs. A five-act play followed, naturally a tragedy, *Arthur de Bretagne,* which was in the baggage when the young man, twenty-one years old now, went where all good Frenchmen go, to Paris. There the external environment was healthily cruel. The play was read by a professor at the Sorbonne. Was read by an actor. By a drama critic. It had faults. Was rejected. But—the dramatist was in the city he wanted to be in. The French world swirled and swished around him. He met old friends. They were studying to be doctors. That drama critic had given the dramatist incidental advice—why not?—and he enrolled in the College of Medicine.

Good chance and bad chance occur in every life. No two persons do the same with them. Some suck out every advantage. Some collapse at the first mischance. Magendie was the great physiologist of France, recognized the quality in Bernard, and Bernard the quality in Magendie, worked for him, then with him, differed with him, left him, returned to him. A single thoroughly understood relation like that (no man has more than one or a few) would light corners in both their lives, but to thoroughly understand would not be easy, not with those two, not easy for the psycho-analyst, not for the literary analyst. Bernard's reserve remained. It may have changed some in appearance. It was formidable. It always had been. It would be difficult to define. (He himself gave out nothing.) Plainly the reserve, whatever it completely meant, was always somewhere at the back of his power. Magendie, when his own death was approaching, predicted for Bernard, if he could not bequeath, and conceivably may have been satisfied not to bequeath, the Chair of Physiology at the *Collège de France.*

Much of Bernard's work was finished before that appointment. Everyone in the fields of medical science knows the work. The reports, articles, eventually books, many translated, are extant, and they reach around the world. His public addresses are. His popular writing, and he did some, can be found in back issues of *Revue des Deux Mondes*. Despite all, one feels shut off, more than from most human beings, and what shuts one off—one keeps wishing one could more exactly identify it. One sees a painted portrait of him and one thinks, ah, a lock of the door is turning, now one will walk in on the scientist, the thinker, the pharmacist-Beaujolais-playwright, find the cue to that bearing, to that statement in the writing so invariably poised, but the next instant the lock still is locked. One knows, too, annoyingly, that there is a key which almost fits and one wishes one had the leisure, possibly the fortitude, to file at it.

The laboratory in which he worked would to us in rich America seem dingy and poor, a place where a worker all winter keeps a running nose and wears a muffler and a tall hat. Discoveries could nevertheless be made under the tall French hat of that day. His laboratory zeal continued uninterruptedly, on frog, on rabbit, on dog. Experimental medicine was in a triumphant mood. The dogs were tied down. No anesthetic. When the blessing of anesthesia came, late in Bernard's life, he used it. A friend related how the master, whose hands were in the open abdomen of a dog that was howling mournfully, turned with benignity to ask the friend to wait a bit, went on with the surgery. Two centuries earlier that dog would have been nailed down, alive. In those old days dogs did not suffer pain at all.

He married late, thirty-two. He married unhappily. He married a woman who disapproved of vivisection, presumably did not want howling dogs so close to her house (the laboratory was close), which might make her sympathetic, but as one knows her better one thinks that even howling dogs may be acceptable. Two sons were born. Both died in infancy. Two daughters lived. When the parents eventually separated, the daughters clung to mama, neither married, had the pleasure of her meticulous bookkeeping, later joined her in establishing an asylum for stray cats, doing as it were penance for papa. To Bernard the separation, and the essential separation that may have begun twenty-five years earlier on the marriage night, could have meant, though he would not have required this, that nothing now stood in the way of the last chapters of some of the most far-reaching experimentation and meditation any biological scientist ever achieved. He led his own field, physiology, into several of its modern directions, placed several milestones, still unmoved, on the long road that leads from Thales and Democritus and Hippocrates and Aristotle and Galen and Vesalius and Descartes and Harvey, to us. Meanwhile he had annoyances. His wife was one—of course he never spoke of that. His colon was another. Not the politics of his time, even though it included the debacle of the Franco-Prussian War, he appearing to have none of the patriotic fervor that boiled in his fellow countryman Pasteur. And stupidity wherever it blundered in was an annoyance. Facts, facts, facts, but they remain mostly outside, could be mistaken, but the existence of a hard inside core cannot be mistaken. One still wishes to file at the key. One would accept help. One would like to hear the sounds with which he swore, that surely was not often, or see the look, that must always have had an overlay of politeness. Inscrutability was indubitable. Integrity was. Nobility. Loneliness—of course he never spoke of that either. His labors continued, discoveries that would be modified as most discoveries are but some that would stand, at least up to now, as he left them. He defined carbon-monoxide poisoning. He proved that the liver stores sugar in the form of a starch, glucose in the form of glycogen, converts the glycogen back to glucose as the body needs it (the understanding of the steps awaiting a subsequent evolution of chemistry), secretes the glucose directly into the blood, internally into the blood, this being the discovery of the first internal secretion, therefore the beginning of today's spreading knowledge of the endocrines. He made probings of the entire digestive tract beyond the stomach. Studied and characterized the similarities and dissimilarities of plants and animals.

It was Claude Bernard baptized the laboratory frog, gave it the name of the Old Testament personality Job, and with that baptism something for a moment does come through the platinum plating, a laughless humor.

He studied, studied, studied. Not anyone ever gave to physiology a greater thrust. Not anyone ever gave to the blood more total meaning. He proved the highly important fact that the size of blood-vessels can vary and that nerves can control this, accordingly would have been in a position to allot to two signs of emotion, blushing and blanching, a simple mechanism: vessels open, vessels close, blood flows in, blood flows out. When, late in his life, Pasteur wrote to him eulogizing his achievements, pointing to the logic in his physiological manner of thinking, as in the discovery of the glycogenic function of the liver, Bernard, reading that eulogy, said he had had a paralysis of control of certain of his blood-vessels, that he was blushing all the way up to his eyes.

In the course of many concrete discoveries there came that generalizing one, that idea which, silently or not, pervades physiology and medicine, of a constant internal environment maintained by a myriadfold balancing and rebalancing. It was his crown. It had put into the long road one of those rare turnings that display suddenly a new country.

He sickened. He had had spells of sickness over years. He was more than sixty. Was aware of his age. It is possible he felt discarded, as even the great old may feel, though he would have been cautiously discarded, his eminence what it was. He still from time to time delivered his lectures at the *Collège de France.* Made his trip to Saint-Julien, where he had visited year after year, on this occasion stayed longer than he planned, allowed Nature time to cure him if she could; his own medicine could not. Nature discarded him too. He was in pain, no complaint at that, except why was there pain when it was of no use? He died. Some have found it possible to be enticed by that perennial question, did he or did he not at the last call a priest? It is true that the Abbé came from the dying man's room, but Paul Bert, great physiologist too, was there, reported that Bernard throughout the Abbé's visit had been in coma.

Claude Bernard—the Abbé—Paul Bert—somewhere two spinsters clinging to mama—our lives in summary all are peculiar.

It was 1878. All France formally mourned. Zola modeled a character after him. Flaubert, powerful French novelist, and Turgeniev, powerful Russian novelist, were two other tall men with tall hats and unapproachable mien in the streets of the Paris of that day. It is comforting to know that Bernard in his later life had had the close friendship of a married woman, his five hundred letters to her preserved, that his quiet eyes could and did light upon a younger person of the other sex slipping in on one of his lectures.

# HOMEOSTASIS

## A Restatement

Years later, at Harvard, Walter Cannon, American physiologist—he died in 1945—restated Bernard's idea. He reshaped it. He brought it into line with his own experimentation, his thinking, his developing concepts of the involuntary nervous-system, his developing concepts of pain, hunger, fear, rage, and in so doing he brought the idea into line with the physiology of the time, helped thus to bridge Bernard from the nineteenth century to the twentieth.

Cannon coined the term *homeostasis*. Bernard's constancy became Cannon's homeostasis. Bernard's fluid-around became Cannon's fluid-matrix. Both men, of different quality, different intellect, different intensity, believed that all systems of the body contributed to the body's constancy, but it was especially the involuntary nervous-system, the autonomic, that Cannon thought controlled the constancy of the fluid-matrix. His work invariably returned there. Cannon was a worker. He never stopped. He had a steady mind. He was a provocative formulator. He was a polemicist, possibly innocent of being, but he was.

His term *homeostasis* flooded the literature. At any meeting of physiologists, biochemists, pharmacologists, pathologists, every roomful might hear the term several times an hour. A generation heard it; another generation hears it. Physicians soon began speaking of getting their sick back to homeostasis. Pediatricians pinpointed the steps of homeostasis in the newborn. All categories of medical scientists every year for at least a quarter of a century would be finding new places to use the term, new experiments to use the idea. A psychiatrist might describe all of his work as a guiding of the disequilibrated mind back to homeostasis. He might go so far as to equate Cannon's homeostasis with Freud's pleasure principle. A few, and this seemed remarkable, equated the stability of the fluid-around with the stability of mental tensions. An occasional one described an external and internal mind environment, related this to an external and internal body environment, then spoke of the four at once, which did not necessarily make everything clearer.

Even politicians and sociologists rolled Cannon's term on their tongues. This would have pleased Cannon, this drawing-in of the social sciences. He always welcomed social implication. He was homespun in look and in mind. He wrote: "It is pertinent to observe that these swings from right to left and vice versa, which produce in society as in well-developed organisms a trend toward a middle course, are possible only in a democratic form of governmental organization." Democracy = homeostasis. A man gets up in the morning, eats breakfast, goes to vote, goes to the office, goes to lunch, returns to the office, returns home, kisses his wife, snaps on the TV, all to maintain the items of family and nation in a happy constancy like that in the fluid around his 1,000,000,000,000,000,000 cells.

Another term began sweeping the literature—mechanism. The two met and wooed and wedded—homeostatic mechanism. It was a wedding that time would render dull, dull, dull. The study of the single constancies, like body temperature, like body pH, like blood sugar, other components of the blood, Claude Bernard's internal secretion (he named only the one), drew investigators this direction then that. Cannon too, but in his experiments he kept returning to the emotions, their signs, their nervous-system relations, nervous-system controls, chemical controls. He was not a chemist. He was a physiologist who throughout his career sought to fit into Claude Bernard's idea his own branching homeostasis, though he probably did not think of his career this way. Life around him meanwhile was more and more mechanized. Physiological laboratories were more and more mechanized.

# DISEASE

## Large Swings

Imagine Cannon's swings swinging too wide. Imagine a healthy body trying to keep in harmony with an unhealthy external environment. That environment may be minute, bacteria, viruses, may be huge, a sick nation, a sick era, or a sick body-part that has become essentially external to the workings of the whole.

Imagine what was a healthy heart now beating with decreasing efficiency against increasing friction of blood moving through narrowing blood-vessels. It was a balanced pump, but it has too long been pumping blood into out-of-balance conduits, and fluid is beginning to collect in places it should not, in amounts it should not, finally a pint in the chest of a child, a quart in the abdomen of a woman, the fluid-loaded tissues not yielding to the physician's effort to dehydrate them, ankles bulging over shoe tops, alabaster-white sacs under the eyes, body-mechanics troubled by swelling and load, body-chemistry troubled, feelings troubled, the person troubled, his wits trying to assist his body, only interfering. The swings are swinging wider. Breathing is too fast or too slow. Blood-pressure soars or tends to collapse. Extremes are becoming both effect and cause, and still the swings are swinging wider. The body and the mind with frank agitation are straining to bring extremes back to mean and the wife of the sick man is shy about speaking to him, glances furtively toward his bed, hopes he will not lose his temper and commit inadvertent suicide. What she thinks she is seeing is her impatient sick husband and what her doctor could hardly make her believe—even if he did not have a sneaking doubt of it himself—is that she is seeing a disturbed machine with bumps and knocks striving to get itself running smoothly again. At the start only one machine-part was tipped, and tipped back by the effort of another machine-part, but the tipping spreading, becoming a complicated zigzag, part supporting part, the sober supporting the drunk till the whole neighborhood staggers, till the physiology is everywhere fighting excessively to make up for not fighting efficiently.

Textbooks do not express this quite this way. Textbooks are not yet prepared to present a sick man as corrective swings lying in bed or perambulating around a lunatic ward, physiological and psychological zeal striving for homeostasis. Tradition is strong. Disease continues to seem so-and-so-many signs plus so-and-so-many symptoms, with the corrective dynamics, what the body is striving to do, definitely not in the foreground. Claude Bernard and Walter Cannon, their ghosts, could be considered as still around and goading physician and experimental worker to the not necessarily easy task of searching ceaselessly for the rebalancing factors in an imbalanced body and head.

The pattern of imbalancings and rebalancings is the sickness. That is what everybody on Linn Street says sand-blond Harry has, a knocking machine bringing in its age-old devices to reduce as far as possible its too-wide swings. Through a long past, Galen to Osler, disease was described as the sight, sound, smell, feel of the enemy, and goes on being so described, though more and more the descriptions include body, brain, mind, striving for constancy, obviously striving at the point of assault, less obviously over the entire hinterland.

If the strains persist, if cells change in their molecules, that is biochemical injury. If biochemical injury persists, if cells die, not enough newborn, organs damaged, that is pathology, may be recognized as such by the physician's fingers, seen under his microscope. Disease is a process aiming at recovery, partially destructive, its intent equilibrative, slanted toward health, but, missing that, accepting the downslope toward death.

# DETERMINISM

## *"L'Idée directrice"*

The first time in our lives some preacher, probably not of the cloth, assured us that everything on and beyond the animal'd earth was an eternal balancing, rebalancing, outside us, inside us, in the large, in the small, in the solar system, in the universe, in the cells of the liver, in the cells of the brain, all expressible in curves running through each other, Claude Bernard's constancy keeping itself constant, Walter Cannon's homeostasis keeping itself homeostatic, nothing else, nothing deliberate, ever, the person seemed mad. Then we got used to the notion and joined the chorus.

A plan *after* the house is up, sketched by the draftsman from the finished thing, is how science understands plan. Plan for a creature that stayed in the sea, one-celled or billion-celled. Plan for a creature that came onto the land. Plan for a creature between sea and land. Plans, plans, plans. Science observes what is before its eyes, wide-open eyes, sometimes eyes that have not lost the dimension of wonder, but human eyes that see a human world, helplessly focused by the techniques, instruments, hypotheses, fashions of a human mind in a human era. Difficult for any of us to accept this though we know it. Difficult to make our historic sense operate except between the covers of a book. Difficult to believe that science can really only find what it is looking for. Science does accept and even insist that the human being is an inescapable part of the human experiment, states this, incorporates it in its principles, in its experimental techniques. In its writings, its seminars, its classrooms, it meanwhile scrupulously and rightly refrains from the personal. Also, it never allows the suggestion of purpose, though like a good doctor it permits each effect to have its cause, that effect to become a cause, and thus on forward or backward.

The man of science is a determinist—things are as they are because they are as they are. Everything can be brought within that enclosure. Everything can be mathematically examined, related, and accounted for. Alive body, alive brain, alive mind, and the dead, what we say are the dead, that should be the alive chemically transformed, are only logical steps in the blind grinding of the universe. Step, step, step, astronomy, physics, chemistry, cytology, and, on the top step, man, the thinker, chews the cud of his thought. He thinks, and not only thinks that he alone on this planet thinks, but that his built-in human-brain manner of thinking is the only manner. His dog thinks, yes, but about a bone. Man does not think about a bone, not all of the time. Man is vain in the presence of the other animals. Man is confident whenever the facts are not just too grim, and even when they are grim quickly recovers his confidence. Man in this and other respects is the eternal optimist, eternal pragmatist, adolescent, contemporary, but underneath is also the eternally haunted.

Science's overall determinism is strengthened by each man's individual determinism, the result of his experience. He is so sure of this that one must sometimes wonder if he is not unsure. Sometimes he vents himself in a burst of annoyance: "Vitalism!" Sometimes contempt: "Religion!" Sometimes: "Metaphysics!" Sometimes a seeming observation: "Nothing is supernatural." Another observation: "No man-type intelligence behind." Making certain of the last: "Nothing behind." Correcting: "Nothing that is not science's oyster." Professorial: "All life can be genetically and environmentally explained."

Anyone would be able from these premises to go on to no freedom-of-the-will. Whatever was mistaken for freedom-of-the-will would on examination reveal that it was in fact determined. Safest for science to allow no freedom-of-the-will, as safest for the church to hold tight to it, because were science ever to let that go, that window open, anything might fly out or in, even what might embarrass science, for though science often has embarrassed the sentimentalist and the church, it still might be embarrassed in its turn. Not likely. Not soon.

Claude Bernard stated his determinism coldly, but he too appears not to have been at ease, called something *l'idée directrice,* the directing idea. What did he mean? That has been haggled over. It could be that an insistent thinker cannot do otherwise, his thinking at a point getting tired, giving up, suspicious that his data may depend only on whether the year is 1969 instead of 1869, more cause-and-effect facts having accumulated, the heap bigger, the meaning the same. Physiology to Claude Bernard was the study of the constancy of his *milieu.* The body was the result of a multitude of constancies. It was a tender tremulous tremendous adjusting, internal environment to external, external to internal, organ to organ, tissue to tissue, cell to cell, reaction to reaction, microsome to microsome, chromosome to chromosome, helical DNA to helical DNA, submicroscopic shape to submicroscopic shape, Democritus' atom to Mendeleev's. The constancy of the internal environment can be considered the shadow of universal adjustment falling across the thin thread of life path.

Behind physiology lies, if you like, evolution, the theory. Near to one hundred years ago while Bernard was developing his theory Darwin was developing his. Our planet is attractive when two such eminences push up their heads in the same brief period. We all believe in evolution. We took it in with our barley water. We believe as Darwin believed. The years since Darwin have added mainly detail, not conception. Lamarck, who earlier than Darwin pondered evolution but differently, entered the nineteenth-century mind, departed from that mind, has somewhat re-entered the twentieth-century mind, while Darwin just stayed. You believe in evolution. I believe in evolution. How ridiculous either of us would appear to himself alone in his room at 3 A.M. if he found himself suddenly not quite believing, though he would not be bringing back angels and miracles of course. Only he would be not quite convinced, as he easily is in daylight, that he himself directly observes evolution. At 3 A.M. he could see William Jennings Bryan and that courtroom in Tennessee trying that schoolteacher, and laugh with satisfaction to think how far he has come, think what a distance this unprecedented half-century has advanced him. By selection from spontaneous variations have been evolved the species. If someone for an instant feels this difficult he is quickly reminded of the vastness of postulated life on earth, three thousand million years by current calculation, and, to earliest man, seven hundred and fifty thousand years, or two and one-half million. Given these spans anything could occur, would occur. Notwithstanding, on some days in some moods someone not in the least Tennessee-trial-oriented catches himself staring at the gaps. While science explains he stares. From amoeba to python to house-mouse to house-cat to tree-shrew to chimpanzee to Fidel Castro are such engaging leaps. From archeozoic to proterozoic to paleozoic to mesozoic to cenozoic are a fatality softened somewhat when he remembers again that man's mind is still an instrument that divides, cannot do otherwise, divides, divides, divides.

Behind physiology and evolution lies, if you like, astronomy. Again by adjustment there has come a universe that did or did not expand, a universe that did or did not arrive by the big bang, but, either way, a universe incomprehensible except as we think not of it but of our measurements of it, a universe achieved from nothing, a universe that continues to be achieved from matter and anti-matter, but however achieved is magnificent, produced the night sky which we have beheld from when we first looked up, the Great Bear and the Southern Cross and to the right someone shrouded carrying Mars as a crystal lamp, and all the visible stars and all the invisible with mathematical infallibility each where it ought to be.

Behind physiology and evolution and astronomy lies, if you like, physics. Down into the atom goes every high-school son of every PTA mother, sees the particles when the atom is smashed, the few or the scores depending on the manner of smashing, the year of the smashing, sees electrons revolving, sees cosmic rays hitting the atmosphere and producing muons, sees neutrinos passing straight through the earth, sees one after another of the particles found by the physicist's instruments, or by the physicist's reasoning, each particle with statistical infallibility where it ought to be, and there is better than a hint that these are but the heralds of hosts of symmetry still to come.

It leaves us with this: symmetry is everywhere, even behind any ostensible physical or chemical asymmetry. Order is everywhere, even in ostensible disorder. The random never is random. Plan is everywhere—plan after the house went up. The structure got there without mind, without forethought, without beginning, without end, a derelict, and the courageous blueprint of science was worked out by the draftsman subsequently.

Nevertheless, there still is left in the riotous sceptical close of the twentieth century an occasional soft temperament. This one clings to the discredited other plan—plan *before* the house went up. Fool! He requires Olympian motive. He too took in his with his barley water. The blind grinding operated upon him too but left him incapable of imagining a primordial mass-energy from nowhere, then the particles, the waves, the atoms, the molecules, the molecules that grew larger, the molecules that replicated, the living cell, the orchid, then Shakespeare, then Beethoven, left him incapable in some hours of not asking if this sequence could ever, in spite of the inevitability of all the successive evolutions, have occurred without an antecedent deliberate something-or-other. Fool! A bit of colored glass from his youth distorts his vision. He gropes, he sputters, he blurts out an incoherence and someone nearby says with disgust: "Teleology." Teleology might be the reverse of determinism and the meaning of both might change with the winds of time. Purpose could even be old-style hell-and-heaven purpose. Granted that this is unlikely. Granted that none of it could be on a human scale. Granted that it would not be understandable purpose. To say that something is not understandable is not to say that it will never be understood, or that it is supernatural, nor need the speaker be crossed off as benighted antiscience, Protestant, Catholic, Jew, Buddhist, sun worshiper. He all-helplessly has the not-understandable as part of the inherited organization of his mind, so let him be. Or, he has the not-understandable as a construct that the organization of his mind caused him to place in his mind, so let him be. Possibly he is stupid. Possibly not. Possibly he is arrogant. Possibly humble.

Should he be humble—this is to be recognized—he might perform even such an absurdity as fall to his knees not because of an outer commandment but because his having-been-born, his having-lived, his having-entered-into-confusion, his having-to-die, continue so to overwhelm him that he is shoved to his knees by his own inner compulsion. He does not call it mystical compulsion, because he shrinks from being smiled at, even by himself. Not only can he not put away the idea of not comprehending, he is ridiculous enough to debate his outmoded thought: to wit, a man's power to reason is built upon his experience, and his experience upon the manner in which his body and brain were built. The cockroach, *le pauvre*, comprehends its world, yea, but comprehend the world, nay. To the outmoded one it might seem that what science has taught him best is that body and mind were from the start beset with limitations, these assuredly leaving some, conceivably much, of reality beyond his reach. His human head is always getting in between. There is no escape. His human head places its own limitations. The limitations in any case are fact. He may speculate beyond his limitations but he does not know beyond them. A silhouette is cut from the unknowable by the knowable, the existence of the knowable proving the unknowable, or so it could be argued. It would not follow—this too is to be recognized—that everyone's god must wear a white beard and a white robe, be made in man's image, think man's thoughts, float in a white cloud over the blue Italian source of it all, the ancient sea. Anyone's god (*god* is a man-made word) just might be, and might be past all theological or philosophical or scientific reckoning, be, as some alchemist of the Middle Ages would have considered, timeless, matterless, a literal from-nothing, a literal is-and-was-and-will-be, an inscrutable for which an artist might daub in a term like luminosity, a luminosity that some intelligences block, some centuries block, and others, intelligences and centuries, let through. Anyone's god just might be. Then God would not be dead as Nietzsche said and as the 1969 schoolboy and his teacher repeat in unison, and would have been glimpsed through a crack forced open by a science that did not mean to force open anything, yet might seem realer than if glimpsed by faith alone. Even in 1969, and even to a science that eschews first causes, a particular first cause just might be.

"It might? And upon what ground does a man then stand?"

"Upon uncertainty—an alert uncertainty that continues in all things to try to find certainty with no hope ever of finding it—where man has always stood."

"And be satisfied with that?"

"Be satisfied with that."

# V

# MUSCLE

———

## Ivory Until It Moves

# KINDS

## The Silken Cord

By weight a man's muscles are some 40 percent of him. On some days they feel none of him, on a bad day, all of him. He has 434 individual muscles. He eats muscle, as do other animals, but his he eats fried, broiled, fricasseed. If he is a biologist he studies muscle.

There are three kinds of muscle, with three appearances under the microscope, and three responsibilities.

*Smooth. Striated. Cardiac.*

The first—*smooth*—does its work in bowel, blood-vessels, pupil of the eye. The second—*striated*—fills the world with acts, enabled Galatea to step down from her dead ivory, enabled laborers to raise St. Paul's from the ground of London, Notre Dame from the streets of Paris, the Sphinx from the sand of the desert. The third—*cardiac* —is right now quietly or noisily beating in your chest, and in Job's, the frog's.

*Smooth* looks smooth under the microscope, is laid down fiber next fiber next fiber, fine fibers uniform throughout their length. The fine fibers look frail, but when they join their strengths they can generate great force, as a cramp in anyone's belly suggests. Scrotum, nipple, prostate, rectum, vagina, gullet, bladder, hair follicles, various valves, walls of corridors, of tubes like that from ovary to uterus, the grand uterus itself, all of the unwilled movements of all of these are the work of smooth muscle.

*Striated* looks striated under the microscope, its fibers striking one for their sturdiness, and for their gay barber-pole-column appearance. The fibers have alternate light and dark bands. With careful lighting, a film of striated muscle may strike one as a more delicate fabric than any one has ever seen.

It has been thought that the more the striation the faster the performance. Striated causes the slight tension that is always over the body-wall, that gives the body its feel of vibrancy, feel of life, which is much of everyman's beauty so far as he has it. Beauty was hardly in evolution's idea—but how do we know?

A medical student in an idle humor dressed a skeleton in a white dissecting gown, and at a distance in the late afternoon the skeleton might have been mistaken for the head nurse, but, a yard closer, the illusion dissolved; the slut simply had neglected to put on her muscle-slip.

Striated wrinkles a man's brow, turns his frown into a smile, tightens his eardrum, grinds a chop between his teeth, kicks a football.

*Cardiac* has striation too but its own kind, one fiber everywhere bridging across to the next fiber, the ordinary light-microscope quite sufficient to show the bridging. By the bridging the heart knits itself into a single mass, that beats. Any fragment beats. If the heart of a frog is cut out with a scissors, as has been done in a hundred colleges, then shredded, the shreds beat. The heart of a hummingbird may beat eight hundred times a minute, ours seventy or eighty, quicker in the expectant debutante, slower in her no-longer-expectant bachelor uncle. When Job is not out marketing and not at home worrying his heart beats approximately thirty-six.

# CONTRACTION

## *Lifts the World*

Muscle contracts. That is its genius. It shortens. Some muscles shorten slowly. Some shorten rapidly. After a muscle has shortened it may immediately lengthen, so that it can shorten again. There are accompanying electrical events. There are accompanying chemical events.

When a muscle shortens it pulls on whatever it is hooked to. Say that is two bones. The two join to make a hinge. When the muscle shortens the two are pulled toward each other, the hinge is closed. Then the muscle relaxes and the hinge opens. The door opens. In the green frog hinges close-open, close-open, close-open, and that is the frog hopping from stone to stone across the brook to the mud of the other bank. In the green cadet hinges close-open, close-open, arm lifts, hand salutes, band plays "Star-Spangled Banner."

Jan Swammerdam—1637 to 1680—devised an experiment in which he suspended a muscle in water in a glass flask that had a narrow top. He marked the level of the water. Then he pinched the muscle's nerve, and the muscle contracted, got fatter, but the level of the water stayed where it was. So, when a muscle contracted, though it changed shape it did not change size. Today it is found that there is some change of size but so slight that Swammerdam would have scorned it. Nevertheless, it indicates the direction of today's muscle-thinking, indicates the refinement.

The calf muscle, the large muscle at the back of the lower leg, is called gastrocnemius, frog or man. For the movement of the body of either there are fast fibers; for simply holding a position, slow fibers. That muscle puts the power into the hop that all day sends the frog across the brook toward town. In a man it makes possible, as Oscar Batson stated it, the springing step, tiptoeing, toe-dancing, raiding of pantry shelves.

Imagine a frog's gastrocnemius fastened at its upper end to a stationary support, at its lower end to a lever. Wires from an electric battery touch it, and a switch is in the circuit. The muscle is electrically stimulated. It contracts. It lifts the lever. The tip of the lever rests against the smoked surface of a revolving drum and therefore writes a curve. That is the simplest laboratory situation.

In the above experiment the muscle was not prevented from shortening: that is, it was allowed freely to lift the lever. In another experiment the shortening is prevented, the muscle held fixed, and when it is electrically stimulated all it can do is build up tension inside itself. That also can be measured. Both situations occur in the daily life of frog or man. A two-hundred-pound man with a twenty-pound pail is walking along, the muscles carrying the pail have inside them the tension, otherwise the pail would fall from the man's hand, but the muscles are not shortening, else the pail would not stay down where it is. However, now the man decides to lift the pail, and for that the muscles must shorten, and thus the one type of contraction passes over into the other.

Amputation stumps of World War II made possible human experiments imitating frog experiments. The levers were attached to the stumps. Other recordings were of the muscle's electrical response during action. Other recordings were of the amount of oxygen a man breathed in when he was doing either of two kinds of muscle-work, going upstairs, going downstairs; spoken of as positive work, negative work.

It has been calculated that the gastrocnemius of a human sprinter is capable of lifting six times his weight. A calf muscle lifts six men! The sum of the muscle-fibers in a human body has been calculated at a quarter billion, which seems many, but six men also is many. Were that quarter billion rigged to pull at one time in one direction it could lift twenty-five tons. Muscle does thus become the silken cord that lifts the world, and a phrase becomes a fact.

Muscle is classified as an effector. It effects. It does something. Any body-part that thus does is an effector. Man orates, hippopotamus idles in the shallow hot river, frog hops, the chameleon produces color, the firefly fire, and light-producing effectors range up and down among earth's creatures. Electric eel, electric catfish, electric ray produce an electric shock and the body-parts concerned are effectors. Some electric fishes produce voltages great enough to stun an ox. Inside our body at its quietest are uncountable movements, some visible under the microscope, like the lashing of the cilia that provides the force that paddles the ovum to its destination, or the beat of vacuoles, or the crawling of crawling cells. Then there are the contractions of those 434 individual muscles with their quarter billion fibers.

# MUSCLE-MOLECULE
## *With Millions in Parallel*

Picture again that calf-muscle of the frog. With a scalpel it has been dissected free of the frog. Imagine that you have done it. Now take a needle and, carefully, there where the tissue splits naturally along the length of the muscle, tease it apart. The total muscle is constructed of bundles of fibers. A bundle is a fasciculus. Fasciculation is a contraction of such bundles just under the skin, one after another, spurts of contraction, twitchings, happens normally with a tired eyelid, but can be a sign of nervous-system disease.

Now, tease the fasciculus into its fibers. A fiber is still large enough to be seen with the naked eye. Tease a fiber into fibrils. A fibril requires the microscope. Tease a fibril into micelles. Tease a micelle into filaments. A filament is the smallest unit known.

In this stepwise fashion you have left behind you your naked eye, then the ordinary microscope, then the electron-microscope, and have traveled that road that leads to where you must place your faith in molecules.

The pursuit of the physics and the chemistry of the muscle-molecule, and the relating of it to contraction, have not only not ceased but the pursuit is filling in with an abundance of experimentation and often bewildering hypothesis.

Years ago a contraction was spoken of as an explosion, too slight to make a noise. For an explosion there must be a saving of explosive. That rude comparison is not entirely out-of-date. When a muscle relaxes, it would be saving; when it contracts, spending. The saving and the spending have both been subjects for chemist work. Wing muscles in insects contract hundreds of times a second.

Years ago a contraction was spoken of also as the recoil of a spring. While a spring was being stretched it stored energy, when released, spent energy. This did appear gross for an affair of molecules though on and off the work of some student of the muscle-molecule has encouraged one to think of a spring. We know so much more of the shapes of molecules and how molecule-shape may go hand in hand with molecule-work.

It has been generally accepted that there are two kinds of muscle-filament, that the two kinds lie parallel and overlap, one thick, one thin. The overlapping occurs at the ends. On the instant of contraction the overlapping is greater, the filaments sliding past each other, telescoping, which could explain the shortening, and could explain also the changing microscopic picture during the shortening. What happened in the small might thus have helped to anticipate what happened in the large. The evidence from X ray and electron-microscope accounted increasingly also for the light and dark bands seen in striated muscle with the ordinary microscope.

If in one's mind's eye one now added submicroscopic to submicroscopic till it became microscopic, then added microscopic to microscopic, then macroscopic to macroscopic, one might once more see that frog hopping from stone to stone across the brook.

A Hungarian scientist working with the chemistry discovered that if two muscle-proteins (about 20 percent of muscle is protein and the rest mostly water) were brought together as a thin jelly, and this dropped into a juice expressed from muscle, the jelly contracted. It got stiff after it had been pliable. This was an early successful stride in the advance upon the nature of the chemistry of striated and cardiac muscles especially, and upon their phases of contraction.

Twenty-five years and more have elapsed since the Hungarian did his time-withstanding study. Recently, he has appeared inclined to laugh lightly at this work, to speak of it as almost trivial, and to concern himself with music and philosophy, as Plato concerned himself with music and philosophy. He would laugh lightly at such a statement too.

One of the Hungarian's muscle-molecules possessed a double talent; could contract, could excite the drawing of energy for the contraction from the high-energy source. It was a contractile chemical (had that energy), and could excite the contractile chemistry (an enzyme therefore). A link within the molecule supplied the energy (is supplying it for you now as you read), and the production of that energy costs the same as the production of any energy. This energy was drawn finally from afternoon tea and cake, and the energy was employed in the washing of diapers, in the electron-microscope examination of muscle filaments, or, if you are a frog, in hopping, hopping, hopping. The frog's tea and cake would be pond water and insect.

The story of muscle-contraction is a modern story, is based on earlier stories, is part of a serial written by successive authors as they appear and disappear, every now and then a particularly brilliant installment. All plots and clues and episodes fuse at last into one grand tale with a title, *Contraction 1969*. On Times Square on New Year's Eve far within the visible scene are thin filaments sliding past each other, and millions or billions or trillions of molecules interacting while great muscles get shorter and longer. Happy New Year!

# VI

# HEART

—————

## Regular Day-and-Night Deliveries

# FOUR CHAMBERS

## *Pump*

Long back when each of us was a child, in some stillness, some dark night it may have been, he discovered this thing that beats in his chest, this fact. It would be years, if ever, before he could think such a thought as that this was just a muscle contracting. He put his hand there. . . . The engine was in him. . . .

A schoolteacher, later, showed him a drawing of a heart. Another schoolteacher, later again, brought a pig's heart from the butcher shop, told the class their own was like that. She sliced in. It was lined with red velvet. From then on he knew that his blood moved over surfaces that were smooth, and, later again, he might have thought it out, that gentleness besides power was built into this pump.

Not all animal-hearts have four chambers. The turtle has three, the frog three, but dog and cat and monkey have four, like us. Two of our four are on top side by side, called atria, formerly called auricles, still often called so. (Auricle=little ear.) Two are on the bottom, ventricles. (Ventricle=little belly.)

The walls of the chambers being muscle it seems plain to us (not plain to anyone before William Harvey) that when a chamber contracts, the blood in it is squeezed, and any heart-door that was ajar is slammed shut, and any that was shut is thrown open, and the blood rushes past the open door, the valve.

Thereafter the blood continues on the postman's appointed round. From every part of the body it is collected in the right auricle → is delivered from there to the right ventricle → thence lungs → left auricle → left ventricle → out to all the zip-code areas. That is one round, one cycle. Collected from the parts of the body → dropped into the right auricle → dropped, driven, sucked into the right ventricle → delivered to the lungs where it enters dark red and departs light red → to the left auricle → dropped, driven, sucked into the left ventricle → out to the suburbs → back to the heart. Servetus, twenty-five years before Harvey's birth, had described the routing through the lungs, also knew of that change from dark red to light red, and because of this and other knowledge was ordered burned to death by his friend John Calvin. Calvin was only doing his duty, serving God.

The wall of the right ventricle is thin, if held toward the sun the light comes through, and it can be thin because it delivers only in and out of the lungs. The left is thick, must deliver from toenails to scalp.

Behind breastbone and ribs, somewhat to the left, in a stout sac that prevents outward bulging during the beat, and with pads of fat cushioning it, hangs the heart. From no more than its shape and its hardness the student might suspect the economy that lets this pump with such mail-clerk regularity meet its day-and-night appointments. In the laboratory an experiment will long be over, and still the dog-heart be beating, and in a hospital ward a poor cancerous woman because of the stubbornness of her pump finds it difficult to get where she is resolved to, back into the warm earth.

Through ages of change on the planet, through mammals, birds, reptiles, this organ has had the same seeming simple design, the same seeming simple action, and the least that can be said: you have something long-thought-out pumping inside of you, brother. Thump, thump, thump, goes the pump in the chick in the egg on the second day after hen and rooster came to a decision, a quiet thump in a chick, and the intelligent hen listening over her eggs hears nothing, though the chick might well hear his mother out there cackling. In the bit of transparent slime that in a few days will be a living snail is the thump, thump, thump, quieter still, no human ear able to hear it.

Three thousand million beats beat in a man who gets old, hundred thousand or so in one day. By heartbeats a life is not brief. Rarely is a beat missed; if a few in succession are, the body drops like a stone.

How much heart-work is done in a life? By one melodramatic arithmetic each ventricle puts out 150,000 tons of blood, by another, raises a ten-ton weight ten miles. For the cardiovascular expert to have taken off a minute to make such a calculation brings his mind down toward ours, helps ours climb up toward his, starts us thinking of the gambles some surgeon today risks with the heart, then—do not ask why—starts us thinking of stars and moon and the beginnings of the universe. It could be no more than that in our mind the heart that had thumped, thumped, thumped, now just ticks, ticks, ticks, is frail, is not threatening at all.

A Greek twenty-three hundred years ago watched an animal embryo before its heart had formed, could see, there where it would form, a point that paled then flushed. Harvey two thousand years later watched a point that paled then flushed, comprehended that the paling foreshadowed a heart emptying, the flushing its refilling. Paling-flushing, emptying-refilling, the auricles always ahead of the ventricles. This fact of auricles always ahead of ventricles was one step in Harvey's logic. All steps in that logic were fought over. In some detail they would continue to be fought over, often only because fight is a pleasure. Yet, a child who had had to hear, dinner after dinner, of riot or basketball, might get weary and fall asleep early, and then if in the stillness and the dark of his bed he again put his hand there, what leaps at it is the apex of the ventricles. He turns on the light. Lies quiet. Looks. Over his chest a sleep-evoking rhythm comes and goes. He can hardly know that that orderly rhythm is not quite orderly, that within every few beats one is larger than the next, that one smaller than the one after. Years again, he has become a freshman medical student, and he stands over the exposed heart of a dog, its brain guillotined off, its sensations gone, its pump pumping, and his eyes try to hold on to what is before them, as Harvey's eyes had tried and could not quite, lacks the concentration or whatever it takes, and the student lapses into a cool mathematical wonderment. The heart gives the force to the circulation and every cell of our body depends on it.

## ARTERY

### *Road Out*

From the pump go the pipelines—go the day-and-night deliveries.

A man's chest has been laid open, his heart beating, a meditative heart-surgeon looking in, all of his face except his eyes masked. That scene would frighten a child, fascinate a philosopher, send a modern-art sophisticate back to his studio. (A mighty sight it is up at the top of the heart.) An intense man, the meditative surgeon, but quiet, compels himself to be quiet, and lucky it is for his patient. The scene is an old one for him, little puzzlement, but great puzzlement for that medical student permitted in for the first time on the unspoken condition that he will efface himself.

The Middle Ages called the *great artery* the aorta, and the word means to heave. The aorta takes off from the left ventricle, sweeps up then back then down, makes an arch, arch of the aorta, after that drops almost straight through chest, through abdomen, the vertebral column behind. From the top of the arch branches go to the head and brain. Below, branches go to the lungs, sac of the heart, diaphragm, liver, bowel, spleen, pancreas; right and left a stout branch goes to each kidney; branch, branch, branch, at last those distant ones that deliver to a man's toes.

The eyes in the mask glance toward the arch. That draws the eyes of the medical student toward the arch. Pipelines are coming, pipelines are going. So much of us is pipes.

Possibly the surgeon is not quiet. Possibly if we could see inside him he is not, and even luckier for his patient to have a disciplined worried surgeon with an enforced quiet who anticipates exactly where the knife will go next, knows that were that knife to idle an instant this sleeping human being might sink deeper than was in the plan. And what difference would that make in eons? None. But we do not live in eons, not a surgeon, not a poet, surely not a patient. We live today-tomorrow, and none of us is satisfied to be what he seems to be, a comma on a page in a stupendous history.

The wall of the aorta is elastic. It is resilient. The walls of all the arteries are resilient, more so in the young, which is why a child can dash from her father and he can only puff after. But were there nothing but elastic tissue, were there no tough inelastic framework within that elastic tissue, the artery like a rubber tube inflated too often would blow out. The aorta has been studied for that fabric, studied for its place in a geometry of pipes, for the way a viscous fluid leaping into it would stretch it, for the changing demands put on it. Harvey knew of the changing demands, knew that the blood was thrown in, sometimes larger amounts, sometimes smaller, sometimes a fast circulation, sometimes a slow, sensed the problems as perhaps no one ever had or perhaps ever will. The seventeenth century as well as the twentieth century must have excited the mathematician and engineer who are inside of every cardiologist.

Cross-sections of the aorta have been sliced out, fitted into an apparatus where they can be subjected to stress, the nature-made substance pitted against the man-made, sometimes the surprise expressed that nature still can do better than man.

Each heartbeat throws in the blood, the arteries yield, then pay back that yield while waiting for the next beat, hence the stream despite the pump's intermittency moves smoothly.

In the suburbs, at a point, the arteries take on a heavy circular muscle. When this contracts the stream at the point is narrowed. Blood can by such contraction be diverted from here to there, from where it is less needed to where more needed. That contracting can be too tight and too continuous, emotion among the reasons, and the heart accordingly pumps against a useless narrowing, does useless work. Evolution built the system flexibly, but not for useless work, and a man to his astonishment, if he has time for astonishment, may leave his job unfinished.

The branchings of a good-sized artery were counted. It was the mesenteric that delivers to the bowel. It has fifteen main branches, one thousand eight hundred and ninety-nine bowel branches. There are then the branchings of the branchings. Each branch has less space than the artery it came from, but the summed space of the branches is always greater than the space in that artery. More space but the same quantity of blood, so the stream must be moving slower and slower. Furthermore, many speeds are in each branch, one speed along the walls, another at the center, and every manner

of speed in between, making a situation that the statistically inclined can develop as far as they like. A few moments ago, the single
swift stream of the *great artery,* now the slower and slower streams
of those smaller and smaller branches spreading out over the countryside, giving one a gay feeling, all that blood hurrying toward
unnumbered and unnamed creeks and rills.

# PRESSURE
## *Pushing Deliveries*

A mosquito has blood-pressure, and if some insect physiologist has
not yet gotten around to measuring it, he will.

In every creature high enough to have a circulation there is friction between vessels and blood, also between blood and blood, so
there must be pressure to keep the stream going. Five quarts of
blood or thereabouts must go the round-and-round in us. Five quarts
must be squeezed through all the narrow passageways and brought
back to the heart. Imagine the heart to stop, the vessels to vanish:
blood-pressure falls to zero. No force + no resistance = no pressure.
Of course, the circulation adds no unnecessary resistance, as the
rich throw nothing away. We are never surprised to know again that
the body operates economically. With so much to be done, if
more than just enough were spent on each beat, those persons
destined to wear out because of their hearts would wear out sooner.
Nature does not wish us to wear out sooner. Has her reasons, we
ours. Does not wish us to wear out later either. Should one of us
decide to be shrewd about it and sit on his posterior holding a fishpole and alternate that with driving a ball over the green, or, in
Harvey's England, "to ride comely: to run fair at tilt or ring: to play
at all weapons: to shoot fair in bow, or surely in gun: to vault
lustily: to run: to leap: to wrestle: to swim: to dance comely: to
sing, and play of instruments cunningly: to hawk: to hunt," et
cetera, our beloved spouse might say, afterward, that she never
could understand why a man who had taken such care of his health
could have passed on so soon.

As with Harvey's Englishman, our blood-pressure goes up when we do exercises, when we have a fright, when we have an appointment with the administrator. Odd to consider that the week-end-tanned administrator can in five minutes send up what evolution kept down through five hundred million years.

The first taking of a blood-pressure was a century after Harvey, in 1732, by another Englishman, Stephen Hales, a clergyman, who fastened a mare, as he stated it, "alive on her back and exposed the large artery in her thigh about three inches below her belly." She was anyway "to have been killed as unfit for service." (Must not spoil a good horse.) The Reverend put a tie around the artery, inserted into it a tall glass tube via two brass tubes, and made the plumbing connections with the windpipe of a goose. He said: "The design of using the Wind-Pipe was by its pliancy to avoid the inconveniences that might rise if the Mare struggled." (Mare must not struggle.) So, perpendicular stood the glass tube. The Reverend untied the tie. First with a big leap, then with small and smaller leaps the blood rose and reached a height of more than eight feet. "Perpendicular above the level of the left ventricle of the heart." What it meant was that a mare's arterial pressure lifted a column of blood eight feet. A man's can five feet. (Must not inquire what became of the mare.)

No one expects a man to be as compliant as his horse and when a physician takes a man's blood-pressure, which no physician in Harvey's or Stephen Hales's day did, nothing is nicked except possibly the man's plans for his winter vacation.

Each of us has had it done to him. A narrow rubber bag (in the long-used instrument) is fitted inside a cloth bag, the combination wrapped around the upper arm. A rubber hose connects rubber bag to a U-shaped glass tube filled in part with mercury and etched with a scale. Human pressures are in millimeters of mercury, the mare's was in feet of blood. Rubber bag, rubber hose, U-shaped glass tube lead one into the other, so pressure in one is the same as in the other. A rubber bulb is for inflating. The physician locates the artery in the bend of the elbow, listens, hears nothing. Begins inflating. Up goes the pressure in the bag. Up the mercury in the tube. The artery begins to flatten under the pressure. The friction between moving blood and vessel-wall increases. A noise. Finally, the artery is flattened altogether. No noise. Now the physician lets air leak out, and when enough has and the pressure in the bag falls below the highest pressure in the artery, some blood slips through (as it did under Harvey's finger when he was performing his tourniquet experiment), and there is a click. The physician reads the mercury. That is the highest pressure in the artery. More air leaks. More blood slips through and the quality of the sound of the clicks changes. Then the pressure in the bag drops below the lowest in the artery. No noise. The physician reads. He has now a high reading and a low, a systolic and a diastolic pressure. That word, *diastolic,* came from medieval Latin, and that came from Greek, and the English of it was not used until the year Harvey was born. Systolic was used the next year, 1579.

Blood-pressures are routine in our lives. Anyone not interested may be a fool, but if too interested is sure to be a fool. Descartes's machine-of-the-body runs best when it is running without our thinking. Human beings think too much. A giraffe does not. A giraffe's systolic is somewhere about 360, a man's 120. A tough job for the giraffe's heart to lift her blood twenty feet above the ground, and should she lean over to slip on her shoe with her teeth, as some girls do, there ought to be a drastic shift of pressure in her delicate skull with swimming of her demure thoughts, but that does not occur because she has valves to prevent it. Lucky for a giraffe's mind to have a giraffe's body under it. Lucky for a giraffe's life that she does not every hour worry that her machine is going to break down. Lucky giraffe.

# CAPILLARY

## Gone Underground

The thin walls of the vessels meanwhile have gotten thinner and thinner until they are thinnest of the thin, finest of the fine, one cell thick, or possibly one should say one cell thin. A delicate membrane supports those cells.

And we have here arrived at the true capillary.

It is a cell or cells rolled into a tube. Difference of opinion arose as to whether there were or were not infinitesimal passageways also leading in and out of the tube. These would be sieves to filter molecules. Since the capillary was pieced together of single flat cells, microscopic paving stones, some researchers were of the opinion that there was a cement between the stones, a mortar, and that that was where the deliveries in and out were made, the deliveries required by the tissues, the deliveries produced by the tissues. The idea of cement is out of favor today but one can see the picture, crowds of molecules with their contained atoms being picked up on either side, pushed or pulled through to the other side. It is a most active transfer. It is busier than any gross movement of blood along the channels. Within the blood there is the busyness of its molecules, a busyness to fill one with awe, though to some eye the blood may look quite placid.

At the time of the research on the cement this was thought to dissolve, the molecules slip through, the cement re-form. The technique was essentially to dye the cement and watch it chipped away by the traffic. It turned out not true. Day after tomorrow someone, no doubt, will be directly watching the tumbling molecules, and right now with the electron-microscope the content of a capillary looks as large as a lake.

Most times the capillary has struck men as a frail tube. The name comes from the Latin *capillus,* means hair.

Latin is the source of most of the vocabulary of our body. Aubrey comments on Harvey's Latin. "He wrote a very bad hand, which (with use) I could pretty well read. He understood Greek and Latin pretty well, but was no Critique, and he wrote very bad Latin." Aubrey says that Harvey's *De Motu Cordis* was "doone into Latin" by a Sir George Ent.

Blood moves slowly in a capillary. When the second-hand of a watch has leapt once, blood has entered a capillary, loiters, and when the second-hand has leapt again this blood has departed that capillary. (On the average.) Some of its fluid will in that time have leaked through the wall, the blood-cells be more packed. The picture is always different. With injury the packing may be so great that any novice can see it.

The routing of the blood—in detail—has been from artery to

arteriole, to precapillary, to thoroughfare capillary, to true capillary, to prevein, to venule, to vein. The length of the capillary was one fiftieth of an inch. (On the average.)

It is evident that the destination of the entire system is the capillary. Heart and arteries and veins adjust their work so the capillary can get done its work. Thither the heart is pumping. One might say that the rest of the circulation is the servant of the capillary.

August Krogh, genius, Dane, physiologist, Nobel Prize winner, did some of the loveliest studies in this world on those small vessels. He chose muscle capillaries. He showed how they open and close. Only a few will be open when the muscle is at rest, perhaps fifty times as many open when the muscle is at work. By-products of the work decide this. Imagine an open one: blood flows, the by-products are washed away, the open one closes, the by-products accumulate, it opens. Myriads obey this type of regulation. There are other types. Everything in us is regulated. That could stifle our soul—not one molecule is free. Krogh would poke the needle of a syringe into an artery heading toward muscle capillaries, inject, kill the animal, slice out a piece of the muscle, harden it, cut microscopic sections at right angles to the capillaries, the open ones visible because filled with the injection material. So he would sit there and count. So he would prove, in numbers, the startling difference between rest and work. A bit of muscle the diameter of a pin might contain seven hundred muscle-fibers, two hundred capillaries, and the two hundred all open during work, all closed during rest. In the human being with his anxious intellect the opening could occur when he was merely thinking work.

For that reader who likes things laid end-to-end the capillaries of his body, connected, would make a hairbreadth pipe reaching several times around the earth.

Harvey knew that there must be connections between the two sides of the circulation. That fact troubled him, but he had not the instrument to let him see. He spoke of pores, as Galen had. Those pores are our capillaries.

Too bad that Harvey never saw a capillary. He would have needed to live only three more years.

Now contemplate the tip of your wiggling minimus, your little finger, and the thought of what goes on there may make you laugh outright, and not for the funniness but because your finger will have become one of the world's great traffic centers, a jammed Champs Elysées with its jammed sidestreets.

# VEIN

## Road Back

The same industrious worker who counted the arteries that deliver to the hills and valleys of the bowel, counted the veins coming out of those hills and valleys, from the smallest tributaries, the venules, to the large collecting mesenteric vein. Millions drain into thousands, thousands to hundreds, hundreds to one.

Veins are different in texture from arteries, less stiff, more yielding, abler therefore to hold variable quantities of blood. During the rhythmic increase and decrease of space inside a lung there is an increase and decrease of space inside the lung's veins. They follow the lungs. It is because of that quality in veins that they can serve as a reservoir for blood, this blood always of course on the move from the reservoir's inlet toward its outlet. More than half of our blood, 65 percent, may be in our veins.

The cells of the inner lining of the finest veins can pull apart, leave tiny channels through which there could be a passage of molecules in or out, which the electron-microscope lets us see.

Miraculous in an animal embryo to follow the assembling of veins, each as if it knew what it was about, never as if it were obeying a blind learning of a blind past. Everywhere we get glimpses into the ancient history of veins, through vertebrates, birds, turtles, snakes. Harvey watched veins in all of these. Everywhere the developing body throws beams into the corridors, and what is taking place in the full womb of a woman often has been regarded as a kaleidoscopic résumé of a story that runs off to the beginnings of the world, many installments in that story, a chronological sequence, no doubt revised, revised, revised, but always kept under one cover.

On the blood's return-trip no more force is wasted than on the way out, just enough left at the last to spill the blood into the right auricle, zero pressure, or nearly. It must be more than that when a heart is enfeebled and its work more difficult. Harvey was not surprised that when a vein was slit its blood should flow out with only mildest pulsing, slightest pressure. His mechanical sense told him it must. From head and neck the blood needed only to drop to the heart, from legs and arms needed to be hoisted.

Hoisting is assisted in three ways: by valves in the veins, by muscles that contract around the veins, by the breathing chest.

Harvey demonstrated. He bandaged the arm of one who, as he said, was lean, whose veins were large and stood out, a man who had been exercising and his blood racing, and whose valves looked like knots. Anyone can perform Harvey's demonstration. Let him place his forearm on a table, identify a vein toward his hand, put the index finger of his other hand on that vein, press, then with the free thumb milk the vein in the direction of the heart, till he has milked beyond the next valve, has emptied that stretch of vein, then lift his thumb. The emptied vein remains empty. The blood does not slip back. The valve prevents it. Thus by valves the stream is helped to go forward because when the muscles around the veins squeeze them, the stream cannot slip back. Veins and venules lie everywhere, remind us again of the mammoth detail everywhere packed into us. That contracting of muscles to push the blood forward is called muscle-pumping. It operates day and night, helps the day-and-night deliveries, sweeps the river along, the child's river, her cat's river, the cat's mouse's, because we all are in this game.

So, the three kinds of muscle-tissue labor for the circulation. At the start it is the heart—cardiac muscle. Along the course it is the muscle in the vessel walls—smooth muscle. Wrapping the veins and contracting on them it is the muscle of the skeleton—striated muscle. The more powerful the contractions the faster the flow. Then there are the breathing muscles—striated also—that suck the blood upward into the huge collecting vein, the vena cava, from where it is dropped into the right auricle to be ready for its next round.

# HEART-SOUND

## *Din of the City*

The heart has its valves. When they bang shut, the flesh of the heart vibrates, the flesh around the heart vibrates, the flesh of the chest wall, and from there the vibrations go out to the ear of the physician. His machinery codes them, the code travels, is decoded, and he hears a heart-sound, hears—no one knows just how.

Those vibrations from the heart may instead be written on a moving surface, leave a readable report, a phonocardiogram.

Our body is by no means noiseless. Everybody remembers some purple occasion when he tried to keep down a noise, and it would not be kept down, insisted on rising embarrassingly in the general direction of the dignified moon.

There is talk of the new music, as of the new arithmetic, and one can easily imagine some new composer bringing in the noises of the heart, a tom-tom-tom-tom background beat, broken with the siren of an ambulance, bustle of the Receiving Ward, scream of a child, burned. A Hospital Symphony. Some composer may have done so. One cannot as easily imagine the fracas of doom that Beethoven— had someone insulted him by asking him to place his ear on a naked chest where the physician places his—might have fabricated of those rhythms and timbres with the warm resonance of the hollow of the chest behind them. Notwithstanding, the textbook says that the sound is *lupp-dup*. Day and night *lupp-dup*. Only *lupp-dup*. An unfeeling *lupp-dup*. A sophomore medical student seriously listens, seriously hears *lupp-dup*. He listens below the nipple, moves his shiny stethoscope, per instruction, rib by rib up toward the neck, under the arm, localities that last week he might have thought too remote. No one forgets the half-hallowed three minutes when he had his first important fever, and the family's doctor listened.

First heart-sound—easy to hear. Second—easy to hear. Third— not easy. Fourth—difficult.

Harvey compared the movement he saw when a horse took a

swallow of water, and the sound that went along, to the human pulse that he felt and the heart-sound that went along. Harvey was Harvey. Harvey was so sharp, and his statement always so vivid.

Valves—those doors that are both silken and sturdy, and efficient—swing between auricles and ventricles and between ventricles and outgoing arteries. So there are four valves, two left, and two right. As the ventricles fill with blood, the valves between them and the auricles first float, then when the ventricles contract and pressure mounts, bang shut. At the same instant, almost, the other two valves, toward the outgoing arteries, swing open and the blood rushes past. The combined tones—muscle, valves, blood— make the first heart-sound. Most of it is those valves banging shut.

When the pressure of the blood-columns in the outgoing arteries (they are columns now indeed) mounts above that in the relaxing ventricles, the valves, that had swung open, bang shut, the blood-columns vibrate, and the other two valves, between auricles and ventricles, sweep open. The combined tone—muscle, valves, blood —makes the second heart-sound. Most of it is those valves banging shut.

Should a valve leak, the normal sounds are altered, murmurs are added, sometimes even entirely replace the normal sounds, and the murmurs occur because particles of blood are thrown in sundry directions by the leak in the valve, abnormal directions for a normal heart, and the sounds are abnormal. Time, place, quality define the murmur. The physician puts his attention on all three, time, place, quality, conceivably does so partly self-consciously in order to make sure that he is concentrating, which would help him in his thought to shape the character of the trouble and what its degree. The physiologist states the physiological facts impersonally, but the dignity of a cardiologist ought to bend sufficiently and usually does to indicate that he knows how bitter this abnormal mechanics may be for the mind of the man waiting there for a specialist's appraisal. Maybe the specialist will look gently reassuringly toward the man and say: "My appraisal is that your situation is not black and it is not white."

A physician's trained ear may hear exceedingly faint sounds. It is quite routine for him to hear the heart of the fetus in the womb.

A heart built badly originally, or damaged afterward, may still allow for long survival. No physician may ever have listened to it, nor it ever have called attention to itself, so no one ever knew it had a flaw, this discovered only when the poor tramp came to post-mortem. Meaning that for years there had been abnormal sounds unheard by human ear, though it is possible that in some stillness after a period of strain, as resting from a vigorous dance when he was young, or shoveling snow on a January midnight when he was old, the man himself caught himself reflecting on the foible life, and could not have recognized that it was something in his machine was starting his reflections, if it was. And his ear did unmistakably hear his abnormal heart-sounds toward the close when he was propped in a hospital bed and his head beat the pillow to the rhythm of his pump.

For years a venturesome surgery increasingly has done almost what it pleased with the human heart. Over the world this has made regular front-page news. Until 1950 surgeons did not enter the heart, accomplished only what they could without entering. Today surgery in and around the heart has almost lost its venturesomeness. Birth defects are corrected in an infant boy or girl, a damaged valve in an eighty-year-old lady, the great pipelines directly patched or replaced, and heart-sounds had their small part in the development of this surgery, especially back at the beginning.

All diagnostic techniques will have been brought in and all professional heads put together before that curious morning when surgeon and colleagues and assistants, the so-called team, are each in his proper place in the operating room. The surgeon makes the initial cut with what appears flourish, but there is no flourish, no waste. He is the senior of the team. He is a meditative man, intense, quiet. The surgeon bows his head as he stands there, but he does that outside of the operating room too, a habit with him, though one could fancy it was to indicate the need of humility in any truly intelligent human being, which would not preclude the possibility that underneath the bow he is possessed of pride and style and explosive impatience of fools.

The cut looked simple, and one could fail to remember this man's long experience, and to remember the long history of such cuts on the part of the long line of eminent surgeons. One might have expected blood to gush. It did not. It was not given the chance, every ready-to-bleed vessel being clipped as quickly as it opened its small red mouth. There followed several cuts, layers of cuts, finally a swift cut through the muscle right down to the ribs. Then the ribs were wedged apart, with care but with confidence, the chest laid wide open. The lung seemed impatient and intrusive. It was pushed aside. The heart was exposed. The heart appeared indifferent to the proceedings. It kept at its business. Beat. Beat. Beat.

Boldly, now, the surgeon clamped a clamp straight across the left auricle. It was a big clamp as mechanically gentle as a big brother, and the heart still appeared indifferent. One might fancy it was putting on a performance. Instead of all this it might early in the proceedings have been sidetracked from the circulation, as anyone who can read knows, a man-made pump substituted, oxygen bubbled through the blood as it passed along, the man's brain never for a moment deprived, and his own heart brought back into his circulation at the end of the surgery. The beat can be grossly halted, whipped up again at the end of the surgery. For a brief time cross-circulation was tried, blood, that is, delivered from the blood-vessel of a donor, not without danger, and for various reasons relinquished. Experimental work meanwhile is keeping pace in many laboratories. One idea has been to build an artificial heart, implant that, make it take over the activities of the damaged or the completely disabled living heart. (Bring in Jonathan Swift again for the right words, the right surprise, the right quality of irony.)

In this present patient, this eighty-year-old lady, the valve between her left auricle and left ventricle has become scarred by disease, the passage narrowed, her life limited, and increasingly in danger. She and her surgeon had talked the matter over, realistically both sides, and they decided on this morning.

This surgeon has now come to the point of threading a purse-string of stitches along the line of that clamp he clamped across the left auricle, then a second line of stitches, on the chance the first might break. (Suddenly and without looking up he spoke peremptorily to a nurse, an unanticipated severity in his voice, and one respected him only the more for that.) He sliced off the flesh above the clamp. Took away that clamp. Pushed his index finger down into that hole, into the heart. The wet tissues hugged his finger. He was inside that eighty-year-old lady's beating heart, and immediately one thought of his finger as being just there. The entire operative field danced. Deeper into the ventricle went the finger. Now he appeared almost recklessly to thrash it side to side, knew how, ripped scars, widened the passage. Then he coolly drew out the finger, replaced the clamp, washed away the debris, tightened both purse-strings, reinforced with stitches all the cut layers of flesh, closed the hole in the chest. His finger had been down where the first heart-sound is made.

Natural to think now of the slim poetic consumptive René Hyacinthe Laënnec, the Frenchman who in the early nineteenth century invented the stethoscope and died of his disease at forty-five. One might have thought of Keats, of Mozart. Of Laënnec it is told that he wore a tall hat and that another consumptive, not meaning to, spit on the hat out a hospital window. Laënnec looked critically at the spit, commented that the poor fellow had not long to live. Laënnec was possessed of extraordinary hearing. With unaided ears, and standing well away from the sickbed, it is reported that he could hear the tones of a patient's heart, as a rare ear (of the blind sometimes) hears falling snow. Napoleon meanwhile was thundering across Europe, earlessly.

# ELECTRO-CARDIOGRAM

## *Walkie-Talkie*

Electricity is in the beating heart. One should fancy plus-signs and minus-signs on two sides of the membranes that wrap the muscle-fibers, the same as with other muscle-fibers, and indeed with nerve-fibers. Everything is wrapped and the wrapping important.

The continuous electrical forces that are always coming off the heart are not easy to follow, not in one's mind either, though the total of them, the summing, the overall wave of odd shape is not difficult for anyone to see. This wave repeats with each beat. It is there to be recorded any hour of any day or night of any week, a regular irregularity. It is a slow wave as such things go. This electricity excites the muscle-tissue of the chambers of the heart, and the chambers contract, and the blood goes its rounds. From cycle to cycle to cycle there is this odd recurring electrical wave, and it is a fact about the beating heart that Harvey did not know; if he had it would surely have occupied his persistent mind.

The fraction-of-a-second-to-fraction-of-a-second recording of the electrical sweep is not only possible but it is clear, we think. There are mountains of records. They heap up everywhere. Interpretation of them is not as clear, because though events in a single muscle-fiber can be interpreted reasonably with principles of physics, events in the whole heart are far from as reasonably interpreted. It is consoling, though, that if they were more fully interpreted we probably would not learn a corresponding amount new about the heart. Meanwhile the clinical usefulness of the electro-cardiogram, the ECG, is manifest fact. The useful research that it has inspired is fact also. True, clinically the ECG may not give assurance to a man as to life or death. Every newspaper reader has read of someone who died an hour after an acceptable record, and of another who lived on and on with a bad one.

The instrument is the electro-cardiograph. The record is the electro-cardiogram. (Telegram. Radiogram. Cardiogram.)

An electro-cardiogram is today part of the most casual medical examination. It helps a diagnosis, may clinch it. The record can show what and where in the heart the damage is, show it five-minute to five-minute in a hospital bed, or show it year to year in the regular checkups, so tell whether the damage is at a standstill, whether progressing, but always more secure as a record than as a prediction. The clinical usefulness often reaches beyond the heart to chemical upset in the body's internal environment, the so-called electrolyte state, clarifying problems there. We may remind ourselves that the horse-and-buggy doctor often treated the heart skillfully, and had no ECG, but undoubtedly he would have understood heart-illnesses better had he had. Also, he would have followed better the results of his medication, kept closer alongside the improvement or failing of a heart.

Conventionally the electro-cardiograph is wired to the patient, or subject, at left wrist, right wrist, left ankle. Three electrical outlets. The arms and the leg that lead to the wrists and the ankle are to be regarded as flesh cables, gross electric lines, because flesh in electrical respects is a dilute salt solution, a conductor. It is not electri-

cally homogeneous but is sufficiently so. The heart also lies sufficiently near the center of the trunk, is itself a conductor besides a
producer of electricity. From it electricity radiates in all directions;
radiates to the pubis, thence to left ankle, to instrument; radiates to
left wrist, to right wrist, to instrument. For electro-cardiographic
purposes the heart lies, then, at the center of an equilateral triangle,
one side of this triangle being a line drawn between the two
shoulders, the other two sides being lines between shoulders and
pubis.

Toward these lines, the force and the direction of the heart's
electricity is aimed; and from this emerges the three standard leads
handed down by the father of electro-cardiography, Einthoven. He
was a great Dutchman. He visited America, impressed everyone.
His original recording instrument was one of high sensitivity, the
string galvanometer, too high sensitivity to be widely practical.
With that instrument he began his long study of the complex electrical potentials created in the heart of a living body, a study that has
gone on ever since, but has not appeared to contribute proportion-
ately to our basal understanding; not remarkable, not disappointing,
but simply as expected, considering the undeniably difficult nature
of this small pulsating private world stuck in the other immense
revolving world. Einthoven's scheme had soon to be modified, and
there were modifications of the modifications of the modifications.
The all-present computer is today brought in and yet no instrument
or combination of instruments is equal to the intricacies. And the
Einthoven triangle itself, or any other triangle, becomes only an
approximation whenever one thinks of the problem as the physicist
would want to think of it.

Lead I. Lead II. Lead III.

Lead I picks up electrical differences between right and left
shoulders, giving information mostly, as one would anticipate, about
the top of the heart. Lead II, differences between right shoulder and
left ankle, information mostly about the heart's right border. Lead
III, between left shoulder and left ankle, the heart's left border. This
is a loose summary. Any instant in any record in any lead is the
resultant of many electrical additions and subtractions and neu-
tralizations. To calculate that exactly continues impossible, but
ingenious thinking has been stimulated and ingenious techniques

developed, and the physicist type of physician-mind has been given substance for its theorizing. Physics itself, the science, will probably not be enriched.

Electricity can be tapped with two electrodes. It can be tapped from the two ends of any side of the Einthoven triangle, or variant of the Einthoven triangle. It can be tapped by one electrode. For a time it seemed that one-electrode tapping would revolutionize electro-cardiography. This tapping can be from near the heart, or far, scant or much flesh intervening, from inside or outside the dog heart, from inside a single chamber of a human heart. No two chests are alike, and that has added handicaps to the game. Harvey's mischievous baboon always has another idea. He has accumulated the electro-cardiograms of many animals, crab to lobster to elephant. The last would have interested Harvey. Indeed, who would not be interested in the ECG of an elephant?

The electricity spreads through the heart, excites the successive chambers, passes through walls and partitions, but does no damage, and that is because it is everywhere neutralized, and because the voltages are small.

Patiently and slowly from the data of many electro-cardiograms an electrical heart has been pieced together. It is an electrical ghost of a heart. The word *ghost* possibly is justified because our heart as idea is to our mind always somewhat filmy, always somewhat remote, always somewhat strange, not the fact, the idea. The ghost could of course come alive, be projected onto a screen, make dramatic as well as visible to the uninitiated something of the electrical turmoil that goes on in a heart. This was done first years ago. It was a physiologist demonstration. Doctors were sitting in one room and the man who owned the heart was lying off in another. One of the doctors remarked that the feelings inside himself were as pulsating as what was pulsating in the other room.

Today electro-cardiograms are recorded photographically, or written on wax paper with a heated stylus, or by a moving stream of ink flowing from a tiny glass cannula, are like any taped news. One can rerun them. Recently volunteers have worn miniature electro-cardiographs fastened to their chests, and the autobiography of a man's heart could be writing itself while he was writing another with a pen, writing the events of his day. He might do this while speeding thousands of miles above the earth, the records accumulating in a small station far below on the earth.

Vertical lines ruled on the paper make it possible to read the contours, the fractions of waves in fractions of seconds. Horizontal lines make it possible to read the heart's voltages in ten-thousandths of a volt. These voltages are ten times greater than those of the brain. The twentieth century is surprised by none of this. There is no mystery. That recorder fastened to the chest is essentially a walkie-talkie. It records heart-cycle after heart-cycle for as long as there is tape and for as long as the pump keeps pumping.

Any moment of any record is meanwhile rising above or dropping below a neutral line, or staying on the line.

Electrical activity begins in the auricles, as Harvey essentially knew, and the sign of the activity is a small upthrust in the electro-cardiogram. Whether up or down does in fact make no difference, the physician able to reverse the written directions by simply turning a dial. This upthrust normally is rounded, called $P$ wave. It registers the electrical activity in the auricles; on behest of this electrical activity the auricles contract. There follows a neutrality, a level line in the electro-cardiogram, a pause in the heart. During that pause in the heart the activity would have been moving from auricles to ventricles, then into the muscle masses of the ventricles, the sign being again an upthrust but a different shape, a spiking called $QRS$, and thereupon the ventricles contract. There may be no $Q$ in the $QRS$ or no S. After the S, again a neutrality. Again a level line. A pause in the heart. The electricity now would have been moving from $S$ to $T$, where a rounded upthrust registers the electrical preparation of the ventricles for the next cycle. Sometimes there is a $U$.

All of this occurs for every beat, and a human heart may beat 150 or more times a minute, a mouse heart 375.

But any telling of this story that is as old as the hills, though of course nobody heard it till Einthoven came along, may annoy us for two reasons: that it is too simply told, that it is too complicatedly told. Both annoyances are justified. What is worth picturing is a heart that has passing through it these regular electrical sweeps, a heart beyond believing in a universe beyond believing. But if you prefer, keep thinking of the heart of a slaughtered beast in the slaughterhouse, put into a butcher's refrigerator, ready for your stomach. We are all, a little uncomfortably, in that slaughterhouse together. It smells. But that is not what bothers us.

That record—the ECG—is a report of electrical events producing mechanical events and the result of chemical events. Most of this is perhaps not yet interpretable. Meanwhile, we are probably staring at the record with routine acceptance, as Harvey would not have, but we may also be staring with confused bewilderment. The electro-cardiographer is doing the practical thing. He is bending his head over a record that he has spread across a console table and with cold-sober logic is analyzing an illogical subject.

# FAINT

## *Peak of the Journey*

Every minute a pint and a half, depending on the cardiac output, is delivered up a man's neck toward his brain, less toward *her* brain, she smaller. Two carotid arteries deliver four-fifths of that blood, two vertebrals one-fifth. Carotid comes from Greek, means plunge-into-sleep, the arteries getting their name no doubt because squeezing them in the neck caused unconsciousness.

In the skull is the postman's top-of-the-hill. Up there is evolution's masterpiece, has the royal right to one-seventh of his blood, gets it, a still larger fraction of his oxygen. If not, it quits, is apt to suddenly, and since one of its jobs is to maintain the great ape's erect posture, the great ape collapses. The cause of a faint is an abrupt drop of blood-pressure but there are causes of the cause referable to every part of the body and the mind. Faint, and something is revealed about you.

A man has been sick, has had enough of it, enough of bed, starts toward the bathroom, but his belly muscles because he has not been using them are flabby and do not give sufficient buttress to the blood-vessels inside there, and the muscles of his legs are flabby and do not give sufficient buttress to their vessels, also these are low down, have the weight of the columns of blood on them, so overfill. Too much of this man's total blood is in consequence virtually stagnant, lost to use, and if in the bathroom he adds some ordinary act of slight cost, brain-blood drops, and should the man be lucky he will have eased himself back into bed, if not lucky will be lightheaded, his vision blur, his power to answer questions blocked, a mutter, an oath, and down he sinks to Mother Earth.

A tall lady entirely in good health but on this afternoon the department store is too public, too crowded, too stifling, the tall columns too tall, the shunting to her brain too slow (it needs to be only by a trifle), and down sinks a tall lady, her customary poise made a mockery of, poor lady.

That well-known cadet at West Point standing still in the ranks, that perfectionist soldier, does not move one unnecessary muscle, may wish he had, his capillaries and veins given less than the necessary prop from his contracting muscles, the one-seventh of his blood not available for the top, and down goes a soldier and not a shot fired. He will have a headache tomorrow.

Blood sloshes freely in our veins, those who know veins tell us, and tell us that though veins have been studied for centuries every year they present new problems. They were doing that in Harvey's day. He observed veins, did his neat experiments with them, reported them thoroughly in his immortal book.

A hot bath has dilated the vessels of a bather's skin, that glows, the river basin of his body enlarged by the dilation, some of his blood also virtually stagnant, lost to use, and the bather does what he ought not in the tub, faints. So many ladies and gentleman for so many commendable reasons say their adieus in bathrooms, bounce from the bathroom straight to heaven, that it is advisable on entrance into such a place always first to turn on the lights and look behind the door.

A stuffy summer day made an elderly woman's skin aflame, her blood delivered in quantity to the surface to drop off heat, the rest of her body stinted, her brain stinted, and the climax was like the other climaxes. People past middle age may feel a bit giddy in June, never mention it, their circulation troubles explainable, their psychological troubles not as explainable. A medical student fainted in the operating room. Among the causes were erect posture, standing still, operating-room temperature, strong emotion. At this particular operation both assisting nurses happened to be pregnant, and pregnancy affects circulation, therefore when the two nurses saw the medical student faint it struck them they had an even better right to. That made three. Everybody spoke of it at lunch.

A pale mailman was threatened by a chained dog. He started to run and, if he had, his blood would have been hustled toward his heart by the contracting muscles of his running legs, but there was a gate that had locked itself. The mailman could not run, his blood was not helped toward heart and brain, and down went a mailman. The dog could not figure what had happened. Emotion may cause a person to run, as the mailman intended, also cause a person to stand still, and stand still for several reasons, as the medical student, the cadet, the nurses, and the reward of all may be that their frightened heads go to join their frightened feet.

Regardless of the cause of the faint the body lies supine or prone or on one side, curled up like a fetus, as people enjoy saying, dropped from the vertical to the horizontal, brain brought down from the top of the hill to the level of the rest, making the heart do less lifting, and the feet brought up from the bottom of the hill, also making the heart do less lifting, the faint becoming a cure, as it always has been interpreted, a cure of a symptom whose causes are many and sometimes hidden. A person who is supine to begin with is apt not to faint. A healthy student who is supine in an experiment can only with difficulty be made to faint. William Harvey never fainted—one suspects.

# CORONARY

## *Point of First Delivery*

The heart must eat and drink and breathe, rid itself of waste, keep its appointed rounds through snow, rain, heat, gloom of night, keep in repair, do this in health and not be stopped by any little sickness. Like the other blood-rivers those of the heart rush in (coronary arteries), go underground (coronary capillaries), wind their way out (coronary veins). Galen, that ambitious Greek practicing in Rome, gave them their name, coronary, means the making of a circlet, a crown; and Harvey, choleric Englishman practicing in

London, recognized why the animal has them. Galen no doubt knew too.

That word *coronary* is streaked with mortality. It is a word for which the bell tolls. Long or short notices, usually on the next to last page of the first division of the *New York Times,* with a photograph of a seventy-eight-year-old gentleman taken when he was forty-eight, dwelling on his virtues, is every day the public proclamation of disaster in the coronaries. Our grandfathers and our fathers did not speak our sophisticated speech; every person in the street could not have said that a clot in a coronary was a coronary thrombosis, a block, a coronary occlusion, the dead spot resulting, a coronary infarct.

Coronary thrombosis. Coronary occlusion. Coronary infarct.

New Yorkers read those titillating terms the morning after, then turn to the book reviews and the notice on last night's play. Possibly the seventy-eight-year-old if he could send back his message would assure us that it was an acceptable way to bow into the night. He had laid out his dressiest evening jacket, next his bed, laid himself on the bed, naked as he was, anticipating the luxury of ten-minutes before cocktails, and so they found him, a cautious smile on his face—was not sure that he was yet seeing straight all things celestial. Ladies suffer coronaries less frequently but they gossip brightly about those who do—their mutual friend had had two busy days in Washington, returned Wednesday night, Thursday morning sent for his breakfast of coffee, orange juice, toast, two slices of bacon, rang for the cook, sent back the breakfast, did not want it really, nothing wrong with it, that afternoon admitted he felt ill, not very, nothing alarming, but agreed to go to the hospital for a routine check-up, rang violently for the nurse at 3 A.M., she off to the telephone for the cardiologist, and while at the telephone heard his body plump. He was a big-bodied man.

To calm anyone foolish enough to immediately need calming: a person may survive his first attack, be felled by the second, third, fourth, even fifth, sixth. When it is sixth, that is the Lord on High absent-mindedly pulling legs and wings from a previously happy enough insect. But the event may also be as caressing as an unexpected kiss—a bit of clot delivered straight to the mouth of a large coronary, plugging it neatly, all flow stopped on the instant, no oxygen, so adieu Coney Island, miniskirted ladies, the thumbed volume of Keats.

What a fascinating planet we do live on. What charming partners to dance around with, black domino death, our own variegated particolored mind.

Blood driven from the left ventricle into the aorta made a first delivery just beyond the aortic valves, into right or left coronary, what the bulk of that delivery was, depending on the blood-pressure, on the size of the channels, on how tight or less tight the heart-muscle was squeezing the channels. It is when the heart is relaxing, in diastole, that the flow is rapid. That word, diastole, was Greek, became Roman, became English.

Leonardo's drawings of the heart place the coronaries correctly anatomically. It is possible that Harvey saw those drawings. Surely Harvey saw everything that everybody was seeing in Italy, in his Padua, and upon that in its degree would have based his experimenting and his observing.

As to the amount of blood to the heart, it is greater than to any organ except thyroid, kidneys, liver. As to coronary arteries, they are pencil-sized; the books say pencil-sized, so they must be pencil-sized, except there are so many sizes of pencils. As to capillaries, the heart has twice more than other muscle. As to veins, 70 percent of their blood is poured into a lake at the back, most of the rest into the right auricle, some trickles into each of the other three chambers. After that it is the postman's routine round: right auricle, right ventricle, lungs, left auricle, left ventricle, city, suburbs. Every molecule not delivered and used, every molecule not sacrificed to wear-and-tear, not sequestered, not lost to bleeding, returns to the heart, travels, travels, travels, then 3 A.M. (it was 3:15 actually) and next morning that notice on the next to last page of the first division. A man getting on, as we say, is apt with his coffee and orange juice to hurry to that page, bring back the long-ago when he saw Maude Adams in *Peter Pan*, heard Caruso sing in *Carmen*. Ladies at breakfast still read the wedding announcements.

If a fluid thick as blood—to bring this properly to the laboratory—is in a cadaver forced into one of the two coronaries, it does not appear in the territory of the other; if thin as water, it does; which could argue that one side of the heart has no immediate blood-connections with the other but that connections might be established if given the chance. And this could argue that there is less likelihood of attack if a block has developed gradually, allowed for opening or channeling. An eager surgeon might not wait, might build a channel of a morning, TV and *Newsweek* telling their readers that such hearts are revascularized. Surgeons have joined vessel to vessel. Surgeons have encouraged vessels to grow into the heart from outside the heart. Have shut off neighboring vessels so as to divert their blood to the heart. Many ideas have blossomed in surgeons' and researchers' imagination, are blossoming, will blossom.

That gradual opening might also suggest compensation for the gradual closing as we get older, evolution seeing the chance to get us older still, a little. Everything is done, by evolution, by Nature, by man, by science, by the American Heart Association, to keep the pump happy, keep it in repair, keep it pumping, pumping, pumping, we lunching, sleeping, hating, lechering.

A distinguished counselor on the heart used to boom out in public lectures that if stricken in his next sentence, please no colleague rush up with an oxygen tent. Enlivened his lecture that way. It also gave him a transition to the subject of control of coronaries. He always had thought that lack-of-oxygen was the key to opening, so naturally if one of his coronaries got plugged he would want every other one open, would want no "wretched, rash, intruding foole" to kill him with oxygen. That old counselor loved life, and having lived long enough knew too he could change his mind. "And had he ever said that lack-of-oxygen opened coronaries?" He never had. Nor was he contradicting that in other places in the body not lack-of-oxygen but presence-of-carbon-dioxide seemed to do the job. That booming counselor, like the other counselor Polonius, is today "most still, most secret, and most grave."

A closing word. Everybody knows angina, tormented his uncle, then killed him. A sudden severe pain, feeling of suffocation, called also *breast-pang*, occurs where coronaries are narrowed and stiffened, the attack itself excited by some sudden blood-demand during physical effort, or some strong emotion which tends vessels to spasm. Angina worried the horse-and-buggy doctor, who treated it with nitroglycerine under the tongue. Worries our cardiologists, who, among treatments, have sometimes implanted an electrode in the skin over the chest, and put an electric switch in the patient's pocket. He could control his bouts, somewhat. Hippocrates and Imhotep two to three thousand years ago had their bouts with angina. Our century experiments with it as with everything. Some years ago that sac around the heart was fashioned into a tunnel, the tunnel joined to the outside of the body. In Bozo, the dog. When everything was healed the experimenter at his pleasure reached something sharp through the tunnel, jabbed. No pain. Harvey knew heart-muscle gave no pain. Next, a thread was looped around the left descending coronary, not a whimper from the dog, but a definite whimper when the thread was jerked on. Bozo then might lift his left foreleg as a man in an anginal attack protects his left arm, suggesting the pain and the reflexes of angina to be triggered from the wall of a blood-vessel, but they are triggered from other places too.

Experiments come and experiments go. Observations come and go. Years back it was reported that we human beings were selected for this terminal notoriety well in advance, at the time we were pieced together as embryos, soon after boy meets girl, when the chromosomes line up. Thirty-four percent (that was the figure) of hearts had right and left coronaries balanced, 48 percent right dominant, 18 percent left. The last were the relatively doomed. However, in our existentialist day everybody carries doom around with him. Nevertheless, he goes on seeing his doctor, bribes his doctor to let him have one more round out at the golf links, two more rounds with the planet around the sun. He does give up cigarettes, it is true, smokes a pipe with conspicuous innocence, puts his hat on devil-may-care. Bless him! He is doing his part to maintain life's race-track character.

# VII

# BLOOD

---

## Five Quarts

# WORK

## Red Wine

A nosebleed left a red spot on a girl's white handkerchief, a French girl, so old so young as they all are. She looked accusingly at the spot. That blood came from inside her—we are all always ready to burst at some seam. Life is filthy. She muttered *vin rouge,* shrugged. The color was not right for *vin rouge.* This she noted with no amusement, with distaste. A tubercular's hemorrhage on a March morning in front of Saint-Sulpice was one of the impressive sights of this world, and the man went the whole way to proving that blood had important work to do, died. He had come to Saint-Sulpice to pray. A researcher, a fastidious man to be doing what he was doing, young, collected blood from a horse that he had shot in the head, slashed open its neck veins, the opened faucets running, overflowing the containers, the red flood inching out across the stable floor, the emptied horse lying in the shiny lake. (Not a speck was on the researcher.) A cut with a penknife may make a finger seem not red-veiled, but blood through and through.

What is the work of blood?

To be literal, it is to provide the 1,000,000,000,000,000 cells of the body, if a human being, with a pantry, and with a place for piddling. They reach their hands into the pantry, satisfy their needs, drop their wastes, and the products of their work. 1,000,000,000,000,000 cells—streams of blood running through—yet our body can strike us as so small when we put on hat and overcoat, walk out over the vast heath alone.

It seems malapropos to call the blood an organ, but it is an organ, also called organ, the organ of the blood. Its molecules move all directions. They move quickly. They mix. They equalize.

Possibly that is it—the blood's work—to equalize. First to equalize its own content, as it does moment to moment, strew its molecules around, scatter them inside the body, scatter a comparatively small number beyond the body. Equalize the body's water. Equalize the body's acid-alkali. Equalize its heat. Generally maintain the body's constancy, Claude Bernard's constancy.

Odd to consider a man dipping water from a cool spring, drinking, the water entering him and his blood, some of it promptly departing again in the sweat of his brow, joining a cloud on a hot afternoon and as rain contributing to the level of the Pacific. Nonsense! Of course. But in its crazy way, true. What lately was part of the outside world, and soon will again be part of the outside world, has had temporary residence in the blood.

# CONTENT

## *Winetaster*

Forty-one million five hundred thousand leeches were imported into France in 1833. Each bloodsucking leech when its great moment came drew its sample, disgusting winetaster. That sample was not for laboratory analysis, not for study of the blood's content, the heyday of leeches having passed before the heyday of laboratories. As many as fifty leeches might suck from one human body at one time though usually it was only five or ten. Physicians were themselves sometimes called leeches. This morning, 1969, thousands of physicians have drawn samples, or left written orders for interns, or nurses, or senior medical students, various sizes of samples, various techniques, various body parts. From the neck vein of a five-day-old infant that at the prick of the needle opened its grandmotherly eyes, squalled, closed its eyes, complained, and the syringe filled with blood. From the ear lobe of an eight-year-old boy who jerked away his head and a scarlet bead hung off his ear. From an athlete who unannounced fainted—the skinny intern was having his moment. From a shackled woman in an insane asylum, a large syringeful. From the blood-forming bone of an elderly gentleman, a special sample. One of the world's blood-scientists regularly takes such samples from human blood-forming bone, is up in the morning before the wards are awake, starts his day at 4 A.M., rows out into Long Island Sound and there does his thinking, and if someone is in the boat with him he may relate again from the beginning how an old doctor guided him into his career by saying over and over: "Look into the blood."

A still more special sample is drawn from the right side of the heart of a woman lying with propriety in the dark of an X-ray room, a thin "spaghetti" tube having been inserted into her arm vein, pushed until it was in her right ventricle, and from her also a large syringeful. A lean cardiologist is watching.

Each of these techniques has been a first step on the way to investigate an item of the blood's content, normal content, abnormal content.

How many human beings throughout history have made human blood offerings to their deity! How many purges throughout history! Napoleon I. The latest war. The latest murder. The Dubliners recount that throughout their affair the river in Dublin ran red with blood. The mind wanders to the realistic face of that French girl, wanders to the old doctor talking and advising his protégé, then years later, the protégé famous and in his rowboat at an hour that should excite inspired conceptions, as he thought it did, the blood in his own body merely so-and-so-many quarts of molecules.

A sample will be spun at high speed in a test tube, the cells thrown to the bottom, the fluid stay on top. The fluid is the plasma. The cells are red cells, white cells, platelets. Plasma + red cells + white cells + platelets = blood. That seems too ridiculously simple in the light of what the professional hematologist knows of the blood's complexity, but the division is useful, is practical, and the terms are in many mouths as one statement of the content of this *vin rouge* that is the body's nourishment and defense. The innocent-looking platelets, in recent years the subject of wide study, are important in the defense.

But the physician may have wanted a different statement, wanted a blood-count, a red-cell-count. He jabbed the ear of the eight-year-old, put the tip of a pipette into the drop, and the blood went up the pipette. The winetaster who tastes the wines of France also uses a pipette. The physician took from that eight-year-old's red river a measured amount, diluted it in the pipette, a measured dilution, needed to be careful because any error here would be multiplied in the later steps, and blood-counts have always the possibility of a 10 percent error. A dab of the dilution he placed on a microscopic slide etched with squares of an exact size, square millimeters. Then over the dab he laid a thin glass, producing in that way a film of an exact thickness. With the low power of the microscope he now counted the reds in several squares, struck an average, multiplied to get the number of reds in a cubic millimeter, multiplied by the dilution to get the number in undiluted blood, finally jotted down that eight-year-old's red count.

Five million four hundred thousand reds is average for an adult male, half a million fewer for an adult female; 7,500 whites is average; 250,000 platelets. Huge numbers these may seem and are, or are not. In the living body the numbers of microscopic and electron-microscopic items can be grasped only by persons who habitually think in numbers and do not allow themselves the luxury of imagining. Blood is about 55 percent plasma, 45 percent cells.

If the physician had let that drop just drip from the ear, it would have clotted; after a time a fluid would have appeared pressed from the top of the clot, the serum. This is a clear amber. Clot plus serum would be still another statement of the content of that French girl's *vin rouge*.

A chemist would wish a chemist statement. Even the most enthusiastic chemist would recognize that the items he knew and his way of thinking of them would always be up-to-now, that it would be different next year. All of the body's construction materials are in the blood, all the demolition materials, solids 20 percent, water 80 percent. Carbohydrate, fat, protein, salts, water: that would be a gross list. An eminent American chemist died and left his list: water, oxygen, carbon dioxide, hydrochloric acid, other acids, alkali in cells and plasma, protein in cells and plasma. At that time it was thought there were two proteins. Later four. Later thirty-five. How many in the mind of *le bon Dieu*? Probably He would not trouble to keep track. The increase to thirty-five came during the blood-letting of World War II, a cheap fluid then.

Xanthine is in it, hypoxanthine, creatinine, amino acids, ammonia, ugly names that get friendly to those who deal with them every day. Metals: iron, copper, sodium, potassium, calcium, magnesium, cobalt. Products of digestion. Products manufactured by the body. Foreign items. Lead gnawed off a toy train. Arsenic contributed by an impatient legatee. All molecules, atoms, ions, all of earth and air get in, and the content of the stars. At least, it is believed that the big universe from who-knows-where to who-knows-where is built of the same materials. Stars in our blood. Poetry, you say. Our body operates on prose, you say. Whether you are wrong or right, the blood is a compromise between what can be packed in and a fluid that can be pumped. It must contain everything that comes and goes to and from the cells of the body, and yet it must flow. This *vin rouge* is not thick, not thin, just right. Do not smile.

# PERNICIOUS ANEMIA

## *Poor Vintage*

Blood can be sick. Pasteur said of the wines of France that they could be sick. Much about healthy blood has been learned from sick blood. "Look into the blood," the old doctor said to the young one by way of helping him to become a research worker.

A person may be born with sick blood, a sick gene, or the blood may just get sick; for instance, some atrophy of the stomach. The person has blood-production trouble and this can be confined to a single so-called formed element: to wit, the red cell. The red cell begins its development in the bone-marrow but in this sick blood the development does not go the whole way. The red cell does not reach a normal maturity.

In the year 1929 there was a physician who was sick in this way, and every day he took a sample of his blood, studied the red cells under the microscope, saw how the number of the immature ones was increasing, and because of that sign, and because of other changes in his body, in his nervous-system, had some masochistic satisfaction in watching the grim specter approach step on step, while every other physician in the city watched him with that unsuccessful furtiveness that most of us watch the fatally sick, until one day he quite unostentatiously died. His disease was pernicious anemia. In 1929 pernicious anemia was no better understood than cancer in 1969.

A smooth sore shiny tongue, a yellow pallor of skin and lips and of the white of eyes, loose bowels, a cringy stomach, a growing enfeeblement, peculiar sensations in fingers and toes, tinglings, numbness, possibly disturbances in locomotion, all of this in a person middle-aged or older would, today, make the dullest physician suspect pernicious anemia. How that disease inscrutable then was followed till its cause had been found is one of the vivid chapters in the history of the blood, in the history of the body, in human history.

An American research worker was inducing anemia in dogs by bleeding them for long periods of time. Anemia is a reduction either in the number of red cells or in the amount of their coloring matter. The worker always bled the dogs short of killing them. His research had to do with the effect of food on blood production and one fact noted was that if liver was fed to the weakened dogs they improved. Two other American doctors took that hint, fed liver to patients suffering with pernicious anemia, and they improved. This was the following year. The improvement was dramatic. Several years elapsed. It had long been known that in pernicious anemia the stomach did not properly secrete hydrochloric acid, this state called achlorhydria. With surgical removal of a stomach for cancer there is also achlorhydria, since the stomach secretes the acid, and there is also a severe anemia. Still another American set himself to that. Feeding acid did not help. Feeding a variety of foods did not. Feeding juice from the stomach of a healthy person did not. However, if a healthy person was to eat a mixed meal, or a meal of beef muscle, allow this partly to digest, have that removed by suction and eaten by the sick person, that would help. The blood picture would change. Or, if the sufferer were to eat the beef and then swallow the juice of a normal stomach, that would help. Or, if the beef were incubated with normal human gastric juice and eaten. Any of these procedures would be appetizing to a dog.

Evidently, there was a factor that it was necessary to eat, and a factor that it was necessary for the stomach to manufacture. If those two conditions were met, something was formed that made a something absorbable in the intestine, this then entering the blood, traveling to the marrow of the bones, that then did the important work of maturing the red cells. What was absorbed in the intestine might instead travel to the liver, be stored there, to be let out as needed (an exceedingly small amount), and this explained why eating of liver cured or improved patients. Eating stomach ought to. Was tried—hog stomach. It helped. Hog stomach had both factors, that necessary to eat, that necessary for the stomach to manufacture if the absorption was to be normal.

More years elapsed. More research. The food factor turned out to be a reddish chemical, a ton of liver yielding only twenty milligrams of it, vitamin $B_{12}$. It had cobalt in it. Cobalt was one of the metals in those original oceans back at the beginning when the vast line of the living was beginning its odyssey. More years elapsed. More discoveries. More interpretations. What was plain was that the pernicious anemia patient had been cheated somewhere, was a chemically deprived person. Proper treatment made up for this. Vitamin $B_{12}$ may be the most powerful therapeutic agent known. Within hours the blood picture changed, within days the marrow changed, and the physician could say that his patient would be seen on Thursdays at Rotary again.

A happy history, the whole of it, because of the step-on-step march of the discoveries, a march that has kept on and has led and is leading in still other directions. A warming history to an American because so many of his countrymen were part of it. Within a few years secrets hidden through centuries slunk from the caves and our knowledge of the organ of the blood had a renaissance.

The working-out of the structure of the molecule of Vitamin $B_{12}$ went far to winning one Nobel Prize in chemistry. The winner was a woman. That was only a few years ago. Nobel Prizes have rarely been won by a woman, three altogether in chemistry.

## RED CELL

### *Color in the Wine*

Five million four hundred thousand red cells were in each cubic millimeter of blood, on the average.

*The Life and Death of a Red Cell.*

Born in the marrow. To study that birthplace the physician thrusts a short hypodermic needle into the breastbone, draws out a specimen, calls it taking a sternal puncture. Breastbone = sternum. The patient does not like it. In the old days, before night baseball and Sunday golf, when human beings had to be more robust, they cracked bones between their two hands and feasted on the marrow.

The physician examines that specimen of marrow under his microscope. This amounts to an examination of the manufactory, investigation of the winery. In the early fetus the winery may be the liver, may be the spleen, in the later fetus the marrow generally, in the infant the marrow of the ends of long bones, in the adult the marrow of flat bones like the sternum. Throughout the active part of its life a red cell has no nucleus, but throughout its maturing it does have, and if a nucleus is necessary for reproduction and if the living are dead unless they pack the earth with offspring the adult red cell is dead. Wrong however to call it a dead sac. Each year we find that it has been responsible for some additional kind of important labor, living labor. Its shape is a disc. In our blood-stream discs seem dancing everywhere, but under some circumstances they pile up like dollars, the single dollar palish yellow though quite loaded with pigment. A rough life this disc leads, bumped by the crowds, squeezed with thousands into one end of a narrow capillary, then squeezed through, then squeezed out the other end, this repeated and repeated, a subway rider's life. Nothing that did not have an elastic framework could stand it. The disc is concave top and bottom, which has engineering value, gives the red cell (erythrocyte) the largest possible surface for the passing in-and-out of molecules, that is its vocation. Also, the concave top and bottom give the disc the capacity to swell and shrink, which it may do several times a minute, has one size in the arteries, another in the veins, this the result of a continual back-and-forth of its physics and chemistry. Gases come in, go out, water comes in, goes out, salts in and out, the wonder is that the poor sac survives as long as it does, though finally like all cells of the body it breaks down. A by-product of the breakdown is bilirubin, bile-red, that enters the blood, travels to the liver, is expelled from the body to the earth mostly in the feces. This expulsion happens methodically and quickly. With jaundice not so quickly, the bilirubin stays longer, seeps through blood-vessel walls into Bernard's fluid-around, appears in the skin. 35,000,-000,000,000 is an estimate of a man's red-cell population, 10,000,000 born every second, 10,000,000 die every second. When a red cell has made the round-and-round times enough it begins to leak and continues to leak until nothing is left but its thin wall, that is called a ghost, in the blood-stream resembles a ghost. The turnover of reds birth to death has been calculated from that color that leaks.

Several methods measure the turnover, the life span. One is to introduce red-cell construction material in identifiable form into a human body, tag the material, wait for the marrow to use it, at intervals draw blood, note the first appearance of the tagged, its later disappearence. This would be the span. About twenty days. *The Life and Death of a Red Cell.* Born in the marrow with a nucleus, keeping its nucleus throughout its maturation, about four days, during which time there are three or four cell-divisions, it meanwhile creating color from its nucleus, then getting rid of its nucleus. Now it is a graduate red. It stands at some busy corner. Is pushed off into the crowd. Travels with the crowd. The hurly-burly of its life never stops. Batted by the mechanics of the circulation, by the chemistry of the body, by the ceaseless adjusting to keep Bernard's internal environment constant, so, naturally, it develops wrinkles. It gets old. Its resistance gets low. Then, pushed one last time, it goes straight toward a scavenger waiting for the ailing. That scavenger, a cell, is an important character in the drama of the blood, and in the drama of body-defense.

## SPLEEN

### *Wine Cellar*

Decrepit reds are unsentimentally disposed of. (Something of us is always being disposed of.) Excellent funeral arrangements exist for this. The spleen is a chief functionary. It is mortician and graveyard—one of the morticians, one of the graveyards. Liver and marrow are graveyards also. Bits of graveyard are everywhere in us, gutters to die in. About once a minute with a rhythmic motion the spleen heaves. (Something of us is always heaving.) A bloody locale, the spleen. It has partitions and between the partitions is a pulp held in place by thin fibers. That makes for corridors, recesses, corners to turn, and there is the slowly moving blood. Dark in there—if a light were lit all would be deep red. And venous sinuses, channels for holding vein blood; these are distensible, able to hold more and less. Red cells are pushed between pulp and sinuses, and it has been thought that when they are pushed through pores just too small for them the decrepit cannot survive, and go the way of all other flesh. Bilirubin, bile-red, saturates the scene and flows off via the bile duct into the intestine. Thus liver and spleen are not only neighbors geographically but they work together.

Everywhere is the scavenger—a Dickensian character.

A small organ, the spleen, weighs 150 grams, 5 ounces. In an adult human being it is five inches long, lies left of the upper end of the stomach, left of the pancreas, above the left kidney, in a bend of the large bowel that appears to be bending around it. It alters during work. It may be abnormally enlarged by disease. In one case eight pounds of spleen were drawn out through his incision by a surgeon, that spleen having required removal because, among other reasons, it had been abnormally destructive of the platelets of the blood.

Besides being mortician and graveyard the spleen is a manufactory. It manufactures lymphocytes, may pass lymphocytes back and forth between it and the bone marrow. In the fetus it manufactures red cells, therefore not surprising that later in an emergency it may take up that work again. It manufactures chemicals concerned in blood manufacture, we believe.

So, the spleen works hard. Everything in us works hard, only the total creature sometimes does not, has to force himself.

The spleen besides all else is a store for red cells. It might be pictured as a dim cellar of the body, many red cells, and the red cells pictured as thin-glassed prettily-shaped pale-tinted minuscule bottles slowly moving between the partitions, bottles of *vin rouge*. A wine cellar? A special wine cellar? The thin-glassed minuscule bottles are slowly moving, one might almost say floating. A deep-down dank wet wine cellar. Rathskeller. An expansible and shrinkable Rathskeller.

Joseph Barcroft and some others were en route to South America, to Peru, for the purpose of studying the changes in a human body when it climbs a mountain. To pass the time at sea they made routine measurements of the volume of their blood and the depth of its color. (Barcroft always had to be doing something, may also have had to force himself, but his work looked like play.) Both the volume and the depth of the color rose as the ship steamed into tropical water, into the Panama Canal. Manufacture of new red cells could of course not occur that fast. Somewhere in the body there must have been a reservoir that could release them. This steaming into tropical waters first called Barcroft's attention to the possibility of there being blood-stores, and an organ that he immediately suspected was the spleen. (Barcroft was the same who wrote so engagingly about how small the amount of hydrogen in Bernard's internal environment.) The various blood-stores that he speculated about were skin, liver, lungs, and the spleen.

Barcroft knew that carbon monoxide is a poison because it suffocates red cells, chemically prevents them getting oxygen, makes them therefore unable to perform. So he placed rats where they must inhale carbon monoxide, then at intervals killed the rats, found that the poison did delay getting into their spleens, "a lag of about half an hour." Unpoisoned reds would be in the spleen after there were no unpoisoned ones in the general circulation. Therefore, a rat without a spleen ought to have a smaller chance of survival because the reds in its blood-stream would be quickly poisoned, and no storehouse from which to draw unpoisoned ones. That proved true. The reverse also, if the poison once got into the spleen it was slow to get out.

Having concluded that the spleen stored reds and released them as needed Barcroft was pleased to learn of experiments that tested the effects of exercise. During exercise more reds are needed because more air is needed. Forced swimming is exercise. If rats are stalked, hit on the head during forced swimming, their spleens ought to be smaller than if stalked during sleep. That proved true also. A smaller spleen was found in hemorrhage and in emotion. In emotion the body was spending in all directions, so red cells were needed. A dog's spleen shrank, Barcroft found, if a duster on which a cat had lain was brought near the dog's nose, shrank to half if the cat mewed, to a quarter if the dog chased the cat. When a dog has a mission it fortifies itself from its Rathskeller.

Someone may have wondered how Barcroft could know the size of a spleen inside the living body. How did he measure it?

He had several ways. One was surgically to loosen the spleen without damaging it, bring it outside, stitch it under the skin. Another way was surgically to fasten metal clips to its edge, the clips blocking X rays, and Barcroft making the measurements on X-ray plates. He estimated that a dog's spleen is able to store roughly one-fifth of all the dog's red cells, 2,000,000,000,000 of 9,000,000,-000,000, and also can contract and push them out. A human spleen has less smooth muscle than a dog's, probably does less contracting, less storing, but it does contract. Barcroft stated that surgeons at operation find the spleen smaller, which one would expect because surgery includes anesthesia, hemorrhage, emotion, consequently more red cells needed, and the spleen would be letting them out, perhaps pushing them out. A surgeon asked about this admitted that when down in a human abdomen he occasionally gave the spleen a slap, and it got smaller.

# WHITE CELL

## *Creeper in the Wine*

The white cell is different from the red, different look, different life, different work, is a member of a smaller population, seems an individual often, as the red cell never does. Leukemia is a cancer of whites.

Seventy-five hundred is the average count, but the range wide, four thousand to ten thousand. This means that for every six hundred reds or thereabouts, one white.

The whites are not a single type. A type has its own birthplace, rate of growth, transitional forms one to another, longevity, graveyard. Lucky the whites do not have to pronounce their names. Polymorphonuclear neutrophil. Polymorphonuclear eosinophil. Polymorphonuclear basophil. Lymphocyte. Et cetera. It used to be thought, simple lymphocyte. Is thought that no longer. Then, the monocyte. The monocyte is a phagocyte. Other whites are phagocytes, too, and a phagocyte takes into itself bacteria, debris, particles, et cetera, disintegrates these, sends them onward over the path of everything.

Various echelons of experts study the whites, know them in detail, sometimes tedious detail. Experts are experts—one says that, then quickly feels guilty, because one remembers leukemia; no years of even dull effort would seem too much when one thinks of that group of frightful white-cell diseases, some already comparatively curable, and if the cures were truly successful all cancer might be attacked from that stronghold.

One of the birthplaces of the lymphocytes is the lymph node, and these take their name from that birthplace, as some Europeans have taken their names from their birthplace, as Leonardo da Vinci did. Other birthplaces are bone marrow, connective tissue, spleen, thymus. The large lymphocytes transform stepwise into the small lymphocytes, so that there are not merely the two old types, large and small, but an almost continuous series of types.

The whites that are born in the bone marrow, grow up there,

reach maturity, leave the place of their birth, make the Grand Tour, die. The marrow serves also as a reservoir for whites.

Some whites creep in our blood quite as if they had ideas of their own. Millions and millions of private lives. The number flowing or creeping in us may be different in an hour, different tomorrow, decidedly different if we have a cold-in-the-head. Surely they give us another picture of ourselves from that returned by our mirror this morning in the bathroom. Creepers in the wine. The polymorphonuclear neutrophil will appear to climb a pole, think better of it, continue off onto a blood-vessel wall, think better of that, go to something floating in the stream, hang on to it, leave it, keep that up, a sluggish microscopic acrobatics.

The eye bent over the eyepiece at the top of the barrel of a microscope may see crazy sights.

Life-span of the whites? May be brief, may be long, for one type some say three weeks, some for the same type say sixty hours. Either way, we need not be hematologists to be curious at this massive birth and death. (The massiveness makes us wonder again about cancer.) For another type, one week, two weeks. For the lymphocyte, some forms, the life-span was once thought hours, now is thought years.

A member of a type has been fished out, grown in the laboratory, watched from birth to death, its span directly clocked.

The polymorphonuclear neutrophil has a nucleus that when it gets older adds lobes, up to five lobes in the very old, a fact useful in establishing the life-span. It is possible with this type to stimulate a population explosion. So a count is made of the old, the population explosion stimulated, the old allowed to die off, the young allowed to get old, the census-taking continuing, till the young have gotten old, have been watched all their lives, and the life-span thus arrived at. The old are puffy, lie about, climb less, creep less, but add lobes, hence recognizable. The span is about three weeks.

The white cells' graveyard? For the final rites the lymphocyte returns to its birthplace, one of its birthplaces, the stuff of its nucleus being preserved for making new lymphocytes, what remains dumped into the *vin rouge* to stock it with protein. The graveyards of other types are spleen, liver, bone marrow. Scavengers wait in each of those localities. Scavengers are always cleaning the streets of the circulation. Scavengers are often quaint characters. They are cannibals, and the cannibalism can be followed with the microscope. It entertains a man. It also distresses him that even while looking into a microscope he is reminded of the melancholy way of the world.

# VOLUME

## ". . . *So Much Blood in Him*"

Last century two decapitated criminals were bled. Previous to the decapitation they had a sample of blood taken from them, and it must have seemed a tantalizing tiny sample under the circumstances. It was diluted one hundred times to provide a standard for color comparison, afterward. Then the decapitation. Then each criminal was given the same consideration. All the blood that would was allowed to flow from his body and head, then his blood-vessels were washed, his tissues chopped up and the color soaked out of them. Washings and soakings were diluted until the dilution had the color of the standard. This, the quantity of it, gallons, was divided by the number of times the standard had been diluted, and that figure told how much blood those miserables had had in their bodies during their lives, and since all of us are sinners the figure probably applies to us too. The method was not very inaccurate. It proved our blood to be one-thirteenth of us. Five quarts go the round-and-round. Less in women than in men, and varying man to man, woman to woman, but for a seventy-kilo man is apt to come close to five thousand milliliters; a one-hundred-and-fifty-pound man, five quarts. The method was spoken of as direct, as it was.

There are indirect methods. A known portion of dye is injected into, say, our Judge. Citizens ought to know about the *vin rouge* in their Judge. Time is allowed for the dye to color his judicial plasma, four minutes or thereabouts, long enough to mix but not long enough to leak into the tissues. A sample of his blood is drawn, spun in a centrifuge, two layers form as with any lesser blood, plasma on top. The color of that plasma then is matched with the colors in a set of color tubes, and this tells straightaway how many times the injected dye has been diluted. If three hundred times, the volume of the Judge's plasma would be three hundred times the volume of the injected. Plasma of blood is to cells as fifty-five to forty-five, but only approximately, must be determined for each person; once determined, eighth-grade arithmetic gives the answer, the volume of blood of our Judge. The dye is special. It must not, as said, and does not too much leak into the tissues, stays in the vessels, is taken up by the plasma, not taken up by the red cells, not taken up by the scavengers that loiter in spleen and liver and other places of what is called the reticulo-endothelial system, has a color easy to distinguish from most body colors. One would suspect the discovery of such a dye required patience. It did.

In our radioactive day there are radioactive methods, the red cells tagged, the protein of the plasma tagged, and the methods seem beautifully accurate, now. All methods have errors, but if everybody studying blood-volume uses the same method, knows its errors, the errors will at least be the same errors, and some day someone will stand up at a meeting of the American Physiological Society in Atlantic City and burst forth with some fact that is truly valuable and truly new.

We are more comfortable in our minds if science is practical and anyone might ask if there was practical use in determining blood volume.

A person has a hidden hemorrhage, is bleeding somewhere inside him, a peptic ulcer possibly, when there might be free blood in his abdomen, as dangerous as if he had been mugged and slashed in his neck by a thug. The hemorrhage caused an immediate fall in blood volume, but the red-cell count did not change because though he had lost the blood there had not yet been a diluting by water seeping into the vessels from the fluid-around. So there was less volume but the blood itself still unchanged, the red-cell count unchanged. Soon the water does seep in. The blood-volume is re-established by the diluting, but the diluting has driven down the red-cell count. So, had there been a hemorrhage, and had the other signs and symptoms not been sufficient for a diagnosis, knowledge of these shifts in blood-volume and red-cell count might have saved a life.

Lady Macbeth was so impressed by the volume of the blood of murdered Duncan that she began to talk of it in her sleep. "Yet who would have thought the old man to have had so much blood in him?" It was only 8 percent of Duncan.

# VIII

# LUNG

## Rock of Earth in Raiment of Silk

# AIR

## Nothingness Around

At the beginning of the world—how out of date that sounds—Jehovah created the heavens and the earth, and the earth was a rock, a small rock among the many small and large rocks that were out there, all the nearer ones lighted by our sun. Rather a cold rock, His earth, His chosen rock. Over the shoulders of it He therefore laid a raiment light of weight, a thin chiffon, almost nothing, air. Within that, generation upon generation would live their lives. The chiffon would stay where it was and what it was, would seem not concerned with the generations, nevertheless help make them what they were, keep them what they were, maintain their constancy, maintain the harmony that was within each even if not between any two. Mountaintop or down where the rivers flow into the sea man and hawk and ass and jackal nibble at the molecules that are the air, a number of kinds, and in each are their atoms, and in the atoms whatever is in atoms and will yet be revealed to be in atoms. Of the molecules, hundreds of times more are oxygen than carbon dioxide, two thousand times more are nitrogen, and some rarer molecules sprinkled in. Then there are water molecules, many, and the glory that Joshua saw of an evening when he looked "unto the great sea toward the going down of the sun" was owing to water molecules. Joshua breathed-out more of these than he breathed-in and so do hawk and ass and jackal. As to the proportion of the kinds, that is the same on mountaintop as down where men slosh in wet sand at the edge of the sea, but as to the number, say, in a bucketful, this is greatest at the level of the sea, fewer and fewer as a man jogs up toward that mountain above mountains where Jehovah dwells. No molecules in heaven. Too high.

Plainly this nothingness at which man and hawk and ass and jackal nibble is not nothing. It is something.

Each kind of molecule breathed-in by Joshua in his day was the same as that breathed-in by us in our day and by men of every nation before Joshua, because their bodies were the same as ours, or near enough, and must have been put together of the same materials, and must have taken in the same materials to stay what they were. A maximum of oxygen entered at each breath, a maximum of carbon dioxide left and of nitrogen about the equal entered as left, Joshua's body no more than ours being able to make use of breathed-in nitrogen, which merely went out and in. One hundred and fifty times more carbon dioxide, about, went out than went in, most of it having come originally from the cells where the hell of the body is always smoldering and producing carbon dioxide.

But how could one know accurately the quantity and kind of molecules farthest in? One could not if one insisted on being literal. Molecules do not wait to let themselves be counted or measured but continually move. Nonetheless, an old and honored way of measuring added learning. A long rubber tube was fitted with a mouthpiece. Near the mouthpiece a side-tube led to a sampling flask. The man who was allowing himself to be studied in this fashion breathed-in, breathed-out, kept doing that, then on command breathed-out with full force, the last of that to come out of him, from that farthest-in of him, was trapped by the sampling flask, analyzed. So one knew. The nothingness meanwhile paid no attention. Was the same chiffon. After the man had thus allowed himself to be studied, he just went on breathing, methodically, humorlessly, Jehovah having popularized the habit.

# BELLOWS

## *Tides In, Tides Out*

Imagine a single breath having started. It is scraggy Abel we have been watching. Abel's diaphragm and the other of his breathing muscles have started on the next contraction. The Old Testament has always left in us the feeling of the primordial, where thoughts of an ancient function like breathing belong.

Abel's diaphragm is a thin strong muscle-partition between chest and abdomen. Below the diaphragm are the masses of his liver, a dome on the left, a dome on the right, the right higher, its cupola up near the nipple. The single breath advances. The contraction of the diaphragm advances. The domes flatten, the floor of the chest lowers, and in consequence the total space in the chest increases top-to-bottom. The rib muscles at the same time contract, and the ribs that were directed downward sweep like hoops to the horizontal, and in consequence the total space in the chest is further increased, but now it is front-to-back and side-to-side. So, top-to-bottom, front-to-back, side-to-side, more space in the chest in all directions, more space for the lungs to expand into, which they do. They fill out the chest. They always keep right up with the chest's movements.

After the contracting of the muscles (inspiration), there is a corresponding relaxing (expiration), less space in the chest, less space in the lungs. The lessening could also have been not merely passive but active. For this the contracting of still other muscles would help, though usually the expiration is passive. Usually there is only the rhythmic contracting and the rhythmic relaxing—a sucking-in and a letting-out.

What we (and Jehovah in the wings) are witnessing is the quiet breathing of our daily lives. In—out. In—out. A mighty taking-in it would have been if all the takings-in since Genesis had been gathered together, since the world was young, before Jerusalem of the Jews. In—out. In—out.

At a moment no different from any other, and for no reason therefore, the regularity of in-out makes the mind think machine. Anywhere in the body the mind at a moment may think machine.

During the taking-in the entrails are pushed down, and to make room for the entrails the belly wall relaxes its tone, its tenseness. It bulges. Nothing in Plato's lower belly is to be allowed to interfere with this body's craving. In—out. In—out. A bellows truly. When a bellows has opened there is increased space in it, and something is drawn in. P-f-f-. Or, for him who prefers it spoken this way, the weight of the raiment laid by Jehovah over the earth presses something in. P-f-f-f. Or, for the few who are convinced that the learned should speak learnedly only, there is created a gradient of pressure from outside to inside, and the something coasts in. P-f-f-f. The infant opens its bellows twice while the midwife who sits like a watchdog by the crib opens hers once. She opens hers sixteen times a minute, about. Breathing is slow in sleep. Fast in fever. Smooth and silent on most days but not smooth and silent if there is threat to the source, if Jehovah's raiment cannot instantly get in or out, when the machine goes wild, and whether the machine is hawk or man it gulps as do the bodies of fishes pulled up from the sea and dropped on the sand. Whosoever has witnessed a surgeon laboring inside a chest to remove a cancerous lung never forgets it, the surgeon as busy as a sailor on a ship in a storm, never forgets either the sight of that lung when it was hoisted through the grievous wound, three lobes if on the right, two if on the left. In man the lung is free of the wall, in ravenous bird is rutted into the wall, deep grooves where the ribs are.

When a lung is inflamed and swollen it has less space in it. When a failing heart lets blood dam in a lung it has less space. When we lie down and the entrails press up against the diaphragm from below, less space. A sick man breathless when lying down has more space when helped to sit up, breathes then more easily, his abdomen lifted off his lungs. (This is really no figure of speech.) A nuisance to be breathless even for the short while that short while lasts. A woman in the final third of pregnancy may be breathless, a happy breathlessness if this is her first-born and her lord sympathetic, admits that a woman in travail is a woman in travail, as he would be apt to if his maleness requires a family. By the time of that final third of the pregnancy much extra has been seeded and has grown big in Plato's lower belly.

# ALVEOLUS

## *Entry*

The ultimate place where the molecules of the air of earth come and go is the alveolus. That is far into the lung. That is truly the shore, the border, there where the great outer meets this single inner. "And he brought them to the border of His sanctuary." The alveolus is the sanctuary, small, and there are many. Through their walls the gas molecules pass in, pass out. We are unaware of the passing, have usually neither the pleasure nor the unpleasure of our breathing. Someone said that the infant however takes big breaths and feels the air running all through its healthy body, and that this is one of the primal pleasures of life. Possibly that pleasure would have encouraged the infant to expand chest and lungs to the uttermost and thus body and head would appear to have been emerging together, physiology and psychology.

Jehovah did not bless every creature with lungs. The one-celled living its life in pond water keeps in good health though it has available only the oxygen dissolved in the water; indeed, it can be larger than one cell, just so its girth does not exceed one millimeter. If more than one millimeter, whether living in air or in water, it cannot get enough oxygen by such physics. Man of course cannot. No bird of any sort. Oxygen never would reach the tissues of man's body and that was the reason Jehovah laid in tubes. These start at mouth and nose, continue through windpipe, which divides, right, left, those divisions divide, so on, narrower and narrower, all the windy passages together constituting the airways of man or hawk. At the narrowest is that final entry, the alveolus, the sanctuary. A cast was made of the airways of a human lung and it looked like a tree from which the storms of autumn had blown off the leaves. If the eye follows along those narrowest divisions it comes to what is called a bronchiole. This divides, and its divisions divide, and at that point there are ducts, then a sac, and the sac pouches, each pouch has outpouchings, and those terminal outpouchings are the alveoli. There the gas molecules enter the body and exit from it.

Each alveolus is microscopic, but, because there are so many, great surface adds up from all the walls of all of them. Each closets a man's most intimate air. There the gas molecules of the air of the alveoli are forever balancing with the gas molecules of the blood. All our lives a light breeze is playing there. In—out. In—out.

Imagine one alveolus. It has thin walls. Outside it are those other thin walls, of the capillaries. Blood is flowing in the capillaries. Through the two sets of thin walls the breathed-in molecules go from alveolus to blood, and the breathed-out leave blood, leave alveolus, leave lung, become part of the nothingness around. Around the jackal, too. We and the jackal are on this cold rock together and ought not to forget it. We have gasped together. Aristotle did not forget it. Thoreau did not.

The total length of the capillaries in the five lobes of the two lungs of a man has prompted mathematical embellishment; it comes to one thousand miles. The wall-surface of the alveoli prompted similar embellishment; it comes to one thousand square feet. To arrive at that latter figure, first the total number of alveoli had to be counted, as far as this was possible, then multiplied by the wall-surface of one alveolus, which is measurable. What it means is that a veil of capillaries one thousand miles in length and a veil of walls one thousand square feet in area are the essence of the lung lobes of a man. In order to achieve the one thousand square feet, back in the womb in the embryo the beginnings of a lung everywhere fold, or appear to, fold, fold, fold, and at last when all the foldings are added together there is a vast surface, a veil of walls of vast surface—surface for all the molecules of the gases to pass in and to pass out. Surface to meet the body's needs when it lies down to sleep. When in the fresh of morning a woman rises too late and rushes to the well for water. She breathes hard. She feels the distance to the well greater than it used to be. That vast surface, blood on one side and air on the other, is only just sufficient, everything in her body laboring hard, her diaphragm contracting, her ribs sweeping to the horizontal then dropping. All of this laboring is registered in her mind, that also is laboring. "And Jehovah-God caused a deep sleep to fall upon the man, and he slept; and he took one of his ribs, and closed up the flesh instead thereof: and the rib, which Jehovah-God had taken from the man, made he a woman, and brought her unto the man. And the man said, this is now bone of my bones and flesh of my flesh: she shall be called Woman, because she was taken out of Man." That rib, called Woman, because she rose too late and rushed to the well for water, barely could get in the air to do it—the air to do it was exactly the air necessary to do it.

# CAPACITY

## *"What Ye Have Need Of"*

Esther. Joshua. Ruth. Nehemiah. Moses. Each of those grand characters without cease breathed-in a pint of air, breathed-out a pint, maturity to sepulchre, took what each had need of from the immense and awful nothingness, gave back to the nothingness. Scraggy Abel too. A pint, approximately. In—out. In—out. The animals too each according to its needs. In—out. "The wild asses stand on the bare heights, they pant for air like jackals. . . ." While my neighbor prays out loud, while my other neighbor whispers sensuous sounds from The Song of Solomon and will not let his mistress sleep, while the Hebraic scholar intones in deep tones The Law, in—out, a pint. If the Lord Jehovah has said that the enemy hordes shall fall upon the mountains of Israel, shall be left to ravenous birds for food, and if the soldier of the Prince of Rosh heard Jehovah and was filled with fear, more in, more out. Fear like strife increases ventilation. If at the end of a usual in, Esther keeps drawing in until she has drawn in all she can, two quarts will have been added to the pint. Esther is a goodly-proportioned woman. If instead at the end of the usual out she pushes out all she can, it will be something more than another quart, and yet even then some will be left in her lungs, impossible to push out. (All of these and other portions of our breathing have quantities, and the portions have names, as a maximum inspiration following a maximum expiration is named the vital capacity.) In a stillborn, air may or may not be found in the lungs of the little corpse; will be found if that one ever took a breath. Important sometimes to establish whether it did, whether air was ever in its lungs, whether the scribes should write

into their ledger that this creature ever or never lived, on the chance
that the crazed mother was not in agreement that No. 4,932,769,432
ought to have lived.

But something more cheerful, what would be the capacity of a
soldier of Moses blowing into one of the two silver trumpets for the
calling of the people to the temple or for the mustering of the
armies? How much could he blow out? Four and one-half quarts. If
his Esther were by his side, five and one-half. Years later he might
look back at that golden time of his youth when he did blow into one
of the two silver trumpets, Esther by his side, and might say it was
all there was to life.

## PLEURA

### *Like a Coat Without a Seam*

At the sacrifice the lungs of the slaughtered beast sometimes balloon
through the torn-open chest, and those lungs resemble the lungs of a
man, each covered by a tough membrane, the pleura. Pleurisy is
inflammation of the pleura.

The pleura has two layers. Someone might think it is like a coat
without a seam, but it is more like a paper sack that has never been
opened. Its inner layer is fitted close to the surface of the lung, and
there where the great pipelines for blood and air arrive and leave
the lung, the inner layer appears turned back to become the outer,
this now fitted close to the inside of the chest and to the top of the
diaphragm. No space is between the two. There is no actual pleural
space. During breathing the two rub over each other, oiled by a
lymph.

In the fetus the lungs fill the chest, and the chest moves, but only
the slightest movement. Then the infant is born. The chest expands
sharply, the lungs expand with the expanding chest, air is sucked in,
or say pushed in by the weight of the nothingness. (Atmospheric
pressure.) The two-layered sac accordingly is stretched, and to the
end of that human being's days the lungs will always be expanded,
and the pleura will always be right in contact with the inside of the

chest and the top of the diaphragm. The pleura will be more stretched during the breathing-in but stretched still at the end of the breathing-out, and one fancies these moving parts forever trying to return to the idleness of before-birth, never able to, compelled eternally to travel with the chest's eternal movements, hurryingly in the soldier of the Prince of Rosh, feebly in aged Rachel waiting among the sick.

A rip through the pleura alters this. A rip could have occurred from outside or inside. A knife stab or a crush wound might cause a rip from outside. Blowing night after night against the resistance of a horn might cause a rip from inside. Occasionally a rip occurs in what appeared a healthy person, but was not quite, a sudden piercing pain announcing the rip. Previously there would probably have been a blister in that layer against the lung, the blister getting larger and larger, then some strain, it may have been no more than a deep breath during a physician's examination, and the blister burst. Air now leaks out of the lung and seeps between the two layers, creating a space between them. Now there is an actual pleural space. At the breathing-in some air leaks from the lung into this space, torn flesh prevents it getting back at the breathing-out, so it accumulates. If it accumulates swiftly the person dies, but it is apt to accumulate slowly, eight hours, ten hours. The character and the place of the sudden pain would have told, to a person who knows about such events, what happened.

A surgeon while operating had sudden pain, fell down, a fellow surgeon knew. (Spontaneous pneumothorax.) With a syringe he drew out the accumulated air. Had that not been drawn out, the lung could no longer have followed the moving chest, would have been stopped by the cushion of air, giving the lung a rest, but rest that did its owner no good. The lung might even have been pushed across the chest till the other lung also was hampered, the heart, that hangs between, batted back-and-forth with each breath, a "dancing heart," the man or woman shortly joining those who went before, who sleep, the priest says.

Tuberculosis formerly often required that a lung be given forced rest. The tubercular must breathe, so he must have working lungs, but if only one lung is sick a surgeon can cut a window through the ribs on that side, air rush in and collapse that lung, it stay collapsed for as long as necessary, have its rest. (Open pneumothorax.) Instead, the surgeon might push a hypodermic needle between the layers of the pleura, inject air, this pressing on that lung, collapsing it. (Closed pneumothorax.) If watched with the X ray that lung still moves but small movements. Gradually the injected air is absorbed and more air must be injected, this repeated possibly every several days. A remarkable thing has been done: some of the earth's raiment has been transferred to the pleural space where air ought not normally to be, but saving this man from death, as air once let him come alive.

# HEMOGLOBIN

## *Man's Blood and His Green Pastures*

Oxygen molecules, carbon-dioxide molecules, nitrogen molecules, a sprinkling of molecules of rarer gases, water molecules, at every breath go in, go out, and this over every mountain of Israel and every other mountain and every valley of the earth for a long time. Billions of oxygen molecules enter the body, travel in the blood, arrive at the cells. Billions of carbon-dioxide molecules leave the cells, travel in the blood, leave the body via windpipe, mouth, nose. That traveling, in the red cells, of oxygen and carbon dioxide depends much on a pigment that has iron in it, hemoglobin.

Our grandmothers did not understand about hemoglobin, nevertheless were worried that we might not be getting enough iron, so convinced us we liked spinach. All that spinach! Though the body needs iron to manufacture hemoglobin, it may even be that it cannot use spinach-iron. Most of the iron it uses comes from the breakdown of red cells. In the many atoms that make a hemoglobin molecule only four are iron.

Hemoglobin can be called a pigment that breathes. Jehovah probably synthesized it by creating the right two genes, then allowed the genes to take over. This occurred on the fifth day, the day when the waters were let to swarm with swimming fishes and the air to fly with birds. "And there was evening and there was morning, a fifth day."

Another prime color worn by our earth since Genesis is green. "He maketh me to lie down in green pastures." Plants draw from the sun and are fruitful because of green. Astronomers have shrunk the sun but it is still out there and hot and big enough and overabundant with energy. "And He saw that it was good."

The green chlorophyll is a small but complex molecule. Photosynthesis is the process that enables the energy of the sun in the presence of chlorophyll to combine carbon dioxide and water, release oxygen, produce carbohydrate, which are first steps up the evolutionary ladder to human life. With chlorophyll, and sun, and air, plants create themselves, and we eat them. Mars has a green, mosslike green arriving and departing through the Martian  year. Mars has little or no water, the thinnest of atmospheres, carbon dioxide in it, and up over Mars during the Martian day there shines the sun. If this was the same in the olden times, and if this is the truth now, then we would understand the scant oxygen in the raiment that Jehovah laid over the shoulders of Mars.

Genesis rewritten would read: sunlight of heaven, elements of earth, a green pigment, and there was oxygen, and there was life, and there was a plant. On a later day among the seven days there was an animal, and the animal ate the plant. "I have given every green herb for food: and it was so." Still later there was that woman, and the man and that woman sat down to supper and ate plant and animal both, and it was so. The man was an early man, and he made this and he made that, fire and tools, and he wished directly to use the fire of the sun, and he was skillful with tools, yet nothing quite worked to his satisfaction, so he was a probing tentative man.

Chlorophyll contains the metal magnesium. No one knows just why magnesium was the chosen, magnesium for the plant, iron for the red of the blood, copper for the color of the feather of the bird.

Hemoglobin is heme plus globin. Heme gives the color, and upon it depends the taking up of oxygen. Heme is ancient on earth, may be as old as half the earth's calculated life-span, if one can say life-span for what was through so much of the span dead. Before hemoglobin, before chlorophyll, heme was already part of the minutest life.

The hawk has hemoglobin. The jackal has. The ass has. The lugworm has. We have. Heme apparently is the same from species to species, but globin varies, so different animals can have different hemoglobins. Within one species of sheep there were found three separate hemoglobins, and by their hemoglobins alone would the members of that species be assured some degree of individuality. A quarter of a century ago a sensitive student seriously asked: was the hemoglobin in the larva of the fly that lived as a parasite on a horse the same hemoglobin as that inside the horse? It was not. This led to the study of these pigments in the larva, in the muscle of the horse, in the butterfly! Which curious logic is mentioned to show how hemoglobin and related pigments may be considered to have pulled the researchers with gay strings by their noses. The books and articles on hemoglobin would fill shelves.

Abnormal genes produce abnormal hemoglobins, that produce abnormal bloods, that produce abnormal bodies, and we say of these people that they suffer inherited blood diseases.

Hemoglobin is the bulk of the red cell and when the hemoglobin begins to change, even though it was born entirely normal, we might say that the red cell is getting old and soon will die. A molecule alters and a cell alters, gets old, and if enough molecules alter and get old, a man's breathing alters. "Now the days of David drew nigh that he should die."

We already know that besides the veins and other small vessels of our skin, and besides the number of the red cells in the vessels, the state of the hemoglobin in those cells helps account for our complexions, paints us, the rose of our lips, the peachblow of our cheeks, the shades within the bright colors of the healthy and the pallors of the sick. The size of the load of oxygen that the hemoglobin at a moment is transporting explains the scarlet of arteries, the blue of veins. The high pink of the corpse, anno Domini 1969, brought to the morgue from a garage where the unhappy one let the engine keep running, is owing to hemoglobin combined not loosely with the oxygen of fresh air but tightly with molecules of the exhaust.

What a rich rugged drama, in Act One to have created the earth, in Act Two to have made it come alive, in subsequent acts to have bestowed the dowry of frailty upon that life, incorporated the life and the frailty of a human being in a woman, over her countenance powdered this fickleness of color. John Donne wrote unwittingly of hemoglobin:

> Her pure, and eloquent blood
> Spoke in her cheeks, and so distinctly wrought,
> That one might almost say, her body thought.

Pigments of life, green for plant, red for man, much alike in the pattern of the atoms of their molecules, the variants often comprehended, an immense chemistry comprehended, our little old earth is both strange and sweet but a bit childish to be so preoccupied with two colors. "They shall be full of sap and green," saith the 92nd Psalm. "The life of the flesh is in the blood," saith Jehovah. The old doctor two thousand years later said: "It's all in the blood. Look into the blood." John Donne wrote wittingly: "I observe the physician with the same diligence as he the disease."

# OXYGEN TRANSPORT

## *Out to the East*

The journey of oxygen can be lung-to-bowel, lung-to-brain, lung-to-hand, lung-to-foot, lung-to-anywhere-in-the-body. One might say outward bound—it is outward from the lungs. Eastbound. Whether on mountaintop or down in the valley, that oxygen would have been pressed on by the molecules around it, some pressed exactly into Elijah's nose and mouth. Thence it would have gone into an alveolus in Elijah's lung, through the wall of the alveolus into Elijah's blood, journeyed in the blood, reached a capillary, dropped off among the cells around that capillary. What was unused journeyed back, vein, heart, lung, there joined the fresh load being picked up. Blood moves through a lung capillary in seven-tenths of a second, or thereabouts, slow.

The alveolus would always have been replenished from the raiment over mountaintop or valley, over Jerusalem, over Bethlehem, over the villages or towns that sprang up everywhere after Genesis. More oxygen is in the alveolus than in the returning blood, so oxygen tumbles from more to less, downhill through the two thin walls, dissolves in the blood.

If the body had to depend for oxygen on the dissolved, it would need seventy-five times as much blood, and a rabbi would look like Jonah's whale beached. Each oxygen molecule does indeed dissolve, reaches the wall of a red cell, gets through, finds a hemoglobin molecule, the one destined for it, unites with it. A not-tight union. Then again there is the journey outward from the lungs. Eastbound.

How much oxygen can be loaded in the lungs?

This depends to begin with on how heavily the raiment over Jerusalem or Bethlehem is pressing. The more the total gases press, the more the oxygen presses, the more loaded. At sea level, 98 percent is loaded. At all ordinary levels where each man thinks he chooses to dwell, 98 percent.

Loading depends also on how acid the blood. Low acid favors loading. In the lung the blood rids itself of carbon dioxide, therefore of carbonic acid, therefore is less acid, therefore loads more oxygen. While carbon-dioxide molecules jump out, oxygen molecules jump in. No oxygen molecule could have anticipated the wonder of the journey when lately it was on the mountaintop and looking straight into the face of the moon. Jehovah "formed man of the dust of the ground, and breathed into his nostrils the breath of life; and man became a living soul." Jehovah breathed oxygen into those nostrils. Jehovah went about that world of Israel and Judah with a voice of thunder, like Homer's gods of old, or like any human being given authority, scolding and threatening and smiting with His sword and promising rewards.

# CARBON-DIOXIDE TRANSPORT

## Back to the West

The other direction. Westbound. Cells → fluid-around → capillaries and veins of returning blood → right side of the heart → capillaries of the lung → alveoli → mouth or nose → mountaintop or valley.

The journey of carbon dioxide employs several types of accommodation, also depends much on hemoglobin.

Stepping-in and stepping-out of any accommodation must sometimes be swift. Whenever an action is slow in a test tube, swift in the body, one suspects a chemical whip, an enzyme. Here the one is called carbonic anhydrase. Whips the union of carbon dioxide and water inside the red cell, instantly. Carbonic acid is formed, instantly. Bicarbonate steps out, instantly. For balance, another ion, chloride, steps in, instantly.

Everything on the rock of earth, in the firmament, in the body's rivers, in the red cells, in the fluid-around, must balance. Balance, balance, balance. A fuller understanding of the stepping-in and stepping-out makes the travel habits of the carbon-dioxide molecule less mysterious. To state it in a parable: the same carts, the red cells with their hemoglobin, refitted somewhat, transport both molecules both directions, but more oxygen eastbound, more carbon dioxide westbound, release oxygen out among the cells where it is needed, carbon dioxide in the lungs where the body can get rid of it. All the types of accommodation, and there are many, must also balance. If a molecule finds one type occupied, it accepts another. The entire arrangement seems not the blind grinding of the universe, which it is, but conscious, and quite contemporary.


# REGULATION

## *Order Is Beauty*

Order is in that in-and-out. Order is everywhere in the act of breathing. Order, strict order, is beneath all the surface of the living person, who in his mind sometimes seems in such disorder. Order is in all that Jehovah does.

In each breathing-in and each breathing-out, order. In the quiet of evening, in the harassments of day, deep down, order. In the machine everywhere, order. In the next breath of that old sinner caught at last, order. "I will gasp and pant," saith Isaiah. This ordered rhythm of breathing is affected by all that the body does, by changes in the blood-flow through the lung, by changes in the composition of the blood. The rhythm is tailored and retailored, fitted to this one body's constantly changing needs, more than any man knows. Always and at the end, order. Act fits to act, part fits to part, the total creature fits the total world, and here somewhere in the middle of that total world is this animate spot, breathing. In—out. In—out.

As with the beating heart, so with the respiring lungs, all appears softness with precision, and man and jackal breathe without knowing, even in the excitement when Jehovah rains brimstone and fire upon Sodom. If the student pages backward through evolution, everywhere is softness with precision. If he watches the mother hawk standing at the edge of her nest and looking down into the open mouths that also appear to gasp and pant, softness with precision.

Breathing is affected from beyond the body, is affected from within the body, to the end that the body shall always have the exact molecules in the exact quantity. Attempt to change that. Even if you are son or daughter of the wealthy Philistine, better not attempt. Better bow your head and admit you are driven by ancient law. Or, try to deny the law, to convince yourself of the law. Hold your breath. Keep oxygen out, keep carbon dioxide in, use up the one and pile up the other, and though you are firm in your resolve, a next breath will crash through. Pinch your solemn neighbor's wife— watch her breathing. Watch his. Watch yours when you run to get away from his wrath. Take a long drink of the new wine and it will be followed by a long breath. Take a tramp over a rough mountain and your breathing will be enormous but steady beyond your expectation; you may never even think of your breathing. Feed upon the extract of the poppy and your breathing will grow faint and fainter. If the mood is upon you to sing loud the song of the drunkards, your breathing skillfully will avoid getting in the way of any note of that song. We sigh. We groan. We moan. We dispute. We partake of green herbs. We partake of "whatsoever hath fins and scales." Breathing is affected by all. Allow a young girl to walk dreamily into the village, a robber to be affrighted in the forest, a horned owl to hold a struggling rodent in its grip, the breathing of all four—girl, robber, owl, rodent—will keep pace with what each is doing.

A student years ago isolated one limb in a mutilated animal, disconnected the blood line to that limb, then with his own hand worked that limb back and forth on its joint, and the breathing changed. A similar passive working of one limb in an unmutilated human being caused a change, and here no blood line was disconnected.

A chief drive of breathing is carbon dioxide, trifling in the nothingness and abundant in our body. But there is a combination of known drives, and there may be unknown. Known and unknown keep regulated this highly regulated force.

Carbon dioxide, manufactured in every region of the body, must be kept in balance with that in the lungs, and the percentage in the lungs remains remarkably constant. When it rises, breathing rises, ventilation rises, the extra blown off. The machine self-regulates. The body must have the oxygen, but carbon dioxide is at least largely at the controls. When oxygen gets critically low, lack-of-oxygen no doubt takes bigger responsibility, and one can perhaps safely say that a main drive is adjusted to other drives, the result being the perpetually modified rhythm of the creature's breathing.

The question often asked of the heart is here asked again. Whence this rhythm; whence these heavings, whence these waves that on occasion swell to billows? Is the regulation all within the parts that deal with breathing? Is it the steady push of a gas, or an acidity, or a lack-of-a-gas that produces an intermittence? Is it an even flow of energy into the breathing muscles, regularly interrupted? Is there a piling-up of tension to a point of breaking, then a break, then again a piling-up? Is it? Is the nature of all body rhythms essentially the same? All earth rhythms? Is a colony of cells inside the controlling system and known to be in command of breathing—is that itself rhythmic?

Two skillful workers once upon a time took a goldfish and with a knife trimmed away everything except the colony of controlling cells. Harsh on a glittering goldfish. The fleck of tissue lay before the workers. They connected a registering instrument to the fleck, and, lo, a rhythm came from those cells, its timing the same as the goldfish's breathing gills. Conclusion? A rhythm exists in the cells that command breathing, and upon that the other rhythms of the body exert their effects, as do the rhythms of heaven and earth. To be sure, that was a silly goldfish and one cannot expect to find in a goldfish "the tables of stone, and the law and the commandment."

# MIND SIGNS

## *High and Low*

Joseph Barcroft took a hand in these problems also.

He investigated upon human beings, upon himself, always had a healthy recklessness, once remained twenty minutes in an atmosphere that had two hundred times more carbon dioxide than ordinary air. Another investigator was with him. Both agreed that they would not care to repeat the adventure, their intellects badly confused, Barcroft's wife not knowing what was wrong with him, kept him in bed. He was vague for days. His thinking on all sides was disturbed. From similar adventures, years of them, a principle emerged: if ever the body gets too much or too little of anything, the mind shows the signs first.

Too much or too little of the gases of the air can be easily arranged. Too much carbon dioxide can be directly breathed from a tank, and too little follows from the taking of a succession of big breaths. The big breaths sweep the normal carbon dioxide from the lungs, hence from the blood, hence a more alkaline blood, hence lower calcium, and lower calcium may cause convulsions, and convulsions mean trouble in the brain. At the same time the vessels of the brain are narrowed, less blood flows, less oxygen goes to the cells, hence a blunted mind, a faulty memory, dimmed eyes, dimmed senses, the ear not hearing the ticking of a watch brought close to it. The taking of the big breaths may become a habit, give a person satisfaction (the psychiatrists come in here), soon the breaths seem not big enough to him, so he takes bigger, appears an addict, suddenly faints. That habit may date back to something else, the physician may decide, to a memory, a face and the eyes that went along with the hard breathing of a dying mother. Possibly the habit dates still farther back, a young child sucking in big breaths, distracting his young mind, filling his sister with amazement; he never approved of her. Or even farther back, before he was born, some alteration, some mutation in some ancestor.

Instead of too much or too little carbon dioxide there might be too much or too little oxygen. Whoever has gone up a high mountain remembers how exalted he felt, how in his mind he became quite careless of the commandments; meanwhile the valleys below were too beautiful to believe; then when he returned again to the edge of the sea and read the poetry he wrote up there, he saw how giddy he must have been. (Like poets who write poetry under drugs and read it afterward.) If the too little oxygen had continued too long his friends might have found him giddy permanently. Anno Domini 1969 a jet plane puts a man up there in the twinkling of an eye and the oxygen in the pressurized cabin may have failed. Up there the proportion of oxygen did not change, but the pressure did, dropped, accordingly the pressure in the alveoli dropped, and in the blood dropped, in the tissues dropped, in the cells of the highest parts of the nervous-system dropped. Mind signs begin at ten thousand feet or thereabouts. Mountain climbers and all the newspaper readers who climb with them have heard how in high places the mind slows, the power to think weakens—the oxygen cylinders right where they could be pulled over and used for resuscitation, and not pulled over. Should the person decide to stay in the mountains, not at that killing a height but high, he would be taking bigger breaths day and night, and because of that the form of his body would be altered. His chest for instance would get barrel-shaped. The hemoglobin in his blood would increase. Chest and blood would be rebuilding on a new and grotesque scale so as to nourish the brain that underpinned the mind. That blood would be thicker. It would need more red cells. More would be born and mature. Such thick blood would be harder to move. This would give trouble to the heart. If the pressure of the atmosphere were to drop still farther, say to one-fourth or one-fifth of what it is at sea level, there would be pain and the mind signs might be severe to the point of unconsciousness. The color of the skin would be redder and darker.

Did Jehovah teach an angel to fly, and then for the sport of it teach a man to fly, or was His method simply to create a human mind, leave teaching and invention to it, while He turned His invention to the making of another galaxy or quasar? And might that be a fundamental question?

Physicians have had experience with too much oxygen. In these situations body signs appear. From the start the airways of the lungs may be irritated, plugged with mucus, the man suffer what seems a chronic cold-in-the-head. The lungs may be damaged. The brain may be, and in consequence the posture and the locomotion may show signs. Also there may be twitching of lips. Spasms. The most distressing as the most alluring of the signs, however, are those of the mind. Inattention. Blurring. Blackout. Physicians speak of oxygen as a poison and it can be for animals. Barcroft wrote of the death of two larks which were placed in a chamber where there was excess oxygen. First, both became excited (human beings become excited too) and this was followed in both larks by convulsions, a quick death of three hours in one, a slow death of several days in the other. The story as told by Barcroft is scientifically interesting, humanly distressing. Larks belong in the pure air but it should not be too pure, and not too much oxygen. We are creatures of earth and air but made to function best where the composition of the gases and their tension in lungs and blood are not far from what they are at the level of the sea, whence we came.

Methinks I behold Jehovah among the galaxies. Methinks I am present at the original thrust whereafter the stars ran off to play. Methinks I hear a rumbling. Methinks there is a great toil. Methinks a body and head are in the making. Methinks that in the tissue of that body and head the molecules of air are balanced with those of blood. Methinks there is a lone songbird singing in the firmament, and on the ground a lone Adam in a deep sleep put upon him into whose nostrils there had been breathed the breath of life, oxygen. From the nostrils there issued molecules of carbon dioxide. Two kinds of molecules, neither seen by human eye, smelled by human nose, tasted by human tongue, finely divided in the raiment around the earth, and what a difference that has made to you and me since it has let us stay a while among the living, Moses too, and Socrates, and Christ, and Descartes, and Claude Bernard.

# CHEYNE-STOKES
## *Trouble*

Trouble cometh to hawk and jackal and ass and to the man of understanding heart who at year's end cut his beard and the hair of his head, and other men's. Trouble cometh to the barber. A third of a century later it still would be possible in the mind to hear the solemn sound of his breathing that night he lay in his last trouble, and his physician sat and watched him, then walked away. Awesome that breathing. He breathed so for two days, then they prepared him, embalmed him, put him in a coffin, and it was three days more before they carried him off, had trouble with that too, the stairs so narrow. Seemed a long time to a child on that cold February noon till finally they started slowly toward the burying place, where were all those ribs and all of them owing to that one original rib.

The words that now follow were written a century and a half ago, anno Domini 1818.

"For several days his breathing was irregular, it would entirely cease for a quarter of a minute, then it would become perceptible, though very slow, then by degrees it would become heaving and quick, and then it would gradually cease again. This revolution in the state of his breathing occupied about a minute, during which there were about thirty acts of respiration." The words were John Cheyne's, one of the goodly company of Dublin physicians whose glory would shine through the generations. John Cheyne does impress us. Besides being the first to leave a careful description of this breathing—periodic breathing—his timing of it comes close to our timing. The breathing must have been known everywhere in Israel and Judah. Hippocrates, physician on the island of Cos, said of it that it was "like a person recollecting himself." But no one before John Cheyne carefully described it. Indeed, uncounted physicians, uncounted members of families from tribe to tribe could not have missed hearing it echo through some house. The departing was making his concluding suckings from the nothingness. He took a breath, took another bigger, another bigger yet, stopped, no-breath.

That is the one type. In a different type there are small breaths in place of the no-breaths. The house may be so still. During the big breaths the departing may have waked and babbled a bit and been quarrelsome, then during the no-breaths was already not of this earth. The watchers were thinking: will it start again? If it did not they could go to their own beds. That, then, is the pattern of the breathing. When the cycles are long it is apt to mean a failing left heart, when short, brain damage. Physicians know the facts. Once, he that is lying there will not start again. "And now that he lieth he shall rise up no more." The earth will not miss him. It is part of any day. No mourning. No weeping. No black cloth. No cease in the earth's conceiving of fishes and prophets.

If the physician is standing by a bed set apart from other beds in a place for the sick, and if the physician says of a man's breathing that it is periodic, he means that in every cluster of breaths there is such waxing and waning. One breath is larger than the breath before and so for a number of breaths in succession, then each smaller than the one before. It is a large rhythm rising and falling and carrying along the smaller rhythms of the single breaths. Or, a cluster of unusual breaths is followed by no-breaths, this followed again by the unusual breaths. In the breathing chronicled by John Cheyne, called Cheyne-Stokes, there are always or nearly always the no-breaths and usually the waxing and waning. On a mountain-top a similar breathing may be normal in sleep. A dog may need no mountaintop but breathe after this manner in front of an open fire in a house low down where the rivers run into the sea. The newborn may. A man in good health can bring it upon himself of his own will. He takes a number of big breaths that somewhat wash his lungs of carbon dioxide, therefore the drive to breathing no longer as strong as it needs to be, the washed blood of his lungs goes to his brain: his breathing stops. As soon as it stops, from all parts of his body carbon dioxide gathers, and in all parts oxygen is used up. He may get blue from lack-of-oxygen, and some have thought it was this that started him breathing again. A breath. The lungs again are washed, somewhat, the carbon dioxide lessened, somewhat, so the breathing stops a second time, not for as long, and thus on until all has become what it was before the man did of his own will take the big breaths. This manner of explaining lays all to the chemistry.

Another explaining lays more of it to the ancient controls low in the brain, those cells known to have charge of breathing, the effect of level upon level, the effect of hierarchy.

During sickness the sensitivity of these cells may be depressed. The probability is that their sensitivity is altering in health too but sickness alters them seriously, their state helping to sink the man into that final trouble that affrights the watcher because a common pattern of the living creature has become an uncommon one.

As for the solemnity of the sound echoing through the house, that arises from man having heard this breathing just before the awesomeness of death for as long as any man can remember. What a man hears depends upon what he has heard, upon the way the generations of his life have prepared his ears to hear the Psalms, the fall of rain, the thunder. A man may not hear the snorings that crash through rooms a whole night long, may not hear the noises of a living city, but the breathing of the dying he hears. The physician witnesses this solemnity a hundred or a thousand times. He knows it by ear, by sight, by understanding, anticipates when it will begin. Then one evening he is thinking: now it will begin for me. The physician's hour. Jehovah is nearer. More likely, to be sure, the physician will not be thinking in the least of his own breathing that is already sounding through his house, which was his father's house before him.

# IX
# DIGESTION

---

## It All Works Like
## a Hungry Dog

# TUBE

## Beyond the Lips Is More Than Teeth

In a zoo at midnight a good-sized live pig was pushed inside the door of the bachelor apartment of the python. The pig was nervous. After the conventional preliminaries that began with squeezing the holy breath of life out of the pig, the python dislocated his jaw, which is part of the ritual for any of his better meals, and in went the pig, slowly. Whereupon at his upper end the python bulged. Next morning he bulged less and farther down. The pig was melting. Life was making life out of life as the pig advanced through the great snake's digestive tube.

Man gets in his pig without dislocating his jaw and next morning does not bulge, not much. Man also makes life out of life. His digestive tube begins with teeth in the vestibule; he flattens his lips over them, puts a socially acceptable façade over the ravenous.

His tube has many diameters, wide, narrow, valves at strategic points, turns and twists named and unnamed, a purse-string double muscle at its ulterior end. It is the scenic way at his private amusement park. The stations are the following. Lips—the doors to the outside. Mouth—the vestibule. Pharynx—the passage through the throat. Esophagus—the gullet. Stomach. Duodenum, jejunum, ileum—small bowel. Caecum—the word means blind pouch. Ascending, transverse, descending colon, sigmoid, rectum—large bowel. Sigmoid means S-shaped. Rectum means straight.

Thence out into the sunshine or the warm rain.

Flesh-eating animals have a comparatively short tube, four times the body's length, vegetable-eating a comparatively long, twenty times. A distinguished expert claims the figures inaccurate because the measurements were made on the relaxed bowel of a corpse whereas they ought to have been made during life with everything unrelaxed. Man's tube is on the short side, establishing him as the brother to the lion, the tiger, the other carnivores.

# FOOD

## *The Daily Bread*

Besides our lunching upon the spring lamb, the flying bird, the tender tip of the young asparagus seasoned with salt and pepper, we are ourselves the lunched upon, after a while. We are edible compounds. We eat. We dream. We think. Better that we do not think too much on what we eat. Better that the rabbit does not think on what it eats, especially when it eats my neighbor's tulips, because if a rabbit employed its meager thirteen grams of brain for thinking, it might not have any brain left for eating, and after a while we never would see a rabbit eating tulips, and that would be too bad. But even those who do not care about rabbits or tulips teeter-totter between lunching and being lunched upon.

About a hundred years ago it could for the first time be said with confidence that there are three fundamental foods: carbohydrate, fat, protein. The human body needs the three. Lacking them it sickens and may die. Later, water and minerals were added as necessaries. Later, vitamins. Later, trace amounts of this or that. A sturdy pot roast (once part of a complacently grazing living cow) plus a seasoned salad (from the pregnant soil) plus a dish of velvety raspberries (that yesterday shimmered like dewdrops on a shrub) meet all requirements of body and brain. Yesterday shimmering and living. Today? Well . . . Tomorrow? Living again, just changed their address.

Digestion is the processing of food from the state in which it is devoured to the state in which it is delivered out of the digestive tube into the blood.

Milk is our earliest food. Some shiver to remember it; some righteously drink it all their lives. Milk contains the three, carbohydrate, fat, protein; also minerals and vitamins. Cow's milk has too much protein, not for the cow, but for us, and in cow proportions, and since human infants do not grow as fast as the calves that are hurried off to market, and since protein is the stuff of growth, there is the practice of diluting cow's milk for our babies, never completely satisfactory because, diluted, the protein still is cow protein and still in cow proportions.

Eggs are high in protein, fairly high in fat, very high in calcium, a good supply of vitamins—everything ready for those fledglings whose necks we shortly would have wrung. Fish is high in protein, calcium, vitamins: salmon and mackerel are high in fat, whitefish and sole low in fat, proportions that we do not think of for their value to the fish because we do not think of fish as swimming in the sea but as hooked out of the sea and eaten on Fridays. A year or two ago there were fourteen million applications for fishermen's licenses, a year or two before that, only seven million, which with fisherman's luck could mean twice as many lives hooked. Meats differ in the amount and kind of fat and protein—we avoid also at lunch to think of rib roast as flesh off ribs. Cereals have a coating of an indigestible substance called cellulose and again a career that began in a happy living wind-blown plant. Flour has protein. When soaked with water it makes gluten. Add yeast (living) and this acts on the carbohydrate, forms carbon dioxide, which blows up the dough. The pan is pushed into the oven. Heat cracks open the starch granules of the flour, soluble starches and near-sugars run out, and the surfaces exposed to the greatest heat are caramelized and browned, and that is your big sunny loaf of bread. Big and sunny and for the moment dead. Vegetables and fruits, both living, have almost no fat. Potatoes, living, are high in carbohydrate; bananas higher; prunes higher; raisins higher; desserts high, often high also in fat, sometimes high in protein, as cocoa, sometimes high in calcium, as chocolate. (Whoever needs to should jot down the facts.) Alcohol may accompany the meal. If beer, it is comparatively high in carbohydrate and digests slowly.

All of the delight of all of the above reaches not much farther than the mouth, reaches the point where we think, isn't that nice, and the next instant think that something agreeable has gotten away from us. As the delight disappears farther into the digestive tube and from the mind it leaves a callous sense of well-being that persists until drowsiness supervenes. Drowsiness is absolution for the sundry murders just committed.

# DEGLUTITION
## *A Swallow in Three Acts*

A parrot's beak can crack a Brazil nut. A dog's grinders are claimed to build up a pressure of three hundred pounds. A debutante's molars have perhaps a third of a dog's power. The debutante's jaw drops. It lifts. It protrudes. It retracts. She swings it from side to side. In short, she chews. Her lips return the mouthful to where her pearly teeth can get at it a second time, a third time, any number of times depending on her mood or the gristle in the steak. At last, she has the mouthful mounted on the top of her tongue and it has a name now, bolus. The bolus is moist: dry food, as when our guilt is heavy and no spit in us, is difficult for the mouth to sculpture, and there are other difficulties. An unsuspecting bolus sits on her tongue, waits.

Then the trap is sprung.

Swallowing starts.

What happens is without division but a physiologist of the nineteenth century described swallowing, deglutition, as occurring in three stages, so we all do. Will that need to number everything 1-2-3 ever let the human mind alone? The first stage carries the food past the anterior pillars—flesh pillars one on each side of the entrance to the narrows that lead from mouth to throat. The second stage carries it through the throat, pharynx; the third, down the gullet. A hunk of beef, a jigger of whiskey, a safety pin, all go the same way. Since the diameter of the tube varies, it is not surprising that when a person bent on self-slaughter drinks carbolic acid, that was fashionable before sleeping pills, some places are burnt, others not.

The tongue deserves a word. During chewing it operates like a masher, also executes the sudden thrust that drives the bolus backward. The tongue is muscle, largely. Muscle is operating all along the way. The teeth are clenched, by muscle. The cheeks are pressed against the teeth, by muscle. The lips are sealed, by muscle. The bolus is rammed against the debutante's pink palate, by muscle.

Trigger zones deserve a word. Some are on the palate, some on the uvula, the tit that hangs in the midline, and these zones pick up the news, telegraph ahead that the bolus has started down.

Not far down is a treacherous crossing of roads. Air and food must both be gotten into the body, air through the two entrances, nose and mouth, food through one, mouth. Air then continues over Route Windpipe, food over Route Gullet. Engineering problems were bound to arise. Food must not slip back into the mouth, not into the windpipe, not into the nose. Accordingly, the mouth is closed by tongue and pillars; the nose by lifting the soft palate, bringing together the pillars, fitting the uvula in between, and contracting the top of the throat; the windpipe by jamming the voice-box against the base of the tongue, pulling everything forward, and thus opening the gullet behind so that food can get in. The V-gap between the vocal cords is sealed, by muscle. Breathing has stopped except for the slight start of an inspiration which creates suction in the now air-tight gullet and—zip, bang, slush—down plummets the mouthful from the soft-eyed cow. For fluids swallowing is quick, seconds. For breakfast gruel not so quick, still seconds. For solids slow, sequential muscles of throat and gullet sequentially pushing the solids along. An infant swallows as perfectly at birth as a man after seventy or eighty years of patient rehearsing.

# ENZYME

## Busy As Fleas

Through ages there were evolved molecules to speed up digestion and other action. They were whips. They were harassments. Fleas. Protein fleas. No chemist, no geneticist, no scientist would call them fleas. Call them giant molecules that activate, yes, macromolecules, high-molecular-weight molecules because of whose evolution life lives. Many proteins are such fleas.

Enzymes.

The scientist calls them enzymes. He says they make or break covalent bonds. That may not be easy to understand. But, they make a molecule accomplish at low temperature, such as our body's temperature, what without enzymes it could accomplish only at high temperature. That would be easier to understand.

Twenty years ago it was said that one thousand such enzymes had been isolated. Later it was said that twenty thousand to thirty thousand were operating in our body, that that many submicroscopic chemical activators were entering on cue to keep our chemical drama moving. Oxidation-reduction drives, it was said. Each enzyme had the same general talent, to speed what otherwise would run too slowly, to fit itself in among busy molecules and make them busier, to excite some chemical step in the living. Step. Step. Step. When then one considers that there are such steps everywhere along the assembly lines, that a single organ has been estimated to do a hundred distinct jobs whose rates are all determined by these specialist molecules, that combinations of them operate close together, that there have been mutations and mutations, as there will be, hence enzymes sometimes only slightly different from each other, one begins to understand somewhat how food entering at the mouth and its residue leaving at the lower gut is able between those two points to keep at reduplicating the fibers in that calf muscle that raids a pantry shelf, lifts a foot to the curb on Times Square on New Year's Eve, puts force into the debutante's grinders while a special enzyme in her glistening spit is chemically reducing one class of the foods that the grinders are grinding.

One asks then, anticipating the answer, what is it that truly breaks up the deliciously dressed goose? What changes goose molecules to debutante molecules? Spaghetti to the dog Paddy? What decomposes, recomposes, with such nightmarish regularity? What minces the chewed till it is finally so small and finally so different that it can slip into the substance of the Chairman of the Board of Trustees or the writer of valentine ditties, or into the humble cockroach that eats and performs zigzag maneuvers to escape and drops drippings and writes nothing? What makes our digestion chemically move? What makes our body's life chemically move? What keeps the machine-parts chemically meshing, turning?

The chemist says enzymes, calls them enzymes. He says much about them. He says that they may start an action one direction and when it has gone far enough reverse the direction. He says that that usually takes two. He suggests that the myriad of such changed directions might provide one definition of life. They provide the acceleration in the drive that maintains the constancy of the internal environment. They keep each special cell at its special task with a zeal that could make it think, if it thought, that the Host had His eyes on it individually. They are the spur. They give the all-around push to that "sum of the forces that resist death" which would be too slow and therefore not powerful enough without them. To the hard-headed science-minded mid-twentieth century they might well seem the essential accounting for that gigantic organized labor that, nevertheless, even at the short distance down from an airplane, appears to have achieved no more than a patchwork of variegated coloring sprinkled over brown-black earth.

Far back before the dawn of the living world indiscriminate molecules were floating around. Then some changed, underwent something equivalent to a mutation. Many such mutations with evolutionary selection of the successful mutations. Many such. Many. Many. Among them a gene-molecule that made an enzyme-molecule. Many genes. Many enzymes. Many protein molecules that could speed some recurring chemical action and, in final consequence, the world took on a new face. It woke up. It was no longer merely a planet. Enzymes everywhere were speeding action. Were receiving what amounted to instructions as to how to speed. Today we emphasize and are interested in a class of molecules that give the instructions, nucleic acid molecules, a specific one, DNA, and the instructions, the messages, are carried by a specific other nucleic acid molecule, RNA. (Man often talks like a parrot in order to conform to the rules of the associations of scientists whose prisoners he is.) After a while, in any case, there was Job, the small green frog, and there was you and there was I. Myriads of enzyme molecules were in and about us, inside every cell of us, each under silent orders, eon-orders, the heavy hand of time on everything. Enzymes were harassing food molecules. And all of it was to the inexplicable end that each of us would inexplicably do his labor and live out his life in a total of chemistry that inexplicably was carrying out a mechanic destiny. At least, so it might seem to anyone who for two minutes lifted his head off his chemical bench. It might not seem so if he lifted it for three minutes.

Platinum the lovely metal placed in a test tube of chemicals, as any high-school student knows, speeds action. The platinum is not used up, adds nothing to what goes on there, just changes the rate. A substance that does this is a catalyst. Some have thought that a catalyst merely attached itself to the surface of what it acted on, but others, that it always took part in the action, disengaging itself when the action was done, and others, that this differed catalyst to catalyst. The last is probable if for no other reason than that complexity is the way of the world, and of every detail of it.

An enzyme is such a catalyst. An enzyme changes the rate of actions without itself being used up. It is an organic catalyst, or, in this region we are talking of, a living catalyst. For years it was believed that the existence of the organic depended upon the existence of the living, but this is no longer believed and the facts are indeed just the other way.

As for the enzymes of the digestive tube they begin in the mouth and follow one after the other, each with its fixed labor, right down to the tube's nether end. A special enzyme stands behind each special action. Indeed, it stands behind each step in an action. This step could go on without this enzyme, but not fast enough, not joining with a multitude of other steps, actions, so as together to create this that breathes, that runs, that sits, that listens to a heart beat, takes a blood count, watches a peculiar type of breathing, a peculiar moment of digestion, then goes home from the laboratory, scrubs its face, and rides all dressed-up to a party in the evening.

Formerly it was taught that an enzyme was vital, this implying it had a touch of the supernatural, but today the word *vital* has been outlawed, all pleasure blotted from it by the irreverent chemist who has dared to purify and even to crystallize about a hundred enzymes. The first was in 1926, by an American. A crystal is a lovely thing but never impressed one formerly as alive, which was old-fashioned, and indeed today it has become almost easy to imagine a crystal in some brilliant concatenation of crystals to produce the bubble life.

Enzymes of the human body work best under the conditions of the human body, the dog's under the conditions of the dog body. Drop the temperature inside Paddy, the Irish terrier, and her dinner will not digest. Contrariwise, raise the temperature enough and her enzymes will be destroyed, she with them. Take an enzyme from an acid locale where it was evolved to work, say, the stomach, place it in an alkaline locale, upper bowel, it will not work. So on. The sweep of the enzymes, the march of them side by side in the cells of our body, each keeping its place, each doing its job, driving, timing, accomplishing, this disciplined army of submicroscopic fleas is so amazing that one finds oneself contemplating with a new respect one's next-door neighbor with his wife and three children, the chemist, and the neighbor next-door to him, the geneticist, who, blind though both sometimes may be, add finding to finding and end by achieving what one could think achievable only by the most inspired insight. Admitted, the chance is that their kind of achievement never remains achievement.

Emil Fischer long ago fancied any enzyme and what it acts on as key and lock. A particular key for each particular lock. What the enzyme acts on is called substrate, and the substrate is believed to move around the enzyme, find a position on it, an active center, one or two or three of these, attach itself, speed reaction there, the substrate broken down, products produced, the accumulation of the products presently delaying or even stopping reaction, the enzyme always released for further work. Three Frenchmen found how the active centers of an enzyme might interact with one another, produce something that performs a different kind of work, and they indicated as parallel to this the rearrangement of the hemoglobin molecule when it is oxyhemoglobin and when it is reduced hemoglobin. Their discovery brought them a Nobel Prize.

In our mouths the substrate would be food, and our saliva, as said, has an enzyme that acts on one of the three basal foods. The products produced by that, along with the food which has been chewed but is still undigested, are carried onward in the tube, where other enzymes continue the digestive work. Meanwhile that enzyme back there in the mouth goes on being methodically freed, more of the basal food methodically attached, products methodically produced, and this roundelay repeating and repeating, in you, in me, in Paddy the Irish terrier, in all debutantes.

A final idea, or only another idea, is an enzyme inhibitor, a molecule that blocks the action of another molecule. Consider an enzyme inhibiting a step in digestion. Digestion is blocked. Wrong chemicals collect. May damage the creature. May kill him. In that case the inhibitor would have been a poison. When scientists in World War II were scheming chemical warfare they found inhibitors that could kill in that way. Whereupon scientists of the enemy, whichever was enemy, searched for and found inhibitors of the inhibitors. An enzyme might be poisoned also by ordinary drug-store poison. Cyanide in the gas chamber kills the doomed because it poisons enzymes and the cells of that body cannot use the oxygen brought to them. If a man is poisoned in the gas chamber or at his wife's dinner table, he thinks it personal, which is unreasonable.

Let a mathematician go on with this account. He should have no trouble now making clear how enzymes can build one man unlike another, in his anatomy, in his histology, in his thought. A smart idea, enzyme, occurred to Nature on a cool morning when for once she had slept well. On another morning she invented the fleas, those we swat, those gigantic constructs that annoy that still more gigantic construct, Paddy. However, they do keep Paddy's mind active, interrupt her day-dreaming.

Gene. Enzyme. Flea. Paddy. Man. Verily size does seem the essence of nothing.

# FISTULA

## *Portraits in a Corridor*

This that now follows ought not to be read with the book by the side of the luncheon plate, because—well, it ought not.

An unpleasant word, *fistula*, comes from Latin, means pipe. Here it means any unnatural connection between the inside of the digestive tube and the outside of the body. In Pavlov's laboratory in

Leningrad one might see a succession of fistulas, the following all in the same dog, according to one of Pavlov's co-workers. A salivary fistula—the duct of a salivary gland brought out through a slit in the dog's cheek. Next, the gullet cut across—two openings, each brought out and stitched into the skin of the neck, and if that dog was fed ground beef, a moment later the beef soaked with saliva flopped from the upper opening into a dish, a farcical situation, called sham-feeding, a caper played on the dog, the beef never reaching the stomach unless the saliva-soaked was re-fed through a funnel into the lower opening. Next, a small-stomach fistula—a tenth of the stomach separated from the rest and that tenth given a surgical exit through the abdominal wall. Next, the other nine-tenths given an exit. Next, a small-bowel fistula—a loop of bowel lifted up and sewed into the dog's belly, and an exit to the outside cut through the belly-wall and the loop. A pancreatic fistula—the duct of that great digestive gland brought to the outside. All these fistulas were in a single dog, a healthy dog, a contented dog, if Pavlov's co-worker's memory was accurate.

In a human being there sometimes has been need of a surgical fistula, as where lye was swallowed and scarred shut the gullet, or boiling clam chowder, and therefore food for the rest of that person's life needing to be fed via a fistula made surgically into the stomach. Rarely, but sometimes there has been an accidental fistula.

The idea to use such for digestion study like all ideas was young once. It was the dewy morning of June 6, 1822, at Michilimackinac, Michigan Territory, or Fort Mackinac, a combination trading station and army post occupied by United States troops after the War of 1812. The population was French, American, Indian. A youth that morning shot himself in the left side, a mistake, the muzzle of his musket only a yard from him, the wound not killing him though "literally blowing off integuments and muscles of the size of a man's hand, fracturing and carrying away the anterior half of the sixth rib, fracturing the fifth, lacerating the lower portion of the left lobe of the lung and the diaphragm, and perforating the stomach." His name was Alexis St. Martin, a name that would live long in the romantic history of digestion. Alexis was employed by the American Fur Company as a voyageur, transporter of furs and men across the north and northwest. About twenty-five minutes after the shot a

surgeon, William Beaumont, arrived, dressed the wound, pronounced the case appalling, the stomach "pouring out the food that
he had taken for his breakfast." The following day there was fever,
difficult breathing, Beaumont gave Alexis a cathartic, bled him,
believed he would die. Five days later the wound was sloughing.
Eleven days later the sloughing was extensive, the wound appeared
healthy, the fever subsided. Four weeks later the appetite was good,
Alexis' digestion satisfactory, bowels moving, but the opening into
the stomach did not heal.

Here was an inquiring surgeon's chance. Here one man could look
straight into another man, and Beaumont looked. He looked every
day of the week, whenever Alexis was willing, which was not
always. "I had opportunities for the examination of the interior of
the stomach, and its sections, which had never before been so fully
offered to anyone." He looked for what actually was there. "I had no
particular hypothesis to support." Watched water pour in at the top,
remarks that this was "a circumstance, perhaps, never before witnessed, in a living subject." One catches Beaumont's enthusiasm.
With a glass rod he touched the inside of the stomach, recorded. He
said that the lining had "a soft, or velvet-like appearance." He
recorded that changes in the lining accompanied changes in Alexis'
mood, that with "fear, anger, or whatever depresses or disturbs the
nervous system—the villous coat becomes sometimes red and dry, at
other times, pale and moist." He drew out samples of gastric juice,
sent them for analysis to the best laboratory of that day, was as
scientifically enlightened about all of this as anyone then could be,
began it in 1825, continued till 1833, published his findings in a
book entitled *Experiments and Observations on the Gastric Juice
and the Physiology of Digestion*. A reader can open that book at any
page and be rewarded and amused. Beaumont made his patient his
complete responsibility. The patient became a declared pauper.
Beaumont housed him and gave him his daily bread and had to keep
an eye on him because the pauper regularly got bored at the proceedings and absconded. The covenant bound Alexis to "serve,
abide and continue with the said William Beaumont, wherever he
shall go or travel or reside in a part of the world." (Sounds like the
marriage covenant.) Alexis did not respect covenants and did not
enjoy the travels of an army surgeon. Letters are extant, the surgeon

trying to imbue the patient with the scientific spirit, and the patient drunk. Why not? With a hole in your stomach?

The coolness to use a seriously wounded man for digestive study, the experiments, the observations, the book, are the most unique and may be the most far-reaching achievement of any American who has studied the human body. Those interviews with the stomach are dignified and innocent. They are an investigation into body and head of the highest animal, a prolonged series of clinical procedures carried out in the spirit of 1969, reports on the relation of gastric juice to food, to fasting, to the hour of the clock, to the state of mind of Alexis, which means, to the state of mind of a man who owns a stomach and properly forgets that everything in his daily life affects it. But the science of medicine should not forget, as it has not, a whole large division of it having concentrated on the effect of mind on body, to the point of illness.

# STOMACH

## *More Than a Storage Bin*

That great baby, man's favorite pouch, that he always speaks of slightingly as he does of the beloved, lies between the lower end of the gullet and the upper end of the small bowel. Food comes in at the top, cardia, called so because near the heart, goes out at the bottom, pylorus, Greek for gatekeeper. A mobile organ, rubs gently against its neighbors down there in the dark, lined on the outside by a silken lining, on the inside by another silken lining but pleated and that has the open mouths of thirty-five million glands. Some stomachs are small. Some seem so because of their worried state. Some seem dropped because they dip down into the pelvis—this is perfectly normal. If only one could say that a steerhorn stomach went with a steer but it may go with a dove of a man. The stomach usually is J-shaped, and no matter what shape it may change with its minute-to-minute mood, or life, that can be hard. The first layer of lunch is spread around the inside wall, automatically relaxed to be receptive, a second layer inside the first, a third inside that; meanwhile an animated luncheon conversation is going on all around the table, all around the stomachs. Starch digestion, that began in the mouth with that starch-digesting enzyme of the saliva, continues below in the center of the vat. But why do we say vat? What is the hidden self-consciousness that tempts us always to speak jeeringly of our great reservoir? Reservoir it is. The digesting it does could be done well enough farther along; stomachs have been cut out and nothing too bad happened. Protein molecules start their breakdown in the stomach, that has the enzyme pepsin to attack the large protein molecules, split them to small. Large to small is the pattern for all digestion. Pepsin requires an acid world and the hydrochloric acid coming from the glands of the stomach is strong. William Beaumont knew about the hydrochloric acid, though how it is produced he did not know, and we today still are not sure. The stomach has the enzyme rennin that curdles milk, cannot in the strong acid of the adult stomach but can in the alkali of the infant where curdling is important. (Everything is arranged.) Some fat is broken down—fat slows digestion. While the attacking molecules and the food molecules are engaging each other in chemical combat, muscles of the stomach wall are acting on the softening lunch. A nagging. A maceration. A trituration. A pushing forward. A steady pressing on the late filet mignon or goose like the pressing on a toothpaste tube. Nothing that once gets into the digestive tube is allowed to loiter. Periodically a wave travels in the direction of the pylorus, that is an easygoing gatekeeper and permits food, chyme now, to pass through and onward into the small bowel with almost no argument. Yet how the pylorus is opened and closed we again are not sure. Water is in and out almost as fast as drunk. Beaumont watched water. Food when in chunks delays stomach digesting, and emptying. High alkali in the small bowel delays. Sight and sound and smell of food hastens. The arrival of food in the stomach with its consequent distension hastens. So does the arrival in the upper bowel.

Fear and gloom may stop all of this and gaiety excite it, the stomach's inside color changing, as Beaumont observed, the acid changing, the quantity of total gastric juice changing. Medicine today on every side stresses the effect of emotion on physiology and every wife knows that her husband's rages may transmute an excellent dinner into an indigestible one. "If only that man would lie

down for half an hour before and after dinner—not get so upset."
They who have watched the inside of dogs' stomachs would say that
wife was giving her husband sensible advice.

# GASTRIC ANALYSIS
## *Authorized Prying*

To obtain a sample of gastric juice requires a fine finical procedure;
some persons say it is nauseating. "Everything came up, nose,
mouth, eyes, but I kept it in!" That was a medical student proclaiming his success. He had passed a stomach tube into himself, as
medical students do (would have preferred to pass it into someone
else), passed it into his mouth, might have into his nose, where high
up it would have turned a corner, gone down his throat, continued
conventionally into gullet and stomach. He railed, reeled, regurgi-
tated. That distinguished expert mentioned earlier said that people
swallow stomach tubes as they live their lives and if they make a
great fuss it tells the physician something he ought to know. One
especially long tube arrives at the stomach and continues, duodenum, jejunum, ileum, continues, continues, is assisted by the
physician and by waves of muscle contraction, continues, continues,
until it has been known to appear on the West Coast. Logically to
complete that picture the two emergent ends might be tied in a knot
and thus a human body strung on a rubber clothesline, which is
stated primarily to remind the reader that a tunnel does run through
him, connects outside with outside, as Midtown Tunnel connects
New York with New York.

Stomach tubes are used every day of the week to obtain samples
for gastric analysis, samples farther on for other analyses. Sampling
of human cavities has gained a lively place in the last decades, all
the quiet places washed out, the washing scrutinized with the
microscope for anything anyone can think of, helping diagnosis and
treatment and sometimes prolonging life. Tubes are pushed into the
windpipe to obtain washings from a lung. Into the sinuses of the
head. Into the genital organs of the male to obtain prostatic wash-

ings. Variously into the female to obtain vaginal, uterine, tubal, ovarian washings. Into the rectum, and up. Washings have long been obtained from the inside of the female breasts—one still resists the notion. Into any veins. (Except those of the ovary that are said to have a sharp angle.) Stomach tubes, to return to those darlings, are used if a person has swallowed what he should not and now would not, to feed patients who cannot otherwise be fed, and to find whether hydrochloric acid is where it ought to be, or is not, and how much. The tubes are always long enough and with some to spare—stated to allay the nightmare of the other end slipping in. If the gastric juice happens not to be flowing freely it may be encouraged with tea and a cracker, which is good; a swig of alcohol, better; an injection of histamine, best; effective, that is. Commonly several samples are collected, curves plotted, in this way the juice of the single Johnny defined, and when reports come from ten or a hundred cities and villages the juice of the typical Johnny defined.

A toy balloon may be fitted to the end of the tube and the combination swallowed, watched with the X ray, and when the balloon reaches a desired point it can be inflated via the tube, a writing device hooked in, and communications begin to come back out. A hungry stomach contracts, as we all learn, and records of hunger contractions are printed in all the textbooks. Even two balloons a distance apart have been included in such a scheme, the two inflated, a portion of the tract thus isolated, sealed from the rest, and the content of that portion sucked out. Even three balloons, the first maneuvered into the small bowel, the second between bowel and stomach, the third left in the stomach. Happy day! Whatever has been learned from this is a tribute surely to the investigative calm of scientist and subject.

# VOMITING
## *Arrangement for Bringing Things Back Up*

From time to time there occurs a reversal in the direction of digestive events. A household matter. In the easily embarrassed adolescent years to have it occur may depress one for a week, but thoughtful old age will sit above and watch, even achieve a disinterested admiration for the mechanics, coupled with a touch of annoyance at a speck that did not miss the nightshirt. It was 4 A.M. A fever was burning. The man had been attempting to get rid of a cold-in-the-head by washing down a pill with a swallow of milk, and the return was so fast that he thought petulantly of Wimbledon and Forest Hills. Just had time to get his red beard out of the way.

Meaning?

A human body was undoing something done to it, or a mind something done to it, or even on occasion a sick man was wishing to impress his nurse. Whichever, the digestive system was performing an age-old act. Nero and his Romans knew it well.

The preliminaries often are hesitant and full of false promise. A spasmodic breathing usually starts them. A dog starts them by eating grass, and in a dog the human being can observe the mechanics unemotionally, the wise bitch interfering with nothing, simply letting everything happen. The mouth fills with what is mostly water, some of this running down into the gullet, watering the walls; in a man his mind now goes definitely along with the mechanics. The gullet relaxes. The stomach relaxes, except its exit that vigorously contracts and stays contracted. Thus there results a long limp pouch that begins at the mouth and is shut tight below. The diaphragm lies on one side of the pouch, the abdominal wall on the other. At an instant, diaphragm and abdominal wall contract, the pouch is caught in the wedge, squeezed, the contents going the only way they can, up and out. This is the body's technique. As when everything is in the conventional direction, windpipe and nose are automatically sealed off, mouth opened. Two incidental facts. If the abdominal wall should be paralyzed, rare, the act is not possible, one side of the squeezing wedge not able to perform. If the stomach has been surgically removed the act is possible, bowel instead of stomach caught and squeezed. In an experiment more than one hundred and fifty years ago a pig's bladder was substituted for a stomach, the act induced, studied; the conclusion one can draw is that investigative curiosity ran rampant in that day, too.

# BOWEL

## *Twilight of a Morsel*

Every morsel of every lunch in every nation of every continent of the seven continents is at this moment journeying to its twilight, and over the same scenic way, mouth to gullet, gullet to stomach, to small bowel, to large, to the Matterhorn or wherever.

The small bowel, human small bowel, any small bowel, is temperamental. Often it has been described as a narrow long factory. In that factory one manufacturing process follows another, the processes overlapping, muscles supplying the automation.

A surgeon will incise a human abdomen and loops of small bowel bulge through his incision. Whoever sees this sees it forever, that belly full of bowel, everything appearing to want to do something and not knowing what, push some direction and not knowing where. This can be observed at one's leisure if a researcher will incise the belly of that tulip-eating rabbit, fit a window, as is done, the rabbit allowed to recover. Looking in at that window the eye recognizes movements. After a while it recognizes types of movement. A type that travels as fast as five inches a second—called rush. A type that propels the food methodically forward—peristalsis. A type that gently rocks sections of bowel to and fro—pendular. (Marked in the rabbit.) A type that over a length of bowel causes a sudden series of indentations that remind anyone of a string of sausages, then in a few seconds each sausage divided at its middle to produce a new string—rhythmic segmentation. An internist relates that through the delicate abdominal wall of a delicate woman he has been able to peer straight at the motor handling of her food, especially if surgery had reason to thin an already thin wall, as happens.

Inch-long strips of small bowel frequently have been experimented with. A rabbit has been killed by a blow on the head, a yard or so of bowel scissored out, flushed, sliced into strips, a strip suspended in warm salt water with oxygen bubbled in, then the spontaneous movements of the strip recorded, and their variations.

A man was hanged at San Quentin, and human bowel similarly scissored out. It proved us all rabbits, rather.

While the muscles of the digestive tube are supplying automation the glands of the digestive tube are supplying juices. Special juices for each food. Special enzymes in each juice. Per day, the total manufacture of digestive juices may be two gallons, more than the total blood. A single salivary gland has been known in five minutes to pour an amount of saliva that weighed as much as the gland—saliva is the first digestive juice. Then come stomach juices. Then small bowel juices.

A small-bowel gland may look as if a microscopic finger had been pushed from inside into the bowel wall, the resulting pouch lined by one or more special kinds of cell, each manufacturing its enzyme or chemical, this welling up where the finger was pushed. Sometimes the finger has side-fingers. Sometimes the cells are lumped and the juice flows from a duct. As to the cell itself, when at rest there are granules in it, these visible under the microscope, and when the cell goes into action the granules dissolve and the juice oozes through the cell's wall.

Besides the small-bowel juices from the millions of tiny embedded glands juices are added from two huge outlying glands, liver and pancreas. From liver comes bile, from pancreas, pancreatic juice. Per day, the manufacture of pancreatic juice is between a pint and a quart. When it has been boiled and brought to dinner, some persons eat pancreas as sweetbread, like it. In life it lies behind the stomach. It manufactures enzymes for all three foods, carbohydrate, fat, protein. Some enzymes do not act at once but wait on other enzymes to strip them, literally. A molecule strips a molecule and that does not seem fantastic to a chemist.

Concerning bile, many know it as gall, and know the gall bladder, had an old girl friend who lost hers. Previous to that sad event her bile was stored in it—hung under her liver like an eggplant and held an ounce or so. Bile is let onto the digesting food through a duct guarded by a sphincter, a circular muscle. If the bile is not under pressure enough to push open the sphincter, it is diverted to the gall bladder. Tension in the sphincter varies. Show a dog a bone and a pressure as low as sixty millimeters is able to push open the sphincter—the bone has made suggestions to the sphincter via the dog's thoughts and feelings. Contrariwise, twelve hours after a meal when chyme (that partly digested food dropped from the stomach) has advanced beyond the first part of the bowel and no bile necessary, a pressure of one hundred millimeters may not be able to push open the sphincter, and after a fast, three hundred millimeters. While in the gall bladder, salt and water were drawn from the bile, may concentrate it ten times. Bile helps neutralize acid from the stomach. Bile helps the handling of fat, all aspects, emulsification, digestion, delivery to the blood. Per day, the manufacture of bile is between a quart and a quart and a half.

One has the impression that every part of the digestive machine is forever humorless. Humor, indeed, stops with the last mouthful when nature becomes grim and each morsel marches obediently to its twilight. At the end of the ileum—lower right abdomen in the region of the appendix—the passageway is guarded by a different sphincter and there the erstwhile meal goes from small bowel to large. Thus has the gravied goose, after humiliating pummeling by muscles of the digestive tube and insidious disintegrating by juice, been reduced, reduced, reduced. *Sic transit gloria goose.*

# X RAY

## *A Man Will Photograph Anything*

A timetable for these events would be a handy thing for a man to carry around. It would include stations along the road, hour of arrival, hour of departure, illustrated by scenic photographs in black and white, X-ray photographs. The timetable would keep a man informed on what was happening when.

Years after William Beaumont was dead, that other distinguished American physiologist Walter Cannon fed a cat a substance opaque to X rays and at intervals took photographs. He was a raw medical student at Harvard. That was the beginning. Soon a like opaque substance was fed to men and women and boys and girls and babes in arms from Tokyo to Moscow and on to Tokyo again. If a tumor was leaning somewhere or an ulcer had etched a crater, this would be seen, but if the tumor or ulcer was simply a persistent idea, as often, this would not be seen, and nothing is such a holiday as a non-fatal diagnosis. Henceforth the diurnal doings of man's digestive tube were as vulgarly exposed as everything else in his private life. Man is peculiar among the beasts in that he owns a camera and has a passion for snapping it.

For his brilliant discovery Walter Cannon, though he added other discoveries, never received a Nobel Prize.

The subject drinks what looks like chalky milk flavored with chocolate, if anyone likes chocolate. The X-ray specialist has his timetable and is ready, snaps. Or he just watches. He sees the black and white against the ocean green of the fluoroscope. Gruesome sometimes, funny sometimes, that opaque bolus waiting on the top of the debutante's tongue, then diving down and starting its creeping advance toward limbo while on her tongue the next bolus waits. (Without benefit of X ray a parallel sadistic satisfaction may be had in the zoo watching hay go down the neck of a giraffe.) Occasionally the X-ray specialist hurries the advance by kneading the belly wall with his fingers. He may imitate the digestive process more exactly by coating actual food with the opaque and in the dark watch the meal ride by. The first of it leaves the stomach almost as quickly as it enters, the last not under four hours, fourteen hours approximately to that gate between small bowel and large, thence a climb up the right side (ascending colon), a crossing (transverse colon), a sliding down the left (descending colon), followed shortly by a reappearance in the sunshine or the warm rain. Twenty-four hours for the total journey, or forty-eight, or less or more, even a week, even two weeks in the stingy with a penchant to constipation. All of which knowledge and much associated knowledge has been bequeathed to our civilized society without the nicking of any man's belly or any rabbit's or cat's or dog's, and still Walter Cannon never received a Nobel Prize.

# ABSORPTION

## *Faithful to the Last*

But now, the true good of this all, the human intent, the animal intent, the digestive intent? It is absorption. Up to now the processes could seem to have been for the sake of digestion, but nay, brother, they were for thee. It was the getting of everything into the thee.

When the raspberry from the shrub, the sugar from the maple, the lard in the cake (from the nice white pig), prefaced by the sandwich with the thin slice of ham (from the other white pig), had been divided and divided, there arrived that fated moment when every molecule would have the grand privilege of becoming a part albeit humble of a man. Billions of molecules. Think of the dull places they might have gone. A swamp in Georgia. The dark side of the moon. The right hind leg of a mosquito. It worries some people where they go after death.

Absorption is the passing from the inside of the tube, through the wall of the tube, into the blood, into the cells of body and brain and, for all we know, into the mind.

How?

That how is far from fully clear. Generally it is probably the same process, but detailedly it is probably a different process for different foods in different parts of the long tube. In each part it is the chemistry and physics of the food, met by the chemistry and physics of the wall, indeed by the chemistry and physics of the parts of the wall. The wall is cells. There is the entry into the cells, a membrane. There is the substance of the cells. There is the exit from the cells, a membrane. Much study of the so-called kinetics of this movement goes on in Oxford and Cambridge and Chicago and no doubt Stockholm and Bonn. A molecule might, and does, enter the wall by interacting with a molecule in the wall, perchance advances by acting with one molecule after another, climbs out by a similar kinetics on the other side, enzymes paddling everything everywhere along the way. Or there may be just a passing through holes,

Harvey's pores, though no one would say it that way in 1969. At the selfsame instant at the selfsame spot a variety of food molecules may be using a variety of transports, and it seems bewildering, and from all that is known is bewildering, and bewilderingly fast. Water molecules also. Salt molecules also. We are as busy as mice even when we are quiet as mice, and so are the mice.

Protein, a large molecule, is broken down to amino acids, small molecules. Carbohydrate, large, to glucose, small. (This action on carbohydrate formerly was thought to take place all in the lumen of the tube, but is known now to take place also in the wall.) Fat, large, is broken down to fatty acids and glycerol, small. It is the small that are absorbed. Fat might slip through by simply dividing into fine fat, might chemically disguise as soap and leak and sneak through, might chemically produce a molecule that dissolves in water and this slip through with the water (there have been and are many ideas), might use that trickery of steppingstone to steppingstone, molecule to molecule, slip in as fat, slip out as fat, even be seen afterward as fat in the lymph stream, the lymph spaces, the blood of the great vein of the liver.

Everything in the digestive tube that is valuable to a gentleman or a lady or a dog must get into his or her hungry cells, lung, eyebrow, thumb.

A tithe of absorption can take place at the entering terminus, the mouth, as nitroglycerine for an aching heart is absorbed when simply laid under the tongue. Some absorption takes place at the other terminus. Much takes place in the large bowel, much in the stomach, especially water and alcohol, but most in the small bowel. If a small bowel is cut open (in that rabbit) and its inside wall lighted and magnified, juttings become visible, thousands, villi, and during absorption these are in continual motion, a slapping all directions, a panorama concentrated on business. Some scientists have thought the slapping to be a way of bringing a maximum of bowel to the broken-down food molecules, and unquestionably the villi do present a surface far greater than if the bowel were smooth. Other scientists have thought the continual motion was to pump food-laden blood and lymph out of the villi to make room for more food-laden blood and lymph. The villus appears most nearly a cone overrun with blood-vessels, and one realizes that lumen and vessels are closer than one thought. Food and blood are closer: that is, world and creature, spaghetti and liver.

A man does not eat his dog—that occurs to one. Why? What makes a man fastidious at that crossroad? Grenfell, the explorer, was able to bring himself to eat his dog, and was praised for it, for his judiciousness in saving his own life. Perhaps the rest of us would do the same this noon except that we can see there is enough roast beef and chicken. One is disgusted with man. One is sad for him, too, always pulled between the Sermon on the Mount and something that can be tracked to how his particular conscience defines self-preservation. Anyhow, man does not eat his dog.

# DEFECATION

## *Reflective Interlude*

Dust to dust after the right dust has been saved—that would be a condensation of this total story.

Over the funereal final scene the bile casts its color. Red blood cells break down, their hemoglobin remainder is green (biliverdin) that goes to the liver and adds a red (bilirubin), those two hues tinting the dust that changes in color from day to day.

The chyme that hours ago left the stomach was transformed into what arrived at the end of the small bowel, continued to be transformed, passed to the large bowel, continued to be transformed, passed onward, hindward.

What passed into the large bowel was apt to be swept along in a wave several times in the twenty-four hours, the sweep beginning often in the middle of the transverse colon, thence into the descending, into the sigmoid and beyond—out of the digestive tube and out of the pelvis and out of the body into the world. When sufficient had collected in sigmoid and rectum, and the pressure there had reached forty or fifty millimeters of mercury, a man might find himself reflective and sensible of a fullness somewhere. This might be followed by a wish. A stupid man might look as if he had a thought. However, the sweep needed not go sedately from thence to thence for there to be the wish. A single peristaltic wave might shoot whatever it was into the beyond, the wish come right along, or the wave might have begun farther back, where small and large bowel join, encouraged by something still farther back—breakfast! It is a common way for the body to salute the day.

The ulterior end of the digestive tube was locked by that purse-string double-muscle, the inner muscle responding to the laws of pressure, the outer to man's command. Assistance was given by that same abdominal force that assisted at childbirth, as Galen back in Rome already said, the force being in the same direction, down. The pressure can rise to two hundred and eighty millimeters, twice the arterial blood-pressure. Figures for it are to be trusted because they have been scientifically registered with scientific instruments in scientific experiments, on medical students, no doubt. Small wonder with such pressures inside our bodies that blood vessels may burst inside our skulls, where it is not a good thing. The bathroom remains the most dangerous room in the house.

Defecation postures are similar and different and familiar over the scale of animals. A dog, having circled in the well-known manner, thrice perhaps, assumes temporarily the posture of the kangaroo. A horse practically pretends it is not taking a posture, but its tail gives it away. Man sits. Man is the thinker.

# X

# METABOLISM

## Tallow to Flame

# DEMOLITION

## Meeting House

So there are the products of digestion. They are absorbed through the gut wall. Chemistry is implicated, physics implicated. The molecules are delivered to the blood, to Bernard's fluid-around-the-cells, go through the walls of the cells, into the cells, to their different destinations in the cells. Some to the nucleus. Some to the nucleoli. Some to the mitochondria. The ribosomes. Other structures. All of this is fast. All involves transport—important to 1969 research. Each structure has its job, does it. The result is life. The tallow has burst into flame. The tulip has burst into flower, or has gone into the tulip-eating rabbit, the rabbit into you, if you eat rabbit.

In the cell—it can be plant cell as well as animal cell—the molecules take part in demolition, the tearing down, and in construction, the building up.

All comes under the head of metabolism. That word is from the Greek, means change.

A seed grows to a sapling, a zygote to a baby, a baby to a man, the man falls in love with a woman, stubbornly stays a man, as his riding horse his riding horse, the innumerable shifts of energy that make those possibilities possible are metabolism. Man, woman, horse, horsefly that bit the horse, bacteria and viruses that infected the horsefly, toxins manufactured in the bacteria and the viruses, toxins coming out and killing a physician or a politician, all are metabolism. The infinitude of change forever taking place in us, and in those who inhabit air and earth and sea with us, is metabolism.

"What else is there?"

"Between us, Friend, there is much."

That word—Friend—did not just slip into the last line, was put there, because for the next pages we are Quakers.

Allow the exaggeration.

Our body is the Meeting House. On this First-Day-Morning the flesh-and-bone Meeting House walked to the brick-and-slate Meeting House, and on this First-Day-Evening will walk by the stone wall around the graveyard. A graveyard has always been an excellent place for preparation for the coming six days, for speaking with one's neighbor, smiling occasionally, laughing never.

Metabolism in a graveyard, one would think, must be at the lowest, or none, but then one remembers the grass, the insects on the leaves, the birds picking off the insects, one bee buzzing loud during The Silence, the whole line of this ravenous world depending on metabolism: grass, insect, leaf, bird, bee, buzz, Quaker.

Human body and human head are recognized as having their billions of cells, and in the cells, billions and billions of molecules, in the molecules, billions and billions and billions of atoms, the innumerable actions of all of these played off in pools (another chemist word), metabolic pools.

Each pool, each molecule, each atom is no more an individual than a Roman soldier in a Roman phalanx. Each is the result of change. Each survives because it fits the design, Friend. Everything under the stars must fit the design.

Two overall methods are employed in this study. First, the measuring of an item given off by the total body, or an item taken in by the total body. Second, the analysis of a detail down where no microscope or electron-microscope can see. Two methods. First, the total. Second, the detail.

# CALORIMETRY

## *Is Hot*

Heat-given-off is a way to measure the total.

Heat flows from every hotter part to every cooler part, then flows to the usually cooler air outside of the body, where it is measured.

A space is walled in. If the experiment is to be upon a man he is no more cramped in that space than he would have been in a roomette in a Pullman. He has a bed. An electric fan. A telephone. A window. This roomette has been economically thought out. There may be a bicycle hoisted on a metal frame, the wheels not touching the floor, so that the man can pedal and his body still stay in the same spot, he do exercise by the clock, so-and-so-many revolutions against so-and-so-much friction in so-and-so-much time. The friction can be varied. How much heat is given off? The amount is a measure of the metabolism of exercise. Or the man may lie on the bed, not move, and the amount is a measure of the metabolism of rest.

Biochemists have studied the energies of the camel in the desert, chick in the egg, man in the roomette. They call the roomette a respiration calorimeter.

That calorimeter can have no heat leaks. There can be no loss or gain except as due to the chemistry of the man pedaling, or lying. Double walls, cold air forced between them to keep a steady temperature, and water flowing through coiled pipes in the walls, the temperature of the water read as it enters and as it leaves the pipes, its amount measured by an ordinary water-meter, and from these data the heat-gain for any period is easily calculated.

Our body is always hot, relatively, even when doing nothing. Who has not at some time entered a freezing empty theatre and waited for the audience to arrive to warm it? Who has not had a fever and felt hot, nervously read the thermometer as it rose, not paying much attention when it dropped. These flesh-and-bone machines of the Almighty have played many roles, Friend, and one role has been that of the furnace. A heat bill often has been made out against a body. In the roomette, some heat comes off into the air as water vapor, which condenses, is absorbed in sulphuric acid, the gain in the acid's weight calculated, and this included in the heat bill. Such an investigation can be for a night, for twenty-four hours, for minutes, and, when completed, the researcher can have the balanced-ledger satisfaction of the banker because in energy respects the body is as precise as an ideal bank. The ledgers balance. Income and outgo at every point balance. The law of the conservation of energy is as true for this nutshell of Hamlet as for the physicist's universe.

# BMR

## *Needs Air*

Oxygen-consumed is another way of measuring total metabolism.

How much oxygen does the body require, the oxygen eventually used to burn our food? How much is taken in for that? The common BMR, Basal Metabolic Rate, aims at determining this.

It measures the amount of oxygen breathed-in over a period of time.

This keeps the body as warm as the healthy body needs to be, and supplies the energy to get done the body's work. Had that food been burned directly (a fire) the energy would have been released at a burst, wastefully. For economy it must somehow be trapped, be deposited in so-called chemical bonds, high-energy phosphorus, and paid out to meet the body's expenses.

When a physician takes the BMR his subject needs to be basal. That means, his muscles slumped almost as in sleep, his heart pumping at a necessary minimum, his lungs filling and emptying at a minimum. This basal level could be forced lower by drugs. Whoever is taking the BMR keeps in mind its intention, to find how much oxygen this quiet body consumes. He takes it in the morning. Before breakfast. No exercise. No extremes of temperature. No hard words with wife or secretary and not even on the day before, if that can be arranged. A minimum of emotion. A peaceful morning following a peaceful night. The best place is a hospital bed though a half hour of bed-rest is adequate. More may be too much. A man awake in bed thinks, may think with feeling, and that can get his body unbasal again, could cause a 20 percent alteration in the BMR.

A typical test is known to most people.

The physician makes his patient comfortable, he says, pinches his nose shut with a clip, stuffs shut his mouth with an air-tight mouthpiece, the man on his back. If he has fallen asleep the nurse wakes him because the BMR of sleep is ten percent too low. Too low also after a tranquilizer, and too low if a part or all of the thyroid gland has been surgically removed, or reduced by disease, or if too little thyroid gland was there to begin with.

The subject lies. He breathes regularly. In the old-type instrument, that in principle is the same as the new, the oxygen comes from a bell that falls and rises with each breath, a pen attached to the bell and its movements recorded on paper wrapped around a rotating drum. That produces a tracing. Since each breath removes some oxygen from the bell, it drops. Meanwhile the carbon dioxide and the water manufactured by the body and breathed-out are gotten rid of inside the instrument. (A burning tallow also produces carbon dioxide and water.) After six to eight minutes of such breathing the clip is unpinched from the nose, the mouthpiece jerked from the mouth, and by the slant of a line drawn with a ruler through the falls and rises on the tracing it is simple to calculate oxygen-consumed. That figure is then corrected for sex, age, sea level, temperature, and the new figure referred to printed tables where oxygen-consumed is expressed as heat—heat-produced per square meter of body surface.

Why body surface?

That used to be answered glibly, more surface exposed, more oxygen-consumed. It is true that oxygen-consumed when related to surface does turn out to be much the same for large animal or small, elephant or mouse, the smaller always having a relatively larger surface. A mouse bulk for bulk, ounce for ounce, has greater surface than an elephant and burns relatively more intensely, as a mouse should. Shape enters in. A tall thin person has more surface than a squat and if the two weigh the same the tall thin pound for pound burns more, like the mouse. Any animal (mouse, elephant, snake, guinea pig, man) has an irregular surface difficult to measure, and there are other difficulties. Notwithstanding, the day-to-day hour-to-hour life-to-death drives that play on the BMR, particularly the glandular drives, thyroid, pituitary, have become better understood because of what has been learned from the method.

The physician expresses the BMR as a percentage, so-and-so-much above or below the average for people of the same sex, height, weight. For the infant just out into the world the BMR is low, in a few weeks is twice what it was, in two years may be as high as it ever will get, drops again in the old. The old burn less, therefore eat less, and when any one of them eats more we think we know an additional fact about him, his mind no longer in control. Usually the BMR is higher for male than female. During the years of the possibility of conception a woman's stays steady, drops at the change-of-life. A single cigarette may raise the rate nine percent. Coffee raises it. Tension. If a patient is afraid of her physician, worried by her illness, by the fancied dire meanings of the test, by the modern-art waiting room, by the bill her husband will get, and if during the test a riveting hammer is rebuilding Doctors Building, that BMR is worthless. Intellectual activity on the whole has not much effect. Emotion has. A high rate is +70, a low —40. An Englishman, a Frenchman, an American, a Russian, each may behave as exaggeratedly as each thinks the other does, but their oxygen-consumed and heat-produced do not behave exaggeratedly. The living machine pays not much attention to nationality, race, status.

# ISOTOPE

## Neon Light

The purple foxglove with its thimblelike flowers long provided the digitalis the physician prescribed for the ailing heart. A special foxglove, however, for marked-molecule purposes is grown in a greenhouse. Sun pours through the windows of that greenhouse. Carbon dioxide is piped in. The carbon dioxide is not the ordinary gas breathed-out by animals and human beings and breathed-in by plants, because its carbon atoms are marked. Each is distinguishable. It is labeled. Consequently any molecule of which it may be a part is labeled. The foxglove in that greenhouse grows like any other, only the digitalis extracted from it, if fed or injected into man or animal, can be followed as it travels through the body. Those carbon atoms, so to speak, are carrying neon lights, and the detective-chemist follows the lights.

Two types of isotope—stable and radioactive.

The stable exist in nature. They are marked during their chemical evolution. In their chemistry inside the living they act like the ordinary atom, but they do not have the same weight. They have the same numbers of protons and electrons but different numbers of neutrons, and their weight distinguishes them. They vary as siblings. Tin has ten siblings. In water the usual hydrogen atom weighs one but there is another that weighs two, and the water molecule that contains the heavier hydrogen weighs heavier. A third water molecule is heavier still, its hydrogen weighs three, proving that the fluid the bathing beauty bathes in is a complicated mixture, a fact that she ignores, sensibly.

The radioactive isotopes are marked by their radioactivity. They emit radiations. They have an unstable nucleus. Some radioactive isotopes probably originated toward the beginning of the universe, some are artificially prepared. A reactor prepares them. The bombardment of a natural element (in a cyclotron or synchrocyclotron or linear accelerator or other atom-smashing bombarding equipment) introduces radiations into an element, it then immediately emitting radiations. It may radiate more actively than natural radium and the scientist counts the radiations. He is thus able to follow the molecules containing the radiating atoms in brain, bone marrow, placenta, even in the fetus inside the mother. It is like fingerprinting. Because a radioactive isotope would be steadily losing radiations it may be thought to be dying, and is indeed described in terms of its half-life. One carbon isotope has a half-life of 21 minutes, so any study made with it must be in a hurry, another a half-life of 5,570 years, so why not put off till tomorrow? Isotopes today are packaged, shipped in trains, stored in sealed rooms, catalogued, priced in expensively printed brochures, sold off the shelf.

Since marked molecules in their chemistry act like their siblings but can be tracked, the habits of the living can be studied inside the pools of the living. The chemist can step from isotope stone to isotope stone through the swamps of the body. Carbohydrate, fat, protein, vitamins may be watched in their metabolism.

# CARBOHYDRATE

## *Coal in the Bucket*

Carbohydrate is eaten at every meal, often between meals, Bissinger's chocolates, pretzels with beer, a snack at midnight when we are more asleep than awake—but we are wide awake on the scales next week.

Carbohydrate is one of the three fundamental foods: carbohydrate, fat, protein.

During digestion carbohydrate is broken down in mouth, stomach, bowel. Glucose results from practically all such breakdown. Glucose is a small molecule. It may nevertheless in itself be sufficient to supply bacteria with their entire food needs, carbon needs, supply carbohydrate, fat, protein. In us the glucose is delivered through the lining of the bowel into the blood, thence hither and thither through the body, finally into the substance of some hungry cell. A chain of enzymes remakes that glucose molecule into intermediate molecules, these combine with others, combine with phosphorus, produce high-energy phosphorus, and this when called on can accomplish at the low temperature of the body the high-temperature work that must be done.

Glucose supplies two-thirds of our body's energy, is our readiest fuel, coal in the bucket.

Our economical body does not allow glucose molecules to float around with nothing to do, so converts them to glycogen. That was Claude Bernard's discovery. The glycogen can be stored, principally in liver and muscle. Glycogen is a starch. Glycogen is a large chain molecule, branched, the links in the chain being glucose molecules, and when the body needs glucose, so-and-so-many links, the right number, are released into the blood, carried, picked up by the cells.

When all the energy is spent, all the work done, all the glucose and oxygen used, there is left, world without end, carbon-dioxide and water. Only in the liver can this action go to completion. In muscle it goes part way, to an acid, this later converted back to glycogen. The carbon-dioxide and water are dumped out of the body via lungs and kidneys.

A well-fed man has been found to have three to four ounces of glycogen in his liver, one storehouse, half a pound in his muscle, the other storehouse, fuel enough for the body for a day. Whenever in this chemistry oxygen is not immediately required it always is eventually, as for the burning of any tallow, the end-products always carbon dioxide and water, as any tallow.

Everywhere in us sugar is burning. The burning is a special burning—that of the living. Put in a few lumps please. Into the tea.

# FAT

## *Coal in the Bin*

Fat or the fat-like—the lipids—we also eat at every meal. Fat and what metabolism makes of fat reaches into all the systems of the body. Fat is in each one of our 1,000,000,000,000,000 cells. Fat is part of the permanent structure. But, it also is fuel. It also is storage.

During digestion fat is broken down to fatty acids and glycerol. Fat = fatty acids + glycerol.

Those fatty acids are burned eventually to carbon dioxide and water, supply heat and energy. That useful word *burned* must of course not be read in the everyday sense. The process is not that direct, the molecules are not sped up to the speed of a bonfire.

Of the fat we eat some remains undigested, travels the mileage of the digestive tract and is dumped out of the body in the feces. It might of course just be carried in the blood as microscope-visible fat. Even eye-visible—after a fatty meal. In hyper-fat states the blood looks milky. The destination of a single fat molecule is a single cell. This cell has been calling for fat. The fat arrives, enters.

Socrates was fat, they say.

By the time the fat molecule has been entirely used up, it has gone in and out of the bowel, in and out of the blood, in and out of the liver, in and out of the blood again, into a cell, and enzymes everywhere spurred the journey. Glucose is coal in the bucket; fat is coal in the bin. This fat is not merely dropped into the bin but chemically fused in, huge amounts often. It is a savings account that the body can draw on much longer than on glucose. Fat yields more than twice the energy, more than protein, and can keep yielding energy after the other foods have quit, wherefore starvation can be mercilessly slow, the victim live sixty days, longer, die when the last fat has been burnt, the end of the tallow truly. Among the fats are special ones, so-called unsaturated fats, that the body cannot survive without, cannot manufacture, must eat. As for the stored, much is in the bowel neighborhood, a great apron hanging down in front of the abdomen where it contributes to insulation. Joints are protected from jarring by cushions of fat. Kidneys rest in beds of fat. Half our fat however is under our skin, as everyman knows, prevents heat loss from our snug inside world to the indifferent outside. Isotope fat (marked molecules) sometimes has been fed to an animal and its wanderings in the body traced. Soon the fat stores show themselves sprayed with the marked molecules, meaning that the newly manufactured was moving into the stores, and that in the stores was moving out. The quiet banks and shoals are not quiet at all.

# PROTEIN

## Stone

Protein is the third of the three. Protein is nearest to flesh, but carbohydrate and fat and everything else is working in Robert Hooke's small cell.

Protein is more than half of the solid substance. For long years it was the most prominent in research, then lost first place, now has it again.

By one estimate the cells of the mammal were considered able to build a million different proteins. Each would have been built originally by evolution using mutation, and our chemistries while we were alive would have kept repeating, molecule, molecule, molecule, worn down in one way or another, rebuilding, reduplicating, replicating, every moment guided by instructions coming from the cell's nucleus, rapid as measured by our clocks. A minute in that world would be a long time.

Protein like the other two foodstuffs is eaten at breakfast, dinner, supper. During digestion it is broken to amino acids. Amino acid + amino acid + amino acid = protein. A chain. It may be an enormous chain. Only in recent years has something been known about the three-dimensional shape of the protein molecule. Different amino acids in different numbers and linked in different orders make the molecule, but for any one molecule the order is fixed, the amino-acid sequence fixed. Each link is in its place, each chain folded in a place or places. Such a molecule might be fifty or more times bigger than some other big molecule and still not fall within the seeing power of the ordinary microscope. Electron-microscopy and X-ray photography and the computer have been necessary for the establishing of the architecture of such a molecule. Keystone is the word used for protein, building-stone the word for amino acid. All twenty amino acids may be in a protein, or fewer, and since the twenty, if there are twenty, can be arranged in any order, the mathematician will have to tell us the staggering possibilities.

We eat it, digestion breaks it up, the amino acids get through the gut wall, circulate in the blood, which is a river flowing with amino acids besides everything else, the amino acids picked up where an old cell needs repairing or a new one needs building. A sick man has been kept alive with an infusion of amino acids introduced into one of his veins, which should not be surprising. Protein demolition occurs in stomach and bowel. Protein building occurs in the cells everywhere, instructions always given by the nucleus, the local architect. Ten of the twenty amino acids are essential, like those fats, meaning the ten cannot be manufactured by the body, must be eaten as plant or animal, we once more doomed to eat the living or die.

Amino acids are in the blood that enters the liver, but none in the blood that leaves the liver. Where did they go? They are stored there, or converted to blood proteins, or torn to harmless by-products, especially urea. Urea leaves the body in the urine, and that way the amino acids from which the urea came are temporarily removed from the scene of the living. They will come back again somewhere to be part of the living. Everything does. Nietzsche's eternal recrudescence. If a liver has been experimentally cut out, no urea.

Nitrogen being the core of the amino acids, it is the core of protein, core of flesh. A body needs that nitrogen. It is critical. Much is. Every chemical grouping in those pools or flowing in that blood-river is critical to its degree.

Besides the maintenance of the tissues of the body, and their growth, protein can supply us with glucose. No wonder it has been called the perfect food—perfect for a dog, not quite for a man, a dog's metabolism encouraging the table manners of a dog, fill up on meat, whereas a man to get enough energy from protein would have to fill up with an amount that would choke his engine.

Enzymes are proteins. Enzymes whip along the building of proteins. So, protein drives protein.

The actual manufacture in the cells takes place in the ribosomes, small bodies, small factories, many of them, visible with modern microscopes. The manufacture strikes one as a jerky Chaplin film, atom jostling atom, but all of it over in a minute, literally. One amino acid is squeezed in here, another squeezed out there, the giant molecule throwing itself together in a way that would be cockeyed except that everything is, or nothing is.

Amino acid welded to amino acid welded to amino acid, each action of itself simple enough, unit to unit, monomer to monomer, and then there is the polymer, and that may be a giant. Protein is a polymer. The amino acids are bonded together, the links spoken of as peptide links. Two peptides are a dipeptide. Three are a tripeptide. Many are a polypeptide. Protein is a polypeptide. Sometimes this is drawn out straight as in hair. Might have many lengths. Might have the shape of a globe, an ellipsoid, a helix. Might be twisted. Might be one chain. Might be a number of chains. Might be a combination, helix and no-helix. To risk another figure of speech, the protein molecule is not the stone of the Meeting House, but the stone of a colorful Gothic cathedral. Our body is an infinitude of cathedrals. Impossible to overstate our body, possible of course to state it wrong. Each protein molecule is three-dimensional, has height, width, depth, and this grand dignity do the vulgar digestive juices methodically demolish.

If one extricates oneself from terms, and if one remembers always that the chemically complex is the summing of the chemically simple, and the simple open to intellectual attack, to understanding, then, despite that the summing is fabulous, and occurs with fabulous speed, it does help somewhat to fathom how green grass can become rust-colored cow.

A fascinating further fact is that that grand molecule is in perpetual flux, as the fat molecule. The cathedral drops out a stone here, a stone there, a new stone stuck in here, there.

# NUCLEIC ACID

## Blueprint

It was thought that our heredity must be locked in proteins, that they alone were adequately complicated. Other molecules were too simple to separate pig from tiger, tiger from man, Chinese from American, one Bostonian from another Bostonian.

There were difficulties.

So, it was a relief when another class of chemicals came to the fore. The nucleic acids. They had been discovered as early as 1869, were disregarded. Had been found in the nuclei of cells, had been refined to a white powder, contained phosphorus, to the chemist were not at all like proteins.

However, they also were built up of simple units, nucleotides. A group of amino acids produced a protein; a group of nucleotides produced a nucleic acid. Needless to say nothing is as simple as a row of words.

One chemist-geneticist and soon many were drawing the nucleic acids from their relative obscurity, positively to the front of the stage. It was well toward the middle of the present century that it began to be realized that the nucleic acids had the key, that they could explain the deposition, then the transmission, of much of life and of our individual lives. A new idea of the chemical basis of heredity loomed. There was a wild rush.

One idea was that there must be a mold, a precise mold, called template, and that into the slots of it, so to speak, the nucleotides were fitted, or fitted themselves. In the nucleotides were their smaller fragments. All of this was restlessly in all the cells. It was the presumed, then the proven genetic material of the cells. It was desoxyribonucleic acid, DNA.

How did it code the history? How did it reduplicate itself?

Some chemists became convinced that the molecule was in the form of a helix, a spiral, held together by chemical bonds, the bonds of various binding strengths. X-ray photographs of crystals of DNA were studied for the distances between the circlings of the spiral, and the angles these made. Chemist models were fitted and refitted. It became a passionate game. The chemists were trying to find the nature of that template. A few were convinced that it was a continuous sequence of nucleotides around the spiral. Molecules were picked up from the blood-stream and poured into that mold. When the template was filled, the completed product dropped out, this all in seconds, the template left behind to be refilled. Thus did our heredity roll along. Here were the orders for our lives and the lives of our progeny.

That helix quickly planted itself on the tip of the tongue of the world, not the chemist world only, life a spiral, in fact a double spiral, the one spiraling one direction, the other the opposite direction, a double helix.


# VITAMIN

## *Mortar*

In the eighteenth century while Americans and Frenchmen were writing political documents, Englishmen and Japanese were experimenting with nutrition, on sea-going ships. Captain Cook made a three-year voyage, included fruits and vegetables in the diet of his sailors, lost four men, phenomenally low when a ship was a pesthouse. Lord Anson in 1740 had three-quarters of his crew down with scurvy, then the dietary acumen of the Scotsman James Lind and the forcefulness of Sir Gilbert Blane got a ruling through the British Admiralty that all sailors must take lemon juice. Scurvy withdrew like an apparition come daylight. In the nineteenth century, Dr. Takaki, a young Japanese, persuaded his government to send out two ships, identical lengths of voyage, identical crews, the sailors of the first ship to eat their customary polished rice, those of the second to eat meat and vegetables. One hundred cases of beriberi (pain, irritability, paralysis, swellings, failing heart) flared on the first, sixteen on the second, and each of the sixteen had stuck to his polished rice. Later it was proven that beriberi could be cured after it was started if the diet was supplemented with an extract of the polishings, or if the husks were eaten, these containing whatever it was. Later still, in animal experiments, rats were fed pure salt, pure carbohydrate, pure fat, pure protein, and the result was dead rats. Also, the offspring did not grow. Pure foods lacked what natural foods had. A rat must not be too pure.

Experimentation soon ran world-wide.

A Polish scientist gave the name, vitamin, spelled it with an *e*, vitamine, under the mistaken belief that it must be an amine. One American worker more than others lengthened the list of vitamins. Clinics defined the deficiency diseases. Everything was better defined, vitamin sources, methods of assay, vitamin metabolism, year after year.

These molecules do not get to be part of the body. The body cannot manufacture them. They must be consumed. They do not supply energy. They are necessary in the chemistry.

There are twenty, as far as is known, and the chemist uses all of his chemist intelligence to find what each is doing, what factor or co-factor it is, what co-enzyme, how it enters in and comes out from among the reacting molecules in the private swamps we all are.

A woman visited the clinic with heavily pigmented skin, ulcers in the webs between her fingers, numb spots on the under side of her thighs, bed sores, other agonies. What she had been living on exclusively was beer, calories enough but lacking vitamins. If one sets about it in earnest one can achieve vitamin deficiency right in the middle of our sophisticated cities. In unsophisticated Australia, in the interior where the aborigines slice off an end of coconut and drink the juice, there is no deficiency, the juice loaded with vitamins. In experimental deficiency the lack can be restricted to one vitamin, but in daily life there is apt to be a lack of several at the same time, the diseases often therefore not fitting the textbook. The absolute amount of a vitamin that a human being or an animal requires has no doubt a level, but the level often is not known, and there is reason to believe that the level differs animal to animal, person to person. Below the level, deficiency; above, nothing. The excess is cast off. That last is not true, for instance, for vitamin D, manufactured in the skin by exposure to the sun, where an excess may cause kidney stones and chalky deposits in the arteries.

If the chemist finds what chemical work a vitamin does, and if that vitamin previously had a letter of the alphabet as its name, it can now be given a chemical name. Pteroylglutamic acid. Para-aminobenzoic acid. Cyanocobalamin. Some letter-names are however less confusing and remain.

When an embryo at a period of its development has via its mother been deprived of a vitamin there might be malformation. Chemistry of the unborn has a timetable increasingly understood.

The following would be a list of vitamins. It might as accurately be read off a package of dog biscuits, since God's dogs require vitamins too and in right amounts if they are to be watch dogs and not corroded tramps that scratch at back doors.

Vitamin A—necessary for the health of tissues in and around eyelids and eyeballs, for a healthy retina, healthy lens, to keep off night-blindness, necessary for the nervous-system, for normal epithelium, tissue that lines the surfaces of the body and its cavities and forms glands, and to protect pregnant mother rats from having offspring with small brains, harelips, cleft palates. Man-oriented man ought at least occasionally to admit his debt to the rat, at least to the white rat. Liver and the oil of liver contain vitamin A, so potent in polar-bear liver that human beings can be poisoned by it, but absent if the polar bear has not been chewing the green and yellow parts of plants. Oil of liver and other lipids of the animal kingdom contain the vitamin, while the vegetable kingdom has the chemical forerunners, the precursors, the vegetables fed by the sun, getting their quanta of energy, and the animal eats the vegetable, man eats the animal, saves his eyesight.

Vitamin B—a complex that began as one but that proved to be many, up to ten, among them thiamine, or $B_1$, necessary in the metabolism of numerous plants and bacteria and all animals, including pigeons, lack of it producing an enfeebled pigeon with an unsteady gait, its head jerked back, its body twisted, and though oxygen is coming into its lungs it behaves as if it were suffocating. When given thiamine, that is the royal touch.

Pantothenic acid—necessary to keep a rat's hair from graying, important to the rat. Pantothenic acid is found in plants and animals, is of ancient metabolic lineage, we needing always to get enough of it, as we do. Riboflavin or $B_2$—keeps up a man's weight, keeps the angles of his mouth from cracking, keeps off raw scaly bleeding lips, an inflamed tongue—is found in yeast, milk, peas, grain, particularly the germinating.

Vitamins, listed this way, could begin to loom as large as anything in the world.

Nicotinic acid or niacin or anti-black-tongue factor staves off a disease suffered by millions, with weakness, depression, diarrhea, scaly skin and thick tongue, no appetite, no clearly identifiable ill health either, known around the world as pellagra. In our country in the southern states during the depression there were every year thousands of pellagra deaths, not recognized as nutrition deaths, the mystery about them as great as the vitamin mysteries of the eighteenth century. Later they were no mystery at all.

Cobalamin or $B_{12}$—cobalt in its molecule, necessary to human life, contains the intrinsic factor that prevents pernicious anemia. Ascorbic acid or vitamin C—prevents the loathsome scurvy. Vitamin E—prevents sterility. Vitamin K—prevents capillary hemorrhage.

Bacteria have been knitted everywhere into the story of higher creatures. $B_{12}$ was synthesized in the bodies of bacteria. Some bacteria have gone on killing us, some have fed us. In an experiment an animal has been kept alive without bacteria, though normally all lead infected lives. A human infant may be born sufficiently sterile, but then in a few hours the game begins, and the bacteria begin providing vitamins, the parasitic guests begin repaying the host who is supplying them with a warm wet roomy place to live and breed. In cows, ruminants, bacteria are killed in the upper bowel, where absorption is active, disintegrate there, the ruminant easily absorbing all the vitamin needed. In us, the hordes are in the lower bowel, where absorption is less active, but our machine usually absorbs enough if it is working efficiently. Nicotinic acid, biotin, riboflavin, inositol, choline, thiamine, vitamin K, come to us in significant amounts because bacteria are multiplying in our bowel, die, and our body benefits from an extract of corpses. The administration of an antibiotic may halt an infection, but halts also the bacterial luxuriance of our bowel, fewer corpses, less vitamin.

# LIVER

## *Kitchen*

Procter & Gamble, who wash the world with Ivory soap, have the inmost part of their chemical combine in northwest Cincinnati. Busy around the clock. People who live in that city sometimes feel the urge to drive that direction at night, a colony of steamy buildings softened and rounded by the night, here and there light-bright yellow windowpanes, looming conduits with the dimensions of Roman aqueducts, railroad tracks running through and beyond. An iron fence encapsulates the whole. The air smells. The air tastes. Interlocking chemical processes proceed with clocked speed, and many clocks and processes at the same time. Here there is no need for fancy. The facts are fancy. Everything begins more or less with tallow and ends with molecules bottled or boxed or wrapped or barreled, glycerine, fatty acids, soaps, supplying a thousand factories, a million kitchens, a million bathrooms. A huge squat heterogeneous chemical aggregate. One thinks: laboratory. One thinks: stew kettle. One thinks: metropolis. One thinks: digestion. One thinks: metabolism. One thinks: liver.

Largest organ of our body is our liver. Wonderfully complex. Chemically complex. Complex in one way in the fetus. Complex

somewhat differently in an old man. A gland, Friend. A gland that
secretes into the blood and into an outflow duct. A gland that has
greater bulk than all other glands of the body added together,
weighs forty to sixty ounces in the adult, which is its dry post-
mortem weight, heavier when wet with blood, only three to four
times more blood in the entire body than in the liver; has a deep red
color to the eye of the surgeon.

In the X-ray room its ghost heaves with the diaphragm at each
breath. It extends across the top of the abdomen, the spleen on its
left, the ribs encircling it, provides a roof for stomach, large bowel,
small bowel. Blood-vessels enter, leave. The physician reaches his
fingers under his patient's ribs to find the liver's edge, decides
whether this one has kept its shape and position, or whether it is
two finger-breadths below where it ought to be. Does it have its
natural texture? Is it too hard? Too soft? Tender? Would this patient
be better off in liver respects if he gave up his fine wines?

A massive chemistry of innumerable reactions steams in that stew
kettle. Gross analysis shows the liver to contain fats and the fat-like,
and amino acids, blood proteins, metals, many chemical species, and
behind them as Macbeth said "A glass which shows me many more,"
plus liver enzymes known and suspected, each assigned to its duty, a
fact that the researcher trusts long before he knows the nature of
the duty. Sometimes an enzyme deliberately tantalizes him. He goes
to the slaughterhouse to pick up a liver, wants it as near normal as
he can get it, therefore stands ready with his pail of ice while the
animal is being slaughtered, lets the liver drop from animal to ice,
hurries to his laboratory, and before he arrives has lost the enzyme
he meant to study. It has degraded into something else.

Samples of alive human liver are routinely obtained by pushing a
needle either from underneath through the belly wall or from above
through the chest wall, a knife concealed in the needle. After the
sampling the patient is kept quiet for twenty-four to forty-eight
hours, a bloody bleeding liver being a nuisance or worse. The reason
for the sampling is to examine for cancer or some other liver possibility, and while settling that diagnosis one way or the other or
now or never the sample is adding its page or comma to our accumulating understanding of that idea in the mind of the Almighty, *Liver*.

Magnificent is the liver's power to rebuild. Some organs have no power, but the liver, even when masses as large as eighty percent are gouged out by accident or in an experiment, rebuilds to its old contours. Fabulous numbers of liver cells regenerate within a day, within an hour. All is always ready. The rebuilding furthermore is with intelligence—no other word fits the situation—because if a dog is starved and part of its liver cut out, the rebuilding is not to the well-fed size but thriftily to the starved size. (Dedicated laboratory dogs do not mind starving.) Respect your liver, Friend.

In the living liver, cell leans against cell, parenchymal cells, the functioning cells of the organ, microscopic geometric form leans against microscopic geometric form. When a block of rat liver is frozen and fine-sliced, the slice viewed with the light-microscope, the cells are polygons, usually with a single nucleus, a wallpaper sameness that heightens the wonder of the unsameness of the chemistry. When viewed with the electron-microscope the cells prove loaded with mitochondria, those microscopic powerhouses, as many as a thousand in a single cell, packed with enzymes, different enzymes, working in assembly-line trimness, each step concluded in a fraction of a second.

The body's defense system has some of its kinds of cells marshaled in the liver. They line its blood-filled sinusoids.

Some hepatologist (that's what he's called) has summed the obligations of a liver to a man. He claimed hundreds, which would justify that figure of speech "metabolic metropolis." It is a place ten times, or ten times ten times, more complicated than could be imagined today. It chemically crashes, chemically creates, chemically demolishes, chemically constructs, chemically breaks down, chemically builds up, is busy with all three basal foods, carbohydrate, fat, protein. Manufactures glycogen from glucose and from other sugars, and no matter what sugar the glycogen came from, the links that it delivers to the blood are always glucose links. Glucose begins being stored in the liver at one blood-concentration, stops at another, precisely.

Digestion broke protein to amino acids, the liver breaks the amino acids, and finally there is left waste nitrogen in a form that can be removed from the body in the urine, chiefly as the small urea molecule. Carbohydrate. Fat. Protein. Enzymes. Near-related and far-related and scarcely-related chemicals labor close together in this Procter & Gamble. But the liver cells are still only those invariant cells.

Then, the liver detoxifies. It leeches the poisonous from poisons. It does this by chemically coupling the harmful into complexes that are harmless.

It takes part in both blood-production and blood-destruction. So, it takes part in the metabolism of iron. Liver and spleen and bone marrow between them hold about one-fourth of the iron of the body. The liver manufactures blood proteins, stores them, slips them into the blood at the rate they are needed, and one good of this is that our blood is prevented from getting too thin and upsetting the balance between thick and thin.

When that great organ the liver fails, the doctor says with cold logic, liver failure, and liver failure may result from any of many diseases, the molecules that a healthy liver rids us of are now piled up in the blood. Upon this fact is based some of the physician's liver tests. These in turn contribute to our understanding of the labor of the liver. Experimentally, entire livers are cut out. Recently human beings, victims about to die of liver failure, have had their blood circulated through the liver of an alive pig, hours of this, several times a week, and the result, that was not life-saving, was nevertheless promising. A person who loses his liver outright dies in a day.

When the liver cells have manufactured the bile it is collected in bile-capillaries, and some of that story we already know. Those capillaries, some say, start right in the liver cells, and they do definitely start between those cells, lead from them into larger conduits, those into still larger, and the largest run off between the great lobes of the liver, finally into the first part of the small bowel. Into it the bile flows intermittently. It helps there with the absorption of fat, a fact long recognized and still not entirely understood. When pressure on the bile is not great enough, as when a dog does not see and has no hope of a bone, his bile does not flow into the bowel, is backed up into the gall bladder, waits. Its free flow, at the proper time, must now not be obstructed, as it could be by tumor or inflammation or cirrhosis. If obstructed it dams back into the blood, the whites of a man's eyes get yellow, his skin yellow, his urine yellow, also his bed sheets, this recognized by physicians from before Hippocrates. The man has jaundice. A severe jaundice makes people turn in the street and look at their old acquaintance. Physicians speak of a creeping jaundice, a bad prognosis.

Why has nature put these labors all under one roof? Why has she given the liver so much to do? Why so many small factories in this huge chemical combine? Capitalist instinct? Merger instinct? Economy of space? Not likely.

An infant's belly may jut because of its liver, the mother worry. The adult liver is most often up under the ribs where its immensity does not show. An infant physiologically and medically is in some respects a different animal from the adult, hence pediatrics. Sickness may make a liver large at any age, the state called hepatomegaly. Sickness may shrink a liver, atrophic liver. Sickness may fill it with fat, fatty liver. Normally, every one of us has liver to spare.

# PASTEUR

## *Le Maître de la Maison*

Now might be the time to hang another portrait in the corridor.

A citizen of the world. A governor of the Meeting House. A highly original student of metabolic and other body problems. Of course, he was not born last year, or the year before, but not everyone thinks that we should just drop all history and begin everything from scratch, as one says. This one's mind had nothing of the vague of some great minds, was earthy.

One human being can accomplish only what one human being can accomplish. That is certain. That limitation gives many human beings who accomplish nothing satisfaction. One human being sooner or later finds that he must join with others if he is to get done what he had hoped to do alone. However that, over any field, art, learning, science, public affairs, there is sure to be waving a flag that proclaims the glory of some single one's inexplicable achievement. Pasteur was such. Much of what we know of the detail within detail of the living we owe to that inflammable irascible grizzly Frenchman who, like Beethoven, had a pailful of the water of genius spilled over his zygote.

It must have been exciting to Louis Pasteur when he discovered that two crystals, tartaric acid, both the same chemical, dealt differently with a beam of light. He had studied crystals. Probably he did not foresee (but how do we know?) that the study of crystals, crystallography, would become a powerful aid in the chemistry of the future. Certainly he did not foresee that the way crystals let X rays (because he did not know X rays) pass through them might have something to do with our understanding of the very foundation of life. As for the tartaric acid, it came from the crust that forms in wine vats, and wines were and are important to France. The one crystal caused the beam of light to be rotated to the left, the other crystal caused it to be rotated to the right. This is not mysterious today. The meaning and the use of the rotation of the beam was indeed beginning not to be mysterious in Pasteur's day, he historically recent, died in 1895, our grandparents his contemporaries.

There was the further fact that a green mold fermented the one crystal, not the other. A mold was living. Fermentation, this induction of chemical change, had been operating from the beginnings of life on earth, had been used by man in his daily affairs for nobody knows how many centuries, but scientist imagination had not been given to it. Now one man, Louis Pasteur, was recognizing that a difference in a crystal's pattern, a pattern of atoms, could determine whether fermentation would or would not occur. It was Pasteur's first intense scientific penetration.

To those around him it could hardly have seemed, either, to have much to do with the human body when he argued that it was the *life* in the yeast made wine out of beet sugar. To him it seemed that chemistry of fermentation was chemistry of life. Yeast cells had been studied long before Pasteur, and their part in bread and wine was known, both bread and wine holding prime places in the human being's daily routine. But in Pasteur's mind it was their chemistry that got the emphasis, as it might today in our mind. We today use the word *enzyme* much more frequently than *ferment*, began to turn the idea in this fresh direction when it was found that the juice of such living cells pounded up and broken induced the chemical change as well as did the cells themselves. It was leading the way, as we repeat like parrots, to the molecular depths. We know that Pasteur's yeast cells were packed with enzymes, which is to know more than he knew but not as much more as it could appear to eyes always popping at the progress of the last six months.

Louis Pasteur, the human being, the brief traveler on this cooling planet, when a schoolboy was merely average as to grades. That happens so often where there is a mind with a deviant talent. Like other Frenchmen he was born with a pencil in his hand, was a competent draftsman. By the time he was fifteen he had produced drawings of persons that let us see how he could look straight into the faces of those he loved and not look sentimentally. When, later on, he did that, looked into faces, were they sometimes transformed into huge concatenations of cells packed with ferments, buzzing with intermediary metabolism? Hardly. Pasteur's intelligence, and that earthiness, easily turned off the fanciful, turned in the direction of what one calls the practical, but from that, in his case, theory might drop like ripe fruit.

He decided, the schoolboy, cautiously as was his nature and in spite of his excitability, that though his grades were indeed average and would let him slip into the Ecole Normale he had better wait. He would delay. The grades were lowest in chemistry, again that irony. He caught up. He took a prize. He was inspired by a teacher, a reason for an occasional teacher to reflect on his career without too much chagrin. He himself did some teaching, but experimental worker is what he was, first, second, last, passionate experimental worker, inspired usually not by some great idea waiting to be cracked open but by some gutter reality, some trouble, some worry of Frenchmen, his countrymen, as when their industry ran into disaster because of the failure of their wines, vinegars, beers, those products of fermentation.

His knowledge of how to work matured as did his knowledge of how much work a human being can do. He had furthermore a mind that did not require authority to be on his side. And he had an eye that could see microscopic shapes.

Human eyes had been seeing such microscopic shapes for a long time, two hundred years, but not seeing them with that characteristic intense penetration. From this time forward such shapes would be seen more and more, and the shapes would be more and more minute, until the human mind, sometimes also the human eye, did literally reach to the shapes of molecules, with their patterns of atoms. Human acumen arriving at the shapes of molecules and the patterns of atoms would become commonplace. And commonplace also to subject them all to measurement.

Pasteur had noted juttings on the one tartaric acid crystal and not on the other—noted this with the microscope. He had studied the yeast cells—with the microscope. He was being led to suspect, then to accuse, then step on step to convict, as troublemakers for animal body and human body, those tiny objects sometimes found somewhere in them, and that were like those tiny objects that moved in pond water—this also via the microscope. (Leeuwenhoek had seen those objects earlier.) So the chemist Pasteur was increasingly the microscopist Pasteur. Those two increasingly worked hand in hand. This scientist—and it makes a difference—not only thought his own thoughts but spoke his own language, spoke of beer and wine as getting sick. Healthy wine had only yeast cells, sick wine had other cells, and that appraisal, sick wine, sick beer, healthy wine, healthy beer, was a physician's language and a physician's manner of thinking. To cap all he was the not unusual Frenchman who loves France. He was a patriot. Anything now could happen. Something always does.

The silks of France got sick. The distress of the silk workers swelled to a national cry and then an international cry, the sickness having crossed borders and spread over nations, and now the chemist and the microscopist and the physician and the patriot joined hands. Pasteur knew nothing about silk. He knew nothing about silkworms. He did not even know about the mulberry. In France the sickness was all far south of Paris, but the voices calling him were loud and compelling, and he went south. He went joyously almost. Soon he knew so much more about every side of the subject than anyone else that he could and did express contempt for the whole tribe of industrialists.

Everyone has heard the story. It is part of everyone's freshman course in the history of science. Pasteur saw, along with everything besides, that there were spots on the sick silkworms, spots on their eggs, and he could separate the unhealthy that had the spots from the healthy that did not. What added first to the bewilderment then to the allure of his five years of study was that the silkworm disease was not one disease, that the state of the external environment in which the worm lived could increase or decrease the worm's susceptibility to disease. And this was Pasteur learning principles of pathology, suspecting that the principles would apply to human illness as well as to worm illness, that what he was learning of this infection would apply to all infection.

Bacteria had become the whip that lashed his genius. The microscopist had become the bacteriologist, and before this bacteriologist was laid to rest he would have sown and reaped an immense portion of this scientific field that he had done so much to create. Others would have begun sowing. Others would have begun reaping. Bacteria would be hunted all over the world. Bacterial diseases would be run down one after another, cures found. Within three-quarters of a century the accomplishments would be staggering, but as staggering almost would be the fact that Pasteur's part of the field, his flag over it, was already beginning to recede, and that other field, that the chemist in him had done so much to set into motion, was catching up and running past. His ferments, our enzymes, and their shapes, subsequently the shapes of molecules and the arrangement of their atoms, and, more subsequently still, the way those shapes and arrangements fit into other shapes and produced action, would in our day come to stand everywhere at the front of science's attack on life. Possibly, on mind.

The picturesque details of this man's journey, much of it in Paris, everyone knows, accurately or not, mawkishly or not, superficially or not. What difference does it make? Pasteur's hatred of the enemies of his country. His affection for his family. His solicitude for his friends. His modesty. His fire. His boyishness of heart. Boyishness of thought. His sentimentality, yes. Skill in controversy—to judge by the amount of it, he did not shun it. His letters in times of controversy were like slaps in the face of his adversary. His first apoplectic stroke, that might have killed him, that destroyed much of the right side of his brain, paralyzed the left side of his body, deprived him of the use of his left hand, made him dependent upon assistants, but his thinking unslowed. His shattering discoveries and his universal acclaim occurred after the stroke in the final twenty-seven years of his life. He lost a child to typhoid, a bacterial disease that would be conquered. Here is a translation of what he wrote at the time: "I give myself up to those feelings of eternity which come naturally at the bedside of a cherished child drawing its last breath. At those supreme moments there is something in the depths of our souls which tells us that the world may be more than a mere combination of phenomena proper to a mechanic equilibrium brought out of the chaos of the elements simply through the gradual actions of the forces of matter." This in his French is as quiet as the Gettysburg address. It had a different humanitarian theme than Lincoln's but was clear like that, full of feeling, simple. The two writings were set down at about the same time.

# XI

# KIDNEY

---

## Shrewd Housewife

# URINE

## Drain Water

We drip, drip, drip, all day and all night, three to four times more by day than by night, drip from the kidneys into the silent pool of the bladder, an underground dripping into an underground lake. One thinks of small boys, nonagenarians, male and female dogs, summer camps, sleeping cars in Italy, hospital end-rooms. The fame of urine as fluid is only next to the fame of blood, ill fame for urine. In the seventeenth century, in tragedy they spoke of urine, in farce of piss. Urine is nonsalubrious. It stinks. Aromatic as it issues forth, ammoniacal after it stands, the ammonia produced by bacteria that multiply. It foams sometimes. It steams sometimes—a horse on Christmas Eve. Is transparent as spring water sometimes. Turbid sometimes. Shiny lemon yellow ranks first among its colors. Urochrome is the yellow, a chemical not well understood, produced apparently from urobilin, which is from urobilinogen, which is from bile, which is from hemoglobin, which means the blood is painting the urine. Every time we urinate we are present at the evidence of the destruction of red cells. The color goes from a colorlessness to a reddish or greenish or brownish black.

All vertebrates, all the loftier beasts, form urine, and these, a distinguished urine-author reminds us, are the same who have the complicated nervous-systems. The maintenance of the constancy—especially the constancy of the human brain—requires this sensitive elimination.

A urological surgeon pushes his fingers through the fat around a kidney, slashes out the kidney, has it in his hand, remarks that its capsule can be stripped off like a glove. Demonstrates. A tough capsule off a tough organ. A pediatrician pushes a needle through an infant's back, on into one of its kidneys, needles out a sample, the infant cooperating.

The kidneys are the urine-makers. They are the body's chief disposal. They are smart. They scrap whatever is bad for the body and brain, and any excess of good. A child goes on a candy spree, sugar rises in its blood, reaches a concentration too high for safety, and the kidneys throw out the excess. A machine-part is bringing relief to the total machine. Paracelsus back in the sixteenth century argued that whether a substance was or was not a poison depended on how much of it. Too much oxygen, we saw, was a poison that caused body signs and mind signs. Too much water is. Too much of anything for one person may not be too much for another. Each to his need. Chemicals entirely foreign to the body may be scrapped. Though ancient and dating back to the paleozoic, the kidneys have the shrewdness when aided by the liver to rid us of a product synthesized by the drug houses last week.

The kidneys drain us. That is what one thinks of first, drainage. But there is that other fact, they retain us. They keep us as near as possible to what we were yesterday. Our correct quantity, our correct quality is saved. Correct glucose. Correct nitrogen. Correct water. Correct acid. Correct alkali. Correct everything. The creature *sui generis* remains behind while his urine, which his mother to inculcate good habits used formerly to say was not a part of him (and she was right), drips off into the rivers. For generations, years, days, hours, minutes, we are scrupulously kept what we are by the scrupulous elimination of what we are not.

Since the amounts and the concentrations vary during the twenty-four hours the physician asks for a twenty-four-hour sample. He wants an average of the day.

In the adult the twenty-four-hour sample contains two ounces of solids, a flexible figure as every physician knows, dissolved in a quart and a half of water, also a flexible figure, as every beer drinker knows. A beer drinker is busier than other men and if he is a big drinker he may be beerlogged for hours, never loses his beer as rapidly as he imbibes it. From the smell of him one thinks his beer must be coming through his skin. On a hot day he sweats more, and, since there is only so much fluid in him, his kidneys are by-passed, so he urinates less. But there is a minimum.

Why beer drinkers are flippant about urine has not been scientifically delineated. Why men are flippant about their stomachs has not been. Why some literary wag sees the possibilities in urine one understands easily enough. In dinner jacket and patent leather shoes on Monday night at the Literary Club he rises deadpan and reads his paper *De Urina.* In preparation he turned pages of histories of medicine, dug out the appropriate quotations, a poem, made a list of the extra-urinary uses of urine, that included, the distinguished urine-author says, the manufacture of medicine, soap, niter. The audience laughs, then the audience settles down, then the audience grows restless, and that could be because after the first five minutes what had been aromatic has become ammoniacal. However, beyond all literature, anthropology, geology, a sample of urine continues to be a doctor's treasure. It was to the Hippocratic school, to the alchemist of the Middle Ages, to the horse-and-buggy doctor of the nineteenth century, to the Mayo Clinic of the twentieth.

Our physician presents us with an empty bottle as we rise to leave his office. "If you'll just return that to me in the morning." How crisp and clean he looks. Clean is our physician, clean is the living body, clean is this vertebrate, this mammal, this higher ape, this human construct, this most intricate construct in the universe up to now. Clean machine. Clean kidney, both as anatomical fact and chemical fact. It scrubs us morning and night. Shrewd housewife.

# NEPHRON

## *Little Helper*

A kidney is built of nephrons. Each is microscopic. One million nephrons in the human kidney. Two million in two. Nephron next nephron next nephron, all packed as nature packs, chick in an egg shell, brain in a skull. Each is a complete urine-making machine. Each labors to wash us. Each may not do exactly what its neighbor does, but nearly. While we at the top are fabricating a Broadway musical, launching a satellite, flirting and following where flirting leads, the nephrons below are maintaining the state of the union. The lowly frog has seven thousand. They cleanse its blood. The elephant has seven million. They cleanse its.

A nephron is a complex microscopic tube about an inch and a quarter long. It has a tuft of capillaries thrust into its shut upper end. That shut end is cupped. The capillaries appear fitted into it by the hand of an artisan, are in loops, fifty loops approximately, with many interconnecting capillaries, some broad, others narrow.

*Glomerulus* is the name of the combination of tuft of capillaries and cupped end of tube. The glomerulus starts the urine-making. *Tubule* is the name of the tube that leads off from the glomerulus. It continues the urine-making.

That tubule is no simple plumber's item, no mere pipe to carry along the forming urine, though it does that too. Today, with electron-microscope, sophisticated equipment, sophisticated methods, sophisticated chemistry, sophisticated thinking, which the kidney always has been able to arouse, we know better how that tubule operates, but still can get confused.

The tubule has a twisted portion that straightens, makes a hairpin loop, twists again, becomes a collecting tube, the walls throughout thinner in some places, thicker in others, the cells of different shapes, different borders, different nuclei, different particles, different chemistry, different work.

Wrapping the outside of the walls are more capillaries, all carrying blood, Bernard's fluid-around everywhere immersing tubes and capillaries, and passing back and forth between them are the products and the by-products of the work of the nephrons, and the raw materials upon which that work is done. It is one of the body's busy traffic centers.

We are told that if all the tubules of both kidneys were spliced they would make a tubule fifty miles long. The importance of this is the huge surface presented to the fluids. To keep our minds in proportion we remember Schrodinger telling us of the English King who stretched out his arm and casually decided that a yard was the distance from his fingertips to the middle of his chest.

At its nether end each tubule is open and there the forming urine forever moves toward the sea. All of those millions of creatures complex enough to have a kidney are forming urine.

From the collecting tube the urine goes via a delta of channels into a gross tube that has a gross name, ureter, and through this the final urine flows toward the bladder. A ureter can be seen with the naked eye. Each of us is blessed with two. They descend from the kidneys into the pelvis, approach the bladder, enter it, and via yet another tube, the urethra, the urine sprinkles the earth. One thinks again what one thought helplessly while trying to contemplate the circulation of the blood—so much of us is tubes.

# MICROPUNCTURE

## *A Sneaker Sneaks In*

How does the single nephron work? There have been thousands of experiments, direct and indirect.

First—the direct.

With a technique conceived by an American a single nephron was studied while it was manufacturing. The technique was ingenious. It made use of the dissecting-microscope. One looked through the eyepiece at the same time that one was manipulating the microscopic field. The tip of a hypodermic needle, attached to the barrel of the microscope, an exceedingly fine tip, was plunged into some part of a nephron, a sample of the urine-water drawn out, this analyzed. One knew therefore what the nephron at that point was doing. An inspector's eye was inside the factory. To go into a factory and look around is not the same as to stand outside and calculate what is occurring in there. This experimentation with the dissecting-microscope contributed to the human spirit, that mostly gives itself credit for nothing but meanness.

Frog and salamander (mud puppy) were the chosen. A frog kidney is not kidney-shaped but ribbon-shaped, silky, frail. First, carefully, it was lifted out to the side of the frog and no blood-vessels broken. Next, carefully, frog and kidney were brought onto the stage of the microscope. Next, a desired spot of that ribbon was gotten into focus. A number of glomeruli, some active, some inactive, would probably be visible in any field. Because red cells are squeezed out of shape as they enter a capillary and recover their shape as they exit, the experimenter's eye is helped to recognize where the capillaries are. Without that microscope, without that procedure, the shaky human hand could never have drawn those samples from the narrows of the nephron.

A quartz tube is beveled to a fine tip to provide that hypodermic needle, this fitted to a glass tube, the two together fitted to the dissecting-microscope and the needle directed, plunged in. The spot can be precisely selected. Guided by the controls attached to the microscope that tip could enter any part of the nephron. If glomerulus, that could be blocked off from the rest of the nephron by pressure applied by a microscopic glass rod. If tubule, that could be blocked for any desired stretch between two glass rods, the stretch cleaned, dried with a frail blast of air, the quartz needle pushed in, a vacuum created, the urine-water sucked out. To analyze such samples, chemical methods of parallel delicacy were required, and in those years were coming in. Earlier they would have seemed impossibly refined, today seem heavy. Without the direct experiments, hundreds of them, on frog and salamander, the present-day edifice of kidney knowledge, so intricate, so reasoned, would have lacked the foundation it needed. Also, might have lacked the faith to go boldly on to make the kidney the supremely understood organ that it is.

# DYNAMICS

## *Housewife's Schedule*

Under the microscope that cupped shut-end with the tuft of capillaries looked like a capsule to Bowman in 1842, so it is called Bowman's capsule. The tuft itself looked like a corpuscle to Malpighi, in 1686, so called Malpighian corpuscle. Outside the corpuscle is space, and into it the filtered molecules go. Subsequently the glomerulus, that filter, was proven not a quite ordinary filter, and debate over it has flared up now and then. Malpighi was the first ever to see a capillary. William Harvey never saw a capillary. Had he seen one it would have filled in the only gap in his argument for the round-and-round of the circulation.

Three membranes compose the filter. The small molecules of the blood are pressed first through the capillary wall, that probably has pores, second through a so-called basement membrane, probably no pores, third through a membrane that is farthest out from the capillary, probably pores. Three thin membranes act as one. Blood-pressure puts pressure on the water molecules of the blood, on the other small molecules too, and through they sail. Blood-cells are too large to sail through. Large protein molecules are too large. Fat molecules are. Therefore these stream on.

With filtration accomplished, Falstaff, my boy, the initial step in your urine-making is complete, and urine always provided you amusement while relieving you of one problem of carousing, to make room for more carousing.

After the filtration, the filtered molecules start down the tubule, and that is the next step. It is a long complicated step, the whole area under debate, stimulating passions and imaginations and the invention often of shrewd methods.

There is now a continuous back-and-forth between the molecules flowing in the tubules, and the molecules building the cells of the walls of the tubules, and the molecules in the fluid around the tubules.

For the filtering there are two million microscopic filters. If all that water were lost into the bladder, thence into the seven seas, the living body would need continually to be replenished by water coming in at the mouth, and each of us would resemble one of those porcelain bambinos at one of those pretty Italian drinking fountains, forever filling itself, forever emptying itself. This we escape, because most of the water filtered via the glomeruli is brought back into the body via the tubules. A water molecule entering at the top of a tubule has 100-to-1 odds against its getting out at the bottom, and getting out of the body. *Reabsorption* is the name for that bringing-back. One-fifth of the water of the blood is filtered while the blood-stream is passing through the capillaries of the glomeruli. Most is reabsorbed, about 80 percent near the beginning of the tubule, about 15 percent near the end. Reabsorption is not for water molecules only. More than 99 percent of baking soda is reabsorbed, more than 99 percent of glucose, 95 percent of common salt. All molecules that should be saved for the body are saved, a perpetual rebalancing of the balance sheets in the different parts of the tubule, and possibly a different rebalancing in the different tubules. Even the gross architecture of the tubules, though generally the same, has differences.

The tubule has another talent, a reverse talent. It can take something from the blood, transfer it to the fluid moving down the tubule. *Secretion* is the name for that. It has still another talent. It can outright manufacture chemicals. Ammonia, for instance. Ammonia has some of the nitrogen that the body must lose, also helps the body lose acid, also can substitute for sodium, save sodium for the body, and with the sodium save chloride, sodium chloride, table salt, verily the salt of the earth. One can believe in the reality of the shrewdness of the chemistry here and everywhere in the body only when one holds steadily before oneself the long time that evolution had to accomplish the shrewdness.

So, the glomerulus filters, the filtered trickles down the tubule, items are reabsorbed and returned to the body, items are secreted and dispatched to the earth, items are manufactured. For the everyday blunt spilling via your bladder, Falstaff, your perpetual paltry private piddling, the term is *excretion*. You always had an elegant feeling for terms, my boy, assisted by the writer Shakespeare. He spoke of horse-piss.

Filtering is work. Filtering requires pressure and the heart supplies it, pays for it. The kidneys pay for the work of the tubules. That makes two costs. The two when calculated and added give the kidney's efficiency. It is low. Might disgust an engineer. But it would please him that all in the universe was still measurable, expressible in numbers, galaxies and star clusters and urine.

CLEARANCE : *Checking on the Housewife*                                                    265

# CLEARANCE

## *Checking on the Housewife*

A man's kidneys are not studied by a microscopic hypodermic needle poked into one of his nephrons. He wouldn't like it. Not even a socially-minded medical student, were he to volunteer on the day before his final examinations, or a prisoner to volunteer for twenty-five dollars and so much off his time, or any other of the ways that the best is brought out in us. For man no direct method is practical. Fortunately there is an indirect method, famous. It was the result of thinking about those three mechanisms: filtration, reabsorption, secretion.

It is called clearance.

A clearance—to get the definition out of the way—is the number of cubic centimeters of plasma that can be cleared of any item in one minute. Animal or man. Plasma is whole blood minus blood-cells. The two kidneys must be imagined as putting their entire effort on those cubic centimeters, free them completely of the item, clear them.

Two samples are necessary: (1) of the urine; (2) of the blood. Then there is necessary a chemical analysis of both to find the amount of the item in both. The urine sample is a ten-minute sample, because easier to collect and then to divide by ten. How much of the item—sugar, sulphate, phosphate, et cetera—is in a one-minute urine? How much in a one-cubic-centimeter plasma? Divide the amount in the one-minute urine by that in the one-cubic-centimeter plasma, and this is the clearance. It is the number of cubic centimeters of plasma clearable into the one-minute urine. Plainly it tells what those kidneys can accomplish. It tells what they can do to the blood passing through them. It states kidney-work in numbers.

With the micropuncture technique the investigator went into the animal kidney. With the clearance technique he stays outside. Clearance studies have been performed in clinics and laboratories around the world, and on many members of the animal kingdom, monkey, fish, dog, man.

About the tubules. Do they substract (reabsorb) from the urine-water filtered into them from the glomeruli? Do they add to it (secrete)? Do they do neither?

To get the answers to those questions it was necessary to discover a substance not reabsorbed and not secreted into the tubules. This substance would be infused into the blood-stream, be filtered by the glomeruli, go unaltered down the tubules, enter the bladder, thence into a test tube. A substance that was neither reabsorbed nor secreted would appear in the urine exactly as it left the glomeruli. Nothing would have happened to it on the way down the tubules. Its clearance would tell us how much of it the glomeruli had filtered, mathematically what the glomeruli had accomplished.

The discovery of this substance came at the same time from two laboratories. It is a starch extracted from artichokes and called inulin. Inulin can be injected into the body, do no harm, not be destroyed by the chemistry, is a molecule small enough to pass freely through the filters of the glomeruli, travel unaltered down the tubules, come out at the bottom as it went in at the top.

A typical experiment would be like this. Say it was on a female dog. She was tied upside down to a table. Had she been a human being she would have signed an agreement in advance, would not have needed to be tied. So, the dog found herself on her back, upside down, four legs chained to four supports. Every now and then she lifted her head to see what was going on, whether a needle was coming. Inulin was infused into one of her veins. After allowing time for mixing, a sample of blood was drawn, and a one-minute sample of urine collected, the dog's belly squeezed to make sure all the urine was gotten out of her bladder. Analysis told the amount of inulin in the urine, and in the blood. Arithmetic gave the inulin clearance. Any substance that the scientist wished to study he would study simultaneously with the inulin. If the clearance of that substance was less than the clearance of the inulin, some of the substance must have been reabsorbed. If greater, some must have been secreted or manufactured anew.

A man's glomeruli, proven by the technique, filter 125 cubic centimeters per minute. The figure seems incredible. It is remarkably constant. Our glomeruli are producing 125 cubic centimeters no matter what we do. But our bladders are receiving only one cubic centimeter. The remainder, 124 cubic centimeters, more than 99 percent, has been reabsorbed. Those are average figures. In twenty-four hours the reabsorbed would add up to a barrelful. Glomeruli are working day and night. Tubules are working day and night. And only a paltry quart and a half is urinated by my man Godfrey or his lady to get rid of their excesses and wastes.

Via the clearance technique the student of the kidney in the twentieth century has won a urinary triumph. He has learned how everyman produces his rain. He has learned how his two kidneys with their two million nephrons operate. And he has learned all of this without poking into them to draw out samples. Dyes have been infused into the blood of patients to test kidney health. Chemicals have been infused to test kidney chemistry. Kidney blood-flow has been calculated—how much blood is flowing through our kidneys as against how much is flowing through the whole of us.

An eccentric scientist might have an irresistible desire to know what the left kidney as against the right was doing. He could exercise that pleasure also. A skillful urologist would have shown him how a catheter could be pushed either left or right into a ureter where that enters the bladder. The ureter is a flesh tube. A catheter is a metal tube. A cystoscope is a larger tube that dilates the opening to the outside so as to let the catheter be introduced through it. An electric lamp is attached to the cystoscope and light can be directed anywhere in the bladder. The cystoscope is one of those tools evolved by man to quench his need to know what is going on in the back yards behind the trellises of himself. Plenty is. Cystoscope and catheter first are pushed through the urethra, which is the all-purpose flesh tube leading from outside into the bladder. The urologist bends his back, cranes his neck, cocks his eye, gets a view of the sunlit landscape, locates the ureter's exit, guides the catheter via the cystoscope toward the exit, pushes.

# RESERVE

## *Thrift*

We have more than enough kidney for most purposes. Every organ of our body has its extra. This is built-in. It is insurance against wear and accident. Masses of an organ can be lost, work capacity reduced, and still sufficient left for life and pleasures and vicissitude. A margin of safety, a reserve, is part of the inheritance of every healthy organ, comes from the genes, and the kidneys are no exception.

For the fetus in the womb, draining and retaining were part of the job of the mother's kidneys. Hers were serving Michael from the moment he was conceived to the moment he clamored out, and his own kidneys took over. Her nephrons were performing for his. Essentially, she was urinating for herself and for him, and nobody thought this untidy.

Throughout the pregnancy Michael's nephrons, when they had developed as far as being nephrons, were idle. They were increasing in number, growing in size, doing no urinary work. The alive cells of the filter portion were changing from cubes and columns, which they were first, to the thin flat sheets necessary for filtering. The glomeruli were crumpled together. The kidney had possibly a tenth of its adult size. In the adult they weigh five ounces on the average. It was only late in the pregnancy that Michael's nephrons were now and then for a second or two roused into activity, to settle back again, as it struck one, into their before-birth waiting. During about a month after birth the number of nephrons continued to increase, each kidney needing time to get its million, its full devlopment.

The two million would be including this human being's kidney reserve that he would need as the years went by and his body was regularly losing nephrons, or need suddenly on one Saturday night when he drank cleaning fluid with carbon tetrachloride instead of gin from what he thought was the last bottle.

Michael's fetal bladder did contain some urine, and Michael's fetal nephrons did manufacture that urine, and the ingredients, especially urea, did appear in the waters of the mother's sac. This would not necessarily have gone into the sac via Michael's urinary machinery but there is evidence it did, that he was amiably urinating a bit down there into that fluid world around him, as a small boy into a swimming pool. If the prebirth urinating has any future meaning for Michael it could possibly be to exercise his developing kidneys, have them ready, because what they are going to do for him is as important as anything in his bodily life.

At birth Michael's nephrons took over entirely, slowly during the first days, rapidly as he got intimate with nipple or bottle, and the two million were and would be ample reserve. The mouse has twelve thousand nephrons, the elephant seven million, and if those numbers had not included mouse-reserve and elephant-reserve mice and elephants would have been erased from the earth. A large fraction of a human kidney now and then has been destroyed in an accident, and soon the remainder was manufacturing as much and as excellent urine as before, the individual nephrons having expanded. A fully grown kidney does not build new nephrons. That reserve often appears redundant, more than ample. Urologists tell stories. An entire kidney has been surgically removed for tuberculosis, tumor, stone, and the person's life after a few shaky days gone on as if nothing had been removed. More serious and dramatic, one kidney has been removed, then a cyst formed in the other, a sizable fraction destroyed, and the life gone on, the reserve reduced. A physician many times in his career watches the progressive loss of reserve during illness, may know the day that he ought to say goodbye to his patient, and does not say it. In a dog one kidney and half the other had been experimentally cut out, three-quarters of the dog's total kidney substance, nevertheless his wastes and excesses did not pile up, even some margin of safety remained. A bitch who lived in the animal yard of a medical college had that much cut out, three-quarters of her kidney allotment, yet experienced no urinary difficulties, her wastes and excesses not piling up, she getting old and fat, a friend of all those dog-lovers. Had she been sacrificed, which is the word used, what kidney was left would have been found hypertrophied (abnormally increased in size), the tubules of the nephrons enlarged, the glomeruli not much changed. It has been reported in a case of human kidney disease that only 10 percent of the nephrons were left, the life comfortable.


# UREMIA
## *Adieu House and Garden*

In spite of the reserve, in spite of the durability, in spite of the evolutionary foresight put into their construction, the kidneys may fail. This may be rapid, may be slow. Kidneys are failing right across the North American Continent, right across the Soviet Union, right across the street in City Hospital. It is a conventional denouement.

Because through life they have been cleaning the body, removing excesses and waste, retaining the correct amounts of everything, renal failure is serious. Blood arrives at the nephrons as usual but these have lost their discrimination, the tissues of the body are left to get along with molecules that should have been removed, and lack molecules that should have been saved, and if this keeps on—disaster. The facts have been accumulated from laboratories and clinics and dingily lit night hospitals around the world. Kidney failure is not a thing to wish upon your friend. A surgeon has extirpated one kidney, after the operation had to face the embarrassing fact that the other was an inherited freak and useless, the patient dead in a week, or if he is lucky and rich he may be here a while longer. Such inherited freaks are rare. A fetus has survived the whole way to birth lacking kidneys, a gene gone wrong, and after birth the mother could do no more kidney-work for her infant; it died. Usually failure has been less dramatic. Usually it has been a late stage in a degenerative or inflammatory process, nephritis. As we know already, high blood-pressure may have hammered the blood-vessels of the kidneys (if that is the mechanism) until the flow through them was so reduced that the remaining nephrons were insufficient, and thus had high pressure stolen "thence the life o' the building!" Shakespeare spoke with luminous words of doom.

When kidneys fail nitrogen is dammed back, rises in the blood, largely in the form of that urea molecule manufactured from protein in the liver and eliminated by the kidneys. It may be twice the usual amount. Three times. Thirty times. Symptoms appear. That sick man's sickness is called uremia—urea in the blood. The evidence seems to be that the symptoms are not caused by the urea. At least, urea can be administered in huge doses with only transient upset. It is not a poison. Crystals of urea come out on the skin like a frost, like flaky snow, the skin itchy. Items other than urea are dammed back in the tissues, acid, potassium, et cetera. We know that potassium can be a poison and injure the heart, kill. Calcium balance is upset. Whatever the poisoning, or group of poisonings, or imbalances, the verdict remains the same, the chief organ for waste removal has had its efficiency reduced too low and this creature will not see another Easter moon.

In recent years a man-made kidney, that celebrated artificial kidney, has been used to remove substances; the sick man's blood run through a "sandwich" of cellophane and plastic, substances leached into a chemical bath. Substances can also be added. The man-made kidney is gauche, is not like nature's, not working with that slowly evolved perfection. Remarkable, though, that man, a born thing, should have been able to build any substitute whatever. There is today an artificial kidney that cleanses the blood of fifteen patients at a time, making survival less high-priced. Hundreds of hospitals have such kidneys. A person has attempted suicide with drug-store poison or she has saved up her sleeping tablets, and her own kidneys would not remove the poison fast enough, the man-made kidney brought in and removed it thirty times faster. Recently there have been artificial kidneys for private homes, like air conditioners, and a Boston girl hurried from school twice a week and did her homework while her blood was being cleansed.

A loop of bowel has been lifted through the abdominal wall to where it could be flushed with salt solutions, the concentrations of these juggled in the light of kidney knowledge, wastes cast out of someone's blood, the moribund kept alive for weeks, exceedingly rarely, of course, and why keep that one alive? In an experiment, the inside of a dog's abdomen, after the kidneys were removed, has been washed with salt solutions, the concentrations similarly juggled, the dog kept alive ninety days. The abdomens of human beings have been washed with solutions, are washed (dialyzed) for various reasons, the persons kept alive for lengths of time. The examples merely call attention to the fact that it is possible for the body to get rid of wastes by other channels than the kidneys. Even natural diarrhea gets rid of wastes. Everyone who reads the newspapers knows that kidneys have been transplanted from one body to another, precarious once, common today, may not be precarious at all tomorrow. There have been thousands of transplantations of organs in the last decade and a half. Liver, lung, kidney, heart, all have been transplanted, only the kidney with proven long-term success.

Normally, lungs and skin rid the body of wastes, but the riddance via those passageways is either not sufficient or not special enough, has not the discrimination of the kidneys. Those bean-shaped organs alone are the shrewd housewives.

When a person was subjected to too much or too little oxygen, too much or too little carbon dioxide, it was brain and mind showed the signs that most concerned patient and physician. So in uremia it is brain and mind. There are twitchings, jerks, weakness, nausea, hyper-irritability, or the reverse, depression, abnormal thinking, stupor, loss of consciousness, coma, and at the end that Cheyne-Stokes breathing that reverberates through the house. We all when we were children and death seemed what-has-that-to-do-with-me knew of someone suffering with uremia, heard the word, dimly pictured urine backed up into the brain, from the bathroom into the throne room. Those two kidneys scrub also the brain.

# XII
# ENDOCRINE

══════

"La Comédie Humaine"

# HORMONE

## *Lighting the Show*

That word *hormone* was first used at the turn of the century, referred to something that happened in the digestive tube, and seemed unique. No one could have foreseen how many hormones would be discovered, how the endocrine-hormones alone would amount to seventeen, how it would be difficult sometimes to be sure what was and was not a hormone.

What happens in the digestive tube?

Acid molecules are carried with the chyme, the digesting food, from the stomach to the first part of the bowel. There the world is alkaline. Chemical action occurs. New molecules are formed, are carried by the blood to the pancreas, deliver instructions to that organ to release its digestive juice. Meaning that a messenger-molecule is manufactured in one place, carried to another, and there it rings up a curtain on a special performance. This discovery was epochal. The year was 1902.

Hormone. Internal secretion. Endocrine.

Those are overlapping terms. Each has its emphasis. *Hormone* emphasizes that a chemical is manufactured in us (or in our pet ocelot), circulates in us, and somewhere excites or suppresses an action. *Internal secretion* emphasizes that a chemical goes from the manufacturing cells straight into the blood, internally into the blood, no outflow pipe. *Endocrine* emphasizes that the chemical goes from manufacturing cells that are lumped in an endocrine gland. One creature may manufacture an endocrine, another use it, as cortisone from the cortex of the adrenal gland of an animal slaughtered in a slaughterhouse in Chicago accomplishes a spectacular improvement of performance in an arthritic actress in Alaska.

*La Comédie Humaine.* That was Balzac's half-bitter title for his shelf of novels. Our body with our mind does in some moods seem not a half-bitter comedy but a half-grotesque carnival. *Le Carnaval Humain.* Each creature performs on the sandlot of the universe. Each creature has within him endocrines playing supporting roles, by day, by night, under sun, under moon.

A tornado rips across the sandlot, flings down a tent pole, a dozen tent poles, broken bones, horror, death, disgust, exaltation, disappointment, bewilderment, heroism, cowardice, and then for an instant even the unsentimental and the sentimental mutter, *Theatre of the Absurd.* Most days, no broken bones, tent poles all where they ought to be, breakfast-dinner-supper, first-toothlessness to last-toothlessness, till one wishes just anything might happen so the crowds would scream. If they have paid their money why shouldn't they scream?

But—take a walk on the sandlot.

Read the words, the signs, over each show, or, first, read the whole list at the entrance. Man printed the words. Let us pretend that God took a day off and fabricated the side shows, the endocrine glands. Which would be nothing against words. Also, if anyone's scruples do not permit him in 1969 to use that equivocal word God, let him feel free to substitute that seeming less equivocal word, Nature. Nature built the side shows, or evolution did, or chain-reactions.

But—the side shows!

Each is important, unless nothing is, if everything is not merely sad and ephemeral, here today there tomorrow, vanishing like the performers. Sometimes a side show has a front and no show behind, a gland and no hormone, often a show and no front.

But—the side shows!

PITUITARY

THYROID

PARATHYROID

PANCREAS

ADRENAL MEDULLA

ADRENAL CORTEX

MALE GONAD

FEMALE GONAD

THYMUS

PINEAL

What names! what a Sign-Painter! What a Manager-Producer-Director!


# HORMONE AND METABOLISM

## *Performers Must Eat*

Each on the sandlot thinks he was born in the most enlightened period of history—that is the length and breadth of his historic sense. It is only when he is tired or when he is depressed, the kind of clear-headedness we have then, that it crosses his mind that maybe all through history someone has been thinking that his time was the best time. One fact may be different nowadays: everybody reads. Every newspaper tells him of what is going on under the clouds of Venus, of chromosomes and criminals, of the order of the nucleotides in DNA, of some chemist synthesizing a first enzyme, or synthesizing a hormone.

Insulin is the long-familiar hormone of metabolism. It processes our food. It is hormone, internal secretion, endocrine. Is manufactured in the endocrine portion of the pancreas, islets there, small islands of cells. In some animals after chemical treatment the islets can be shaken out of the pancreas and studied.

Insulin, the molecule, has been worked over for many years, continues to be, has seventy-seven atoms, fifty-one amino acids. Chemists have torn apart the molecule, and one skillful chemist has put it together again, synthesized it in the laboratory, and this achievement was a cool refreshing spot in this hot half-century. However, the insulin that a man injects into himself is more economically gotten from animals, but it *could* be put together, and that quite properly is a satisfaction to the human spirit.

The first extraction from those islets of Langerhans was in 1921. In the human pancreas there are a million islets, yet added together they make probably less than two one-hundredths of it, the rest of the gland manufacturing the digestive juice—exocrine not endocrine, because it flows by an outflow pipe along with the bile into the gut. As to the pancreas, it lies high in the abdomen behind the stomach. It has an elongated head. To the chef it is stomach-sweetbread. The thymus is throat-sweetbread. At dinner we forget we are eating animals, unless a tactless idiot reminds us, then, as quickly as we can, we shut our minds on the truth and our mouths on the next bite.

Some believe that insulin does its work at the walls of the cells everywhere, that it opens pores, doors, a stage doorkeeper, lets glucose in, glucose then performing its role in the hurly-burly of enzymes that every cell of the body is. The result here is energy for cell work. What finally is left of the glucose slips off scene in a disappearance act, water and a gas, the water out the drainpipes of the kidneys, most of it, the gas out the chimney of the lungs. That was the way things went in dead Hitler, dead Mussolini, dead Messalina. Each urinated. Each breathed. So did each of the twos that walked with dignity by the side of Noah up the plank into the ark before the rains fell, so Dori the cow, so Marcellus the horse, the insulin-famous toadfish, the sperm whale, the sea whale.

It is those human beings who lack insulin, some machine-part out-of-gear and they stricken with diabetes, that have helped our understanding of this endocrine, and with that specific understanding have helped our general understanding.

The first extraction of insulin from the islets was in 1921 by Frederick Banting. He was not a chemist. He was a medical student just out of medical school—if students could hold a portrait of him in their minds day after day it might hoist one of them also into greatness. That first extraction, like much first accomplishment, followed a long line of ideas strung together from a long line of minds. Banting's idea haunted him. He had been thinking of a career in pediatric surgery, was reading surgical journals, and in one read what suggested to him that if the pancreatic duct, the outflow pipe for the digestive juice was tied, the flow of that juice blocked, there might result changes in the cells of the gland like those found in a patient of whom he was reading, who had had the pipe blocked by a stone. Suppose, further, that the islet cells showed no changes? After all, they did not have an outflow pipe. Nothing was blocked. Suppose the endocrine was still in those islet cells? Suppose it was extractable? This was his idea. He had a strong body, stubborn character, clear intellect. His idea was destined to make medical history. It would write a story that all the world more or less would read. Banting persuaded another student to join him. It was summer. The professor of physiology had gone off on a vacation, grudgingly had given permission for the experiments, had no faith in them, but allowed ten dogs and eight weeks. That professor was to have a shock. A shock by cable. Three cables. By the third he hurried back across the sea. Possibly the third contained some hint about publishing a paper, or the professor's mind picked that out of the air.

So, a summer's work for a medical student assisted by a medical student, the pancreatic duct tied off, an impatient period of waiting, a rising excitement, the cells that manufactured the digestive juice degenerating, those of the islets not, an extract drawn from them, and when this was injected into a dog rendered diabetic by cutting out its pancreas, that dog was helped.

Diabetics all over the world were helped. Many of the tangles of hormones and many of the tangles of metabolism were brought nearer to where they might more freely unravel through an indefinite future.

A pediatrician states that he has seen diabetes in a child nine days old and in a man ninety-eight years old, which assures us all of our opportunity. If diabetes persists, the capillary walls are damaged, so, diabetes and circulatory damage, the two catastrophes occurring together, one gene or two genes. Anyway, damaged blood-vessels and a damaged heart may go with diabetes. Clots form in the damaged vessels, fragments break off, travel in the blood-stream, get stuck, deprive a part of blood, and that may have a bad ending. As well not to talk of that. Nothing unpleasant, though, to talk of a mutation in some ancestor, upon whom the practical joke was played in the first place. For the diabetic it comes to the same: narrowed vessels, less blood, the retina of an eye does not receive enough, a man goes blind, or a foot develops gangrene, a line of demarcation forms (we anthropoids love to chatter), a surgeon amputates. Off comes a foot. Amputation always has made an impression on the carnival crowd, especially amputation for gangrene. Unpleasant word—gangrene.

Of every fifty persons between ages sixty and seventy, one has diabetes. Two percent of the population has. Of the two percent ten percent are children. If an identical twin has it, his brother is almost sure to. Of every thousand persons between ages twenty and forty, one has it. Rarely does he get his foot cut off, but he may die. We say that we do not care whether we live or die, but usually prefer to put it off. One of the greatest vascular surgeons of the world assures us that human beings prefer to put it off, or, he states it positively, prefer to keep on. Only the gods on Olympus can afford otherwise, their comedy having no last act. One catches oneself speculating on what Shakespeare was thinking when he quit London and by stagecoach bumped along to Stratford to begin to close his affairs and write his own last act.

Because the diabetic is not able properly to process glucose it rises in his blood, and spills over, as is said, into his urine, hence glucose in his urine, sugar in his urine, the energy tied up in that sugar lost to that man's cells. He feels weak. His brain lacks its principal fuel. His senses dull. Is not hearing as well as he used to. If he has the narrowed blood-vessels he probably has them in the brain also, therefore two excuses for forgetting that he knew you, like the woman forgetting then remembering the elderly gentleman in Mariemont Square. Anyway, misadventure follows misadventure. When the glucose spilled over into the urine, water spilled with it, so water was lost. Old Aretaeus these nineteen hundred years has been moaning across the sandlot that there is a "melting down of the flesh and limbs to wine." Our physicians say simply: "dehydrated." Thirsty diabetic, he drinks and drinks, and for that the elegant word is polydipsia, and for all the urination, polyuria, and for his hunger, polyphagia. Lacking sugar, his cells burn fat, so he loses sugar, water, fat. Lacking sugar and fat, his cells begin to feed on his own flesh, so he loses sugar, water, fat, protein. Round and round goes the merry-go-round. It squeaks.

At a Carnival, while we are watching one side show we sooner or later reflect that all the others are keeping right on with their performances. How can they? The rest of the world is keeping right on. How can it? We reflect that when we close down for good and all, the others will not. How is that possible when we are no longer about? We never quite get used to this universal impersonality. Even the most debilitated ego feels, not necessarily thinks, that there simply must be something personal. However, say a thing like that out loud and the owl perched on the fence immediately hoots. But, it seems impossible, too, that there could be something personal. Only the resolute do not vacillate between those extremes. Only they are convinced, lucky resolutes, impoverished resolutes.

Not surprising with all of this misadventure in his body that the diabetic should be losing weight. He was too fat. Now he is too thin. But he might still be too fat, there being unmitigatedly fat diabetics, stout diabetics, the stoutness accounted for or an accounting sought. In a hospital ward a cadaverous diabetic will be talking to the other kind, brothers in the Biblical sense, trying to laugh at a joke, laugh at nothing in particular, as we all do. Not everything is gay inside the great tent with the September night-blue top. The cadaverous, who is actually a skinny young blonde, keeps starving herself to save her life, though some doctors say that control of diet never did save a life. Every now and then she swipes chocolates when she thinks nobody is looking, defiantly but worriedly eats in the face of doom, as thousands furtively smoke cigarettes, then starts again starving herself. The stout diabetic also plunges into bouts of recklessness, to catch up with his hunger, which he never can. The other day his doctor put him on a new diet, non-fattening and non-appetizing, and it did not cheer him.

Since fats when imperfectly metabolized, imperfectly degraded, as is said, produce acid, the stout man is tending to acidosis, is wasting his minerals to keep down the acidosis, and, chemistry being the heartless thing it is, misfortune still follows misfortune, and he can be forgiven for sometimes thinking he could welcome the promised land of coma and death. In a minute he is shaking off that thought, sets his jaw, smiles, and if it is toward evening in the poor light of the ward, one thinks that through the smile one sees the skeleton of a face.

A syringeful of insulin might have cut across that line. It is doing so this evening in Zanzibar, Capetown, Philadelphia. Did so that July day, the 30th, 1921, when that dog, rendered diabetic by surgery, was cured, briefly, by injections of an extract of the islets of another dog. Since then an injected hormone has, in Zanzibar, et cetera, been restoring human bodies to their normal 1-2-3, prolonging their survival, stimulating the old one, who has gotten quite old and latterly occupies the end room with three others, to relate his tiresome experiences again, and gravely nodding his head when his doctor assures him that his disease is "merely a nuisance."

Yet he does get angry at the needle. A needle, day after day, can on some days infuriate a human being, suddenly change what everybody thought was a saccharine old man. People get out of his way. Surely it was thoughtful of the Manager-Producer-Director when some years ago he arranged that antidiabetic substances could in selected cases be taken by mouth. No needle! It's ridiculous to feel so strongly about a needle. Some years earlier the Manager-Producer-Director had arranged, also, that a particular insulin could be inserted at a point, slowly feed itself into the body from that point.

Man discovered insulin with the help of his undiscouraged friend the dog; then put it into wholesale production, got it from the pancreas of slaughtered animals, ate the rest of the dumb beasts, one beast at a time, and after supper each boy walked his girl in the moonlight, quoted Shelley to her, other couples walking in the same moonlight with the same ambiguities. Sick couples. Healthy couples. Cruel couples.

Protein molecules, phenol molecules, sterol molecules, the United Nations of the molecules, all are represented among the endocrines. The insulin molecule is protein. It is not a large protein relatively. Its weight, its content, the order of its fifty-one amino acids, all have been known for years. Much of the variation between species also has been known. Still the molecule is being contemplated by the chemist eye. It is built of two chains of amino acids, the chains hooked together in a particular way, the amino acids isotope-labeled, the manner in which they are bonded established. A stranger coming into a room of calculating chemists, politely flattering each other, may think them quite wonderful, even if a single one met outside that room does not seem wonderful at all. This two-chained molecule may be the only protein molecule well enough known to give courage to the chemist to contemplate its structure, then from its structure attempt to explain what it does and how it does it. Evolution—natural selection selecting from pools of occasionally mutating genes—fabricated those two chains and the precursors of those chains, and now millions of years later man has ferreted out what evolution did, and has done some fabricating on his own, added brilliance to the twentieth century, and fancies he is well on the road to accounting for a man's body and his soul, possibly the soul also of the Manager-Producer-Director.

There goes the owl again.

Insulin converts sugar to fat, so gets sugar out of the blood. Insulin helps convert sugar to glycogen, so gets sugar out of the blood. Excites the burning of glucose, gets sugar out of the blood. There are other sugar-removing acts, others suspected, no doubt others still unsuspected despite that this whole field of carbohydrate metabolism has been vigorously studied. To persons outside of chemistry, outside of science (there are fewer and fewer of those), the scientist could appear a gray silhouette, the outline of a man seen through a transparence of curves, graphs, tables. Of course he is not that when we look at him squarely, watch him start home at 5 P.M., later give a hand to his garden and to the runway for his dog. Even so, when one is fumbling around at the edges of this crazy busy Carnival, which many of us like to do, it is not surprising that every now and then one plumps oneself down on a Carnival bench overpowered with fatigue just because one has begun to realize the tireless goings-on inside one.

A freshman medical student was injected with insulin. This was a human experiment. It was a sizable dose of insulin. The results were prompt. He trembled. He flushed. He turned pale. He sweated. He could not conceal he was frightened. He was aware of the wrong inside him. He was hungry, as one would expect—eerie in the X-ray room to see a stomach contract with hunger contractions because a hormone has been injected. He yawned—eerie to see a hormone drop a jaw. In this student—the experiment may have been carried too far—there was sudden great weakness. (He was learning something he never thought of about experimental dogs and cats.) Abdominal pain. Mental clouding. Dizziness. If aid was not forthcoming, which of course it was, delirium, confusion, spasm, death. A gumdrop might have been enough to deliver that student safely to his classmates. A solution of glucose into one of his veins would have been quicker. The overall informal conclusions were clear. Tamper sufficiently with the role of those 777 atoms and the routine comedy has an ambulance clanging at the door, turns life to tragedy, gives it a passing interest and the chemist of metabolism a lift for his work, a fresh feeling for what hormones are doing inside us besides what they are doing inside their chemical structural formulae.

Insulin molecules streak through the sandlot. Other molecules streak along, other hormone molecules, air molecules, water molecules, food molecules. An intricate Chekhov drama is unfolding. Impossible to see it all. A plot starts, a second plot starts, a third, the first slows or stops, starts again, hurries, slows, stops, and in every act and scene hormones are busy. The drama is inside the researcher, too, and with the help of his imagination he knows this and it makes his work real, his body and head still more objectively alluring. Still he may not know quite what to make of it all, yet it would be a reason for a talented one to snatch himself early from his bed, keep the lights in his laboratory shining after midnight, chew his fingernails, pull at his coat sleeves, sit at seminars, ask questions for which he really wants answers, bestir himself until he thinks he literally is seeing those sinuous two-chained invisibilities weaving in and out through the Carnival crowds.

# HORMONE AND GROWTH

## *Hercules and the Midgets*

A body must grow. It must grow sizable enough to elbow its way through the Carnival crowds. It must grow upwardly, linearly, push its head above other heads, innocent in the giraffe, innocent in the basketball player, and not innocent in the chairman of the faculty committee who sits at the head of the table and keeps neck and back stiff and straight.

In the beginning—in the dark of the womb—there was no skeleton, then after a time a small skeleton, then larger, bones larger, cartilages larger, chemistry of cartilage and bone busier, chemistry of calcium, phosphorus, Vitamin D, parathyroid hormone, busier.

Over the growing bones were growing muscles, and outside of them was the growing skin, and inside of the growing body-walls were a few guts and other matters, all growing. This deaf-dumb-blind ambition to grow! In the beginning of the world there was no Descartes body-is-a-machine, then in due order a small machine, then larger, still larger, and for every Descartes machine, for every cell of it, every molecule, plans would have been sketched on the chromosomes, instructions delivered from DNA by RNA (so goes the rime), enzymes accumulated, proteins accumulated, and after more time there would have been a bouncy Vice President of the United States, and before him there would have been Moses.

The endocrines concerned ooze into the blood from pancreas, thyroid, adrenals, male and female gonads. And the brain has its control on the endocrines, the parts of the brain generally but not detailedly known, and the how not detailedly known. Finally, there is a special endocrine, growth-hormone its name, somatotrophin its nickname. It supplies overall growth-drive and subsidiary growth-drives. It seems everywhere on the sandlot. Experimentally it is injected into a white rat at time-zero, then the increase of amino acids, that are the building stones of proteins, watched for in the blood, and the body watched, with no end of variants on the experiment. Growth-hormone may, as was thought for insulin, control the cells' entrances and exits, lets things in, keeps things out, is another stage doorkeeper. But it is also superintendent, day-laborer, curtain-raiser, lamp-damper, any functionary that will help expand this gala enterprise. After the ends of the long bones have become sealed to their shafts, and the growth of the body has become stationary, growth-hormone still goes on being secreted into the blood, this continuing into old age, but, irony of ironies, continues to be secreted also through starvation—even the starving are obliged to grow, at least stimulated to grow.

The manufactory for the growth-hormone is the anterior part of that three-ring show the pituitary, that marvelously intricate endocrine gland. The pituitary is fitted into a saddle of bone at the bottom of the skull under the middle of the brain, which saddle has another saddle inside it, of tough membrane, so that the pituitary may seem the most protected organ of the body. It must be important to *Le Carnaval.*

In past days, long past, the pituitary was thought the source of *pituita,* drippings from above down the back of a man's nose into his mouth. Might be his liquefying soul. A performer might wake of a morning and find his soul in his mouth, and that would bother a superstitious actor. They are all superstitious. Catch an actor on a black night over by the fence where the owl perches, he is superstitious.

Besides growth-hormone, the anterior pituitary keeps at least five other hormones in production. Some of these have the job of keeping still others on their jobs, the amounts of hormone determined by the usual teeter-totter: more in the blood, less released; less in the blood, more released. And there has been introduced a releasing factor. The play is interesting and complicated, and grows more so. One easily understands why the anterior pituitary has been called master gland, but it would take from the interest, and the facts, if one failed at all points to remember that all the endocrines relate to one another. That is true more, and that is true less, but it is true all around the ring, and the anterior pituitary gives the dominant endocrine orders, is the Manager-Producer-Director's office. The glands that receive the orders are called target glands, which is a change in the figure of speech, and makes it fair to say that a special pituitary arrow is aimed at a special target, hits the target, in fact can't miss, and the target obediently sends out its own brand of hormone. This is to the end that every body-scene and every body-act shall receive its right cue, right place, right proportion, achieve for the total performance a harmony quite miraculous in this cast playing its repertory on the time-space of hot and cold stars.

At the top of this, towering over everything of the body, is of course the head, the brain, and over that the mind, wherever and however that is.

*Le Carnaval Humain* womb to coffin gives the mind pleasure, gives our feelings intensity. Balzac in his theatre-art was ahead of us, yet despite the detail that his genius wove into *La Comédie* he may have presented life simpler than it appears, more weeded, more cleared-out. It is cleared-out in some respects in the mind of a genius, in many respects in the mind of a Shakespeare. We know that. We cannot always see the details, the course that the work and the life took. But we know. We get confused. We get tired of wanting to know everything. We want to lie down. We want to fall asleep. Pavlov said over and over that our brains are tilted toward sleep, and he was right. We want to sleep and sleep, let the world go by for a while. That is the reason for Saturdays and Sundays.

Long before Adam and Eve, millennia and millennia, there was the growth-hormone, but not till millennia and millennia after Adam and Eve was it extracted from the pituitary of an ox. Two scientists did it. Americans. They injected their ox-extract into a rat and the rat forthwith grew to a great rat, 50 percent increase of weight, not water, not fat, but the virtuous materials out of which God and Evolution and desoxyribonucleic acid joined to make rat. That an ox could do that for a rat! An ordinary dachshund became a great dachshund—a dirty trick to play on a dignified dachshund. Textbooks print that dachshund's photograph, a puffing four-legged square-piano construction not unlike the puffing two-legged upright-piano construction met everywhere on the sandlot. This is the side show that behind separate curtains exhibits giants and dwarfs and acromegalics, the oversized, the undersized, the mis-sized, the growth extravaganzas, the queer characters, each differently determined by its growth-hormone, together with assisting hormones. All are genetic combinations, reducible to chemical combinations, molecules determined by molecules, eventually always one molecule, more eventually, two molecules, two molecules side by side so restless that though they also want to fall asleep they cannot.

Grossly, what the growth-hormone is exciting is the steps to protein production, increase of polypeptides, increase of amino acids, increase of that ordinary element nitrogen. But ridiculous at the point we have reached to call anything ordinary. Nitrogen is not ordinary.

There were rats before Rome and Greece and Egypt, and whichever was the first mother rat, she manufactured molecules of rat growth-hormone, and these alongside the molecules of the other hormones, and the molecules of growth factors, et cetera, were pushed through the placenta from mother to fetus, wherefore the fetus grew, in time had grown a pituitary of its own, this producing growth-hormone and the other hormones, hence on and on and round and round and rats and rats and mice and men. Remove growth-hormone, altering nothing else, the rat stopped growing. Return growth-hormone, the rat started again. It is all silly and wonderful.

What a convenient animal the rat is, and how considerate of the Manager-Producer-Director to give it to man to play with. Now and then late at night in front of one's headlight there scurries a rat across the Carnival highway, but that is not a white rat. That is a gray rat. Or it is a black rat, anyway a free rat, because the Manager-Producer-Director is a liberal, too, thinks of freedom. He wants Republicans and Democrats and all in between but none too far out to either side.

Those two scientists, those Americans, when they got intimate with ox and rat, extracted a growth-hormone that was crude. Today, the extracts are purified but still not all the questions answered, and never will be if the answers must be final answers, science forever only moving from answer to answer.

Interesting answers there meanwhile are. Up and down the ladder of animals the growth-hormones are possibly never exactly the same. Whale growth-hormone and rat growth-hormone and monkey growth-hormone do not have the same weight of molecule, not fatally the same building stones, not fatally the same arrangement of them, yet definite chemical kinship always between them. There is evidence that all growth-hormones have approximately the same core and that it is the core that has the power to excite growth. Animal growth-hormone does not excite growth in man, though man's does in a number of animals. This might look as if effectiveness went downhill through the species. Cow hormone, bovine hormone, is not effective in man, has a molecule twice as heavy as man's, recently has been chemically pared, trimmed with the digestive enzyme trypsin, something obtained nearer to the core, and this used in the treatment of human pituitary dwarfs. Each of us might even conceivably have a slightly personal growth-hormone.

Just the other day, or the day before that, the structure of human growth-hormone was worked out, and that was a sandlot triumph. The lights on the sandlot burned late at night and the monkeys drank champagne. Fifty thousand human pituitaries were collected in one year. Fifty thousand yielded actually not much hormone. Five thousand were used for the discovery of that chemical structure of human growth-hormone.

If in the course of these late considerations someone suddenly felt an ecstatic enthusiasm for the growth-hormone, let him not forget that health, and the other hormones, insulin, thyroid, parathyroid, adrenal cortex, gonads, and the growth-factors, and the state of the world, and happiness and unhappiness, all contribute, also what she ate for supper, because, though perish the thought, the growth of the nose of the beloved does depend partly on spinach.

# STRESS

## *An Act Is a Strain*

Thomas Addison, British physician, one of the immortals, not the highest but well up the line, described a disease, afterward called Addison's disease, where the patient had an irritable digestive tract, anemia, a peculiar patchy darkening of the skin, weak heart, enfeeblement so profound that he might not be able to lift his arm. The first description of the disease came the same year, 1855, that Claude Bernard described the secretion of sugar from the liver into the blood, the first known internal secretion.

"I am dirty but I am washed," said a woman, in the clinic, stricken with Addison's disease. Even the gums around her teeth would make anyone understand why she said dirty. It was more a bronzing. The effort to breathe was almost too much.

To enjoy the Carnival one must at least be able freely to breathe, freely to lift an arm, freely to meet the stress of an ordinary day, and, if one is lucky, freely to throw energy away.

Walk again on the sandlot. Read again two of the signs.

ADRENAL MEDULLA

ADRENAL CORTEX

They are parts of a two-in-one gland, the adrenal. It has a center and an outside, a medulla and a cortex, the two parts for some thrift of nature fitted together, though their work not any less separate than that of the other endocrines, the combination of the two sitting like a dunce cap on top of the kidney. In the human being the adrenal is about one-thirtieth the size of the kidney. In some animals the two parts are in different places.

The following would be a loose way to describe their separate work.

For sudden stress—hormones of the medulla. For prolonged stress —hormones of the cortex.

Professionals speak of stressors. Life dumps stressors on us and on the white rat. Both bodies, ours and the rat's, have built-in ways of meeting stressors. Some call out the endocrines of the medulla, some those of the cortex, some both, some call out also other endocrines, and call out other systems of the body. Often the endocrines seem one system, their interactions always difficult to follow, one drawing in another, the quantity in the blood determining how much, more, less; all help the body to merge into the one that it is, its head included of course.

Stressors we will have to meet until the planet goes up in flame, if that old story turns out true. Specifically, stressors would be pain, hemorrhage, heat, poison, threat, combat, wounds, fracture of a bone, infection, the taking of an anesthetic (ether for the white rat), also the prick of a hypodermic needle, the climbing of a high mountain, any other activity that threatens the supply of oxygen. But moment-to-moment life requires moment-to-moment adjustment to stress. For that experimental white rat it was forced inhalation of ether, a violent stressor, frequently used. Forced swimming to exhaustion or death would be another. A laboratory worker would not be surprised that that last was a stressor because he has seen it. "That animal has had it." Lazarus rising rotten from the dead must have been a stressor. A baby was born at the Carnival when no one was expecting it, police siren screaming, another stressor for everybody, for the mother too, the father too. Birth always has been. An obstetrician states that late in pregnancy the mother's adrenal glands get squashier and larger, are readying themselves for times ahead, for the supplying of greater quantities of hormone.

Throughout history the Carnival has been readying itself. It has been tightening the tightrope, anticipating that the acrobat might break his precious neck, and now and then an acrobat does, sacrifices himself, so to speak, in order that future spectators may have a pleasurable background. They paid their money because the acrobat might break his neck. (Yes, of course, they enjoyed his acrobatics too.) When that kind of stress for our body and our head utterly ceases, we cease. Each of us knows this, anticipates it, has written his will well ahead of time, though he rewrites parts of it, a phrase every day of the week, never allows himself to forget. We believe in our death no more than in our birth, but at the back of our heads we do not forget, always have half in mind the dark figure that stalks round and round just outside the Carnival fence. The dark figure helps maintain the right tension. Life must be lived with the right tension. And it must be lived now. Pleasure and unpleasure are now. The side shows must all be in operation now. Hurry. That voice? That was the owl on the fence repeating now. There is an end to everything. We are on borrowed time. Excellent. Children do not know that, therefore though children are gay and pretty their gaiety is thoughtless and no interest in it. Their bodies go up-and-down but their minds are not matching. Every adult knows. Some adults have the intelligence to enjoy the cost of maintaining the right tension. Despite that, they are apt to avoid looking directly at the dark figure. If they accidentally do, they look away.

The hormones of the medulla are amines, chemically speaking, the hormones of the cortex, steroids. Those of testis, ovary, placenta are steroids also. The medulla produces adrenalins. The cortex produces cortisone and its relatives, each with its actions, each with its name that is a jumble of syllables to the non-chemist unless it happens to get into the newspapers, as cortisone did. The newspapers said that it had made an arthritic woman dance. The newspapers themselves are always supplying our body and our mind with stress, and that is why people read them, and that accordingly is how they are written; if they allow us to laugh for an instant, that is to lessen the stress in order that we may suffer more richly the next instant.

The best-known hormones of the medulla are adrenalin and noradrenalin. The first stimulates carbohydrate metabolism, stimulates the pituitary, stimulates the nervous-system, stimulates whatever goes into action to meet the sudden. The second, noradrenalin, is released far and wide among blood-vessels, narrows them, hence controls their size, hence controls blood-pressure and the quantity of flowing blood. Size of blood-vessels, blood-pressure, quantity of blood must be precise, so the body-muscles may contract properly to meet the stress, and the heart-muscle pump properly. Precise fueling is necessary for the sudden. The medullas of the adrenal secrete also noradrenalin. So the medullas secrete both. And both do other work. The pituitary being stimulated, far-scattered parts of the body are stimulated, ground into action or hurled into action.

As for the sudden, it need not be too sudden. An evening stroll may be enough to alter the adrenalin level of the blood, even the anticipation of a stroll. A police dog after surgical removal of his medullas sat for a motion picture, maintained his composure and would have continued to if nothing too severely stressful happened. Instead of surgical removal, tubes of radium could have been planted in the medullas of a dog picked up (absolutely legally) at the dog-pound, the radium gnawing, accomplishing the removal. While thus treated like a dog, the dog probably has the best of hospitalization. Cancer gnaws too, and are we human beings required to have more consideration than cancer, than Nature? Furthermore, is there anyone able to say that this gnawed dog will not accomplish something for one of our cancer-gnawed senators? So, farewell, little dog.

Had it been the cortexes of those adrenals instead of the medullas that had been removed, the dog would have died.

Thomas Addison tracked Addison's disease as far as the adrenal glands. He said the cause was tuberculosis, but allowed that it might be malignancy or atrophy. Our clinicians and pathologists have gone farther, have tracked the cause to the cortexes of the adrenals.

A galaxy of hormone diseases has in our time been tracked to those cortexes, and what was learned has added itself to what was previously known. Basic-science acumen and clinician acumen have united. One Nobel Prize celebrated the union, and the publicity this Nobel Prize received emphasized the importance of those cortexes to man and animal. As for the diseases they are mostly rare. This does not mean that the everyday work of those cortexes is rare. It is not. We are grossly dependent on them. One group of cortex hormones affects glucose chemistry, they therefore called gluco-corticoids. A second group affects mineral chemistry, therefore called mineralocorticoids. A third affects sex, its chemistry.

One disease is almost the reverse of Addison's, an excess of cortex hormones pouring out of a secreting cortex tumor. Bizarre sex signs appear. A child with secreting tumor may have the look of an adult, and the desire. A woman has hair grow on her chin and other unwomanly places, her voice drop to baritone; she applies at the Carnival office for the job of the Bearded Lady, her body changed almost to a male's, her temperament remaining female, though that may have become male too, partly because she has sanctioned with her mind what Nature has visited upon her body. A man may experience an abnormal virility, or, by the paradox of hormones, be transformed into a hermaphrodite, in which case he is neither sex wholly, or, turning the dial half-around, both sexes somewhat. Everything is Carnival. Everything is sandlot. The trick is to keep seeing the sandlot as sandlot.

If a person is foolish enough to require for his peace of mind exact explanations for everything, he could ask: how has such an excess of masculinizing hormone come about? That could be guessed at. If the pituitary is remembered, and its ACTH, and that the quantity of that cortex-stimulating hormone released depends upon the quantity of target hormones in the blood; if further remembered that there are several target hormones, should there for any reason be too little of one, this might drive up the ACTH, which in turn might drive up the other target hormones, also the masculinizing hormone. This might be the explanation. The chief interest in this may be that it points to the variety of controls that are always interplaying, and lets us guess back and forth in the way the experimental endo-crinologist does. And misses. Admits he has missed. Starts over again.

For years that cortex of the adrenal was the Miss Hormone Gland of the century. During World War II research was whipped up, it was said, by reports that the Germans were injecting cortex hor-mones into their aviators with excellent results. (At that recounting Lazarus was amused, and stank, and mingled with the all-too-human Carnival crowds.) Twenty to forty compounds were extracted from the cortex. Some were steps in the gland's manufacturing process, some were artifacts produced by the chemist while he was cooking and baking. Six were regarded as actually separate hormones. Then there was a seventh, powerful. In time there may be others.

The discovery that the adrenal cortex was one of the glands fired by the pituitary preceded the discovery that the pituitary was fired by a region of the brain, and thus was the possibility of interplaying extended farther.

ACTH stands for adreno-cortico-tropic-hormone. The mechanism of its control may not be clear to the last decimal point, but, as suggested, much has been experimentally and clinically deciphered. As said, when it or its target hormones go up in the blood the quantity released goes down, and vice versa. There is that see-saw. There is that perpetual balancing. It is spoken of as negative feedback. The endocrines are controlling each other. The pituitary is controlling. And the brain is controlling the pituitary, the mind often controlling the brain, and whoever wishes can regard that last as the feedback *par excellence*. Someone immediately said semantics. Someone always does at a seminar. The control of the hormones by the brain seems to be exercised by a different but closely lying region of the brain. The order of involvement then is brain, pituitary, target gland, behavior of the body. It might be more than one target gland, more than one behavior.

Later, one of the hormones was recognized as fired not via the pituitary but by the brain directly. How exasperating! Everything is always insisting on being somewhat different. Nothing is orderly— ostensibly. Nothing is final. Nothing is settled. No peace for the acrobat. "More light and light it grows . . . more dark and dark our woes." But more and more light it does grow. Surely it does. Only one has the illusion that though it does grow more light, there is more Carnival to light. The acreage appears bigger. It is still the same old Carnival of course.

Thomas Addison may not stand forth monumental like Balzac, but stands forth, a slight figure that casts its shadow. As said, the year Addison described Addison's disease, 1855, Claude Bernard recognized the first internal secretion, sugar from the liver secreted internally into the blood. Much is known of Claude Bernard. Much is known of Honoré de Balzac. Not much is known of Thomas Addison. The three lived at the same time. People right around Addison seemed not to know much either. This in itself is of course to know something. He was thought haughty. He displayed doctor decorum. He wore sideburns. There are a number of such biographic fragments. He was shy. That could be because he came of yeoman stock. Those on top in that day, too, probably stepped on those below and built up a hate inside themselves to justify where they stepped, so they might step harder. Addison was a general practitioner. He was meticulous at the bedside, but once he had made a diagnosis he had no interest in curing the patient, which was said also of Freud, probably was not true of Freud, probably not of Addison. Where Addison seems to have been relentless was in the verifying of a diagnosis. This was noted and reported. He never failed to follow the dead body to the post-mortem room. Diseases of the chest attracted him. He introduced Laënnec's stethoscope to England from France, likewise Laënnec's method of examining the chest. For the rest it would seem a routine story: born, worked, accomplished, got fame, got old, married a widow with two children, had none of his own, found himself suffering from gallstones, became jaundiced, was sick, was wretched, resigned from the staff of Guy's Hospital, went to Brighton, died. Probably he died of cancer. This was 1860, around the time of the publication of Darwin's *The Origin of Species,* a book that gave to man's mind its most securely based conception of biological struggle, stress. Addison might never have used the word *stress* but apparently knew the need of meeting it, for survival, knew the facts physiologically, knew them personally. *Requiescat in pace.*

# HORMONE AND SEX

## *The Glow and the Fact*

The femaleness of the female, the maleness of the male, or the other way, the mind-side, the body-side, before puberty, at puberty, after puberty, the twenty-eight days, the fertilizing of the ovum, the planting of it in the flesh of the woman, the ensuing ten lunar months, sex hormones play on every part of that. And how they play! They guide the chemistry. They guide the physiology. They guide the behavior.

In the human body are many mansions and many hormones. All along the flowery or the wilted sexual lanes they wait, lend their different talents, till on some midnight-to-morning they get the obstetrician out of bed and Nature's end is achieved, prince or pauper, president of the Woman's Club or Rotary, Oscar-winning actress, Friar Lawrence, Romeo or Juliet, assassin, arsonist, rapist. If all of these could find themselves at the same performance of *Le Carnaval,* same esplanade, or same seminar, understand as far as anyone can, which is possibly not far, the impressive extent to which each has come alive and grown to be the notable citizen he is by the command of sundry hormones, these controlled by other hormones, all concocted by sundry glands, the concentrations determined by the concentrations in the blood, with parts of the brain intervening, one would like to see that piece of news write itself across their faces. Old age follows the menopause, follows thirty or so years of the possibility of conception, follows puberty, follows the first day at school, first successful standing on two feet, first cry, first kick of a foot of a fetus in the wet dark, follows a mad St. Valentine's midnight or a snowy Thanksgiving before dawn, or just some stupid evening with nothing else to do, and in the church towers the bells are ringing, it could hardly be to celebrate the youth, the gaiety, the prime, the decline, of our forever-burdened forever-contradictory sexual rounds. A few hormones wind that clock, they say. The thing is utterly impossible but let us hear what they say. If anyone wishes to buy an all-day Sunday ticket for this side show, please go that way, next corner, left. Be warned however that once one buys that ticket one may forget mother, father, wife, seven children, the commandments, burning of hell, reward of heaven.

Ensconced in the bone of the floor of the skull is the pituitary. It is fabricating, storing, releasing its target-directed arrows. Three are aimed at the female sex glands, which when hit send out other hormones, whereupon the female body and head act in never-simple ways, the female finding herself haggled over, cooed at, jealously watched, angrily watched, happily watched, or if she is a bluestocking, intellectually watched, suavely turning the tables on him, uneasy when she does, uneasy when she doesn't.

Three pituitary hormones bear up under unbearable names: follicle-stimulating hormone, FSH; interstitial-cell-stimulating hormone, ICSH (male), or luteinizing, LH (female); prolactin. From such reasonably straight-forward capital letters and unreasonably bulky words, the story unfolds, elaborates, culminates (or gets lost) in a polysyllabic chemistry.

Nature must smile at the unsimple designations we attach to the simple things she does, though if she regards them as simple she sees the planet not as we. Wouldn't it be surprising if actually she were a she, and if she did see our affair in terms of the chemist's steroids and the physicist's omega minus?

Three hormones, three condensations of witchery, three shapes in space have accosted other shapes in space, and in consequence more and more shapes accosted more and more shapes, molecules molecules molecules, accostings accostings accostings, performances performances performances, periodically periodically periodically, and then, sometimes unexpectedly, sometimes expectedly (that's planned parenthood), at the crossroad starred on all the maps a new thing, a wiggling wonder, subsequently often a wiggling bore that at once went downhill, though sometimes also uphill, made its appearance on the sandlot.

The first of the three, FSH, directed its arrow, perhaps not as straight as a bird directs its flight between wind-thrown branches and leaves, but straight enough at its target, and on arrival excited the growth and maturing of the follicles of the ovary, especially that follicle in which the egg-of-the-month was housed, nudged it, pushed it out, but did that only when assisted by the other pituitary hormones, the follicles having begun now to secrete a female sex hormone, estrogen. (Estrogens. Everything hereabouts has a way of becoming plural.) This sex hormone traveled through the female body, met, as the sophisticated say, its proper receptors, that were molecules, and this meeting excited the growth and the sweep of the female characteristics, including the kick-pleat of her skirt, scarce an organ of her body left untouched, scarce a single cell probably, all probably or partly regulated by that perpetual teeter-totter that, as of 1969, regulates everything, this-goes-up that-goes-down, the first half of every menstrual cycle dominated, and, if there is a pregnancy, there ensues the mammoth theatre of the endocrines of pregnancy.

The second of the three, LH, assisted by FSH, directed its arrows so as to cause that house vacated by the egg-of-the-month to secrete, to mature, to become tenanted by a reddish-yellowish microscopic side show that for the span of its survival produced its own female sex hormones, progesterone and estrogens. The former worked upon the lining of the uterus, prepared it for the egg, which if fertile caused all female hormones to continue in production, the lining to grow gorgeous, till that midnight-to-morning when the curtain rose on the trillionth matinee. If not fertile, the juggling of hormones continued, the lining sloughed, the lady bled, and that explained to her the twenty-eight days. She nodded, thoroughly understood you, as far as you understood yourself, then laughed outright into your recondite face, went on reading *Time* for news on the latest pill to suppress or to induce ovulation.

The third of the three, prolactin, was a variety performer. Among its performances, should there have been a pregnancy, and after the ten lunar months a delivery, it instructed the breasts, always with assistance from other performers, to secrete milk to stuff the mouth of that yelling parcel, or, should the mother have been a cow, joined with another hormone to start milk in quantity out to the dairies. Meanwhile, in the pigeon under the window, prolactin was exciting the growth of the crop gland, with the production of crop milk, the cells of the lining of the crop having rubbed off and mixed with water, and that mixture pumped into the open mouths of the squabs; a wrestling match between the two, it seems.

Disturbing it was when as college students we first heard that these same pituitary hormones were vouchsafed to the male also. The sandlot seemed bewitched. For the female they produced female witchery, for the male, male witchery, excited the interstitial cells of the testis to secrete male hormone—to help the male keep male, keep swashbuckling.

This was only the beginning. One could have thought one was in a nightmare, the extraordinary names, the extraordinary performances, the pushings-out onto the stage, the pullings-back into the wings, the regular, the periodic nightmarish repetitions of the nightmare, the quality of the scenes played, even to the human mind that is playing them, and such acting. Is any of it true? Is all of it science fiction? Is one going to wake and find oneself back in 460 B.C., a neighbor of Democritus with his atoms that were round in a flame and sharp-pointed in an acid?

These and these atoms in these and these arrangements were for females only, but not quite. These and these were for males, but not quite, always some leaning, some cheating both ways, making it comprehensible why a body could sometimes be given so small a shove and fall so far, why a life in the sex regions might so often be so unsettled, why a mind could so often play so wantonly on a body. Everything was so tipable to begin with, boy, girl, man, woman. But it did make for marvelous entertainment.

One passed in review the side shows, looked at the 1969 chemical formulae, how alike yet unlike, how all might be reshaped from one shape of molecule, an atom shifted here, an atom there, a carbon, a hydrogen, an oxygen, the completed aggregation hustling through a body built of molecules, disappearing into a cell built of molecules, there splitting molecules, joining molecules, the left-over pieces with military regularity trickling into the urine. One thought how symbols concealed while they revealed. One thought how incredible everything, but then one thought, the pulse of the twentieth century in one's blood, why not? Immediately one was restless again—one remembered that this turmoil was going on inside one's own self. These endocrines were fiddling while Rome burned.

There was the male gonad (that side show) located forever insultingly outside of the male, its hormones sometimes believed to have transmuted septuagenarians into eligible bachelors, flattered young men with lower voices, male bravado, the privilege of shaving, given the rooster its comb and its crow and its bad habits. There was the female gonad (that side show) located like some other of the body's secretest secrets in the silliest places, inside in the depths of the female pelvis, hiding there, its hormones keeping the female cycling in her cycles, swindling the hen of wattles but allowing her to cackle. The master gland tapped by the brain (as that had been tapped by the mind) therefore tapped the male gonad and the female gonad, and these tapped the male and the female, who accordingly felt very male and very female, felt exalted, felt depressed, lifted high, dropped low, never at peace until everything was at peace and no cheer in that either. An odd coincidental happening: fat had become deposited in one place in the female, another place in the male, wherefore a male in overalls had the satisfaction of recognizing how sharp his eyes were, how he could distinguish a female in overalls crossing the sandlot at midnight, then all the trouble and sooner or later or never a placenta, that rialto where mother and fetus met, gotten lush, for ten lunar months performed as a great endocrine gland, put into production all sex hormones whatsoever, the greatest Carnival on earth. The thyroid joined in, sex among its talents, added to the excitement, added to the depression. The adrenals joined, viciously sometimes.

# THYMUS

## *Illusion*

For years no endocrine role was assigned to the thymus, in fact no role. Then, increasingly, hints that it was part of the body's defense, some special, some glorified lymph gland. So it would not have been an endocrine gland.

It did however go up and down with the ups and downs of the adrenal cortex, an endocrine gland. It enlarged with adrenal insufficiency. Those two, thymus and adrenal, had some kinship. With the thyroid also, the thymus enlarging in troubles of the thyroid, in goiter, exophthalmic goiter.

See again the list. It may have been a mistake to tack this one on, possibly nowhere else to tack it, possibly illusory, something to do with illusion. One must be cautious. One must remember that nothing here (or anywhere) is as it pretends to be. Time changes knowledge. Nothing is without prank.

It lies, the thymus, in the upper chest. It must be important in the infant because it is relatively large. It lies above the heart at the base of the neck. It continues relatively large up to puberty, gets smaller after puberty, is small and was believed mostly to stop working in the adult, whatever that work was, shrinks, atrophies in old age.

A tumor of the thymus in a high percent of cases is associated with a disease that has been claimed to be startlingly cured by surgical removal of this side show, if side show.

Lately it again seemed side show, important to the sandlot, important still in defense, important as an endocrine in defense, in immunity, et cetera. For a time it seemed to be sending critical cells to the spleen, to the ordinary lymph glands, other places. It had something to do with the rejection of skin grafts, maybe other grafts, maybe big grafts, transplanted hearts, transplanted kidneys.

Indubitably there is still illusion. Indubitably this thymus, with its peculiar cortex, peculiar medulla, peculiar special cells, would for a time produce illusion. It has earned its uncertain place in the Sign-Painter's list. The Manager-Producer-Director always proves in the long run to have been shrewd.

There has been some learning about the thymus of late, and more learning about the adrenal cortex, about the sandlot everywhere, and about the vast and the small, about light-years and angstroms. Endocrine research and other biological research has been moving inside the small, from angstroms to fractions of angstroms, from the ostensibly crude to the ostensibly refined, from gland to cell, cell to particle. We still travel inward, inward, inward. Is that illusion? A sober science surely could not be creating illusion. The owl perched on the fence hoots. It sounded like a laugh. "Oh, owl, it would not know that it was doing that." The owl hoots again. "Oh, man, would it not?"

# INDIVIDUALITY

## *Clowns Are Serious Actors*

A change of costume . . . A change in the make-up of the clown . . . The make-up does not hide him. His message comes through. As clowns before him, he is trying with clowning to express that there is pain but there is glory in being born an individual.

In the next three minutes while the lights are being damped for the night, end of the day, night falling over Carnival, let your body relax, your thoughts wander. Here and there on the sandlot you still see moving figures and it will be darker than this, entirely dark before you lose the clown's false-face white face. No need though for the face. The Harpo-Marx actions tell all.

A fly has it, individuality. Also has six feet. Every fly has six feet, as if flies had been tooled by machine-tools. Nevertheless, if the human eye is quick it can recognize the individuality of a fly, a solitary fly living with you in a room, the rest of the building silent, 2 A.M. A spider has it. An expert on spiders will be disappointed in you if he thinks you do not believe he knows his spiders one and one. Invertebrates have it. Vertebrates. When it is our kind we call it personality. A cat has it—no one will need to convince anyone who has shared a house with a cat. A sporting man counts on it in his English setter, as the English setter in his man. Both have it with such shadow and substance that the two can sit across the table at dinner night after night, or on a couch where there are two butt-polished indentations, and neither ever gets bored. Dogs ought really to be given a longer life. The Manager-Producer-Director ought to write a letter to his congressman.

In individuality as in everything about dog or cat or man or tadpole endocrines are busy. Sometimes they strike one as too rash, too experimental, in a certain lizard, certain jackass, certain you.

Which is not to suggest as one expert did that the individuality of his dear dumb deluded Cousin Mathilda before that night they laid her out at the fashionable funeral home was only so-and-so-many endocrines bound within her lily-of-the-valley skin. A book that was a sensation forty years ago disposed of the individuality without a quiver: pimples to wrinkles, endocrines were it. Like the recipe for a salad, there was a pinch of adrenalin from the adrenals, thyroxin from the thyroid, three pinches from the gonads, et cetera, and the book was popular and had value, but today reads as if the author were Savonarola and spoke from a fifteenth-century Florentine cathedral. Nothing in life gives a man the right to be that confident, to be that exclusive. We were thrust without such penetration onto this crust of a planet, yet, imagining we had the penetration, do we of this hard-headed third-third of the twentieth century proceed so differently from him who wrote the book? Did not he make an intelligent start on what we have not always as intelligently followed? Have we even much changed the way to get to any one of the problems within the problem? We bring in his endocrines, hope sometimes to stumble on a Nobel-Prize-winning new one, or a new idea about an old one, dash them somewhere, inject them, isotope them, do a hundred experiments, surgically remove whole endocrine glands, sew them back at a different place of the anatomy, manufacture synthetic endocrines, dose the nature-made or the man-made into human bodies or animal bodies, test the back-and-forth between any two endocrines, any three, delete one, substitute another, have the skills of the century, but in the deleting and the substituting do we not sometimes overlook some small historic fact, as, say, how far off that crust of planet have we pushed the ghost? Yes, the ghost. Hamlet's father's ghost. And not only a grand ghost like Hamlet's father's but an ordinary ghost, one of those that a few years ago walked by night down Linn Street, turned the corner, turned again, went into an alley, into that house where past the windows up the stairs at the lunchbreak or five o'clock before supper went gentlemen with black derbies, the ghost of your great-grandfather, of the great-grandfather of Mrs. Neumeister who lived on the other side of Armory Avenue, and there would have been the ghosts, if there are, of Mary Dinosaur and Mildred Dinosaur. What would the fifteenth century have thought of pushing a ghost off the sandlot? You say we do not care what the fifteenth century thought?

Very well, the twenty-fifth? By then the ghost may have been reinstated, each body have its ghost again. That ghost will live in its body as in a rented house and get out of its house, say, by committing suicide. Mary Dinosaur and Mildred Dinosaur have given notice by suicide to the landlady, then arm in arm in their enormous dressing gowns those antediluvian sisters reappear as ghosts. That would be a sight!

An endocrine without a doubt can do drastic deeds to the individuality. A hospital clinic has the extremes walk through its doors. An undersecreting thyroid leaves a fattish mucous-looking fellow who nods every time he lays down his cigarette, an oversecreting puts a near-maniac (says his wife) thrashing about in his bed. The near-maniac has a goiter. Flushes. The bedspreads tremble with his trembling anatomy. Talks through the night. Captivates you. Horrifies you. His eyes pop. He breathes noisily. His body wastes while you watch. Not enough adrenal-cortex hormone in a human body leaves a feeble female collapsing ahead of time to her everlasting rest. Too much—a hairy Amazon. Too little of one pituitary hormone sends out onto the sandlot that freak of growth the dwarf, too much, that other freak the giant, or, if the too-much occurs late, the acromegalic. "That fellow," says the head nurse, "has a secreting tumor of his pituitary up in his skull but all he can talk about, his hat is tight!" Too much parathyroid creates a humpback, abnormal curvatures of his spinal column, his body shortened, an ugly walk; too little—twitchings, tremor, convulsion, spasm, possibly spasm of the larynx and a tragedian's frightened exit. A testicular aggressor and a castrate apologist are accounted for or the accounting sought among the gonads. Get one wrong number in the balance sheet of the endocrines and it can multiply into a mistake so large that a policeman in a cruiser spots in that transient something that he is not responsible for and resists pulling up to the curb.

Most of us dangle between too much and too little of everything, feel a righteous eleven-o'clock Sunday-morning pride in being what we cannot help being, normal. How dismally in hormone balance, how at-home, how integrated (that all-purpose word) is the person we absent-mindedly call our friend, how tedious and dull, yet in his twin brother there is running a distinct and golden vein of fairytale fantasy. Through Paris strides Madame Defarge looking at nothing, seeing everything, bold and cold in fury, a type in well-nigh perfect endocrine balance for her type, and, apologetically keeping up with her, her muscular husband, the wine-seller, balanced for his type, watching her, watching her. But neither of them and none of the others of the Paris of that day, and not Mary nor Mildred Dinosaur of their day were all endocrines, any fool realizes.

It is likely that the endocrines do with the individuality what a portrait painter does at early sittings, lays in the broad strokes, leaves it to later sittings, another technique, mechanisms of a different system of the body, or the two systems joining, with the patience of primordial physiology, to pencil in the singularity, the signs of the life-experience, the small hurts, small satisfactions. Even an excellent painter though he may have come near to painting this, may not know it. Nature may not know. If Nature is molecules—well then, molecules may not know. Better let that alone. Better not forget either in the midst of all the molecular uproar that, for example, a slum-organ like the liver may as a total do gross work that has long been recognized, long been intellectually attractive, valuable to our minds for its revelations of the living machinery, but no one has so far followed all of that work down to the molecules, and when someone does, the learning from those molecules will, usually, be added to the total, not the total reduced to the molecules.

It grows dark.

René Descartes in the seventeenth century thought the pineal, perched in the middle of the brain, was the seat of the soul. The soul sat up there while the man sat below on his own seat in his pew in a French church. Latin *pinea* means pine-cone and the pineal is shaped like that, and reddish. Sometimes it has been claimed to have growth effects and reproductive effects, even effects that accumulated through successive generations, the offspring smaller and smaller, maturing earlier and earlier, as they say do some children of India, or some of some ghetto. Claimed also to have effects like a conceivable sex gland, removal of the pineal causing enlargement of the ovaries, removal of testicles enlargement of the pineal, another of those ups-and-downs that can bring satisfaction for an entire afternoon to a seminar of graduate students and their professors. One present-day cult—these were not graduate students nor professors—has assured us that there comes an instant when the pineal bursts and then one sees the past, sees back, back, back, and if one saw back far enough and forward far enough that would be the eternal, would it not? Three hundred years since Descartes lived and died, and those molecules there in the middle of our brain, plus the grains of sand that go with them, have not yet let out their secret, provided they have one, provided the pineal is not a side show with a show in front and no show behind.

However, before it is completely dark, before night settles over the Carnival, it is a temptation to recall certain experiments.

A female mink may display a proper unwillingness toward one male and be quite free with another. Inject atropine, which temporarily blinds her, and that female who has been hospitable to one male takes on all. Her eyes are out of focus and mink is mink. The discrimination of any female, we know, does go down with mink. It does go down with mink. Which could be brought to ions and atoms and molecules and endocrines, but why? Why so serious, child? This is only the Carnival! Come, eat your cake.

# XIII

# DEFENSE

────

## The Investment

# NATURE

## Statistically Correct Number of Defectives

Nature often seems a person. Seems animate. Seems human. Seems against us, for us. Seems not as our science sees her but as ancient myth saw her. Seems testy. Cool. Female. The earth is hers. The creatures are hers, the man-bodies, dog-bodies, cat-bodies. None can escape. All are earth-bound, universe-bound. Each is no more than a condensation of the primordial substance. Each could be broken up, the molecules scatter, their electrons leave and arrive, their nuclei recombine, after an abatement the whole recrudesce, come back again—a silly thought that everyone has thought, then lifted his glass, gone on with the party. In light-years man still will be the methodic murderer and earnestly slaughter his five siblings, or eat roast pig made appetizing by an apple in its mouth at the Christmas table.

Nature is the unmitigated existentialist. As it is, so it is. She is a mathematician. She is a concoction of gonads and brimstone, a breeder and a destroyer.

She is a defender. She has made her investment and she plans to protect it.

She is shrewd in her defense. Never is bothered when she has turned out a poor job or when a good job has crashed into a poor job. True, she does not want too many defectives. She calculates it. If the defenses are right she knows that replication will take care of the rest, she will get her quantity. She wants quantity. Pores over the percentages, worm to man, and to hell with that withered nine-year-old with an I.Q. of 160, a genius but to whose running nose someone else must bring the handkerchief. The withered one is one, adds up to one, let him sicken, let him die. Wastes no emotion on the other either, the football fullback who has broken his back, or that feeble-minded middle-aged contraption that cannot ever be said to have belonged to the human family, not from the womb. Like Vesalius, she wants bodies. Not corpses to dissect, of course. She wants living bodies, more and more living bodies, then if ever things get crowded she will rely on Malthus, cut down overpopulation quantitatively. A good war. Efficient birth control. Planned parenthood. Legalized abortion. Other social illusion. Faster automobiles. Formerly, she used the black death, and in parts of the world still uses famine.

She is ice. She has done well. Can afford to be fickle. This evening she bristles, tomorrow morning will purr, in the afternoon masquerade as a typhoon, and just before sunset will take off her powder blue dawn (her skirt, that is), change to the gored rose that she will sweep around the horizon, undermine the human mind with poetry. However, beneath all change, or just frankly outside, it is scales, talons, fangs, or a seductive smile.

Through it all she does needlework. In earlier days they called it fancywork. A female can think while she does needlework.

Her best thoughts come in the chill of morning when the haggard sleepless look out of their windows and see the edges of the land cut threatening shapes out of the sky, she lolling in a half-sleep following an exactly sufficient whole-sleep. Is pleased with yesterday's accomplishments. No conscience, men say, implying they have. Last night she sipped a champagne cocktail after she snapped off the sky, this morning inhales lungfuls of galaxy molecules, crisp ones, no wet ones.

She is too old for the kittenish mood she is in, flicks open that other Pandora box that contains the anxieties, lets more of them out among human beings, to keep them human.

Abruptly she crosses space, as if it were a room, to a man-made microscope, peers into the eyepiece, sees down there a defending white cell in a drop of blood. Laughs lightly.

Amusing down there below the eyepiece.

Men say *blind* Nature. Men make bearable in talk the insult that affairs on this planet are not of more concern to Nature. She behaves always as if there were planets enough. She sits, starts again at her needlework, has catgut in her sewing box. Without that stitching there would be no surviving living creature. Leaves the fewest possible unpatched places, mends all the holes, until she gets bored, the bitch. The hole was too deep and in an out-of-the-way place, so she let drop the garment. A dead liberal. Dead priest. Dead prize fighter. Badly torn garments she drops. They roll with the planet.

# SKIN

## *Outer Barricade*

A sleeping alligator would wake if one hit it with a hammer. (Alligators in the swamps are shot on sight when seen by a man with a trigger-finger.) A sensitive creature when approached properly, zoo experts say, charming, the female, when she builds her nest, charming when she watches over her young, suddenly fierce in their defense, in a minute so sleepy, never resists the temptation to go back for five minutes more. To bring an alligator mind via an alligator brain from dream-crammed sleep into glaring daylight in a second with a hammer, what an outrage! By a jungle river an alligator sleeps and sleeps, can risk it because besides the built-in defenses that all of us have it has that outer barricade. One lies prone in a bathtub, a small one (eight inches long), the boy of the house having given it asylum, looks lazily up through soft bright eyes at the near-sighted aunt who has come to take her bath—this ridiculous screaming hysterical world. Those soft eyes look through slits in armor plate.

Man has his outer barricade too, his skin, gentle, soft, yet doing rough work. Its outermost cells are flat and, by textbook, dead, a dead layer ready to rub off and blow away. Beneath that, the cells are less flat, less soft, more what one thinks living, more lacking in that special chemical keratin. Horns, quills, fingernails, toenails, claws have the greatest amount of keratin. The enamel of teeth has a substance close to keratin, and enamel is harder than anything a man has, except his mind, and teeth from when they first appeared under the moon have been a first line of defense. Thirty-two. Glistening teeth. Who has thirty-two?

Skin is a defense against assaults of many kinds, cold winter, hot summer. The way it tans defends, the dark pigment rising from the deeper layers of the skin blocking further absorption of the ultra-violet of the sunlight, defending thus the depths from further exposure to that assailant. How permanently damaging that exposure may be has come to be known only in late years.

Skin is a defense against the inside of us, too, defense against inconstancy of the internal environment, imbalancing molecules being let out, ridding the body of them, the skin classed with kidneys and lungs among the excretory organs.

A defense, too, against that other assailant that is everywhere, water. Keratin does not dissolve in water, a reason that a water-attacking world cannot get at us, cannot erode us. A tanner's skin, says the gravedigger in *Hamlet* "will keep out water a great while." And water in the opinion of that gravedigger is "a sore decayer of your whoreson dead body." The gravedigger knew.

For all of these reasons and other reasons we are not astonished when we are told that the total weight of our skin, this Persian rug in which the Persian of all Persians has wrapped us from most ancient days, is three times the weight of our liver. Our hide is three times our liver! Kidneys, lungs, nervous-system appear diminutive. This hide furthermore is greased. A thin sheet of fat lies on it, fabricated by sebaceous glands, and each time a hair moves it squeezes out a bit of fat, and somewhere a hair is always moving. A polite fat—does not get rancid. An aggressive fat—its fatty-acids and soaps can kill bacteria. (The skin normally holds bacteria.) A friendly fat—prevents us drying, moisturizes us, keeps us pliable, causes one human being not to be averse to touching another, helps therefore the male-female situation, and that would be group defense, each two and two with their knives against the knives of all the other twos and twos. A fluidic fat, makes men slippery, as they are. Women are slippery too.

The living machine was perfected by natural selection, tooled to precision by natural selection, provided inside with an array of defending molecules by natural selection, provided outside with this barricade by natural selection, this skin, then, finally, a mind laid on top, or wherever, and all of this shrewdness meant to protect the dab of sometimes complicated nobility within.

# HEMORRHAGE

## *Breakthrough*

A breach in the outer barricade occurs a thousand or a thousand-thousand times in a lifetime. It may be a jagged breach. "They never make a clean slash you can stitch!" That was a snarling intern, weary, in Receiving Ward, and he spoke without looking up from repairing a man's neck. Three brought him in, and no one of them, not he either, knew how the thing got started except they all were drunk. Severed the jugular vein on the left side. Instead, whoever's knife it was might have slashed one of the four feed-lines pounding into the brain, and that might have been fatal, but the fellow was lucky, Irish. It was only a regular Saturday-night card party. A third of his blood spilled out. If the third had not been put back promptly that also might have killed him. If more than a third his chance of being erased from City Directory would have been excellent. It could be that the Receiving Ward does save too many. Hemorrhages usually are less than a third, owing initially to the mind, the defenses in it, vigilance, imagination, the keeping before itself pictures of destruction by imbeciles on motorcycles, teenagers with 32's, Amazons with arrows, thus helping body and head to avoid destruction, and extending the possibility of procreation. Is actually everything to be considered as finally for procreation? Is that Nature's thoughtless thought? But, again, does she really think? As a man thinks? Does she probe astronomically and reason philosophically?

By the book, defense against hemorrhage is general and local.

*General.* If not much blood escapes, the blood pressure does not drop, if much, it does, less push is applied to the circuits by the pump, less blood left in them, on both accounts less escape from the wound, so the drop becomes a defense. If a drastic drop because of a drastic escape, survival depends on whether that which still travels the circuits under the pressure that it travels can supply an indispensable organ like the brain, and, incidentally, the animal go on using all the intelligence within its range, go on maneuvering, go on doing for itself what it can, which a cynic would say was only more irony.

*Local.* The arteries of the slashed area constrict, narrow the sluices, may shut them tight, the hydrants turned down or off. A chemical starts this, is released from the platelets, those particles that looked carefree and useless as they bobbed along with the earnest red and white cells. What grist all of this is for the man-in-the-street-atheist. Sly Nature has built defense against hemorrhage straight into the hemorrhage. "She won't let us die, not quite, until she gets ready." The platelets furthermore clump, glue themselves to the walls of the small vessels, jam the open ends. The inner lining of those vessels at the same time curls up, also jams the cut-open ends. The whole idea is so neat—in that bloody wound that is so messy. Normally, blood-vessel walls and platelets repel each other electrically, thus reduce resistance to flow, free the stream. With injury the electrical state reverses, walls and platelets attract, cling to each other, the flow hindered, slowed. Then, as all the world knows, to all the world's relief, the blood clots. A cork is stuck in the wound. The breach in the barricade is plugged. The blind and intricate and numerous defenders inside the tissues and inside the blood have again an unbreached wall to lean their blind backs into. So says that same man-in-the-street-atheist.


# CLOT

## *First Aid*

Our blood must keep flowing yet must be able to gel on occasion. The gelling must be limited to the wound, to the jaw where the tooth was ripped out, the puncture streak where the I.V. needle went, the never-seen never-known innumerable small injuries of any man's or any beast's life. (The nurse had shoved in the I.V. needle with confidence.)

The gel, the clot, must be firm, a firm fabric. It must not be brushed off on the outside if the breach has been through the skin, but also not be swept along on the inside, not creep along, not get beyond the scene where it is needed. If it does, if it creeps, say, until it blocks some critical vessel, even extends from a man's leg into the cavity of his heart, as happens, still another will have helped that problem of overpopulation.

Clot construction is worth a few moments of anybody's time. Clot construction takes place inside blood-vessels too, because they get injured and need the defense of clot. There is another kind of construction that takes place inside blood-vessels, that narrows them, that blocks them, that could hardly seem defense, and that is indeed believed the chief cause of illness and death in the United States. But that is a separate story.

The manufacture of the clot is so methodic, so cool to its importance, so grim (for the person concerned), when it is grim. It has repeated itself: try to fancy how many times it has repeated itself, or maybe you have already done that, or maybe you do not care to, in which case you may care to when you get older, though it is usually not age that makes human eyes look around.

A drop of blood suffices for an informal observation. During the first minute after the blood has dropped from, say, a cracked-open finger, one sees in the blood, under the microscope, red cells, white cells, platelets. Then miraculously, even to those not inclined to the miraculous, fine fibers appear. They are called fibrils. The material for making them is in the blood right along, waits there. With the ultra-microscope aided by the X ray the molecules of the fibrils have the shape of fibrils. Soon they will have built a latticework, which shrinks, and in the shrinking the blood-cells are trapped, and that is the clot. On top a clear serum rises, its color between hemoglobin and bile, amber.

Every step within that process of clot construction would have had steps within the step, known steps, suspected steps, postulated steps, and all of them timed, small fractions of a second, and graphs drawn up, time plotted against the specific activity of that step. It is all rather gay.

That inducible, that eye-observable clot construction would have been the same (perhaps) inside a wound, and the wound in the end and at that point would have been plugged with the cork, the leak in the dike mended. *Finita!* (If everything else went well.) This creature's body defenses would once more have shown themselves adequate. The mind, recognizing that everything was going well, would have added its body-quieting defenses, difficult those to define, but a worried mind is able to upset any applecart. Some simple soul will feel he must at this point in conscience say that this creature was destined to be saved, that God would have had it so. Alas, the miserable suspicion remains, just averages.

# COAGULATION

## *House-That-Jack-Built*

The fibrils mentioned are made of fibrin. Laboratories have fibrin in bottles. No fibrin is in the man's blood or he would clot to death, but a forerunner is, fibrinogen. The fibrinogen is dissolved in his blood. It is molecules. Fibrinogen molecules are one of the blood's longest and narrowest and heaviest molecules. The liver manufactures them. Rarely is a liver so sick that it cannot manufacture them.

Nudged by an enzyme the fibrinogen is converted to fibrin.

Fibrinogen → fibrin.

That fibrinogen molecule, that soluble molecule, has the form of three nodules joined by a thread of insoluble fibrin. What the enzyme has done has been to shave off the nodules, and that left the fibrin. Molecules of fibrin are linked end-to-end and side-to-side, produce a tough framework, a matrix. This jams the cut vessels. Also, it is this latticework carpentered in this way that shrinks and traps the cells and forms the clot, and the clot is the endpoint of this defense. But also, obviously, there must be mechanisms afterward to remove that clot. (Again, a separate story.)

The picture here and so far is a visual picture, which is still dependably what we have, but the chemical is underneath, a most, one might say, diabolic chemistry, an evasive chemistry, and the chemists have rudely but also refinedly taken charge. No one can say where it all will lead. However, the strangeness of every moment of it, the complex living body in which it is working, the fact that it should be occurring, that it must be occurring, that it was seeded by evolution and, so to speak, was nurtured by evolution—the true wonder of that is always at the back of our thought.

Fibrinogen → fibrin → clot.

For this critical long-known fibrinogen-fibrin step there is expectedly an enzyme, and this is called thrombin.

Thrombin → fibrinogen → fibrin → clot.

No thrombin is in a man's blood either or once more he would have clotted to death, but once more there is a forerunner, prothrombin.

Prothrombin → thrombin → fibrinogen → fibrin → clot.

This is the scheme at its simplest, its grossest, the classical scheme, and the classical language at its simplest and grossest. There was no factor V, no factor VII, no factor X, no factor XIII, et cetera, at least we have not mentioned them. There are everywhere pro-factors. There are everywhere anti-factors. Those who study the intricacies, who learn the language, are called *coagulationists*. They have swelled to an important part of the world of biological specialists. They are in Poland, in Detroit, in Vienna, in Cincinnati.

Sometimes no coagulationist scheme for the clotting process, if one includes and plays with the chemistry, seems to remain intact for longer than an afternoon. Yet, on the contrary, if looked at over the last decade or even the last quarter century, to the non-specialist the changing schemes often appear not too changed, and when one thinks of the inner magnitude of the body, the changes might never seem enough. Indeed, to follow the changes takes more knowledge than the non-specialist can ever muster, probably. Fibrinogen and fibrin in the last quarter century have increasingly been studied directly, and all on the road from prothrombin to thrombin studied directly, for the chemical concerned, not merely indirectly for its blood-stream effect. A factor, some protein, some lipid appeared needed, was searched for, was found, another needed, found.

All in all there is nothing surprising in that there should have been these discoveries and rediscoveries, these interpretations and reinterpretations. One needs to relentlessly remind oneself how fraught with danger as well as happy possibilities the coagulation process is—also how almost countless are the items in our blood, not only the unnumbered molecules, but what chemical acrobatics they have to perform for our defense. "It is all in the blood," the old doctor said. Everything must be so versatile—nowhere in the immense complexity of the animal is it more evident than here. Everything that occurs could so quickly be final—the coagulated man lie there dead. Coagulated somewhere—his heart, his brain, his leg, merely his poor leg.

Laboratories have prothrombin, too, in bottles. The liver manufactures it with the help of vitamin K, and vitamin K exists in edible leafy plants, and exists in our bowel, where bacteria disport and procreate and fall ill and decompose and in the debris is the vitamin. As of now, four distinct clotting factors require vitamin K. When prothrombin is not available it might be not for lack of the vitamin in food or bowel, but because something went wrong with the liver, or the bile is not flowing out of it properly, or not reaching its destination properly.

Vitamin K → prothrombin → thrombin → fibrinogen → fibrin → clot.

When a person pricked to it by his pride goes to the blood-bank and gives his pint for his friend, his blood is run into a solution. If he asks he will probably obligingly be told that the solution is a citrate, that citrate removes calcium ions, that calcium ions are required for clotting, therefore the citrate is keeping fluid all that bottled blood of the blood-banks. The blood of the blood-banks nowadays need not all be fluid. In other words, that sometimes inconvenient situation has been met too. Calcium ions play their role at several steps of the coagulation process, definitely at the prothrombin-thrombin step. How? The coagulationists are scientists, so admit they are not sure.

Calcium → vitamin K → prothrombin → thrombin → fibrinogen → fibrin → clot.

All hangs by all. True everywhere in the body, fantastically true in the mind, but at least humbly true in the 1-2-3 of clotting. This perpetual intricacy of integration, something new always just discovered, must be tucked in, may be, should be, the chief charm to us of what goes on inside us, so far as we can get beyond ourselves to see what goes on.

Normally our blood has plenty of calcium. Normally it has plenty of everything required for this House-That-Jack-Built, everything to clot us solid, yet our blood does not normally clot us solid, clot in our vessels, for which thank God unless we just happen to have grown weary and decided that everything is monotony, therefore took a lunge that reduced everything to the final monotony.

A person mechanically inclined and chemically naïve might have deduced that his blood is not clotting because it is flowing so swiftly. The great Virchow seventy-five years ago thought that it was the slowing of the flow that was the reason for clotting, and some young coagulationist of the present day might shake his head at how simple the great Virchow was.

Obviously there are other reasons than the slowing. The coagulationists have pointed out many. And, then, there is the big reason back at the beginning, the wound. One could almost have forgotten the wound. Jack built the House in the first place because of the wound. In the wound are chemical details and physical details. Among the chemical is an enzyme, or enzymes, lumpable under the monstrous name thromboplastin, and thromboplastin has classically a still more monstrous forerunner, prothromboplastin. Not monstrous to the specialist. There is a tissue thromboplastin, any tissue, but also a platelet thromboplastin, and they operate in harmony.

Among the physical details there is the electrical charge on the molecules. Some have denied that that has importance. Then there is the physical detail that though in and around uninjured vessels there are no rough surfaces, around the injured there are, and rough surfaces start clotting: thromboplastin is formed, acts on prothrombin, thrombin is formed, more thrombin, more, more, more, faster, faster, faster, and the clot builds.

That faster is astonishing and it may be frightful. In the barbarous days one saw an unpleasant lecture-room demonstration. A Ph.D. would previously have macerated a lung, and now in front of the class would, usually with some introductory pleasantry, inject a few drops of this concoction into the vein of a healthy apprehensive rabbit. Hurriedly inside the rabbit there would have been jiggled together the known and unknown pieces of the coagulation puzzle and a sight it was (gently, brother, gently, pray) in seconds to see that rabbit die, its blood rivers a tree of clot. From the portal vein a perfect cast would fall out or could be shelled out.

Prothromboplastin → thromboplastin → physical situations → calcium → vitamin K → prothrombin → thrombin → fibrinogen → fibrin → clot.

And this still is not all. This road still is not straight enough or it is too straight. In another year this year's description of some stretch of that road is bound to be outmoded, to need updating. The road of evolution is always needing updating. If to accelerate clotting is an advantage to the creature the coagulationists have found accelerators, if to inhibit, inhibitors. Impure extracts of liver inhibit or yield heparin that inhibits, and heparin is manufactured by many cells, definitely by the mast cells. (No matter if anyone does not know what mast cells are.) Drug stores have multidose vials or ampules of heparin. The amount of heparin normally in our blood is small. Its reason for being there is, one would think, to have it available for any tipping toward clotting that occurs in our always vulnerable and always defense-needing body. If a physician infuses heparin into a man (or infuses one of the comparable anti-clotting synthetics) because of indications of clotting, say in his coronaries, the infusion may save his life. Too much heparin may kill him, or paralyze him, start bleeding in some dangerous place, mash nerves. Everything in and about us has its danger, causes some of us to be careful, some of us to be careless, some to be neither, but inevitably to be reflective, each to his degree.

Clotting theory, clotting language, clotting discoveries may appear, as suggested, disproportionately shifty to anyone not caught up in the web of specialist interest, would appear more shifty if he has had time to glance at last month's journals. The remarkable turn of our decade has been the effort to understand all the old steps and all the old factors, and all the new steps and factors, as biochemical events. The chemist has us in his intellectual grip here too, which is another historical and philosophical point of interest. What he shows us is often not as difficult as it is not invitingly fascinating. The mind is always the fascinating. The mind is always imaginably somewhere in among those multitudinous molecules of the blood. Also, the mind is always the overriding defense.

# BLEEDER

## *Flaw in the Material*

"Bill looked so healthy and did his work and you wouldn't have thought there was anything wrong with him, but there was . . ." In earlier years the neighbor would have lowered her voice and added a few sad words about Bill's family. "Bad blood." Everybody knew there were bleeders in that family. The boys. It was said to go back in the family, not all of the boys, but always boys, and the men too if they lived that long, which they didn't usually. The women, though, carried the bad blood. Bleeders were part of the mind of the town. The feeling has changed, the point of view, not altogether, but science has reached even into that small town, affects its thought without its knowing.

Bill's clotting-time (timed with the second-hand of Sir John Floyer's watch) may not have been overlong. It is not in about one-third of the afflicted. A technical complication is that clotting-time does not depend altogether on the sample of blood tested but partly on the way of testing, which seems simple, but that is deceptive. (If Sir John Floyer could have anticipated the consequences of that second-hand on the watch, the good to come possibly from the sense of the value of time, but the bad certainly in being driven by a second-hand, he might have thought twice before he invented, or he might have anticipated and invented anyway, as some of the minds that invented the atom bomb anticipated but went on inventing.)

Bill's trouble might not have been what the town thought, a bleeder. It might have been that his small blood-vessels when cut or ripped did not contract as they should. Their inside linings might not curl up as they should. His platelets might not supply his blood-stream with their coagulation contribution, as they should. Also they might not clump and help plug the small vessels. Or there might not have been enough of whatever the living stickiness that sticks living tissues together. When men rode the sea in sailing ships and were not getting the vitamin C that is in lemon or orange or lime there was scurvy, and with scurvy there was bleeding into the skin, pinpoints or blotches, petechiae or ecchymoses, blood having leaked through the vessel walls, and that would result in another kind of bleeder. Still other kinds. An abnormal liver results in another. One kind crops up among the newborn, their vitamin K not sufficient, and this might have been true for some older member of the family also, either because enough of the vitamin was not being produced by the bacteria of the gut, or because it was not absorbed from the gut, this requiring bile salts, and the bile for various reasons unavailable. Four separate clotting factors depend on vitamin K.

But the spectacular, the almost popular bleeder does owe the dangerous tendency to a hereditary lack. The sufferer has classic hemophilia. It is rarer than its reputation, about one male in ten thousand subject to it, about four hundred hemophiliacs born every year. Or the sufferer is born with a similar hereditary disease not clinically distinguishable from hemophilia, called Christmas disease.

Hemophilia is caused by the lack of a gene that is a defense against bleeding, the lack of an antihemophilic factor, AHF. Like lacking a gene to produce one of the amino acids and resulting in a metabolic disease, a deficiency disease. The blood lacks this defense. It should be in the blood. It isn't. So there is a step missing, a defect accordingly in the House-That-Jack-Built. Bill is suffering from the defect. The clotting-time in his case was much too long. As for clotting defects in general, experts in recent years have demonstrated eight or more clotting defects caused by any one of ten or more clotting factors. But Bill has classic hemophilia. He was born with the lack of the antihemophilic factor.

The lack occurred originally in a male, spontaneously. It was a mutation. That bleeder bled. His sons did not. His daughters did not. But the daughters passed the taint to their sons, and half of them bled. The children of those again did not, and so on, the sons bleeding, the daughters transmitting, till the strain died out. Elsewhere a new strain would be coming in, Nature at her averaging.

Since Gregor Mendel it is known that genes are dominant and recessive. Hemophilia plainly attaches to sex. For anyone who likes to have it stated so, hemophilia is therefore a sex-linked recessive characteristic passed to the affected male via the unaffected female.

The treatment has become increasingly successful. Instead of the repeated transfusions formerly necessary whenever there was an injury and bleeding there can be a regular injection of a precipitate of a concentrate of the AHF, that factor. It is gotten from normal plasma. It may of course let more hemophiliacs reach reproductive age, and they produce more hemophiliacs.

Hemophilia got its turn-of-the-century popularity because of the way it ran through the royal houses of Europe, supplied royal gossip, cousins marrying cousins. "Disease of kings." Alfonso, last of the kings of Spain, was a bleeder, the lack of the gene having been a mutation that occurred among the genes of the family of Victoria of England, and she did, the good queen, scatter hemophiliacs around her. The last Empress of Russia came of that substantial family, and her son, the tsarevitch, was the bleeder who was said to have been saved during his bleeding episodes by Rasputin, the priest, and the method of treatment seems to have been hypnotism. (Human mind peeps out at the side of the curtain even among the mechanisms of blood coagulation.) Rasputin was a historic ugliness. He played his role. Subsequently he was poisoned then bludgeoned, to make sure he was dead. The bludgeoning was with a cane, as we learned recently in our American courts from the man who wielded the cane and is still alive. One geneticist-littérateur has suggested that the Russian revolution with all its sequelae for Russia and Europe and China and us and mankind is owing to a hereditary chink in the armor, the lack of a gene. Interesting. Unlikely. And hemophilia has required no intermarrying royal houses, but it does sometimes require that an afflicted child have a soft-padded bed. In a children's hospital there was one child had an uncontrollable temper and blood that took overlong to clot, or would not clot, those two viciousnesses egging each other on in a pale and haggard boy: uncontrollable temper, hence violence, hence wounds, and undefended blood, hence the possibility of death, hence the bleeding into joints that made them painful and finally useless. Only a Greek dramatist or possibly Eugene O'Neill would have had the language to convey the malice in that fate. The boy is indeed dead now. There are always children in a children's hospital that would have given a Greek chorus substance to chant of celestial perfidy. Little children who cannot live yet cannot die do their part to blot hope out of the blue empyrean.

# HEALING

## *Kept in Repair*

In all hospitals, for all reasons, in all cities, in all towns, they bleed, and she darns, Nature darns, Queen's Square in London, Jacobi in New York, Allgemeiness Krankenhaus in Vienna, Charité in Paris, Sheboygan Hospital in Sheboygan, Wisconsin. For better or worse she darns and they live on. An annoyed intern asks: Why? Tired intern—everybody is tired. With a curved needle he has put twenty-seven stitches into this one. "Somebody clapped something over my mouth when I came out of the bank from cashing my pay check—the gang grabbed me—carried me to the back of the lot—not a light anywhere—six o'clock—robbed me—were making their getaway—one of them whipped out a switch-knife—slashed at my face—I put up my hand—my arm got it—don't know why he wanted to do that—they had my money, didn't they?"

Imagine a smaller wound inflicted in the course of a bad-tempered shaving. The man's face is wet, water and blood flow together, he wipes off the mix, it forms again, several times, a drop of the mix splashes into the washbowl. This happened minutes ago. Over the razor cut there was now a bright gem-like bead, attractive if one looked at it straight. He has dried his face. Along the edges of the wound were redness, swelling, heat, and there was pain, slight pain, a sting when the chin was rinsed with the hot water. Redness. Swelling. Heat. Pain. Those have been the four age-old signs of inflammation, each trifling here because the wound was trifling. While doing its dastardly deed the razor blade irritated usefully two types of cells, blood-vessel cells, connective-tissue cells.

The blade had cut through capillaries, which was bad, but it had stimulated uncut capillaries, and that was some good anyway. Those capillaries had dilated. More blood was brought in. The local defense was aimed at installing fresh supply lines as quickly as possible. And the razor blade had roused, turned up, stimulated (whatever word occurs) blood-vessel cells. These began propagating. Other cells were roused too, began propagating. Old vessels sprouted buds. New vessels formed. New joined old. Channeling began. Blood flowed.

Connective tissue connects. It binds. Fiber lies next fiber with an occasional cell-body squeezed in. That was the sight seen with the ordinary microscope. The cell-body was no more than a nucleus with a thin layer of substance around. It had been economically built, almost shy. Some plasma, some white cells, some fibrin had been left in the cut, the fissure, the rut produced by the blade, and this, besides stimulating the blood-vessel cells, had stimulated the connective-tissue cells. Their nuclei went to work. Branches of cells pushed into that debris in the fissure. Fibroblasts were present. Fibroblasts secrete pre-collagen. (Boil collagen and you make a glue.) Collagen fibers formed. The fissure was crossed and recrossed by the fibers. Dame Nature was at her darning. She was restoring the original fabric. Later the scene would again be fiber next fiber with an occasional cell-body squeezed in. The collagen began too to create the waviness of the scar. This now bulged like a new grave. When those fibers shortened, the scar would sink like an old grave. World-wide. The spinning of the thread was from molecules. The thread was proven with the electron-microscope. It had been used here for the reweaving of flesh, or, if that were easier to imagine, for the self-mending of the garment. Either way when this patch was patched it would be like the rest, thread next thread, fiber next fiber. We fall asleep over miracles.

At the top of the wound the flat cells of the skin had meanwhile laid down more flat cells. They had done this neatly. The new appeared not to have added themselves to the advancing edge of the wound as might be expected but to have been pushed over, slid over like cards. Flat cells from all sides met across that cut. The scar had strong color. When later the quantity of blood in it had diminished, it would be pale.

How long for these events? The original hemorrhage had lasted minutes. The first bridge across the gap was built in hours. The healing was complete in days, or less. Soon it would be impossible to find where that cut was. Bad-tempered men's chins are in continual turnover. Had the catastrophe been a bump on the shin there would have been the purple of blood through flesh, then the borders would have been yellow and green because hemoglobin was converting to other pigments, then a spectrum of browns from the iron of the hemoglobin, then, exceedingly slowly, the world exactly as it was. How fatal everything and because fatal how mindless. The cut went away and we went on. The bump went away and we went on. A vast activity all our lives has been rebuilding us from underneath and we have gone on. One day blood-pressure would drop to zero, breathing stop, digestion stop, metabolism stop, urine stop, thyroid stop.

# PHAGOCYTOSIS

## *Home Guards*

It might not have been a razor cut but a lawn-mower gash. That would have altered the strategy. Myriads of bacteria would have been sown along in, and the defense would have been guerrilla action against those creatures who travel the earth with us, an earth damned from the beginning, as our grandmother taught us.

Three defense-arms.

They overlapped. The first was about as subtle as the Marines, the second just subtle, the third subtle-subtle-subtle. The first—the cellular defense-arm. The second—the chemical defense-arm. The third—the human mind.

As to those fellow-creatures (fair to call them that) the earth teemed with them. The total of them weighed many times more, perhaps twenty times more, than the total animal life of the earth, hard to believe but probably true, figured out by one of those who figure. Some of the fellow-creatures were in the soil, some in the layers of the skin, some in the air. Some were submicroscopic, some microscopic, a few macroscopic. Colossal worms—squeezed into us with a bolus of half-cooked measly pork. Parasites—injected with the sting of a female mosquito, escaped her salivary glands, had the lark of a tour around Manhattan in the coaches of his red cells, his body racked with malaria. (Only imported malaria has been in New York of late years, difficult for the medical colleges to find a case for demonstration.) Viruses—passed from good friend to good friend while they warmly embraced at the Charity Ball. Bacteria—passed from bed to bed in a maternity ward; this was last century, wiped out the whole ward, gave eternal peace to two generations at a swat, left a spectacular chronicle of defeat of the large by the small, raging childbed fever. It would have the distinction of a place in medical history.

The fellow-creatures by joining in their attack on us were only doing the sporting act, taking sides with their side in the perpetual jockeying of the living with the living. If you must be patriotic they were giving their all for their country. Did they not possess it before us? So who invaded whom? They wished to live. Call it wished. Did everything that would enable them to continue. Do not ask questions. Accept facts. Apparently the facts got their anthropomorphic value from the tremendous expansion of man's brain that could bias anything any direction. Accept, but admit that philosophy is fact, too, and the mystical, and wonder, even if Albert Einstein had the pleasure of the pronouncement that the wonder of wonders is that there is no wonder. In context he may not have said just that.

Those fellow-creatures, those small ones, wish to live, but I wish to also, so why shouldn't I wish their destruction? Why shouldn't I take action? I do. My body does. It has that in its blood that defends my life. But what is so special about my life?

Some of the tactics of the small seemed fantastic besides outrageous. As said, we have gotten accustomed to everything, to all the detail of the chicanery. The penicillin-susceptible parasites that the penicillin first killed slowly or not so slowly metamorphosed into the penicillin-resistant. They mutated. The seesaw between the penicillin-susceptible and the penicillin-resistant was only a passing example of all seesaws. Already in the nose of the newborn, hovered over by the nose of the nurse, it began, snotty nose in the nurse, then snotty nose in the newborn, snotty noses in both, then snotty noses in neither, and that over and over. "Fie on't! oh fie, fie! 'T is an unweeded garden, that grows to seed."

Imagine an assault, an infection, established via a packet of bacteria in a kiss. No, via a scratch. No, that lawn mower! Sixty minutes later so-and-so-many million had been multiplied by such-and-such a factor. That's many. Unseen cohorts, trained to Marine deportment, had come to the defense of the host, us. Enemy cohorts were escalating—enemies always escalate. They multiplied, multiplied, multiplied, microscopic bacteria. Combat was on. It had exploded over the whole terrain, was producing effects beyond the terrain, host molecules versus parasite molecules, jungle warfare. For the mind of the host, that was looking down from above molecules (down we choose to think), this was disgusting carnage. Sickness. Sweat. Dryness. Fever. Chill. Commonly there would have been recovery. Commonly would have been an armistice. The casualties, it seemed afterward, would have been in proportions to keep up current life-insurance statistics. To the host it was an annoyance, to the physician, a fee, to the scientist watching from the sidelines, a battle worth including in his folios of battles. Combat has been good for us, the healthy have assured us, spoke of it with moist eyes, said it added strength and glory to the nation, and to the species, and to the individual, whether he survived or not. In any case combat had to be reckoned in. When the species was that with the huge expansion of brain, combat was often indulged as drill, ping-pong, weight lifting, poker, love matches, national and international politics, or merely the tiresome verbal fencing of three ambivalent tolerant intellectuals. Even UN had grown tiresome. One thought: ordinary wars between hordes of anthropoids would one day cease not because they were bloody destructive, but a bloody bore. The majority had come to crave law and order and the police, for the nonce.

In the cellular-arm it would, then, have been cells that defended, and that meant molecules in great abundance, many kinds, carefully arranged, carefully wrapped in a sac, in a membrane. That was your defending cell. Numerous varieties of defending cells. Numerous services in the cellular defense-arm. All citizens were subject to call. The home guards waited till the invader reached their suburb, then bestirred themselves. Others had forever been on the move, insinuatingly. Services overlapped. A phagocyte was any cell of the cellular-arm that ate, phagocytized the foreign. Metchnikoff said so. Metchnikoff was an imaginative student of microscopic defense, possibly pondered the submicroscopic, possibly anticipated the chemical defense-arm, though he could of course not have seen what required an electron-microscope to see, and could not have fathomed what required a patient knowledge of the chemistry of large molecules to fathom, both having been requirements for modern investigation of defense. He was, besides, hard-working, imaginative (and Russian), left us facts, and terms, and original conceptions, introduced vividnesses. Some said he was too vivid, that he was melodramatic, which if it were true, and if it had enraged a few in the past, had lighted lanterns for everybody ever since.

Phagocyte. Microphage. Macrophage. Each is a member of the cellular defense-arm. Leucocyte. Histiocyte. Polymorphonuclear. Large mononuclear. Sinusoidal cell of liver. Cell of marrow. "Fillet of a fenny snake, In the cauldron boil and bake; Eye of newt, and toe of frog . . ." And inside the defending cells were the defending molecules—so it was said, so it was, so it was meant.

In normal times the histiocyte reached out its wet fingers, chemically and physically reached, and into its wet self engulfed what was moving past in the backwaters, a bit of flotsam, some solitary molecules, a benign one, a malignant one. The engulfing looked casual. In times of threat, abnormal times, things were less casual, finally were not casual at all, nothing was casual. Whole divisions of histiocytes swiftly and smoothly and silently on chemical command left their stations for the battle area. Incredible numbers were always in readiness for this, always encamped in the armories of the body, always refilling their decimating ranks. Incredible numbers. Incredible expertness. Incredible is the adjective for our body everywhere.

The bacteria when they broke through the outer barricade, through the gash, asserted squatter's rights, clinched their position by a population explosion, as nations, as races. Local healing was not so quick, not so manageable. Leucocytes packed around the inside walls of neighboring capillaries and veins. A stickiness developed. In some vessels flow stopped. Leucocytes came like eels from all the oceans to the Bermudas and on arrival behaved in accredited fashion, thrust part of themselves through whatever interfering barricade, then more of themselves, then all. (Ramón y Cajal for twenty-four uninterrupted hours watched a struggling leucocyte.) Soon leucocytes were everywhere outside of the vessel walls, were engaging the bacteria, were destroying some, or many, or every one, ingesting them. A battling leucocyte assisted itself wherever it could, maneuvered a bacterium against a surface, some mat of fibrin, jammed the bacterium against that, digested it there, used

the fire-power of its acids or its enzymes. Millions of bacteria, of
course. Millions of leucocytes, of course. Each leucocyte in its
destroying destroyed itself. Gave its life for our sake—merits an
ode. If a bacterium drifted from the battle area into the general
blood-stream it was all but sure to be picked up by the fixed histiocytes. (*Histiocyte* means simply tissue-cell.) Around the edges of the
battle area meanwhile macrophages moved in, formed their own
barricade. A still stouter barricade meanwhile was built farther
out, an abscess wall, of various stone and mortar. (No, not any of
this took place in one's fancy; it was there in fact.) If that last wall
was successful the entire danger zone was walled off, little Willy's
life saved. The macrophages arrived more slowly, were tougher,
brought their lysozyme, their chemical poison into the contest. The
macrophages were able furthermore to multiply after they arrived,
multiply on the scene, which helped. Ceaselessly leucocytes were
digesting bacteria. Ceaselessly the debris was being disposed of by
the macrophages, that grew big and bigger, fat and fatter. They had
charge of the dead and the dying, were morticians besides all else.
An unsavory place. Pus. Purulence. Time allowed, somewhere in
that abscess wall, usually the roof, there was a softening, a lysing, a
necrotizing, a hole. Pus ran out. Sometimes some anthropoid who
was standing by got nervous, stuck in a knife, and the pus ran out
sooner. If the cellular defense-arm had not been too insufficient, the
host not too resistant, the bacteria not too virulent, they would have
been killed off to the last aggressor, the slaughter delightfully complete, and an old man sat by the fire and talked of the days of the war.


# IMMUNITY I

## *Chemical Arm*

A poor white woman's defense was stuck into her with a syringe.
That was when the police brought her to Receiving Ward. It was 5
P.M. Hours passed after that. Then it was 11 P.M. She had waited all
that time, the interns overworked, and when they had a moment
they talked of their military status, or their careers if they escaped
being casualties in some war, or, best, talked to a nurse. Interns in a

metropolitan Receiving Ward have an exciting service, but not easy. At last one of them got around to repairing the white woman's mangled fingers, so just as well that the molecules of tetanus antitoxin had been stuck into her (forty thousand units into a vein, forty thousand into a muscle) at 5 P.M. She has that one bracket of insurance, from the chemical defense-arm.

Instead of cells like bacteria attacking us, viruses may, and instead of cells like phagocytes defending us, the chemical arm may, immune sera may, antitoxins may. How these operate is by no means an obvious chemistry: (1) the capacity to recognize the something destructive come into us, (2) the capacity to neutralize it, to render it harmless. Recognize and destroy. It is one of the potentials of our body more alluring than almost any other. The 1969 scientist concerned with this is sure to be thinking on the molecular level, as is said and said. What were the molecules that built the damageable cell? What were the damaging molecules? What were the defending molecules, the specific globulin molecules? What were their electrical characteristics? Their weight? Their structure? How could that structure be broken down in the laboratory for purposes of understanding it? What is it in that structure that does the defending?

The white woman ostensibly was defended by a liquid drawn with a sterile syringe from a sterile vial. Most of us had thoughts about tetanus antitoxin from as far back as we began to hear people talk. Antitoxin in a vial easily seemed chemistry.

An opsonin belonged to the chemical defense-arm. It was an old wing of the Pentagon, kept its old name, was known long before it was understood, so far as it was understood. It was a picturesque example in a field (microbiology née bacteriology) that teemed with picturesque examples. An opsonin was not stuck into the body with a syringe. The body manufactured it. Normal serum had it. To state the situation melodramatically: a submicroscopic ooze was squirted over an invading bacterium, to ready it to be devoured by a phagocyte when the phagocyte arrived. "I prepare to eat." That is what the word *opsonin* means. Unless the bacterium was so prepared the phagocyte would not touch it. A tasty dressing must be poured over it for what was not its banquet, for what appeared a mechanical destruction, the phagocyte wrestling it and downing it, but that fundamentally was chemistry too. Long before our day a protein was extracted from blood, called leucocyte-lure, to indicate its come-hither quality, and when the bacterium did come-hither it was slain, lured then slain, reminding one of a Grimm Brothers' fairytale or an episode of Homer's *Odyssey*. Such chemicals also helped in the digestion of the slain, everything at the site of the killing cleaned up, put away, neat. No one speaks of leucocyte-lure nowadays but the idea is close about.

So, there are complicated atom-structures circulating in us, some murdering for us, some merely getting the victim ready for the murder, and we because of murder saved from murder. Whether those molecules with specific killing capacities were in us when we were born, or whether we were born with a stem molecule from which all the other specific molecules branched, was and is an unanswered question.

Often our chemical defense-arm might seem to have little to do, like the members of the fire department at the corner, occupies its time with leisurely housework, cleaning up the bacterial and viral world in which all of us are forever immersed as in a dirty ocean or a fetid atmosphere. But let there be a five-alarm fire and every defender can be counted on to toil through day and night, and, also, the fire department immediately and apparently automatically expands its membership. The chief defending molecule was, above, called a gamma globulin, and that was correct, but soon it was clear there were other globulins. In any case, here it is the idea and not the detail that matters.

Among the ways of tracking these molecules the chemist incorporates fluorescent isotopes, tracers that give off a fluorescent light, that dots the scene, and the scientist can view it analytically.

Every defending protein would be different from every other, because the enemy was different, and that would mean many. Every protein would have its gene, though recently it has been believed that each protein might rather be the result of the union of a common gene with parts of any or many genes, which may be the stem idea in a different guise. Also, in the defending molecules as in the infecting molecules there can at any time be mutations, alterations in the patterns of atoms, sometimes favoring infection, sometimes favoring defense. To the outsider all this could easily inspire doubt, could read like Hippocrates's humors or Galen's improvement on Hippocrates's humors. Whatever the later working-out of the mechanisms when danger threatens there are underground chemical units waiting to protect Nature's investment. True too for the flea-bitten Saint Bernard. He too has had a long defensive history, a long evolution of immune devices, to keep him one step ahead of the fleas.

Immunologists—students of the road the living must travel to stay comparatively immune—long have defined for us an antibody as a defending something against a foreign something. The foreign was the antigen. Antibody neutralized it. Anti-egg. Anti-poison-of-snake. Anti-oil-of-caster-bean. Anti-toxin. Anti-bacterium. Anti-virus. Anti-protein. Anti-the-blood-of-another-species. Anti-another-blood-of-the-same-species. Whatever excited the production of the antibody was antigen. Most proteins were antigen. Some carbohydrates. Potentially, any of the large molecules, the macromolecules.

Antigen → antibody.

Varying figures have been offered for the numbers of antibodies, numbers that appeared excessive, then after one had reflected again on the character of life, competitive at every level, antigen-antibody action at every level, the numbers appeared less excessive. How should the layman decide? Accuracy of numbers is anyway for the grocery, elsewhere only helps the mind to see relations, to express relations, to talk understandably. They are important practically to the scientist, though, as we know and he knows, he may use them to hide behind, to hide simplicity.

Some antibodies were patterned eons ago, or their precursors were, and they descended then from generation to generation, but others were patterned apparently yesterday evening at the time of a fever. How antibody is kept going in the body's tissues and fluids, for whatever lengths of time it is, has been one of the questions. How the chemical defense-arm is integrated with the cellular defense-arm has been a question. How antibody could be promptly manufactured in quantity if an antigen that had attacked once attacked again was a question. How antigen injected into a newborn before its immune system was formed could be recognized by the adult, be treated like one of the family, no antibodies formed against it, dangerous therefore, would in earlier times have seemed fantasy and myth. Not that the myth is understood today, but our thinking is against myth.

New terms have followed new conclusions. New terms drop off the pens of experts, fill the air, then one day nobody knows where the terms went. Some terms and ideas stay.

# IMMUNITY II
## Old Military Plan

*Natural inherited immunity*—an immunity that the creature was born with, a resistance to something that might invade it. The baby just placed for the first time in the crib had this inherited immunity. It had been anticipated in the chemistry of that one cell from which this body came. The defending molecules were already in the fluids and the tissues of that baby's developing self, not the formal antibodies yet, but the molecules that could become defending molecules. Sometimes the defenders were bacteria-killing. Sometimes they were enzyme-destroying. Sometimes they were anti-viral proteins. Different creatures had different natural inherited immunities. My dog was not stricken with typhoid and my human associates were not stricken with some of his diseases; therefore he and we dared come near each other, were not afraid of each other, enjoyed each other, were friends. That may not be friendship, not the whole of it, but at least it is an essential for it. Which is not cynical at all because we could hardly be expected to infect each other to death, species-kill each other.

*Natural acquired immunity*—a person had the whooping cough and the antibodies produced in him stayed in him, or they somehow kept repeating themselves in him, and he did not get the whooping cough again. Yellow-fever antibody recognized and remembered yellow-fever antigen when that reappeared. Measles antibody might remember for a lifetime.

*Artificial acquired active immunity*—antibodies were manufactured in a person who had been vaccinated with killed or attenuated bacteria or virus. The bacteria would have been not virulent enough to infect but virulent enough to excite the manufacture of antibody.

Instead, a person might have been vaccinated with something closely related to the trouble-causing antigen. Cowpox virus, for instance, manufactured a cowpox antibody that was effective against small-pox. The person would have suffered a mild attack at the time of the vaccination, antibody manufactured, left in him, he immune for years. This immunity was artificial because he had been vaccinated, and the immunity was active because his body had taken an active part in the production of the immunity.

*Artificial acquired passive immunity*—diphtheria toxin was injected into a horse, or, in practice, first a toxin-like substance, a toxoid, and only then the toxin, the horse thus immunized. It had diphtheria antitoxin molecules in its blood. If the serum of that blood was injected into a man it immunized him passively. Hence the immunity was called artificial acquired passive immunity. Diphtheria antitoxin seems ancient in this rapidly changing field.

Some of the substances of our own tissues could inside us act as antigens. Usually they did not. Usually nothing happened. But something could. The body could attack itself. When it did, the result was an auto-immune reaction. Doctors have thought serious diseases like rheumatic fever and multiple sclerosis might be auto-immune diseases, adding thus to the number of molecular events that could be taking place inside us, their startling character, their vicious character, their ambiguities.

The allergies, the sensitizations that are suffered by human beings and some domestic animals, probably all animals, have been considered antigen-antibody reactions. Instead of being immunized the human beings were sensitized. They were hypersensitive. A person might have been born with the sensitization, then on being exposed to a drug, a food, a pollen, react abruptly and it might be violently, and it looked mysteriously. The signs might be local, great red hives, or widespread, as after a toxin-antitoxin horse-serum injection in someone with an allergic state like asthma. He might die.

Before dying the wretch might have had convulsions, might have half-suffocated (spasm of the smooth muscle of the bronchial tubes), experienced terror. Madame Nature stretched him on the rack for no punitive reason, just stretched him. A similar dying often had been visited upon animals experimentally, called anaphylactic shock, inducible by two time-separated injections of a small quantity of egg white. When one first saw anaphylactic shock in a rabbit one felt, should have felt, an angry unhappiness, then one got used to it, as to Receiving Ward of City Hospital on Saturday night, or being accosted by a prostitute in a London Street, or hell after burning awhile.

Paul Ehrlich, chemist, microbiologist, immunologist, student of this war and defeat or victory of molecules, lived in a time of light microscopes and zoology, whereas we live in a time of electron microscopes and chromatography and radioautography and high-energy physics and nuclear chemistry. Ehrlich could seem a simpleton to some spruce young man at Rockefeller University in 1969. Ehrlich imagined a toxin as a molecule with two hands. One hand had a shape that let it grip some part of a healthy cell, and, while it gripped, the other hand, the free hand, could work the cell, damage it, demolish it. Many cells of course. The cells had their revenge, chemical revenge, produced extra hands, to take care of the future. Those were the antitoxins. They were distributed through the fluids and the tissues of us and other animals, and when the same toxin next assailed a living body the hands took the toxin into custody. This could be retold in protein terms, molecule terms, has been, may not have sounded the same in a retelling that involved chains of amino acids, kinds of globulins, viruses injected into higher vertebrates, sites on the surfaces of antigens inducing the production of particular antibodies, receptor sites, and whatever else, but the essence may have been the same. As for Paul Ehrlich, his theory as here stated was at its barest. It caught the scientists' imagination of the time, caught his, still catches ours, may A.D. 2100 be all wrong, or all right.


# LYMPH NODE

## *Inner Barricade*

The skin was the outer barricade. There was also an inner, complex, multiple. Important in it were the lymph nodes. These were regularly placed defense posts, provided mechanical defense, cellular defense, chemical defense, and in some of the cells was chemical memory; the cells remembered an old attacker, and were ready. There was evidence that these nodes lived long, or might, remembered the attacker for fifteen years.

India ink often has been injected beneath the skin of an animal, the animal slit open, the ink discovered to have left a map of an intricate system of fortifications; the large black dots were lymph nodes.

Every one of us has had a tender lymph node. Has felt its swollen outline. Has the idea of the shape in his finger tips. A human node has been described as having the size of a pea. In the human body these strategic posts would be everywhere along the streams of the lymph vessels, the lymphatics. In them flowed the lymph, slow-moving, appeared indolent, was constituted of that part of the blood that was squeezed out through the walls of the capillaries but not re-collected into them.

The nodes and several related defense structures (spleen, liver, bone marrow) were the fixed fraction of the inner defense. The thymus has recently been discovered to be part of that system too.

In peacetimes a node worked at peacetime jobs, as manufacturing lymphocytes, large and small and all the sizes in between, the manufacturing sped up if the body was invaded. From the small lymphocytes, by a number of cell-divisions spread over five days or so, there were manufactured the plasma cells, and these were shown by clinics and laboratories all over the world to be essential defense. If the animal were to lose them it promptly would lose itself, because they were the final manufactories of the antibodies. Thus the whole scheme stood there at least in outline.

Then the thymus was recognized as a character in the tale.

That illusion gland of the endocrine system sent forerunner cells to the several defense structures of the body, as the spleen, and followed the forerunner cells with a hormone that matured the cells. Forerunner cells and hormone went also to the old useless tonsils and the useless appendix, bringing their uselessness into question.

In wartimes—war of foreign invasion—the mechanical aspect of the lymph node had use, could intercept invaders, as it intercepted the dead particles of India ink. Having intercepted the foreigners it could digest them, kill them with its chemicals. Standing guard at the node's passageways were phagocytes, and we already know what they could do.

Seen from the side of a bacterium the node was a snare, a hazard, a booby trap. In one experiment, ninety-nine of every hundred bacteria that entered a node were destroyed. In another, a node was flushed with bacteria for an hour and not one bacterium escaped into the channels leading out of the node. (Maybe one did.) The node defended against toxin molecules also. And virus molecules were not forgotten, antibody gluing virus molecules together, the mass of them bigger, so that they could not slip into a cell, multiply there, do their destruction.

Inside the cell, furthermore, in all cells was a chemical called interferon. It interfered chemically with invader multiplication. It interfered usefully in the life of plants also, these needing something against the circumambient invisible destructive force in which we all live together, and antibody production did not occur until evolution reached as high as the vertebrates. So, below the vertebrates there was interferon. But we, king of beasts, had it also.

Those lymph vessels drain our scalp, our face, our abdomen, our genitals, our extremities. Cancer might reverse that, might employ those passageways for its own guerrilla-warfare, which was why a surgeon when he removed a breast removed all lymph tissue anywhere in the vicinity, up toward the collarbone, out into the armpits, deep toward the lungs, a surgery that might take five and more grueling hours. The surgeon did similar drastic surgery in other stricken parts, stomach, uterus.

The military plan of the bacteria meanwhile was to join with as many as possible other bacteria, infect the node, tear it to shreds, move on to the next node.

Should the invaders not have been mechanically halted by the nodes, not beaten down by the total cellular-arm of the body, not by the chemical-arm, they were free to swarm. Such a human being or other animal was suffering a bacteremia, bacteria free in the blood, or a toxemia, toxins free. Death might be hard to stave off—not as hard as it used to be.

# A, B, AB, O

## *Bolstering the Wartime Economy*

A night visitor groping along gloomily lit corridors in City Hospital plumps suddenly on big white letters lighted from behind that startle him, but it is a gay sign for a gay place. *Incubator Room.* No one can resist stopping to look through the window. A skinny pink baby lifts a skinny pink leg, a pale sleepy baby coughs, a tomato-red-faced baby lets out one horrific yawn. The three are at the beginning of their long-short journey. The tomato-red looks as if it had high blood pressure already. Opposite the Incubator Room is the Blood Bank, and a nurse is coming out the door, has picked up two pints of blood for a transfusion. Hospitals work hardest in the middle of the night sometimes, but any hour. Families come and go. A worried husband is staring out a window down a night-steps into the night-street. Maybe he is wishing he could run away, far away.

It long had been impossible to replace lost blood, a fact that gnawed at the human mind until a way was found to do it, one person tapped and the tappings added to another person. All over the world thereafter in hospitals, drop, drop, drop, blood entered veins, so-and-so-many drops per minute, the transfusion equipment checked hour after hour. In spite of popularity it continues to feel odd, drain a first body and run the drainings into a second. It is so intimate. "And how do you know whose blood that was?" That also has gnawed some human minds, and the query was right enough. For centuries men had tried transfusion. Sir Christopher Wren who built St. Paul's was one of the early ones. It was tried again and again, man into man, sheep into man, other animals into man, always abandoned for the same excellent reason, death. The bloods, the vibrant fluids, were being barbarously used, but what was thought: merely somehow that they were incompatible. In former days women had fatal hemorrhages when their babies were delivered, especially slow deliveries, happens rarely now. In the nineteenth century an English obstetrician transfused five women who were in bad state, four died, but the fifth lived and the transfusion had saved her.

In the twentieth century came the discovery that accounted for past failures. Bloods were different. One blood could not be safely mixed with some others. However, what was also discovered, bloods did fall (imperfectly but perfectly enough) into four large groups, and that was a practical discovery. It made transfusions simpler but not entirely simple. Madame Nature always holds something back.

Not only was the blood of one group generally safe for transfusion into persons of that group, some crossings were safe. A quick test let a physician know in advance whether the person who was giving the blood, the donor, was right for the person receiving it, the recipient. That much had been clarified before World War I, but the triumphs for transfusion came during World War II, altered war, because now after shooting holes into human bodies with unprecedented efficiency the lost blood could be replaced and everybody happy except the sacrificed young lives.

The manifest cause of the transfusion deaths had been that the donor's red cells clumped in the small vessels of the recipient, blocked the stream, and organs or tissues did not get their deliveries, and if the person did miss dying outright he nevertheless was in trouble. Later those clumped cells would dissolve and that was more trouble.

Clumping and dissolving—agglutination and hemolysis—fall reasonably under the head of immunity, and the discovery of inherited blood groups stimulated the whole field of immunology by lighting up this corner of it.

In the clumping and dissolving the body could be regarded as defending itself against the foreign. Madame Nature, so far as we can understand her, wants the body to maintain its blood nationality, to resist invasion. She keeps strict ledgers, is implacably against passing things from body to body, whether bloods, tissues, organs. It might, not indubitably, be reasoned this way. Living creatures in their long history have had to deal with invading proteins. Parasites have their varieties of proteins. Parasites get into another creature and thrive in him. Other varieties of proteins get in. Red and white blood-cells have their varieties. Best to keep all foreigners out. Madame Nature has made it a danger not to keep them out.

The four original blood groups were A, B, AB, O, or ABO. A

thrifty man has his group typed on a card in his wallet so he is sure
to have everything ready for his auto wreck. Whoever enjoys matching pennies or working crossword puzzles might have an interest in
the matching groups. Blood-cells and blood-serum must be kept
separate in one's mind. Blood-cells have on them, or do not have on
them, A antigen or B antigen. Blood-serum has in it, or does not
have in it, A antibody or B antibody. B blood-cells have B antigen,
but B blood-serum does not have B antibody, because if it did it
would destroy its own cells. This probably immediately makes the
term *incompatible* clearer, and the nurse will not look ridiculous
when she puts a drop or two of serum on a glass slide and adds a
small quantity of the blood-cells of the person scheduled for transfusion in order to see whether the cells do or do not agglutinate.

The traffic in blood has grown as vulgar as the traffic in money.
Blood banks are everywhere. Blood bankers receive and pay across
the counter. (It is indeed hoped that bought pooled blood may be
reduced by donated blood, the danger of liver disease, hepatitis,
less, and it is a danger.) A wife despite twentieth-century obstetrics
loses a unit of blood at the delivery of her baby, the obstetrician
sends for a unit from the bank, transfuses her, and it might be safer
if her husband were the donor, his obligation, and he may indeed
hurry to the bank, pour his pint into the common pool, right the
family account even though he is an anemic small man who can ill
afford the payment. Happy enthusiastic father, tomorrow he would
have passed the obligation on to his mother-in-law. Blood bankers
have a society, the American Association of Blood Banks, with
national meetings. International Transplantation Conferences take
place every year. There is more theory. There is more accumulated
data. There is more effective application. There is an estimated collection of six million pints of blood each year. Every twenty-four
hours every large hospital transfuses blood in every ward, human
lives saved. The wit of physician and surgeon and scientist has
taught them to drop blood out of a container into a vein and a man's
body does not have to wait till his bone marrow has manufactured
the blood-cells, and he in the meantime die. This defense may fail
but it often succeeds, in hemorrhage, shock, burns, clotting abnormality, anemias, before and after surgery. Occasionally, it is
true, one has the impression of an order for transfusion being

written into the hospital chart because no one could think of anything else to do, on the principle that if things are shaky, as well to
have enough of the bank's money right in your own pocket.

It ought from the start to have been anticipated, and it has already even in this brief accounting become evident, that blood is
more complex immuno-hematologically than the four groups. The
living (also the dead) are always more complex, man always wanting to find out something further, finding, still wanting, finding,
which may be crazy but it is man. Besides the A, B, AB, O, there are
more than ten well-known groups and as a result of the linking of
this with that it was said for a time that the figure was three hundred thousand recognizable blood individualizations. If all facts
were recognized, the actual figure could again probably be comprehended only by the mathematician and the astronomer.

# Rh

## *Odd Bit of Investment*

Rh has also turned out to be more than one. When first discovered it
could have been thought essentially a blood group. It had somehow
insinuated itself into the known groups. What was quickly clear was
that it supplied the explanation for a previously mysterious cause of
death of fetuses, of infants, cause of stillbirths, of miscarriages, of
brain damage if there was a birth.

The infant came forth jaundiced, or got jaundiced a few hours or
days after birth, had anemia, had a large liver, large spleen, then
died or did not die but was very sick.

What had happened in the sex situation was that an Rh-positive
father (born Rh-positive, genetically Rh-positive) had at a previous
pregnancy conceived an Rh-positive fetus in his Rh-negative wife.
No Rh in her. While that fetus was in her womb some of its red-cells
with their antigen crossed over into her. Her blood produced antibody against them. This stayed in her. At the later pregnancy the
antibody passed the other direction, into the new fetus. Poor foreigner in the womb, his red cells were destroyed and he got anemic.

His bone marrow labored to make up for the destruction, whipped up its red-cell manufacture, hustled the cells into the circulation before they were completed, and under the microscope one saw the immature reds. The fetus might not survive. More likely he would. In a children's hospital, if he was lucky to be near one, or any hospital, the physicians would start at once on admission to draw his blood, to replace it with blood compatible with his mother's, exsanguinate him, literally, transfuse into him an entirely new blood-volume, red cells that that incompatible serum would not destroy, Rh-negative cells, in exchange for his own Rh-positive cells, like putting a new transmission into an unluckily built car. By the time that infant had used up those transfused cells, his personal blood-manufacturing machinery would have swung into full action, the destructive antibodies negated, his life saved.

The mother meanwhile had had a danger added for herself. She had acquired a serum sensitized with that new antibody, and she dared not be transfused with Rh-positive cells because her serum (as at the second injection with any antibody) would rapidly produce more antibody, destroy the transfused, she die, or she might. Death is wherever life is.

Some psychobiologists choose to regard the fetus, any fetus, as foreign. It is a three-pound resented foreigner. Eventually he is rejected. That is to say, born. It gives to all of us the tone of Greek tragedy. We are all resented and rejected. The interpretation is bizarre if nothing else. It makes us think thoughts larger than one catastrophic birth.

# PHYSICIAN

## *Nature Accepts Help*

There were millions or billions—many anyway—of years between the first fern and the first man. All surviving and non-surviving animals fell somewhere between those dates. No one animal ever paid a professional call on a sick animal, so far as we know. (Not any M.D. animals.) An animal helping an animal get well—that would be milestone on any planet around any star. During the mating season a mate does groom a mate and that sometimes looks like physical therapy; the pigeons were at it this morning on the Medical College roof, a rough-and-tumble in each other's feathers. But a human being would think he was drunk if ever he truly thought he saw Pigeon A fly over to Pigeon B to bring it medical help. Psychological help? It could be, but one always ends thinking it was a human interpretation and hard to be sure. Big-brained baboons are claimed to bring medical help, but one reads the accounts, precisely what did happen, and one ends thinking the same, hard to be sure.

Man brings that help day and night, not only to a mate, to anyone who will pay, or won't.

Human disease is old. Fossil bones prove it. All the kinds of body-defense are probably old. The physician is old. In the planet's tragicomedy, obviously he stepped on-stage only during the last scene of the last act, but in man's tragicomedy he was probably there from the beginning, probably was making his rounds long before there was a first record of him. Hippocrates, the Greek, was merely the first "Father," Imhotep, the Egyptian, merely the first name. The evidence for Imhotep seems to be that he was an architect and an astronomer and a high priest and became a physician only twenty-three hundred years after he was dead.

A dog licks its wound. The physician washes the wound, removes the dead and lacerated tissue, lessens thus at the outset the chance of infection, and should infection in spite of that occur he supplements the body's inherited and acquired immunities from his vials. He trusts his vials, the liquid artificial immunity that is in them, each labeled. And he trusts his syringe. Has reason to.

He knows more than other men both about the defenses built into the living and those he introduces, that fortify a blood-stream with friendly molecules to deal with enemy molecules. Knows about antigens, antibodies, phagocytes, macrophages, blood clotting, transfusion, wound healing, mind healing. Considers how he can make mind-healing help wound-healing. Knows the capacity for repair and growth in individual organs, how far he can count on that power of the liver to reconstruct after injury. How soon after birth he can count on the infant's kidneys to take up their work, take over from the mother's. Knows the protection that is in the essential chemicals in the right foods, how much or how little and what kind a fat man ought to eat, a skinny woman. Knows even—if he has not forgotten and it is not important if he has—the partial pressures of the gases in the air that we breathe via the lungs, at least thinks a moment about it when his cardiac patient is planning a trip to the Andes.

This now requires a different acumen: he knows at what point he should *not* rely on his textbooks of physiology, pathology, medicine, surgery, his journals, but should rely on himself, on his sharpened senses, his premonitions and intuitions, these added to by all manner of humanly useful detail gathered from his experiences. He is alert to such a fact as the day and hour a life is turning irreversibly toward death, sees, that is, when Nature is beginning to desert her investment; or Nature has weighed the situation and intends to protect her investment. In the latter case he plunges, invests all he has, and if Nature is in a mood to accept his money the patient improves. In short, the physician is supporting his hundred or his thousand pieces of immunological information and other information, supporting with his intelligence, his discretion, his manual skill, and his past mistakes.

To the sick man he gives enforced rest, gets him to agree to lie down two hours in the afternoon, or, the contrary, keeps him on his feet, suggests he walk once up and down the hospital hall, tomorrow twice, the day-after-tomorrow up and down the street. Says peremptorily that the patient must stay one more day in the hospital. He humors him out of his consequent bad humor. He baits him, gently. For another patient he reams a prostate, or gets his friend, the specialist, to do it. Or he extracts a splinter. Another friend replaces a heart valve. He himself modifies a milk formula. Quiets a bellyache. Quiets a thought. Quiets a guilt, a shame. Occasionally he launches toward the sky, succeeds in wrapping someone's despair in the gossamer garment of hope. Admittedly, he may damage where he expected to cure. Admittedly, he may accomplish nothing. Undeniably, he has made life last longer for many, and made it more bearable while it lasts.

If he is honest and reads the journals he probably acknowledges why and how he succeeds, what, especially in late years, he owes to physics, chemistry, genetics, to those research workers who may be imagined to have traded for the stimulations of the warmth of the suffering human being the cool stimulations of the laboratory. To say that with less purple, he recognizes what the laboratories of the world have accomplished—for him personally—and he looks toward what they will. He expects the extraordinary. It occurs. All the sciences have joined to make medicine effective. Nevertheless, the physician, if he is adaptable, easily holds his place, does not cringe (or not much) before the more narrowly focused intellects around him. He does not know what they know, does not too much worry that he does not, that he has a less exact but sometimes can claim a wider spread of learning. He prefers to have it that way. He almost deliberately back in high-school, if somewhat vaguely, somewhat self-deludingly, chose to have it that way. He has, throughout, gone on routinely performing the routine acts. He has established the diagnosis. Written the prescription. Pronounced the dead, dead. Signed the certificate. Consoled the family. May have gone to the funeral. Then, at cocktails and in the foyer of the theatre that night, he has done what he could to bolster the dignity of his calling. He has tried to maintain that dignity. He has tried, sometimes, and sometimes falteringly, to maintain its coded and also not-coded rules of conduct. He is apt in consequence to have, but also may not have, a steadily accumulating wisdom. He uses and has used it. He also has used something less: to organize city, state, nation. Sometimes he has thought that city, state, nation were organizing against him, as we all on occasion think; we are human beings, so we think up and down.

The physician is the healing process become self-conscious. He has put mind into the healing process. He does not say this, and it may even not occur to him, but he has put our late-coming human mind into forces as old as evolution. He has allied mind with those blind forces. We all have done that, but some of him has done that somewhat more, and a recognition of this has made, sometimes, a philosopher of him, this doctor. It is a fine good fortune—to be a doctor. It also has sometimes done more mundane things for him, made a traveler of him, made him want to go to find what the rest of the planet thinks of life and of medicine, what the latest medical gains may be, on the spot where they arose. He has traveled to a meeting in Tokyo, in Patagonia, met those other colleagues who sailed on the *Leonardo da Vinci*, those erstwhile students of metabolism who are enjoying a few days of an ultra hotel at government expense. Paracelsus in the sixteenth century went to all the corners of Europe and inquired even, as he said, of the barbers and midwives. Our physician shortly will go to the moon.

Meanwhile the physician will have continued to divide that overall learning into classes of learning, as other professions, will have strengthened medicine, he believes, by narrowing his portion of it, without narrowing himself, hopefully. His technical and theoretical findings he will still have reported in articles and books. Mostly he will have been cultivating his own garden, but sometimes too he will, for an evening or a decade, have been trying to bring together all the gardens, create a single science of what in spite of his trying remains a scattered though now and then an integrated learning. In his unwillingness to compromise the integration, when he is unwilling, as also in his unwillingness to reduce his day to the calendar of a businessman, when he is unwilling, lies medicine's allure.

For these reasons, none very exact, when next you pass an Academy of Medicine on a meeting night and see some not unordinary-looking, not ungentle gentlemen in huddles talking medical discovery or medical politics, one leaning against a wall, one blocking the front door, one draped half around the mailbox, a mushroom of cigarette or other smoke floating over them, all rich but not in the high brackets, no one of them poorly dressed, all scrubbed, particularly the bald ones, their cars out in the street illegally parked, if that sight annoys you suppress your annoyance a moment, remember for that moment that they date from Hippocrates. They may not look it but they date from Hippocrates, and from Imhotep, and Aristotle, and Galen, and Vesalius, and Harvey, and Pasteur, from all who have succored the sick or by their experimentation have supported the physician in his intent to prevent past misery repeating itself in the future, his intent to make human life less pitiable, may make it quite hilarious for a twilight. Much of the past may be embodied in that man, a historic past. He does not often think of this, as most of us, as most men, rarely think of their historic past. Consider though what advantages this one has. Consider what it could do to his mind to live every day near the wonder of the naked human body, that unparalleled machine, but every day also to live near nausea and retching, near misery, to have the difficult privilege of being every day near pain. On that same day he may have too the happy privilege of taking from a human being the worry about his body. Have the happy privilege of that miraculous moment when a newborn literally pops from a womb.

There is the outstanding one—say he is a surgeon. On some frightened midnight we realize that this man is astonishing, that surgery is astonishing, that the economy of this one's talk has been imposed upon him by the abundance of human beings who with their bodies and their minds are leaning on him. He works as the great work, with patience, with intensity, with what swells to be wildness for one half hour. As he has gotten older he has come to resemble the artist. He has grown genially rotund. Last night he made one think suddenly of Johannes Brahms.

# XIV

# NERVE

## Carrying the Torch

# SIGNAL

## Marathon

Third row, aisle seat. The stadium in Athens. The Olympics. In a moment we shall learn who won the Marathon. The crowd is hushed. Now it stands up. Third row, aisle seat, that spectator forgets to stand. He is thinking. He knows he is not tragic, not comic, has not the size of tragedy; at least, though, he is not a coward nor a fool, so knows that he sits alone, and will until that last loneliness in which the greatest chagrin may still be the loneliness. Third row, aisle seat, a good seat, one can see everything, see the excitement in those spectators. He groans with disgust. "An athletic contest." He blurts out that that in his brain is more than a passing of torches, something completely otherwise. He has no idea how to piece it together, body, brain, his single mind, his single thoughts, his single memories, and in his brain his fifteen thousand million units, as in all the other minds and all the other brains. Only the brittlest academician could lightly lay all of this to signaling over nerves, yet every intellect trained to the trade in 1969, were he to look into a human brain, and were he able to look with infinite minuteness, all he would see would be moving signals, nothing else. An eminent Britisher, Charles Sherrington, imagined the signals to become luminous.

Third row, aisle seat, that spectator there wishes he had a humbler system, like that of the cockroach that stepped out of the shadow and was stepped on. Wishes he had a less large ovoid skull for that mad racing to race in. Wishes his small body had not been cursed with his huge head.

Purpose of a single nerve?

Not to make us nervous, of course. A picture that most of us had earliest was of a wire. A flesh wire. Over that go the signals. Another picture might have been of a microscopic track over which there was run a microscopic race. The racer is the signal. It starts, runs with its message, delivers.

An entire nervous-system?

A multitude of tracks over which there is run a multitudinous Marathon. Ten to fifteen thousand million tracks with branch tracks are in the human brain alone. Messages are hurrying everywhere. Each helps to join the north of the body to the south, the east to the west, to blend the total into the affairs of planet, galaxies, universe, the affairs of the inanimate, the animate.

# NEURON

## *Torchbearer*

For years it was thought, by some, that the nerves of the body made a network, all of them connected, nerve-filament to nerve-filament, and over the network the signals raced. The system was built as one, so could do what it often appears to do, operate as one.

The chief defender of the idea of network was Golgi, Italian. The chief opponent was Cajal, Spaniard. Cajal thought the system built in units. Among the human mind's follies and charms is this, that despite having immersed itself in history, having learned over and over that the present usually has regarded the past as ridiculous or puerile or at least quaint, it still can so believe in this particular present, its own. Golgi and Cajal intensely believed, a volatile Italian, a volatile Spaniard.

To Cajal it was units. A unit was a neuron, and a neuron was a nerve-cell with its fibers. The signal raced over the neuron, reached its terminals, and at the end of each there was a cleft, a gap. The electron-microscope in our day can see that gap. Cajal's eye could not. Across it the signal leaps—at least it gets across—in the direction of the next neuron. If enough signals get across enough gaps and add up their voltages the torch is lighted and handed on to the next neuron. Handed-on or not-handed-on, a yes-or-no system. Today, any pipe-smoker on the dinky from Princeton Junction to Princeton without a moment's hesitation would assure us that the nervous-system is built in units.

A unit, a neuron, can be grown. A speck of sterile brain (mouse or hamster or human) or some other speck of nervous-system can be planted in a culture medium such as bacteriologists use to grow bacteria. The speck might also have been put on a sterile glass cover-slip with some of the culture medium, another cover-slip placed over the top, thus the growing directly observed under the microscope. That medium has been stocked with right neuron food, kept at right neuron temperature, and those neurons look then like the nature-grown. Each has its branches. Each has its insulation. If the original speck was taken from the front of the spinal cord, the fibers grow out straight, as they would have in the body, and if from the back of the spinal cord, they loop. One sees there in the culture medium how those units will make their inevitable connections. They were born to go one way. They were born to generate their particular electrical action. They were born to play their small inevitable part in some control of movement, or, if at the head-end of the highest, play their part in some pattern that presumably produces a thought.

In the culture medium other cells grow. Formerly these were believed to serve merely as scaffolding. Later this did not seem true. One well-known investigator insisted that they pulse, beat. This was surprising. Like heart cells, they beat. So there is a slow beat in the food and drink. Beat, beat, beat. Life is different from death. The old Greek, Democritus, sitting with dignity in the stadium, would no doubt have remarked he had understood that fact right along.

Golgi's claim of a network lost. Cajal's claim of a system built in units won. With a vigor that seemed not to weaken as he got older, in more than two hundred articles and up to the last days of his life, Cajal hammered the proof of the neuron doctrine into the thick skull of the compact majority.

Engineering advantages could accrue from a system built in units.

As with telephone, telegraph, railroad, there could be local lines, through lines, lines that converged on large centers like Chicago and Athens and on small centers like Princeton Junction and Charvati, also lines that diverged from those centers. A fraction of the system could be disconnected, reconnected. However, any comparing of the living with the dead has always proved false.

A single unit?

Billions are in our brain. Always a temptation to picture a drawing reproducible in an engraving. Each has a cell-body, and, leading toward that, rootlets, and, leading away, the large branch with those terminal twigs. Rootlets → cell-body → branch with twigs. The junction between any twig and the next unit is often called a valve. Because of that valve—synapse—the signal is forced to go one direction. The overall direction is: rootlets, cell-body, branch with twigs. The rootlets are named *dendrites,* the cell-body *cell-body,* the branch *axon.* An axon may be short, may be long, may reach from a man's spinal cord to the tip of his middle finger. Each neuron operates on its own power, pays the cost for the signal's advance over its lap, and at the end of that lap helps to hand the torch on to the next neuron.

In the living body each neuron is born, endures, dies, in the language of the Persian. It dies privately, or it dies in some great crowd, some mass decimation, as when a blood-vessel bursts in an elderly gentleman's brain, or gets plugged by clot, or is hardened and narrowed and then gets plugged, the gentleman's poise without notice sprung from under him by the trap that is waiting for us all. He fell on some dark street. Or, he put down his head on his desk and the head just stayed there. Or, he had a convulsion in the bright sunlight, leaned to one side, was still. Or, he was lying in his bed to begin with, was found there next morning, and his friends next evening to get rid of unpleasant thoughts said he was lucky to die that way. Hippocrates knew strokes. They happened now and then in the Athenian stadium, a dead or dying Greek was carried out and the games went on.

# GALVANI

## *Thundercloud and Athletic Leg*

Twenty-two centuries after the Athens of Pericles, the year 1737, there was angled out of a womb Aloisio Galvani. (Aloisio, or Luigi.) He was Italian, had the quality of the Renaissance Italian; that is, he was Italian, Italian, Italian. It was close to the time that Tristram Shandy was having the same experience, and likely too that Aloisio's father was as fluttery as Tristram's had been that night in the Shandy kitchen, expectant.

Some envious persons, so many years later, ladies very likely, hinted it was Aloisio's wife, Lucia, deserved the credit for the great discovery, and undoubtedly she and Aldini, Aloisio's windy nephew, were in the room. Three—and a frog.

One of the three held a scalpel. The frog had been skinned. The large nerve that leads to the muscles of thigh and leg had been exposed. Close by was an electrostatic machine generating electricity, and each time the machine sparked, thigh and leg twitched. From Galvani's careful account it might be inferred that the metal of the scalpel picked up the electricity through the air, conducted it to the shank, that twitched. Galvani would write a whole book on the subject. His first article reporting his experience with frog legs appeared in the *Proceedings* of the Academy of Bologna; he was Professor of Anatomy at Bologna. After a while the galvanic battery would be named for him.

As for the scalpel, it needed to be grounded, could not be merely a scalpel held in space, and not enough either for the experimenter's hand to touch the handle, that was bone, but must touch the metal that riveted the bone to the metal of the blade. Two dissimilar metals. Alessandro Volta seized on that. He said it was the contact of the dissimilar metals excited the twitch—physical electricity. Galvani said it was the frog—animal electricity. Between Volta and Galvani there blew up one of the famous debates of science. Neither debater ever vanquished the other. Galvani sometimes was represented by his nephew, he himself too gentle and shy, or some other reason, and represented after his death by his nephew. The debate added fire to the combustible talents of Volta and Galvani. Volta laid wetted cardboard between dissimilar metals, and from that came the voltaic pile, the electric battery, a current that flows instead of sparks, the transatlantic cable (likened later to a nerve), and toasters for breakfast, and Broadway by night; the Sphinx and the Great Pyramids near Cairo that had dreamed on the dark desert for three thousand years would be waked by floodlight; the Parthenon in Athens would be lit like a ball park. From Galvani? Galvani would have begun the electrical investigation of the living. There would follow from that the recording of the rhythm of a million beating human hearts; a steady adding up of the number of hearts with abnormal rhythm that were electrically shocked back to normal; the adding up of the recordings of the rhythm of who knows how many brains; the stimulating for localization during surgery of tumorous or injured or scarred brains; and in colleges and universities the stimulating of millions and millions and millions of frog shanks. Electrical studies would grow more complex, more delicate, the electrodes finer, the electrometers more sensitive. William Harvey "was wont to say that man was but a great, mischievous baboon," and someone much later would say he was a bag of water with items dissolved in the water, and it was in the interval between those descriptions that Galvani introduced his electricity. A mischievous electrified baboonish bag of water—man's mind has gotten him that far. That eminent Britisher of the twentieth century would nevertheless get him an inch farther, would have imagined the electrical flows of the brain to have become visible, and, ah, if all those currents and cross-currents could become visible, what shimmering ghosts would nod to each other on the roads at midnight and each Johnny be as amazed at himself as he has the right to be and seldom is.

There remains the anecdote of the frog shanks. They were intended for dinner, a tidbit for Lucia, who was ill, were hanging by a copper hook from an iron railing, a copper-iron coupling, and they twitched. "Dissimilar metals," said Volta. "Animal electricity," said Galvani. The arguers still were arguing. The shanks were dead. They behaved alive. During a thunderstorm they twitched crazily, big cloud in the sky was inducing electricity in small shanks on earth and at the flash of the lightning the electricity discharged and the shanks twitched. They twitched also in fair weather, but not so predictably, said Galvani. Fortunate that Volta and Galvani were Italian, that they did their thinking heated by their passions, because the point they were making was an important point in an unimportant world. However, the eighteenth century did not yet think the world unimportant, as Athens had not. God still was in His heaven, as the gods had always been on Olympus, and whether the world is important or is not important, electricity, the demoniac thing, and nerve, the life-inspired thread, were promenading together and no man could be sure whither that promenade would lead.

# NERVE-IMPULSE

## Swift and Brief

Up in the skull the demoniac has always been ready to haunt the flesh. The number of nerve-fibers is enormous, the number of electrical events enormous, difficulties of understanding enormous.

Even over a single nerve-fiber electrical events are only comparatively simple.

But the nerve-fiber, the neuron, and the chemical-electrical event that passes along it are approachable. That advancing charge of electrical potential, the impulse, is of measurable length. Each impulse is an infinitesimal fraction of the electrical wildness inside each of our heads.

Over each neuron of the small green frog the nerve-impulse passes too. Impulse follows impulse. Each is a bit of the total, briefer than Shakespeare's brief candle. Notwithstanding, by that brief candle we know when, where, what. No, not quite what.

The nerve-impulse has been studied for years. Physiologists must all have been pleased when the 1963 Nobel Prize for Physiology and Medicine went to three students of this small event.

The gross nerve—the transatlantic cable—is a bundle of many nerve-fibers. Anatomists work with that cable. Surgeons repair it.

Physiologists rip it out of the frog. They also tease out a single fiber. The fiber is wrapped in a membrane. The ordinary microscope cannot see that membrane but the electron-microscope can. Muscle-fibers have a membrane also. It is a close-fitting sheath-gown some ten-millionths of one millimeter thick, and important, because over the nerve-fiber membrane the impulse races.

When the membrane is at rest there is an electrical difference between outside and inside. Textbooks have diagrams with plus signs outside, minus signs inside. Now, let that membrane be touched by electricity, or by a hope, or by a disappointment, and instantly plus signs behave as if minus signs had neutralized them, even reversed them, minus signs outside, plus signs inside. For that instant there has been created an electrical hole.

Beneath the stress of our world, beneath the turmoil of our life, there is this methodic production of electrical holes. Promptly neighboring plus signs pour into the hole, patch it, but leave a hole from where they took the patchings. Pour. Patch. Thus does a hole advance. Thus does the nerve-impulse become an advancing electrical leak. Something like that.

What causes the plus or minus atoms, the ions, to pass in and out of the membrane? What causes the advancing shift? What makes the membrane now let ions in, now keep ions out? That was and is and will be investigated, also seminared, and Socrates and Plato will seem not to have lived in vain.

One species of crab had thirteen times more of one kind of ion outside its nerve-fiber membranes than in all of its blood. While a membrane is in action one one-millionth of the kind of ion that was inside goes outside, and much of another kind, sodium, that was outside goes inside. That sodium needs to be gotten out again. How? The hypothesis is that there is a pump right in the membrane. The impulse arrives, the impulse departs, the ions have exchanged places, the pump pumped, the next impulse arrives, and the total event has required—hold tight to your hat—one one-thousandth of one second.

The bulk of the fiber—as against the membrane that wraps it—is maintaining a busy chemistry. It is doing metabolic work. It is keeping the fiber alive, producing the resting state, the difference between outside and inside, so the impulse can upset it, produce the active state. "We are such things as dreams," and so forth.

A species of squid that goes down to the sea has a giant fiber whose bulk has been squeezed out, studied, and the membrane, the sheath-gown, also studied.

The nerve-fiber, being living, consumes oxygen, gives off carbon dioxide, produces heat. Heat-produced is another such minute quantity, one-ten-millionth of a small calorie for one gram of nerve-fiber. The carbon dioxide is another such minute quantity. Thales, the Greek, the measurer, also did not live in vain.

# VELOCITY

## *Hundred-meter Dash*

How fast is the race of the nerve-impulse along the nerve-fiber? Over the long tracks from fingertips and toetips or the short tracks through the jungle of spinal cord and brain, what is the velocity?

A century ago scientists thought such speeds never could be measured. Over the great spaces of the heavens, yes, but over the small spaces of a living body, no. Six years after Johannes Müller gave this as his opinion Hermann Helmholtz made the measurement. Helmholtz was a quiet man. He freed a nerve in a frog. It was a gross nerve, a bundle of nerve-fibers, and he let it stay attached to its muscle. The apparatus was arranged so that marks could be written on the smoked paper of a revolving drum, one mark when the nerve was stimulated, then, after the nerve-impulse had run along the fiber and excited the muscle to contract, another mark. Two marks. With everything ready, Helmholtz stimulated the nerve as far as possible from the muscle, five centimeters, got those two marks, then near the muscle, one centimeter, got those two marks. A longer time elapsed between the former two than between the latter two—because the impulse had to travel the longer stretch of nerve. He measured the stretch. After that the calculation needed no Helmholtz.

Thirty meters per second or thereabouts was the velocity for frog nerve, approximately a mile a minute. One hundred meters per second is possible in a large motor nerve of the human body, approximately two hundred miles per hour, sixty miles faster than the Bullet Train between Tokyo and Osaka. A jet plane is faster still, many times, a space-ship many many times, and light looks over its shoulder and laughs. The brain is a slow instrument, marvelous but slow, appropriately slow, to keep to the pace of our lives. If our brain were not able to hold back to the contraction-time of our muscles, and to the time it takes our joints to get out of each other's way, we should all fall on our faces as we did when we were babies.

A nerve and a track of gunpowder have often been compared. Ignite the powder at one end of the track and the flame burns its way to the other, that burning having a velocity that also can be measured. Each particle of powder ignites the particle next, the flame remaining steady, does not die out as would an echo that in every instant draws its energy from the original sound source. (The nerve-impulse does not draw its energy from the original stimulus, but from yesterday's breakfast.) If the powder is wetted somewhere along the track, enough to slow the burning but not stop it, once that wetted spot is passed, the burning will continue at its previous speed. It would be much the same for a nerve wetted by a drop of ether.

Anno Domini, 1969, velocity has for years been measured on single nerve-fibers without a muscle attached. A nerve-trunk like that of the frog would have many sizes of fibers. Velocity over thick fibers is higher than over thin. Size and speed bear a strict relation. That discovery was not trifling. It brought a Nobel Prize to America. An almost frightening array of speeds has at every instant been operating inside of you, and the polecat. In a nerve-trunk there may be thousands of sizes, thousands of speeds, hence the possibility of sensitive timing, of averaging, hence flexibility, hence nuance, hence life, hence, imaginably, mind.

Instead of the heavy-muscle lever that Helmholtz used, the researcher today has the most weightless lever in the world, a beam of electrons.

# VOLLEY

## *Not Singles But a Team*

In the living body the nerve-impulses do not travel singly. They travel in volleys. "They come not single spies, but in battalions." It is volleys racing in our brain, our spinal cord, in the far-flung remotenesses of us. If one were able always to conceive of the body as machine, bury it in the mechanistic, see a machine raking garden leaves, everything would remain comparatively simple, but if one is not able quite to do that, everything becomes fantastic.

The fact of volleys was first recognized by an Englishman at Cambridge. The Englishman's crucial experiment included a loudspeaker. Out of the loudspeaker spoke the volleys, click, click, click, for each nerve-impulse a click, more and more nerve-impulses over more and more nerve-fibers. The Englishman could of course amplify the clicks, make them louder, hear them better, by howsoever much he liked.

First, the experimenter converted the electricity of the impulses into flashes of light. Second, photographed the flashes. Third, transformed the photographed flashes to clicks. Fourth, made the clicks permanent on a Gramophone record. And because he had the Gramophone record he could any afternoon play back to himself what had traveled over the nerve of his colleague the cat. Today we tape colleague and cat, with or without permission, yours is mine, mine is yours. The Greek was an amateur in his sense of freedom.

An experiment like the Englishman's had been performed also on the muscle in a human arm, on the membranes. First, the experimenter fitted a metal core into a hypodermic needle. Second, plunged the needle into an arm. Third, manipulated the needle until it had a right contact on some muscle surface. Fourth, brought in the loudspeaker. And out of the loudspeaker dutifully spoke the clicks, each click following the click before, soldiers again. If the arm was bent, more and more muscle-fibers were included, and more and more impulses traveled over each fiber. The clicks themselves were not different, only faster. A literary physiologist stated this felicitously. "The general effect is that of a machine-gun firing at increasing speed but the noise made by the explosion of each cartridge does not alter."

# ENDPLATE

## *Finish Line*

A friend of Claude Bernard sent him some arrows tipped with a gummy substance, curare. The arrows were from South America. They were used by the Indians of the Orinoco River Valley to shoot birds through blowguns. Bernard related how in his laboratory in Paris he scratched a rabbit's back, introduced some of the gummy substance into the scratch, disturbed the rabbit so slightly that it did not stop chewing, shuffled to a corner, settled down, its head drooped, its ears lay back, it turned on its flank, was dead. The death occurred six minutes after the scratch.

Where in the animal's body had that poison acted? In the nervous-system, but where?

Bernard's reasoning here often has been cited as an example of how the scientist at his best comes to a conclusion. For twenty years he returned to the problems of curare, using various animals, and students the world over have performed experiments that paralleled his.

A frog places itself at a student's disposal. Actually, he grabs it, lithe and pliant and slippery. In both legs, right and left, he exposes the calf muscle and the large nerve that leads to it. Previously with a needle he has mashed the frog's brain against the inside of its cranium (pithed is the word), and so threw the frog into a state like that of a man who has suffered a massive cerebral hemorrhage. Next, the student ties a ligature tightly around the frog's right thigh, stopping all blood-flow down the right leg, the nerve not included in the tie, so that nerve still is supplied with blood. At this stage, if the student stimulates the nerve on the right side, that right calf-muscle contracts, and, of course, if he stimulates the left nerve that calf-muscle contracts. Next, the student injects the curare under the frog's skin. Soon the frog becomes paralyzed everywhere except the right leg, which was tied off from the circulating blood, the poison therefore not going to it, but going everywhere else, including the left leg. Nevertheless, if the left calf-muscle, the muscle itself, is directly stimulated with electricity it contracts, as does also the right calf-muscle. Plainly, curare is not a muscle-poison.

Next, the student stimulates the nerve to the right leg, and the right calf-muscle again contracts, even though the poison has been carried to that nerve because it was not included in the tie. Plainly, curare is not a nerve-poison.

When, however, the nerve where there is no tie is stimulated, that muscle does not contract. So curare, that does not poison muscle and does not poison nerve, does poison something between muscle and nerve.

Aristotle would have been fascinated, as through the centuries would have been Harvey, Leonardo, Paracelsus.

Today, that tissue between muscle and nerve has come to be accepted as a special part of the architecture—the endplate.

Bernard in his time could know only that something had been poisoned in the general neighborhood where nerve meets muscle. He could not have called it an endplate. He could not with present-day minuteness have examined the physical and chemical complexity, but he grasped the large facts, was the first to do so, and the pattern of his thinking has been handed down. Wherever a nerve-fiber meets a muscle-fiber, that finish line on the muscle side is an endplate. If enough impulses travel to enough endplates and enough muscle-fibers contract, an ankle turns, an eyelid twitches. Bernard's rabbit died of suffocation because the endplates of its breathing muscles were poisoned and no nerve-impulses reached them, so the animal could not breathe. Death was stirless because the rabbit was not able to stir. It had been deprived of the power.

Curare and its substitutes came into wide use and wide thinking. By surgeons, when they wished to diminish disturbing contractions during an operation. By psychiatrists, when in shock treatment they wished to prevent the convulsions that in earlier days fractured bones. By physiologists, when they wished an animal not to struggle. This special tissue, the endplate, has drawn men's attention to the nerve-side, to the muscle-side, to the baffling detail of structure in both places. Has drawn attention to the mechanism of a miserable human disease. Has drawn attention to local endplate chemistry and physics, and this knowledge has been carried elsewhere in the nervous-system. Indeed, everything even remotely relevant has been and still is investigated with a painstakingness that might seem absurd to the runner of a hundred-meter dash, or to a two-hundred-forty-pound fullback.

# SYNAPSE
## *Runner Touches Runner*

The finish line need not be an endplate, where nerve meets muscle. It can be a synapse, where nerve meets nerve.

Sherrington, the Englishman, gave the name synapse to that junction, and Cajal, the Spaniard, made it a reality.

There is a gap at the synapse. The electron-microscope sees that gap. Messages must get across it. There is delay. There are possibilities of modification of the message.

Many such gaps, such junctions, such synapses, are everywhere in the nervous-system. At some points they appear hopelessly matted-together. But human ingenuity has kept at them.

The physiologist-biochemist-microscopist has stuck a micro-electrode into the terminal part of the neuron. Has stuck it right into that fine gap where neuron meets neuron. Has stuck in two electrodes, one into the one neuron, the other into the other neuron, the two that meet at the synapse. The techniques may have grown commonplace to him. The idea cannot be commonplace to us.

In our human brain are an estimated ten to fifteen thousand million neurons. Each has its dendrites, its axon, and the axon has its terminal twigs, hence almost inestimable numbers of synapses. Hence the matting-together. We could think it was disorder except that we know that nowhere in this compulsive naturalistic world is there disorder. Each synapse has its minute important electro-chemical decisions to make. There the traces of the world upon the brain are written and rewritten, and it could help us to understand how from the multitude of connections the strangeness of a human mind might be built.

The nerve-impulse races over a neuron, reaches a terminal branch. Much immediately occurs. There is the membrane of the neuron. Beyond it is the gap, the cleft, and beyond that is the membrane of the next neuron. At all membranes there are the problems of electro-chemical transport. It is the adding-up of the electrical force of many membranes that determines whether the torch is passed on or not passed on. Yes or no. The runner touches the next runner and that touched runner runs or does not. Many must touch if he is to run. Once he starts running he keeps on to the end of his lap. Once a nerve-impulse heads down the neuron it keeps on for the length of that neuron.

Dendrites → cell body → axon → terminal twigs.

A spray of terminal twigs breaks from the end of the axon. Each twig has an endfoot (not endplate, endfoot). That endfoot frequently has been called knob. Six varieties of knobs, probably. Variety may or may not make a difference, and this is not important for the idea, but it is important that by a definite piece of architecture the contact of neuron with neuron is established. Add another detail. Blebs are at the end of each twig. The electron-microscope sees them. They store chemicals. There is chemical play—ions. There is the play of electrical charges. There is the rise of electrical potential. At an instant there is the dash across the cleft.

Four hundred endfeet may come down on one neuron. Up to a thousand have been counted.

As the region of an endfoot is neared there is increased chemical activity, so that even if the torch is not passed on, the general neighborhood has what could be called a neuron warmth, enough to cause other neurons of that neighborhood to be readier to speed some nearby torch.

The rowdy hurling of a vowel from the hoarse throat of a candidate at a Meet-the-Press, or a senator at a senate investigation, or a rioter prodding rioters, each has beneath it this sensitive neuron stir.

What the researcher does to investigate such a neighborhood may seem clumsy. With a gross electrode he touches a gross nerve in a cat. Or a single fiber. The electrical message leads into the spinal cord, through the cord, is picked up by a nerve leading out of the cord. How many micro-micro-micro-Marathon-miles have been traveled? How many synapses? The delay at a synapse is known from other experiments. The time over short or long stretches of thin or thick fibers is known. Each fiber tapers toward its end and the message slowed. The race from Marathon to Athens leads through crooked-streeted Charvati and other crooked-streeted villages, but also over broad uninterrupted highways. Tomorrow morning the researcher will cut into a new cat. The government supplies the cats. Thucydides would have reported these matters coldly and accurately, Herodotus fabulously.

# CAJAL

## *Observer at the Olympics*

Any book dealing with the synapse, or the neuron, or some large or some small structure in the architecture, eventually also the chemistry inside of that architecture, might overlook a name, but not Santiago Ramón y Cajal. Again and again that name. Usually it is just Cajal. If the frontispiece has his photograph he is dark-complexioned, hair cut close to his compact head, soft tissues of his face laid sparely on the bone, body leaning forward, a tenseness that eludes no camera, definitely a gentleman, a Spaniard. If the book has engravings of his drawings of brain or spinal cord or single nerve-cells, the engravings also are apt to be his. Even in the mountainous output of our time where no stone is left unturned, where every book has illustrations, always the latest, rarely are his redrawn, rarely have they *after Cajal* underneath them, but are reproduced as he left them. References to his discoveries stack up until anyone would reflect on how many men sometimes may spend what one man earns.

A sweating day laborer he seems but neat, a creative mother but male, wildly gay at times, if reports are true, a canny old man at the end.

The place of his birth was a somber village, a lost spot, an island of Aragon in Navarre, the year 1852. His father had to drudge to get his doctor's diploma, to become worthy to heal the sick and practice the art of medicine, as speakers say at medical commencements; then followed the years with the sick, the fees small, the doctor tired when after dark he was through with the day, then supper, then the Spanish nights, then the children to be supported, no compensation to be expected from this one haunted by dreams of being the great painter of Spain, the modern Velázquez, and stubborn about his dreams. Two wills, father's and son's, pitted against each other, the father winning but only because he was the older and had the greater experience, a victory that must be won over and over and never really was won. In the son's memoirs he described that contest. All one can say is that it was not as one-sided as a bullfight; the father tried him in this school, tried him in that, the passion for painting flogged out of him, unimaginative and dour priests keeping him at Latin and whatever boys are kept at for the good of their souls. Regularly the pent-up feelings burst through in escapades punished by starvation, by imprisonment, once a literal prison with bars, for a child. Possibly he deserved the floggings. Possibly he required them. Possibly they drove into genius the discipline necessary for it to survive. What is sure—he got them. The memoirs tell how he saw miracle in the discharge of any firearm, used always to admire his father's gun kept out of reach. He took timber and an auger from a carpenter's job, strengthened the timber with metal, produced a cannon, hoisted that to the orchard wall. Loaded. Aimed. Fired. At a neighbor's new gate. And there fell down on the Spanish earth the wreckage, and on him the bitterness, this in a village that must save every stick and stone. He was eleven.

One reads the robust tale, sees it from the father's side, thinks the son had something of the devil and the father the good heart; then one turns the tale around, sees it from the son's side.

Came a day when the son too got his diploma. *Doctor Santiago Ramón y Cajal.* All things do come to pass. His advance through the medical school had been distinguished because of anatomy. He had his eye for shapes, could draw what he saw, the eye focused not on the Spanish landscape or an hidalgo with a horse, as he had dreamed, but on a dead brain, or part of a microscopic field, a single nerve-cell, many nerve-cells. There the eye stayed. A talent tramped down at one point had pushed up its head at another. Son and father together prepared their anatomic atlas.

Besides being a student of the intricacies of that system of the body that most underlies mind, and an artist fascinated by the living machine, and an artisan who could fabricate the tools to study it, Cajal was a patriot, passion in that as in everything. He was not obligated to enter the army but demanded they enlist him. This was at the time Spanish-American animosity was brewing. Most Americans would be vague if asked what the eventual war was about. Cajal never appreciated America. The army sent him to Cuba, where he contracted malaria, as many did, dysentery on top of the malaria. In those pages the memoirs are not pleasant to read, the mind of the patriot having to rise above and pull up after it the ailing body. He was shipped back to Spain. No one would have trouble imagining that father waiting for that son. Each had used the other to toughen himself. At home he rested. Then he started to work, and started to hemorrhage from his lungs, probably the tuberculosis that frequently followed malaria though there never was a positive diagnosis. Illness does not kill that kind. He recovered his health and up the ladder of Spanish universities he practically ran, soon was Professor of Anatomy at Madrid, his knowledge of the nervous-system amassing day and night because day-and-night was his pace. He invented methods, altered them, flung them aside, took up others, all with feverishness yet with the painstakingness of the painstaking microscopist, cut section after section, stained the sections, tracked the single neuron through the wilderness of spinal cord and brain. All present-day advances in these fields somewhere rest on him, neuroanatomy, neurophysiology, neurology, psychology, even psychiatry, as well as the late-coming molecular neurosciences, all have a greater firmness because of the foundation he put under them. What he discovered in one animal form he verified in others, mouse, rat, rabbit, guinea pig, man, when he could get a man, a dead one. He had the shrewdness to examine embryos at that stage of development when the neuron is small and stands alone like a tree in a white winter. Saw what other men had not been able to. Saw with the ordinary light-microscope. There was as yet no electron-microscope. There was a method of Golgi's, the Golgi silver method, abandoned by that eminent Italian because of its difficulty;

Cajal wrestled with it, forced it to produce. He took his slides to science meetings, struck fellow scientists with amazement, now and then an ass with amusement. When someone could not see what he saw—some pool of cells, some particle in a cell, some path in spinal cord or brain—out came his impatient pencil and with quick strokes he indicated where to look, made the significant visible to slower eyes.

No one has brought to the nervous-system more of the reverence it deserves; no one has been more humble before it, and humble before nothing else. He cleared away the weeds. Actually, there are no weeds, but there is the unexplored and genius knows what of that can for the time be put aside. Without ever saying it, without ever being concerned with it, possibly, he was helping to build that flesh-bridge which would join nervous-system to behavior, to language, to thought. Farther than anyone Cajal comprehended the minutiae of this one base of our being, lessened our bewilderment toward it, and in the respect that brain relates to mind gave mind an earthiness.

His discoveries he published in a journal that he launched, virtually wrote, paid to have printed. Its text was Spanish. The world was not reading Spanish and after various editorial evolution the journal was French. "A bitter thought," he says in the memoirs. With Cajal writing was for the sake either of clear statement of impersonal finding or clear statement of personal recollection, never writing for writing's sake, but he did possess that first principle of high style, went in every sentence straight at what that sentence wanted to say. Drama was in his discoveries, drama in his life, and it can be claimed for the memoirs that they do not smother drama. When he had his microscopic sections ready, his drawings ready, his engravings ready, he wrote. His entire life seems at the point of pencil or pen. Every year we are more convinced that Ramón y Cajal was one of the greatest of all anatomists, Spain's greatest man of science since the Renaissance, the all-time all-world greatest student of the detail of the nervous-system.

He won a Nobel Prize in 1906 but had to share it with Golgi, who probably hated him and whom he probably hated. Golgi's Nobel address was a polemic against the neuron doctrine (a nervous-system built in units) that Cajal's lifelong labor had proven. Such was the atmosphere that year in Stockholm. Too bad. Too stupidly customary. Already as a child Cajal seemed to know that there can be no perfect moment in an extraordinary man's life. In a fool's there can be. Homer's gods up at their Olympic sport always must spoil something somewhat. During his last illness he spattered the wall of his sickroom with ink. Lived with a pen, died with a pen, and the subject of his year-in-year-out thesis was brain and spinal cord and neuron viewed from one end of the barrel of a microscope. The year of his death was 1934.

# INHIBITION

## Handicap

The idea of excitation gives no one much trouble—something excites something. A stimulus excites. Inhibition may be more troublesome, but inhibition is common, operates everywhere in the body. All movement—the way at this moment a reader turns a page—must include a force opposite to excitation. That is inhibition. It long had been studied by the physiologist world, with some word-spinning, some failure, some success, increasing refinement of detail. In Moscow an international conference, November 1963, reported in *Science* in February 1964, was given over entirely to inhibition. The conference was planned to honor the man who one hundred years earlier was claimed to have discovered inhibition—specifically that high levels of the nervous-system inhibit the low. (Inhibition = to delay or stop.) This is true in the hedgehog and the mouse as well as in us.

Excitation. Inhibition.

Both are positive. Both are operating more widely in the body than one a priori thinks. Both operate together: in our chest to coordinate the expansion of our lungs with the movement of our vocal cords; in the circulation to coordinate the heart's beating with our breathing; in every skeletal muscle to coordinate its contraction with its antagonist muscle.

One could imagine two kinds of throttles, chemical throttles, and electrical throttles, depending upon which kind for the moment one is choosing to give one's attention. More than two kinds; at least, more than two kinds of inhibition are postulated, though hereabouts there is murkiness. In any event, the machine employs as positive forces both excitation and inhibition. "A machine which breathes, a piece of animated property," Aristotle said of a slave, making us realize that that sharp Greek intellect thought machine when it thought body; also, that intellect was sufficiently casual about slavery. In our century one physiologist, Pavlov, as we shall learn, considered sleep an inhibition that spread over the brain, a more or less global inhibition.

But the inhibition that has been investigated far and wide, Rocke-feller Institute and Australia, is by no means global. It goes on in every nook and cranny of us independently of every other nook and cranny. Without inhibition Phidias could not have lifted his mallet, Demosthenes not lifted his voice, no twentieth-century white rat run a race. The white rat is a professional. He will run ten miles in twenty-four hours if encouraged—a dishonest word—and during every inch of that ten miles, which is on a treadmill, that rat is employing inhibition, slows something, reduces something to half-action, to no-action, is always inhibiting one part of the machine to keep it out of the way of another. Nowhere is the machine permitted to interfere with itself.

This brake-action (to get it as clear as one can without belaboring it into more of that murkiness) is not simply a stopping. And it is not a sudden drawing-back from, say, an onrushing automobile. A drawing-back is only a reversal of the direction of the engine. The engine was running forward, now is running backward, and, as for inhibition, it is acting in both actions. It takes place inside the forward and inside the backward. It is a decreasing of tension at one point of the nerve-muscle machinery so there may be no opposition at another point where there is at the instant an increasing tension.

All that creeps or walks or flies over the earth requires such in-hibition. The Greek runner of the Marathon requires it, but also the Greek philosopher who sits contemplating Fate, because to sit is to act, is to maintain a posture, is to have some throttles in while others are out. Man-made machines do the same.

In our century another physiologist, Sherrington, considered in-hibition to create just the opposite of the sleep that Pavlov talked about. Pavlov was a Nobel Prize winner. Sherrington was a Nobel Prize winner. Sherrington considered inhibition somehow to create the awake mind, though it may not be real clear what he meant.

Later than either of the above great physiologists, another, a third winner of a Nobel Prize, a clear and graphic thinker, told how in an experiment on hearing he put loud speakers one next another in a row, and found that, if he placed himself near the row, instead of receiving a thundering attack, he simply heard that loudspeaker that was nearest, the others were being blotted out, inhibited.

An early recognition of the fact of inhibition was in the heart. That inhibition was carried via a frank inhibitory nerve, the vagus. The vagus also had excitatory fibers, but each single fiber of it was one or the other, excitatory or inhibitory, the inhibitory dominant for the conditions of the experiment. Therefore when the vagus was stimulated the beat of the heart slowed or stopped. Crabs have frank inhibitory nerves that go to the muscles of their body-wall. These muscles also require at the proper moment to be gotten out of the way of other muscles, require to be inhibited in order that the total crab will be kept in a proper crab-motion. Human beings have no such frank inhibitory nerves to the body-wall, but their bodies have the same need as the crab's, their contracting muscles also must not interfere or be interfered with. For this the gearshifting takes place inside the centers of the spinal cord and brain. Physiologists speak accordingly of central inhibition.

Recently inhibition was recognized as a chemical. It seeped from nerve-endings. Excitation seeped also. Those chemicals altered the situation at the junction, at the synapse. They altered the local electrical state.

Because an advancing nerve-impulse is an advancing wiping-out of the plus-minus difference between the outside and the inside of the membrane of the nerve-fiber, if that difference were exaggerated, which would be hyperpolarization, there would be no wiping-out, or a diminished wiping-out, hence no impulses or fewer impulses, hence degrees of inhibition. When those fine electrodes that were spoken of were placed one inside and one outside of that sheath-gown, that wrapper around the larger nerve-fiber of the squid, it brought supporting evidence.

So, to repeat, largely to keep up the illusion that we know what we are talking about, inhibition operates hand in glove with excitation. It exercises its peculiar force (peculiar as seen through skulls that are no windowpanes) in all parts of all the bodies that creep or run or fly over the globe. The bodies are always proclaiming that they want nothing so much as freedom. Freedom, freedom, freedom. Nevertheless they cling to the earth. Everywhere the earth-bound are checked inside themselves. Everywhere inhibition is a first principle. A sour temperament might have thought of it as only a myriadfold hindering, that even down in the minutiae there is ball-and-chain, that life could be construed as a distributed all-present penitentiary sentence. One can learn in a penitentiary. Learning might be a reason for our having come alive. That might be considered a strange reason.

# REFLEX

## *"Des Passions de l'Ame"*

So many baby fingers have hopped away from so many hot stoves by a compounding of reflexes, accompanied by pain, accompanied by scolding, that it could seem obvious what was Nature's intent for the reflex.

A neuron is a nerve-cell. It has its cell-body and its fibers and, as Cajal taught us, is a unit structure of the nervous-system. The reflex employs at least two neurons. A signal travels to a center, is re-routed, travels to a muscle or a gland. The muscle contracts. The gland secretes. The contracting and the secreting are unit responses, and the total act is a reflex. A performance is more. At the barest a performance is some summing of reflexes. One dapper neurologist defined life as such a summing, and what an impoverished definition that was.

Unit structure = neuron. Unit response = reflex. Unit performance = life.

Zeus coughs. Aphrodite yawns. Those are Olympian performances employing Olympian reflexes employing Olympian neurons. (No mortal can comprehend the gods entirely.) Zeus rises to speak, and coughs. Aphrodite disrobes, and yawns. Aphrodite is not unlike the naked beauty Praxiteles, the sculptor, yesterday morning, summer 340 B.C., hired as a model for his Aphrodite of Cnidus. Yesterday evening Praxiteles went to the stadium, watched the glistening runners, last night sat in the theatre of Dionysus at a revival of Sophocles' *Oedipus Rex,* watched the actors, watched their bodies within their masks, watched the reflexes. Praxiteles is always about his business. Praxiteles is a visual animal.

A reflex can be described a different way. Something strikes a living body, the mechanical energy of the stroke is transformed to chemical energy, to electrical energy, signals start racing over nerves. Torches in many places break into flame. The brain glows— to that imaginery eye looking into it. The race of the single torch over the single neuron depends always on energy generated within the chemistry of that neuron. Torch follows torch follows torch. Signal follows signal follows signal. Neuron meets neuron meets neuron. Something enters the nervous-system, something comes out, the in-out being the reflex, and when reflex is integrated with reflex with reflex, that is the performance.

It is sultry in the Athenian stadium. The same spectator who lolled, still lolls, still is bored at athletics. A beam of light from a metal shield strikes his eyes, starts reflexes, causes muscle-fibers inside his eyes to contract, his pupils to get smaller, those delicate reflexes exciting other reflexes, those still others, and in consequence the head of the spectator turns and lifts toward that higher stadium of Olympus. His tongue mutters: "Zeus thundered, didn't he?"

In Greenwich Village, A.D. 1969, a member of the committee suddenly rouses, looks hopefully across the street into a second-story window, whispers to the not uninterested committeeman next him: "That model stretching and yawning—charming."

An engineer's definition of the reflex is terser, three items: an input, a gearshift, an output. The input could be anything, the setting down of one foot of an Athenian fly onto one hair of the flank of a dog who is sleeping on a marble step. The fly has been buzzing around a pillar of the Parthenon. The fly walks along several hairs. That is too much. That in the nervous-system is summation. The dog, a hound, immediately has Zeus-only-knows-what signals dispatched here and there, gears shifted, and as by a fearful summons a realignment of his nervous machinery, a tensing of his muscles, then the output, that ancient response that in all dogs seems excessive, the scratch, and the multiplication of the scratch, the impatient rhythmic thump. Effect is out of proportion to cause: slight energy applied, the weight of a fly, and, as can happen in the nervous-system, great energy released. Effect is really not out of proportion: that fly has *six* feet.

Everyone remembers his physician testing his reflexes. The physician taps the lower right of the four quarters of the belly wall, and under the skin a muscle leaps up like a toe under a sheet. Taps a wrist and a hand rotates. Taps the body head to heels and jerks followed his hammer. Strokes the sole of a foot. Strokes the palm of a hand. Draws a wooden applicator between closed lips. Blows his breath into a man's eye—impudent. For each of these stimuli there is the mechanical response whose place among the patterns of the daily body anyone can make out. The physician is testing the nervous-system. He may not have the discipline for each reflex to track in his mind the nerve-paths thrown into action. He recognizes the artificiality—this conventional picking of one reflex from many. In life they are always tightly fused one to the other, the body always a total, and this piecemeal testing is partly only to appraise the state of the total.

When Praxiteles of Athens was twenty-one centuries dead René Descartes of Paris wrote the first downright description of a human reflex:

> If someone quickly advances his hand toward our eyes, as if to strike, even though we know he is our friend, and that he does it only in jest . . . we have nevertheless difficulty in keeping ourselves from closing our eyes, which proves that it is not at all the intervention of our spirit which closed them . . . but it is because the machine of our body is so built that the movement of the hand toward our eyes excited another movement in our brain, which sent animal spirits into muscles, that caused the eyelids to close. . . .

That is a literal translation from Descartes's book *Des Passions de l'Ame,* published 1649. The reflex fascinated Descartes, this result rolling off the assembly line, able to bypass us, we not able to prevent it. The twentieth century is practical. It draws a line, adds numbers, concludes that a sick man is an algebraic summing of reflexes lying in a hospital bed while another algebraic summing is standing by carrying out bedside experiments. In the first half of this century much was done to clarify the reflex. The scientist who did most, who acquired great fame, who extracted most knowledge from the reflex, said shortly before his death eighteen years ago that the study of the reflex was exhausted. That gadget, he said, had lost its usefulness. He no doubt meant usefulness to the laboratory, because its usefulness to you and me and any dog and any Englishman's cat, as to all Athenian flies, goes on day and night.

# XV

# BRAIN

===

## Cold Earth of Life

# SKULL

## "Alas, Poor Yorick"

—Hamlet

A medical student wearing a crisp white laboratory gown, a blasphemous fellow, put his finger onto a man's brain, drew away, a cold brain reeking of formaldehyde. It was given him—he could take it home to study if he liked—in the winter of his freshman year by the Department of Anatomy. Some months earlier he had sauntered, a novitiate, into the dissecting room, bodies wrapped in white sheets, on tables that decidedly were not couches, and when the sheets were jerked away, the naked dead lay one next the other, and the tables were parallel and that added something. Pride was all that kept down horror. Nevertheless a few days later this student was leaning his elbow not on a nearby window sill but on Adam's rib. A callousness finally takes care of us all. Even so, it would again be a few days before he could be debonair about a brain in a jar in his bedroom next his bed. That brain is a fact of anatomy, yes, but has an idea attached. *He* had been in there. He is not in there now, no, but something of him was left behind when he vacated, a shirt on a nail.

To pick up a knife and cut into that formaldehyde-hardened haunt would be still another few days. Then, with a sudden viciousness in he went. Probed the cut surfaces. Probed the depths. Turned upside down that assemblage of tissues we call a brain and underneath where the geography is obvious located the obvious. Never cared a damn. Strolled over to the library. Looked at some drawings by anatomists of the Italian Renaissance. Some by Leonardo. One was by Vesalius—Andreas Vesalius—where a skeleton stood in his bare bones as graceful as a dancer, and at his top end in the brain-pan was half a brain, left behind, like that shirt on a nail. Another drawing showed the classic dissecting theatre in Padua. Much occurred in that theatre. Amphitheatre in form, rising rows of seats, steep, and the lectures were by candlelight. Fiery Vesalius. Candlelight. And a corpse.

The medical student dislikes anatomy. Dislikes the smells. They follow him into the street, percolate up through his meals, get into his beer, are under his fingernails. And that eternal memorizing. Anatomy is all memorizing! Each night he adds new terms. Writes new lists. Copies someone else's notes. Primes himself at the least expense to himself with knowledge about that 1,350 grams of brain. Regularly he interrupts his dedication. Gossips on the telephone. Dreams. "Why not dream—soon you are too old to dream!" In that undoubtedly he is mistaken.

Then, one disenchanted evening, three evenings before the final examinations, he recognizes that he cannot waste another hour. He will study this stuff systematically, get done with it once and for all, forget it the hour after the Professor of Anatomy is no longer in a position to blackmail every freshman.

He begins there where the brain joins the spinal cord and works his way up. Medulla. Pons. The rest of the way. Comes out at last on the roof. Roof brain! Cerebral hemispheres! What a relief!

How can this freshman, three years yet till he is a senior, four till he is an intern in the hospital, seven till he builds a house for his bride, be expected to have the maturity to realize that while he is at this drudgery he is also placing for himself the historic and literary and philosophic perspectives of an architecture that is grander than Pisa with its tipsy tower. Grander than Florence, than Rome. Where is the professor could inspire him with that? Where is the professor who if he were himself bursting with that inspiration could afford to let it escape into a classroom?

The freshman is not even thinking of the crasser gains: that this drudgery will help him better understand the controls on the living body, important to a physician; better understand his own body when he swings it into the train that carries him home from the city in the evening; understand the side-to-side sweep of the eyeballs of the man next to him who is staring at the fleeing landscape, the eyeballs looking frantic though the man sits quiet; understand the controls on the beat of the heart of his small daughter when she dashes excitedly down the stairs: "Daddy!" The freshman shrugs away the thought that science and common sense long have considered the brain to be the foundation of mind. That is too obvious. The brain is flesh, the dissecting room is dead body next dead body, and one learns from the dead, and a dead brain has that which a dead foot has not.

Another student at midnight under a lamp has been trying to memorize the Latin names of the bones of the skull and the grooves and the holes where the nerves and blood-vessels during life passed through. He stops. Did that skull lodge a princess? It *is* a female skull? What did she wear at her wedding? Was she white? Was she black? "That skull had a tongue in it, and could sing once." Gentle Shakespeare. Out at the front of the skull there are still some of the half-attached frail bones that gave shape to her face, framework to her smile, the smile with which that day she swept down the aisle of the hushed cathedral. "Let me peruse this face." Different from the two students, a paleontologist this afternoon in an ancient pit found a fragment of skull, and this evening is reconstructing the mind of a man of a clan that foraged lands and ravished women eighty thousand years before Andreas Vesalius was born. Padua was the center of anatomy of all the world in Vesalius's day, because of Vesalius.


# MENINGES

## *"Tie My Treasure Up in Silken Bags"*

—Pericles, Prince of Tyre

The old morgue was at the back of General Hospital. This was across the street from the dissecting room. Recently there is a new morgue, on this side. Families, especially poor families, and they come often as families, approach timidly, ask in which building is the morgue, frightened at what they might be going to see.

The pathologist, the chief, has for years cut open bodies and heads to find why the late renters died. A nimble-tongued fellow, wry, swarthy, always has worn eyeglasses, not afraid of colleagues or corpses, from the start came at the world with an iced outside in the unavailing expectation that no crack of warmth would ever melt through. Nothing about him suggests post-mortem. Even so, he has just turned a corpse over on its face, immediately starts the humdrum day-labor of lifting spinal cord and brain out through the corpse's back. He incises skin, fat, fascia, muscle, in that order, does it with a harsh knife. The arches of bone of the vertebral column he snips, not crashes, wields a neat bone-shears. At the head-end of the corpse he slashes through the hairy scalp, that is bloody in life. Next he saws through the bone of the skull, two plates with a spongy lacework between. He opens the skull from behind, as any coconut. He straightens up. He scans his work so far. Spinal cord and brain in their membranes lie in that excavation in front of him. Splinters of bone and other rubble are scattered about, yet, considering the circumstances, it is a clean place. There is that peculiar smell of recent corpse.

Three membranes—three meninges—cover the entire central nervous-system. It has a triple coat. Each large nerve-trunk goes out for a distance with a sleeve of that coat. The nerve-trunks go in pairs, one left, one right. Of the triple coat the outermost makes one think tarpaulin, and the second and third make one think inner tube. Had there been a hemorrhage the tarpaulin would have been a shade of brown that depended on the time between hemorrhage and death. With a scissors he then clips in succession the pairs of nerves that still pin down the system, then hoists the whole of it in one piece out of the corpse. *Requiescat in pace.*

It is easy at this moment to fancy the nervous-system as some architectural structure of the Italian Renaissance, sturdy and worthy, and from it the mind did somehow rise. Could it be the other way? The physical universe somehow rose from mind, as in one form or another and at one time or another has been thought.

Alternatively, the pathologist might have run his scissors only through the outer of the three membranes, in which case spinal cord and brain would have been hoisted from the wreckage with the second and third membranes still around them. Men are so free with their scissors. They are so insulting to a body dead two hours. The hoisted he drops into formaldehyde, and not until this moment has the bystander, the shock of the scene somewhat faded, realized what a fortress vertebral column and skull are, and what a pliable inner stockade those membranes are.

Those membranes have Latin names: the outer, dura mater (hard mother); the second, arachnoid (cobweb); the third, pia mater (tender mother). The pia is fitted skin-tight and dips into every irregularity of surface of cord and brain, while the arachnoid bridges across the irregularities. Between pia and arachnoid flows, if anything so slow can be called a flow, the cerebrospinal fluid, making a shallow lake in some places, deep gutters and ravines and cisterns in others, with a fine stream of it running down through the middle of the cord, and a good quantity filling the caverns inside the brain. Were one to imagine the brain spread out flat, and imagine it vast, imagine oneself standing like a tourist viewing it all, what a watery empire! Venice. Every narrow canal and plaza has its sentimental story and nomenclature.

The amount of the fluid? It would fill a teacup.

To tap some of it is to perform a lumbar puncture, an LP. A neurologist may perform an LP any day of the week. He asks the sick man to lie sidewise near the edge of the bed, curve his back as much as he can, bend his head forward, pull his knees toward his chin. The last helps further to curve the back and to separate the vertebrae. In goes the needle, a two-inch-long needle, advances firmly toward the cord, into the space between arachnoid and pia, and out comes the fluid. The neurologist measures its pressure. He drains off a sample to send to the laboratory to study its composition. The proceeding sounds simple. A senior medical student was not so sure. He had tried it. Later recovered his composure. "Four of them held her down; not one was able to get in; my first case; in I went." Oh, to be a senior again! To be seeing everything fresh. To have in oneself some of the old knowledge, and some of the new, all of it actually quite new. The student may have forgotten what he learned yesterday, but will have the energy to learn it over again tomorrow, the vitality, sometimes the gaiety, sometimes the cruelty of his private Italian Renaissance.

# SPINAL CORD

## *"Pillar of the World"*

—Antony and Cleopatra

The spinal cord—those membranes around it—hangs in the spinal canal. Nerves lead into it from neck, arms, chest, abdomen, legs, make connections at junction points inside, and other nerves lead out. Highways at the same time streak up and down, to and from the brain. One of the large spinal nerves (such nerves are always bundles of nerve-fibers) that led into the cord of a cat was stimulated with electricity morning after morning by that scientist who wanted to learn about synapses, those junction points, and he picked up the results of the stimulation from another large nerve that led out. The substance of it, what a person would find if he touched the cord, is a soft material, but miraculous as soon as he thought of what it was, nerve-fibers, nerve-cells, other cells that were long thought merely supporting cells, then fluid, the fluid that is in and around all cells, and conduits for blood, and blood. If the cord is damaged and bleeds, because it is fitted in the unyielding bony spinal canal, the bleeding can cause pressure and be dangerous.

The canal is roughly cylindrical, is built in sections called vertebrae. Together they make the vertebral column. Joining the vertebrae are tough ligaments that help hold the column firm, and at the top-end the canal opens into the skull, and in the skull is the brain, that is a soft miraculous material too. The brain does the mental work, presumably, and recently some have claimed changes in the size of the brain when it works. (This was in rats.)

The vertebrae themselves are light in construction but architecturally strong. One vertebra rocks on the next. The freedom of the movement differs in neck, back, pelvis. A thin man may bend forward to tie his shoe, a child bend backward till her hands are flat on the ground, Little Egypt bend all directions for the education of her audience, and a dignified gentleman in the front row bend the least possible, but he manages to see all. Throughout the bending the spinal cord and the nerves entering and leaving are never or seldom kinked or pinched. Between the vertebrae are shock-absorbing cartilages, disks, that occasionally slip, and then nerves get pinched, produce pain, and numbness, and sometimes paralysis.

Since human beings have vertebrae they are classed as vertebrates by the professors, who also are. Thirty-three vertebrae in a human skeleton. Seven are cervical—cervical means neck. The uppermost of the seven supports the head, that world, and is fitted onto the next below in a way that permits the head to pivot, to see around, to keep posted on how the other heads live. Twelve thoracic vertebrae—thoracic means chest. Five lumbar—means loin. Five sacral—*sacer*, Latin, means sacred. The latter five are fused and form the back wall of the pelvis. Four coccygeal—cuckoo. Those also are fused, and abbreviated, sad remembrance of a day when each of us still could wag his tail. How gay we once were! That dignified gentleman before his rheumatism had a column that curved slightly forward in the neck, backward in the chest, forward in the abdomen, backward in the pelvis. Years ago people said: "The young Colonel has a grace." When he was an embryo and his vertebrae were growing, his cord was growing too, but his vertebrae faster, the result being that sections of cord no longer corresponded to sections of canal, nevertheless the sections of cord still numbered as if they corresponded, third cervical, first lumbar, et cetera. A further consequence was that the cord no longer filled the lower portion of the canal and in the Colonel and also in his Corporal toward the tail-end there was no cord, only space filled with cerebrospinal fluid. In that fluid was a straggling downward of nerves, these constituting the cauda equina, horse's tail. The neurologist when he performed the spinal puncture aimed the needle low into that space, because there he would injure nothing critical.

Nowadays the neuro-physiologist may impale a single nerve-fiber, neuron, of the cord. He does it with a microscopically fine electrode. By the electrode he picks up the electrical messages, or the reverse, he stimulates. When impaling, it may be cell-body, dendrite, axon, or the surrounding of these, and he knows where it is by the recordings made via his electrode. In the old days, that can always be thought the crude days, which they were relatively, the experimenter simply sprinkled salt on the raw surface of the cut cord of an animal. The salt was the stimulus. Crude. Yet our glorious new days rest on those old days. And the new days are rapidly adding chemistry. Entrance into the cord, exit from it, areas of the center of it, all are being analyzed. Possibly not too much has yet come of it.

The cord is flesh, but one does keep seeing the fleshless pigtail of a wraith. A braid is hanging down from a brain, a dead braid from a dead brain in a dead head, and the wraith a transparent bluish wraith looking from above at her late embodiment hacked to pieces at this unhallowed hour in this unhallowed place, the morgue.

# DESIGN

## *"The Spring, the Head, the Fountain"*

—Macbeth

The brain is in the cranium. It is within those walls, those Renaissance Italian church walls—let us say of Padua, because Vesalius was in Padua. The brain is fitted neatly in that space in the upper part of our head. At death the cranium becomes a Renaissance Italian mausoleum, a fine one really.

All our life the brain sits up there above our spinal cord except during the conventional eight hours at night when it capitulates and with our body topples to the horizontal.

That space must have the right size and the brain the right size, not be too large, the head not too large, not too small either, space enough for the great nerves and nerve-pools for eyes, ears, nose, for the arriving nerves from all parts of the body, the fabulous forest of connections between the arriving and the departing. As a distinguished American has made clear, space for the brain of the reptile, and upon that the brain of the old mammal, and upon that, the new mammal, the three together constituting the brain of a New Yorker.

Space enough for reserve brain, gray and white matter both, in order that after the wear and tear of youth, including bumps on the head with bleeding, there will still have been enough in the cranium of a Lorenzo de Medici that he would not too shamefully totter to his eternal couch under Michelangelo's three marble figures. Earlier, during the nine months that proud Lorenzo in the dark of his mother's womb was kicking his way toward his restless Florentine life, up in his fetal skull there was the noiseless growing, the noiseless folding, the unequal developing, the differentiating, each step signaled according to the blueprints, to the code laid down for this brain's design. There would be a code for the spider's, if the spider can be considered to have a brain.

No, the brain must not because of illness be too large—a swollen watery brain may trouble the observer somewhat as does the supernatural, leaves an uncomfortable feeling even in a brain-cutter, as one can see in his face.

Any anatomist could make a blueprint of the spinal cord. A blueprint of a brain would not be that simple, though there are plenty of what amounts to blueprints in dissecting guides and other books. These are apt to show us the long-known details, but possibly some recent detail because while the pendulum swings the anatomist adds, a little. The student who on that disenchanted evening decided to cram overnight into his memory every fact about those 1,350 grams, crammed a few that already were not fact because somewhere a brain-cutter had taken his pancake turner, picked up the next slab of dead brain, flipped it over onto a large enamel pan, found a mistake; not often any more, gross anatomy rather a settled science. If a lush brain-tumor had been found, that enamel pan with the slabs would have been sent to the hospital photographer, the photograph filed, and a pennyworth added to the anatomy of disease, hardly to the design of the brain.

An architectural blueprint could instead have been based on the brain's development in the fetal skull, which would have been based on the brain's slow development in the species. That loftiest part of each of us began in a microscopic line of cells. This line was the ground-breaking. It ran in the direction head to tail. The cells multiplied. There resulted a narrow longish plate of cells. To either side of this the cells multiplied faster till there were two walls with a narrow longish trench between. The trench deepened. The top was covered over with cells. Thus did the trench become a tube. Everywhere in the populated world inside of wombs there is this developing of tubes, tubes, tubes. If a man feels no chill at the thought of the amazing facts of any developing of any creature in any womb, and in all the wombs since wombs began, he will chill at nothing.

Down sank the tube. While it did, the embryo's back, the same that later would lean its tired self into a soft chair, was covering over the tube, and over everything in its neighborhood. At the head-end of the tube were the beginnings of the future brain, at the tail-end the beginnings of the future spinal cord, most astonishingly odd shapes at the head-end.

That head-end now divided into three, cavities connecting the three. Development was relentless. Differentiation was relentless.

The first of the three divided, the second did not, the third did. So there were five. Despite folding, despite unequal growth, despite the tight engagement of the baby's head in the mother's birth-canal at the time of birth, nowhere were the cavities obstructed. Latin names were given to the five. *First ventricle. Second ventricle. Third ventricle. Aqueduct. Fourth ventricle.* Ventricle still means hollow, a little belly. During life when the cerebrospinal fluid has been drained from brain and spinal cord, and air injected to replace the fluid, X rays taken, the darker silhouettes, darker because X rays are less blocked by air, produce what could make one think of some ancient relief pieced together by an archeologist.

Arteries and capillaries and veins of the brain have thousands of times had X-ray-opaque material injected into them, been photographed, and those photographs too have added to the design that has been aggregating through the centuries. The aggregating will go on. The naming will go on. In general, the old Latin and Italian names have resisted change.

*Medulla*—small and immediately above the spinal cord.
*Pons*—small and above the medulla.
*Cerebellum*—large and sticking out behind.
*Midbrain*—small, the smallest part, forward of the pons.
*Interbrain*—comparatively large, forward of the midbrain, including hypothalamus and thalamus.

*Basal ganglia*—various sizes of clumps of cells, some close around the thalamus, some at greater distances.

*Limbic system*—an encircling core at the bottom of the brain, reaching toward the ancient and toward the recent, and holding much of our history and some of our hope.

*Cerebral hemispheres*—tremendous, on top, outside, all around, enveloping the whole and always seeming a blossoming.

Lop off cerebellum and cerebral hemispheres and what is left is brain-stem. It looks that. It looks stem. It looks stalk. It could look a stout umbrella handle. Snap open the umbrella and that would be the spreading of the cerebral hemispheres, but not a thin fabric, of course, and more spreading to our fancy than what one sees there in the anatomic stuff hardened by the formaldehyde. In man and in his evolutionarily close relatives there is more spreading, and it is all comparatively new, so the cortex that surfaces it is called neocortex. One can think, if one likes, of those spreading hemispheres as giving contour not only to the human head but to the human mind. The formaldehyde-hardened substance is not easily damaged, the living brain is. So, take up the pretty bambino with care, padre. Nature started this, nurtured this, wrapped this in membranes, boxed it in bone. Care, padre!

# MEDULLA

## *"He Fetches Breath"*

—Henry the Fourth, Part I

From the spinal canal it goes into the skull—from the spinal cord it goes to the lowest part of the brain-stem, the medulla. Spinal cord . . . Medulla . . . The medulla lies just inside the skull, just above the large hole at the bottom of it. An inch of special flesh, the medulla has great age, the ancient invested in it. Andreas Vesalius knew it well. Swallowing, vomiting, articulation of speech and song, the roguery of the tongue: these and more are entrusted to that inch, the control of the beat of the heart, the size of the blood-vessels. A cut through the brain-stem above the medulla in an animal, or an accurate guillotining above the medulla by that supreme executioner, the drunken driver, leaves the medulla continuous with the cord, and there may be no change whatever in blood-pressure, though changes enough otherwise; but a cut below the medulla, the trunk of the body and the extremities deprived of their controls, blood-pressure collapses and the victim goes into spinal shock and may die at once. The doctor is glum when he decides his patient has bulbar poliomyelitis, bulbar palsy, bulbar anything, *bulb* being another name for medulla.

To divide off an inch of nervous-system, to describe it carefully and fully, just an inch, and one inch after another, an inch at a time, helps the anatomist. It helps his concentration. It also helps any student to escape the panic he is bound to feel if confronted with the whole lump of the brain instead of part on part, and even with that help, the Lord knows, he has opportunity enough for panic. It keeps his and the anatomist's mind from wandering though it might not keep his backside from broadening.

Vesalius published the *Fabrica* and the *Epitome* both in 1543. He was born in 1514. So he was done with anatomy essentially when he was twenty-nine years old. He did accurately recognize the brain as the center of human faculty.

The brain-cutter, the modern one, may have reason to examine slice after slice of that inch, the medulla. Being a man who wishes to understand his trade he may then quietly in his mind refit slice after slice, reconstruct the great amassings of neurons, the embedded cables that run up and down. Those that run up have come from scattered parts of the body toward the spinal cord, passed up the cord, passed into the skull, and now are passing through the medulla on to the top of the brain, are bringing information, as Vesalius would not have said. Contrariwise are the lines that come down from the top, through the medulla, through the cord, scatter via nerve-trunks to all parts of the body, delivering information, or orders. On the front of the medulla great numbers of those aggregated down-traveling lines cross, right to left, left to right. This explains why damage on one side of the brain produces paralysis on the opposite side of the body, as was known to old Hippocrates, who walked by the harbor of Cos a thousand and a half years before

Italy came to its greatest glory, in painting, in architecture, in anatomy. And the anatomy was Vesalius.

On the back of the medulla lies a part of the fourth ventricle, a shallow diamond-shaped lake roofed by a thin veil called vellum, and under the lake, in the solid substance of the medulla, nerve-cells and nerve-fibers of many sizes and kinds. Vesalius thought of the ventricles as receiving air when we breathe-in, and from that forming animal spirits, which gave us mind, whereas the animal spirits traveling out the nerves gave us sensation and movement.

Some of those nerve-fibers have an insulation of white fat, some not. A crisscrossing results. The crisscrossing might seem disorderly in the orderly brain but is not. Generations of students of the nervous-system have shown it to contain clear-cut areas that are centers of vital body-control. There are hereabout similar areas less clear-cut as to work and less clear-cut as areas. They make a reticulum—called reticular formation—that we are meeting here in the medulla for the first time and that is extensive. We shall meet the reticular formation again higher up because it goes upward in the axis of the nervous-system, and it also goes downward into the spinal cord. This formation has received close study by that part of the world concerned with these things and new light has been thrown onto some of the ways the nervous-system works. Sleep and waking have become somewhat more comprehensible; some think mind more comprehensible.

Among those clear-cut centers of the medulla is the respiratory center. It too is built of such a reticulum, latticework of cells and fibers. Here the control of respiration is centered, or, here is the central exchange where it is determined exactly how deep the next breath shall go, exactly how fast. Galen, second century A.D., bold vivisectionist, slashed through between spinal cord and brain, paralyzed breathing, knew therefore that this center must be above where he slashed. It was in an animal of the Roman Empire. Galen chose to look at and to study and to reflect on this spot, which is not large, where the nervous-system regulates breathing. Galen chose to spend his life with this kind of fact. Each of us can to a degree, choose the kind of fact he wishes to spend his life with, but he must choose, there being so much, and Galen did choose as wide a range as one man can. He did not know how in that living tissue the control of breathing was managed. The Roman Empire could not yet know. Italy of the Renaissance could not yet know.

Italy was on all sides fascinated by structure, the leaning tower of Pisa, the anatomy of the brain, yet Andreas Vesalius, Professor of Anatomy in Padua, though he did at so many points clarify the anatomy of the living, did not know the respiratory center. That knowledge still must wait. Legallois (1770–1814) localized the center in the lower brain-stem. Flourens (1794–1867) localized it more precisely. A time would come when anatomists and physiologists would seem with a ruler to be measuring off a plot under the floor of the fourth ventricle and saying: "There." Later they would again be less dogmatic and less diagrammatic, or possibly only differently dogmatic and differently diagrammatic. They would think the respiratory center reached over a longer length of brain-stem. They would recognize that its action was less simple than they thought, indeed exceedingly delicate. A cloud of molecules of a gas moving via the blood through the flesh of the medulla would determine that next breath. In this vital part of the brain-stem carbon dioxide would act. Morphine and lobeline and nicotine also on occasion would act. A drink of water, a laugh, a sigh, each would modify the next breath, a thought could, a dream, especially a nightmare, as most of us have experienced. In this nervous flesh of the medulla was the control-tower for the rhythm of our breathing. Much besides breathing is entrusted to the medulla. Much of the ancient in us is guided from there.

## PONS

### *"Come You from the Bridge?"*

—Henry V

Just above the medulla is the pons, second piece of the brain-stem, again one inch. Spinal cord . . . Medulla . . . Pons . . . We are on the way up. All of it has been ancient brain. All of it is roughly the same in all mammals, cat, monkey, man. All of it belongs to the part that has been spoken of as reptilian.

Spinal cord, medulla, pons—this is a conducted tour. Medical students nevertheless sometimes take the tour, neuro-anatomy, only because they must. Vesalius? That was something else. That was different. Nothing like that before or since. Had his inimitable instinct. Was mad. Was wonderful. His great work he finished so early. Then his life burnt out.

*Pons* means bridge. "Come you from the bridge?" says Gower to Fluellen. A broad band of curving flesh builds this bridge. Seen from the front with the naked eye it reaches from one side of the brain-stem to the other, the ends disappearing into the masses behind. The broad band is nerve-fibers. There is a graceful rising from the one side, then a falling toward the other. The full name is *pons varolii*. Under the microscope this inch is more than the side-to-side fibers, has depth rich with structure. Varolius was professor in Bologna. It was a brilliant period of the sixteenth century. He was physician to Pope Gregory XIII, reached thirty-two. Seems a short time to have lived, but time enough to make many studies, most of them forgotten, but this bridge not forgotten. Because of a name attached to a piece of flesh Varolius is another who really did not die, may not have died a thousand years from now, two thousand, and in two thousand even Imhotep, that first Egyptian physician, may have lost a syllable from his name, the type dropped out, disappeared.

Pons is next above the medulla, the next piece, and through it run those same nerve-lines from high to low, low to high. On its back is that same fourth ventricle, that diamond-shaped lake, and under the floor of the lake there is still some of that center for breathing, and of that ancient crisscrossing of nerve-fibers and nerve-cells, the reticular formation. Side-lines enter the crisscrossing. Vesalius was not at a point in history where he could know a fact like that, had not the instruments, must in every hour (he accomplished so much) have kept his young mind open for possibilities, and surely he felt the mystical wonder of anatomy. "In this manner, indeed, the great Creator of things has fashioned our body not only that it may live but that it may, subject to corruption, live properly." He said it that way.

The chief pathologist, 1969, having with his scissors completed the snipping of the pairs of nerves that pinned down the spinal cord, continued into the cranium, snipped pairs there too, and since he was a systematic man, which a pathologist is as apt to be as is an anatomist, he began with the lowest pair and continued to the highest. Because the pairs now were in the cranium they were called cranial nerves, twelve pairs. Galen thought seven. Vesalius also thought seven. A dogfish misses twelve by two. Starting in the medulla, the lowest of the twelve is the nerve that gives orders to the tongue. Then, a nerve that goes to two of the powerful muscles of the neck. Then, a huge nerve, a cable, that distributes its branches to the throat, the windpipe, organs of the chest like the heart, organs of the abdomen like the stomach. A nerve for hearing comes out where the pons joins the medulla. A nerve that controls facial expression. A nerve for sensation in the face. For swinging the eyeballs. For blinking. For chewing. (We have left the pons behind us.) For seeing. For smelling.

When finally the pathologist had entirely unpinned the brain, and had lifted it out of the skull, those pieces that were hanging there like the straggly pipes of the bagpipe were the amputated stumps of those twelve cranial nerves.

A dangerous place to be damaged, the 'pons. If a child has run into the street into a truck, been picked up by the police, rushed to the hospital with that clamor of ambulance that all Americans must train themselves to sleep through, and if a hemorrhage has torn its way into the pons, there are frightful signs, bizarre paralyses. From human misery learning sometimes comes. From deliberately produced animal misery learning sometimes comes too. The pathologist and the anatomist and the physician and many researchers are forever in their minds, with help from their hands, doing what we have been imagining the brain-cutter to do, have been fitting together the crashed or slashed nervous-system, have been doing that for centuries, did it in the Italy of Varolius, in the Italy of Vesalius. It is their business.

Despite the charm of the detail of his body a man may think he has had enough. The suicide who prefers the mouth route rests his revolver against his palate, aims straight back, and even if he trembles is apt to achieve his wish, blows out everything in the neighborhood of the fourth ventricle. A savage wish. He muddies and bloodies medulla and pons. To dissect with a revolver is not real neat. If he uses a shotgun he can aim much more casually, just north, blow off his head as Hemingway probably did, and if he did he probably thought it appropriate, and he probably was right.

# CEREBELLUM

## *"It Is Too Little"*

—Two Gentlemen of Verona

*Kleinhirn* is what the German calls the next part. The word means diminutive brain. What the Latins call it and we call it is *cerebellum,* and that word too means diminutive brain. When anatomy was young, long ago, anatomists enjoyed attaching Latin names to every nook and crook of our body.

Vesalius wrote an excellent Latin.

The cerebellum is a wedge of brain, looks a separated piece tucked under the mushrooming large cerebrum, peeps out. Immediately in front of it is the brain-stem. The casualness of that statement is in order that no one may be frightened off from knowing exactly where he is, even if only in the sense that a sober citizen knows where he is when given an over-polite and exact direction by a well-intentioned inebriate.

In weight the cerebellum is one-eighth of the brain.

If one places one's finger high on the back of one's neck, the nape, inside there is the cerebellum. Three diminutive feet join it to the rest of the brain—peduncles the godfathers called those feet. Through them run nerve-tracks leading to and from the rest of the nervous-system, which means, spinal cord, brain-stem centers, cerebrum. Everywhere in this grand organ of the nervous-system there is a leading to and from. Wealth of connection makes the body what it is. Leonardo, Harvey, Descartes, Vesalius, those men in and around the Renaissance felt this wealth and studied it, and we still study the anatomy and histology, document both. Because of that wealth of connection the human body is able to do the deeds it does, the leopard body also.

The godfathers looked over the surface of the cerebellum and saw that it was everywhere grooved, that one groove dug more deeply than those near it and therefore more definitely separated one lump from another, producing apparently independent organs, small lobes within this small brain, this *Kleinhirn,* and everywhere over that surface they sprayed names that stuck. Indeed, some names were pried off only lately by the young anatomists untutored in Latin and tutored in algebra. The young concern themselves with the work of the parts, the old were fascinated by sheer geography. For instance, a longish roll of brain that runs down the middle between the right and left cerebellum suggested worm, its Christian name therefore becoming *vermis,* worm. It must have been a satisfaction for the godfather who thought of that. To either side of the vermis are the lobes, and the lobes have gray and white, the gray, nerve-cells, the white, nerve-fibers. The white in the depths has gray splashed in it, islands with natty names like dentate, emboliform, globose, fastigial. The purring godfathers had not only knives but words between their teeth.

Here as everywhere in the body one is in some moods reminded, cannot escape the awful fact of repetition, of cerebellum and cerebellum and cerebellum, all the same, almost.

If a thin-bladed knife slices through the cerebellum, on each face of the slice one sees a decoration that suggests tree, *arbor vitae,* tree of life. The young anatomists (also to a less degree clinicians suspecting a cerebellar tumor) find that the parts godfathered by those tutored in Latin are sometimes responsible for clear-cut work, sometimes not, and when not, then the pretty names are beauty for beauty's sake (in a morgue) and the names are apt to be peeled off. That takes time.

The cerebellum often has confused the medical student, the Italian medical student too, because when he lops it off, hardened as it is in formaldehyde, and holds it in the palm of his hand, flips it about for a minute or two, he no longer knows top from bottom, front from back, and that might have embarrassed him in the dissecting room when the professor of anatomy quizzed him, the other students listening. That is only embarrassment to the old-fashioned school-learning, not the new-fashioned. Either way, he will forget the embarrassment later when he is a physician and his gluteals are tired from driving all afternoon seeing patients, or sitting all afternoon in his office doing the same. Flattering to have been the physician to Pope Gregory XIII, but to have had as a patient Elizabetta of Urbino!

# MIDBRAIN

## *" 'Twixt Which Regions"*

—The Tempest

The course upward has carried far and deep into the brain.

Medulla . . . Pons . . . Cerebellum . . . Midbrain . . .

The midbrain is by the eminent expert included as part of the ancient, the reptilian, the brain of the reptile.

The midbrain is forward of the medulla. (This would still be the inebriate giving street directions.) Medulla was an inch. Midbrain is less. It is the smallest division in that original design. Mid-ocean, mid-Africa, mid-Pacific, midnight, mid-air, midsummer, mid-heaven, mid-Italian, mid-Renaissance, also midwife, each word has around it some cashmere of mood. And midbrain has too. If instead of an adult corpse the brain-cutter had gotten for himself a several-month-old dead fetus, and if he had cut open the back of that skull he would have seen four dainty prominences at the corners of a small square, four little hills, colliculi. (The godfathers again.) In the fetus the four little hills loom large, but wait a few months, they will have sunk quite out of sight, which is because the midbrain will have been growing leisurely while the parts around it were growing rapidly. In the adult corpse, to find the four little hills the brain-cutter must take hold of the brain-stem and pull it (gently, pray) away from the brain above. There they are, four. The two below are junction points for hearing—we hear a sound and automatically turn eyes and head that way. At least, that was long the view, has changed somewhat, and will, but it gives an idea of how such a working center might work. The two hills above are junction points for seeing—a sight is half-seen and we automatically turn eyes and head in order to see all. The recent evidence for the somewhat changed view is that these colliculi are concerned more exactly with changes in the *rate* of movement.

Even with all of this description it still may not be easy to imagine just where the midbrain is. Possibly it would help if the dissector propped up the fetal head. That would prop up the midbrain also. The oriental bridge of the pons would then be jutting out, not far but none the less dramatically, and the midbrain would be forward of the pons. (Still the inebriate.) The midbrain is said to have two diverging peduncles. Peduncles = little feet. These now are peduncles of the cerebrum, named so by the godfathers. The peduncles of the cerebrum are more like legs than feet. In their substance run those nerve lines that go from low to high and the other way.

Who would not wish for an hour to have sat in that sixteenth-century dissecting theatre, add that candlelit memory to the memories of his life.

Ask now that the dissector jab into the solid parts of the midbrain. He will come on an island of nerve cells reddish in tint, called nucleus rubor. Another is blackish, substantia nigra. Here again is the reticular formation—in the midbrain now—that core of the brain-stem with its many shapes and sizes of cells crisscrossing. Medulla . . . Pons . . . Cerebellum . . . Midbrain . . . Special work is assigned to a variety of aggregations of cells of the midbrain: to wit, the control of movements of the eyeballs when we look from faraway to near, the control of the movements inside of the eyes when we look from black cloth to a glaring incandescent lamp. Here, again, it is the control of movements that would keep our body foursquare to the earth. And, again, it is to be remembered that the temptation is to make assignments of functions too exact.

The dissector cuts into a narrow canal. That canal is the aqueduct, and through it the cerebrospinal fluid creeps from third ventricle to fourth. Vesalius knew this. We do not know all that he must have known, besides the anatomy, about the function of the parts of the brain. He with Paré, the great surgeon, was at the bedside of Henry II of France, when that gentlemanly King died of a wound inflicted by a lance broken during an encounter at a tourney, the metal passing through the eye and into the brain. Henry II had spells of consciousness and unconsciousness during the ten days that he survived, and the surgeon and anatomist both would have been thinking, doubtless talking, and how interesting it would have been to listen to that talk with alternately sixteenth- and twentieth-century ears.

# INTERBRAIN

## *"And Underneath That Consecrated Roof"*

—Twelfth Night, or What You Will

It was said of Agassiz that he gave a fish to a student, did not tell him what fish, what waters it came from, the student just to find out everything with his native powers, and away with the fish went the student and being a Harvard student returned with a description that surprised the fish. The reader of these pages has not been given even a dead human brain, not a shiny photograph, not a facsimile of Vesalius's drawings, not Leonardo's, not anyone else's of the Italian Renaissance, of the great anatomists who worked in Padua, not any pictorial help for a pictorial subject. If the reader still is reading it can only mean that he is possessed of a patience Agassiz would have admired. The reader has read himself up the brain-stem.

The brain-stem for all its importance to you and me, its subserving the instinctual stereotyped fatal functions, its venerableness as the part of the brain doing most ancient work, its similarity through a long scale of animals proving that evolution has let it stay as it was because it was good, is small. It looks pushed down, looks unexpectedly inconspicuous if one has the chance, rare, to view it toward the bottom of some living human skull during surgery.

Just inside the skull, just above and out of the hole in the vertebral column, was the medulla. The pons was next above. The midbrain was forward of that. The cerebellum was tucked in behind. And now it is the interbrain. Spinal cord . . . Medulla . . . Pons . . . Cerebellum . . . Midbrain . . . Interbrain . . .

It would not be possible to locate the interbrain in a few words, as was roughly possible for the parts heretofore. It might not be possible in any amount of words. The interbrain is at the top of the brain-stem. In it are the controls in the flesh of the primitive, controls for most of the body, controls for some of the head. Two of the interbrain's parts should get attention: hypothalamus, thalamus. Also, the basal ganglia hover close. Also, these parts are fused as structures, more and less anatomically fused, and then unfused, often artificially though usefully, by those who try to find what the parts do. In that dissecting theatre in Padua the fascinated witnesses crowded around while Vesalius was dissecting. Great wild future-seeing genius though he was he might have been yet wilder had he been able through a rift in the Italian sky to catch a glimpse of where his labors would one day lead. But possibly it would not have made him wilder. Possibly he was just born wild. Possibly he was just taking the future for granted and realistically anticipating how it was going to be. Spinal cord . . . Medulla . . . Pons . . . Cerebellum . . . Midbrain . . . Hypothalamus . . . **Thalamus** . . . Basal ganglia . . .

Hypothalamus. In our day the hypothalamus has been actively studied and some new importance found with such regularity that one has ceased to be surprised. The reader, to get his bearings, might do worse than fancy himself a surgeon who had picked his way along the outside of a human brain—a small pale child's with a fast-growing tumor—and come down toward the bottom of it. There he sees the midbrain. He is not concerned with that now, not with those four little hills, but is concerned with two little breasts, mammillary bodies. (Still the godfathers splattering names.) Mammillary → mamma → breast. The mammillary bodies are part of the hypothalamus, and the hypothalamus is, anatomically also, one of the more manifest parts of the interbrain.

So far as the relation of the hypothalamus to skull goes, it is situated immediately above the middle-bottom. Forward of it is the crossing of the two great nerves for the two eyes. And between that crossing, in front, are the mammillary bodies. Weight of the hypothalamus? Approximately one three-hundredth of the total brain. Microscopic content? Gray nerve-cells with gray nerve-fibers in colonies, the latitude and longitude of any one colony so similar from brain to brain that with a man-made instrument and guided only by landmarks on the outside of the skull, assisted by X-ray photographs, it is practical to push a fine electrode into a colony, stimulate it or destroy it, and, from the consequences to the animal, attempt to figure out what work normally was done in that particular colony, what type of control it might exercise over body or head, or both.

A surgeon who is operating in this neighborhood, under local anesthesia, presses lightly, with cotton, but it is still too heavy, his patient loses consciousness. A breath may affect the living brain, so sensitive it is, yet let death come and the pathologist will promptly proceed with his rough work. "Why does he suffer this rude knave now to knock him about the sconce with a dirty shovel, and will not tell him of his action of battery?" Duller men than Hamlet have thought similar thoughts when they saw a brain knocked about on the scrubbed death-smelling morgue table. Instead of the knave's dirty shovel the pathologist uses a long thin clean knife. Down through the dead brain he goes and, at the middle-bottom, goes through the hypothalamus. He lifts away the knife. For a moment he holds the knife poised. Then with its rounded tip he whisks away a stray bit of coagulated nerve, likes things neat.

The thalamus sits on the hypothalamus. In the classic world, from where that word *thalamus* was plucked, it was the inner chamber of the house. In the brain it is not a chamber but is inner. Two thalami. A double ovoid. A right and a left. The two are grossly fused for part of their length, and fused also with the brain-flesh around. Nerve-fibers enter the thalamus from every part of the body. On their way they side-branch into the reticular formation, that core of the brain-stem. A neuro-physiologist called the reticular formation a waste basket because it seemed that everything and anything, all kinds of function, was being tossed into it. The reticular formation in turn sends new lines, new nerve-fibers, to much of the rest of the brain; the importance of this is great, helps pick us out of sleep and keep us awake.

Each thalamus has two sides, has a back, a front, a bottom, and a smooth and rounded top. That top of the brain-stem is the floor of the first and second of the four ventricles, those caverns filled with fluid. Galen described them. Occasionally a hemorrhage has broken into one of the four, the blood leaked into the other three, and when the brain-cutter cut down in he came on the coagulated blood, might be able to cull out a cast of the entire system of ventricles, as that coagulationist culled out a cast of the portal vein. The cast of the ventricles looked like some fossil, some flying mammal, some strange bat with graceful curves and a texture like the inside of a morning-glory. During life the clear cerebrospinal fluid fills the caverns and in it are strewn a seaweed-like material richly supplied with capillaries, a material long thought to produce the fluid, and does, but not exclusively. Harvey Cushing, eminent brain-surgeon, actually saw fluid oozing from those capillaries while he was operating.

From this waterway through our brain we might get yet another idea of the brain's geography, as a visitor might look at a map of the Great Lakes and get an idea of the United States, or sail along the Thames and get an idea of London, or paddle along the Styx and get an idea of the Lower World, or gondola along the canals of Venice, et cetera. Spinal cord . . . Medulla . . . Pons . . . Cerebellum . . . Midbrain . . . Hypothalamus . . . Thalamus . . . Basal ganglia . . .

Those last, the basal ganglia, are scattered about the top of the brain-stem, but also down in it, islands of nerve-cells, lumps of gray, sometimes interlaced with white. What lumps to include among the basal ganglia has, as said, been debated. Perplexing they are to locate. Perplexing to explain. Perplexing names. Caudate. Putamen. Globus pallidus. Amygdala. Claustrum. Like names on Welsh railroad stations, except these are Latin, bestowed by the Latin-studying godfathers.

# HEMISPHERE

## "Grey Vault of Heaven"

—Henry the Fourth, Part II

The roof-brain. We have climbed there now. Top of the house. Cerebral hemispheres. Human cerebral hemispheres with their cortex. These have waxed from mammal to mammal and culminate in man. They have long been regarded not only as each human being's personal crown, but the crown of evolution. To be sure it was the human mind that did so regard them. Vesalius and the anatomists at Padua knew them well. The modern brain-surgeon may think of them differently than did Vesalius, but similarly obligates himself to know them as the anatomist knows them, every convolution, every groove. It is the cerebral hemispheres that the student mostly sees when he looks, furtively the first time, toward the 1,350 grams into which the pathologist is sticking his knife.

In the embryo the hemispheres began as head-end buds. The buds grew. Luxuriantly. Too luxuriantly, one could feel. A weed! This weed would be the instance par excellence of disproportionate rapid growth. Of all brain-parts it grew most shockingly. For a brief time in embryo and fetus those four little hills of the midbrain still loomed large, peeped out behind, but then they were covered over entirely, the cerebral hemispheres with their plating of cortex over-growing everything, and, while they were growing, they were keeping close to the inside of the skull, that was growing also. Those hemispheres were building a dome. They were completing with a dome the edifice of the brain. Man thinks of his own as at least St. Peter's built by Bramante and Michelangelo. What portion of brain to designate cerebral hemispheres is roughly agreed upon. What the immensity of those hemispheres can be in the higher creatures, as against lower creatures, is recognized. And what the immensity in the highest, man, may mean is in some respects recognized also. Gray is on the outside—nerve-cells. White underneath—nerve-fibers. That gray, that plating, plus that white, plus great bands of communicating lines, plus the first and second of those odd-shapen hollows filled with fluid, first and second ventricles, plus the lumps of the cell-masses of the basal ganglia, plus much of the ancient smell-brain or what once may have been smell-brain but is not now though continuing to lie in the brain-pan just above our noses, plus various curvings through the middle regions of the brain—all belongs here. When a hemisphere is surgically removed because of disease, one-half of the barn of a man's skull vacated, whoever sees that sight never forgets it. (Only occasionally does even the surgeon have the chance to see that sight.) Louis Pasteur suffered his noted stroke, lost most of one hemisphere, but appeared, judging by the great work he accomplished, not to have lost any mind. Possibly it would be allowable to speak of the two hemispheres as the vault of heaven. They did invent that other heaven, gave it its shape, the earth-child's vault. Kepler, great seventeenth-century astronomer, believed that it was simply because of the hemispheric hollow at the back of our own eyes that we see the sky as a hemispheric hollow. The heaven, that "vision of fulfilled desire," was a hemisphere up there because of an item in the geometry of our anatomy down here. The eye created the shape that it looked into for an eclipse of the sun by day and an eclipse of the moon by night, the eye aided of course by the cerebral hemispheres, that are so closely related, still inexplicably related, to what the universities think is thought. The whole mind uses the whole brain, no doubt.

# LOBES

## "*Heaven's Vault Should Crack*"

—King Lear

Like a monkey, a man will crack open anything, find out what is inside, crack open a skull, and there is heaven's vault, the brain. He has studied the grooves on the exposed side of it, established landmarks among the grooves, and these have aided him in parceling the brain into lobes, and then, having parceled, over years he investigates into the lobes to learn perhaps what each does for us.

The first of the grooves appeared late in the life of the fetus. It was a grooving to be seen on the cortex, the bark of the brain. Patterns appeared there. The patterns repeated brain to brain. In the fourth month of fetal life that cortex still had had a smooth unreality but then came the grooves, from nowhere, it seemed, shallow ones, deep ones. Say that the Seven Days of the Creation are too amazing to believe but no more amazing and rather like what is taking place in every sleeping fetal head. Is it sleeping? Is that sleep? Waiting to be awaked?

A deep ditch—*longitudinal sulcus*—runs from the front to the back of the brain, dividing it into the two hemispheres. A rugged furrow—*sylvian fissure*—cuts in on each side at about where side-burns begin. A decidedly less rugged furrow—*central fissure*—takes off midway from the top, the crown, runs diagonally and irregularly downward and forward. So, three boundaries: one dividing left from right, one dividing high from low, one dividing front from back. There are others but those three are the large ones. It is they have parceled man's brain into lobes. Man has parceled his brain. In the hope of understanding he has with diligence labeled and re-labeled, has done this in clinics, in laboratories, in seminars, in arti-cles, in books. Those who died a death attributable to something amiss up there have been studied also for the amiss.

Frontal lobe. Parietal lobe. Temporal lobe. Occipital lobe.

One of each of these is on each side of the billions of heads. In late years there have been other regions of emphasis. There has been much talk of that limbic lobe, also of limbic system, supplying a cortex over that old reptile's brain, and it is still down there in ours. It lies mediobasal, as is said, middle, low. Some of what is considered limbic by one student is not by another; some think there is exaggeration here; some think it should be limbic system; some used to think rhinencephalon, smell brain; some, visceral brain; some, thymencephalon, thymos meaning mind, passion, soul. Which all is mentioned only to give the feel of all. It is still a dividing to give a balustrade, something to hold to, encourage us to keep climb-ing, keep trying. Much has been learned. The dead have and do and will enlighten the living.

Frontal. Parietal. Temporal. Occipital. Then the dividings of the dividings. We are understandably concerned. What is closer to us? It could even come to pass that the last third of this century would be proven to have been as consumed with this architecture, with its chemistry, with its meanings, as was the first two-thirds of the cen-tury with hot and cold war and interpersonal relations and inter-racial relations and intrapsychic relations. Not likely.

Frontal lobe—everything in front of the central fissure. Parietal—everything above the sylvian fissure and back as far as a vertical line drawn an inch and a half forward of the brain's rear tip. Occipital—everything behind that line. Temporal—everything below the syl-vian. Limbic—the ring encircling the brain-stem, for twenty-five years thought to deal with the emotions, with things ancient in the mind of the creature.

Road directions would read: turn backward for the occipital, upward for the parietal, downward for the temporal, forward for the frontal, and other directions, and into side roads.

Such scholastic cracking of heaven's vault ought never let anyone forget that the brain is one, that the closest we ever come to one world is one mind, and underneath one mind somehow is one brain.

Franz Joseph Gall (1758–1828) was the first brain surveyor to make surveying fashionable. In the century and a half since his death that surveying never has flagged. Phrenology was started by Gall. Phrenology claimed that the bumps and shallows on the outside of the skull could predict the strengths and weaknesses of the Johnny inside. An observing barber should have been good at this. Phrenology was sure to sell well because easy to understand, the customer for his money had something his fingers could touch. It was A-B-C truth. Gall was driven from Vienna as a quack, honored in Germany, died rich in Paris. Goethe was impressed by him but laughed at him. Flourens had no faith in phrenology but admitted he never truly looked at a brain till he saw Gall dissect one, and Flourens was a great neurologist. So, Gall could dissect. Quackery mixes with excellence often. Grooves on the surface of this flesh had significance to Gall and to Flourens, would get more and more significance, and latterly, as said, chemistry would join in. Two frontal lobes . . . Two parietal . . . Two temporal . . . Two occipital . . . Et cetera . . . The pathologist with his long thin knife and without a blink slices through lobes, grooves, fissures, furrows. But, suddenly, interruptingly, preoccupyingly, there comes forward into one's memory the dark Vesalius who has stood himself in the midst of the gaping crowd in the sixteenth-century dissecting theatre in Padua and is demonstrating on a partly dissected corpse. All his sketches are vivid. In one a skeleton is resting its skull on the back of the bones of its own left hand, the palm of the bones of the right hand meanwhile resting on someone else's skull.

# ARCHITECTONICS

## "A Rich Embroider'd Canopy"

—Henry the Sixth, Part III

Vesalius was dead two hundred years when an Italian medical student noticed something that previously had gone unnoticed. This was the year when across the Atlantic gentlemen were affixing their names to a Declaration, the year 1776. What the medical student noticed was a white line near the tip at the back of a freshly removed brain. He made a report, and with it got himself an immortality, small immortality. Even today when one knows beforehand that that line is there it may be difficult to see. It is white against gray, fibers against cells. A landmark. The general region of brain is where arriving nerve-impulses somehow give to a mind sight. Sixty-four more years, and a French psychiatrist reported that the white line was two white lines, with a dark line between. Later the microscope was brought in to rove over the region and over the entire cortex, that rich embroider'd canopy. It was proven now to be teeming with detail, all previously unnoticed. The cortex of the brain had layers. There was a layer with few cells. . . . A layer closely packed with small cells . . . A layer of medium-sized cells . . . A layer . . . A layer . . . Whole lives would be devoted to scrutinizing that frescoed vault, no doubt sometimes with the secret hope of finding what in the structure of brain might be the basis of mind, that mystery, find what in anatomy might be transmuted into thought and feeling, in order that we might perchance more securely grasp what is at the foundations of us, find the lost language on the buried stone, buried brain. In a single year three independent publications divided the cortex of the total brain into small plots based on local cell and fiber arrangements. Maps began to appear. Cartographers opened shop in many countries. Areas of the frescoed vault were given numbers. Area 4. Area 6. Area 44. One cartographer said there ought to be twenty. Another said fifty-two. Another, more than two hundred. Under the microscope these areas were different and that was very likely because they had different work to do. Types of cells, the patterns they made, the molecules, atoms, ions, though chemistry had not yet really come to birth in this problem, all must be making possible the work, fantastically varying work. Some cartographers were beginning to say that other cartographers had divided too much. Too many maps. Too many small plots on each map. The overall labor—architectonics—had probably taken the anatomist as far as anatomy could. He was making himself stale by bending his head over what had become too trifling. This quintessence of architecture had probably reached the limits of what it could reveal. The secret of the vault must henceforth be sought in other ways.

# XVI
# SENSE

## Message from the World

# ANALYZER

## *City Desk*

An enveloping world of energies strikes our body—strikes the fire-fly's too. "The glow-worm shows the matin to be near." The glow-worm knows and we know, too, that morning has come again.

Monday morning.

Who does not know what the weekend did to his five senses. Or six. Or seven. He could dash out the brains of that eight-month-old brat on the second floor, born with ox strength and the devil and never beaten with the coat hanger. Mother-infant conversation. That flirt! What she looks like when she staggers out for the day, washed-out blonde, cheap hairdo, last week's perfume; lets that blackguard yell for sixty minutes without taking a breath then feeds him. Mark my word, he'll rape his landlady before he's fifteen. And why does she feed him? Because it's 7 A.M. Honest to God. Feeds him because the sacred clock has banged out 7 A.M. Sounds, sights, smells tumble down from that second floor. Slams the icebox door twenty times an hour. And the January wind rattles the windows. Drafts climb in over your shoetops no matter where you sit. Every-body knows Monday morning. Stomach stretched with Sunday dinner. Monday's socks still sensing Sunday's newspapers all over the floor. Muscles numbed from no exercise. Day of rest? In-laws just dropped in—dropped. If only spring and baseball and the violets would come. Damned wind, a beat in it, since eleven o'clock last night. On a Monday your heels jar your head and you do pay the five senses a wry respect. Praised be Jehovah that not any more gets into us than does, that each sense has its limits, its walls, its thresholds.

Through the wet air comes that one still remaining church bell, so close, one not only hears but feels it; sound, touch, pressure, all three are there; and then the bells in one's memory, the bells of history, the gonging towers of Italy, that single gong after they burned him to death at the auto-da-fé in the piazza in Florence, and that bell of John Donne that will toll off the printed page forever: "Never send to know for whom the bell tolls. . . ."

But our walls we do have. We are mathematically walled. Touch, heat, cold, pain, taste, sound, sight, smell, each has a minimum before the attack from that sense begins, each a maximum where the attack shades off into sense-nothingness. Were it not so, this frail construction, this nerve-infiltrated flesh that each of us is, would long since have been banged to death. Alerted body and head must be, perpetually, be peaceless in order to have peace, to stay safe, normal, gay—gay when we are, is in spite of any grey-beard beyond the Milky Way.

The stimulus?

Its characteristics?

Size—must be big enough. Intensity—must be strong enough. Duration—must last long enough. Rate of change—must increase or decrease fast enough. The moon is big, is bright, keeps shining, changes from instant to instant, and for those reasons though it is far away we see it.

A curious fact discovered for one sense, then for another, probably true for all—the nervous-system can of itself to a degree determine whether what reaches the body shall be weak or weaker, strong or stronger, can switch off, somewhat, the brat, the icebox door, the blonde's husband who whistles, can do this not only by an act of mind, which each of us knows his mind can, but by mechanisms inside the analyzer.

Another curious fact—each sense is able to adapt. I am permitted to tighten my belt to the third notch, for a few seconds feel the pressure, then the pressure adapts, and I tighten again, feel again, adapt again, and if I am fool enough to keep that up to the point of pain, the system poorly adapts to pain. I am not permitted to become indifferent to pain.

Receptor?

Anything that receives a stimulus. The real or imagined number of receptors has increased year after year. A receptor is any living instrument, chemical or physical, that picks up any kind of news, from outside the body, from inside the body. It is an antenna. Myriads pick up news. Large populations are packed close together in places like ear, eye, nose. What the receptor does is invites in one kind of energy, keeps out others, lowers the threshold for one, raises it for others.

Analyzer?

Anything that analyzes the sentient world. In 1969 a young neurophysiologist with a micro-electrode in his hand may shiver at that definition. He knows how inexact. Admittedly *analyzer* is less exact, more general, more inclusive than *receptor,* more nervous-system involved. Besides receptor and the news wire that carries the message to spinal cord or brain, there is the often baffling wiring that handles the news inside cord or brain.

The term connotes also why we creatures lug about with us this equipment of radios, transistors, cameras, ear trumpets, to sense a universe that must for the purposes of body and head be ceaselessly analyzed, each sense in its manner, with its limitations, its ends. The higher the animal the more comprehensive the analysis, though not necessarily the slighter the changes picked up from the circumambient. The dog smells well. To the mind of the universe, if that is not wholly metaphor, what could be more significant than the sense enrichment of which each of us is so rightly jealous, but that is also defense. Sheer survival may be the whole evolutionary intent of this built-in machinery. We can detect more quickly, more positively, more offensively, therefore survive longer, which may be philosophically ridiculous but Nature is a practical person.

# SKIN SENSE-ORGAN

## *At the News Front*

Our skin, that outer barricade, is crowded with receptors.

An ordinary microscope easily sees these sense-organs. There is a great assortment of shapes and sizes. One may look at one of them, know what histology and physiology have taught us this one does, remember at the same time how physiology has taught us to doubt them.

A thin slice from the tip of a man's amputated finger is stained, and the sense-organs observed. A bare terminal nerve-fiber apparently performs as one. Another more definite organ looks as if it might register weight, and does. Another reports heat. Another cold. A still more definite one, large, lies almost at the center of the illuminated round field. Appears to be a capsule. Appears to contain a jelly. Sunk in the jelly is a nerve-fiber, the jelly probably protecting the nerve-fiber's membrane in such a way that it may safely be stretched, and, consequence of the stretching, a rise of electrical potential and nerve-impulses starting conventionally toward headquarters.

At almost four o'clock of the illuminated field there is a pacinian corpuscle, famous, a Pacini, described a hundred years ago by Filippo Pacini. Filippo's hobby was electric fishes.

The receptor we are looking at is enormous, two millimeters long, visible without a microscope, has been widely experimented with because of its size and its comparative isolation from other sense-organs. There is space around it. Possibly it is the best-studied single receptor of the entire body. It is found where pressure must be appraised. That would be particularly toetip, fingertip, joint, tendon. A tendon moves, presses, the Pacini is stimulated, the news telegraphed. During an abdominal operation the Pacini is seen everywhere in the gut neighborhood, also in the mesentery that attaches gut and other abdominal organs to the body wall, is seen near the pancreas, where it could be registering the shiftings of the heavy structure, or registering local changes in its blood flow, or both.

The Pacini has been experimented with inside the abdomen of an alive cat. Being built in layers, the Pacini has been likened to an onion, and when the layers are compressed by some movement in the body there is again a rise of electrical potential, threshold reached, message to headquarters. Recently this small instrument has been cocainized in one of its parts and not in another, the parts thus separated for study, a fine electrode stimulating, this in the alive cat. For the stimulating a crystal has been used, the crystal pressed, producing a so-called piezoelectric effect.

One could say that there are tiny onions far and wide just underneath the surface and in the depths of all of us, registering, registering. "We are such things as dreams are made of."

Filippo was twenty-three years old when he described the Pacini.

Many different shapes of receptors are in the skin of that amputated finger. Bulbs. Discs. Cylinders. Beaded nerve endings. Some are more bizarre than others, some merge into others, some become unrecognizable by the merging, dissimilar similarities everywhere. The specialist freely admits that he often is not sure just what he is seeing let alone what is being reported, and whether the report from one part of the body by the same type of receptor is the same as the report from another part. He states what he sees, and waits.

Look now at your finger, or at a finger in your mind, see that teeming inventive surface. See those toy-telephone-receivers everywhere waiting to get a ring, some possibly waiting a lifetime and never getting one.

The skin of a millionaire's abdomen may have only a few poor hairs. Poor naked millionaire. However, each hair is sunk in a moat, and the moat is crowded with receptors. This is the kind of all-present fact that one sets oneself to imagine only on the second Tuesday of the third month of the seventh year at midnight. A first millionaire's hair, a second, a third, is giving notice of the approach and departure of a humble spider; but it could be a black widow, or a malaria mosquito decorously walking on stilts. A millionaire's shaved chin and a lady's shaved shin should then be less alert after a shaving than before, and so they are. Whiskers around the mouth of a rat bring in news from distances twice the diameter of the rat's head. Deprive that rat of whiskers, it dies. It needs its whiskers. The competition of life is hard and each of us needs all the sensory wealth he has. Poor millionaire—cannot pile up a bank account of receptors. In insects, the tip of one receptor comes up through a hole in the insect's hard shell, and a smart male child pulls and pushes at that tip, explains to his father how the contraption works. (A female child might not trouble to explain, or explain better.) Some airplane pilots wear belts of electric vibrators around their waists so as to keep themselves alerted via their skin, it by itself not being quite enough. The pilot needs this additional information.


# SKIN PATH

## *A Message Goes to Headquarters*

The message must get from the person's skin to headquarters, to the central nervous-system, to the brain. An energy—from near or far—meets the skin. A receptor is altered. It makes its particular use of that energy. (Always many receptors, of course.) The message travels inward over the skin path. It travels in a monotony of nerve-impulses that are faster or slower, never larger or smaller. Click. Click. Click. The Morse code has dots and dashes and a changing rhythm, not the nerve code. The single nerve-fiber in the bundle of nerve-fibers does have its own size, hence the impulses over it their own speed, hence what goes over a bundle of nerve-fibers would have a multiple, a complicated advancing time pattern. Also, there are the different anatomic routings. The patterned message does finally get to the top of the brain. Possibly that patterning has not much to do with the result.

But somehow there is the sensation. The mystery of that remains what the mystery of sensation is, has been, possibly will be.

Touch paths, heat paths, cold paths, any of the skin paths have a length that depends on the body's geography. A path may start farther away than a man's ankle, travel in a trunk line up his lower leg, upper leg, pelvis, spinal cord, cross to the opposite side (can cross higher), thence through the brain-stem, reach its destination on the outside of the brain, the cortex. The sensory cortex. That destination is a point in a strip of that cortex, of what might rightly be considered a most marvelously frescoed surface, the strip being about midway between the front and the back of the brain. Maps of the sensory cortex have been produced in three ways: (1) In human beings, successive points have been stimulated with electricity during surgery, the patient under local anesthesia and able to state what and where he is feeling. (2) In an animal, a small area has been damaged, the animal not able to state but observations made on what it does that it did not do before. (3) In an animal again, a monkey, the skin has been touched at successive points with a pin, or with a feather, the electrical consequence picked up from the strip, from the point that corresponds, the point recorded. A map made in that manner has taken twenty hours, the experimenter needing to keep hustling because at the end of the twenty hours the monkey was dead. The maps produced in these various ways indicate that the maps produced in man correspond to those produced in animals.

An experimenter studied the cortex in a pig and found that the snout does the message sending. The four chubby short legs, the barrel belly, immense though these are, had no space allotted, or none was discovered. Farmers verify this in their own way, claim the snout is the most intelligent part of a pig, that the pig digs, explores, eats with the snout. Also, the pig is sensitive in the skin of the snout, and farmers apparently know where to deal with a pig if it gets piggish. Pigs in Scandinavia like everything in Scandinavia look clean, and if a red cherry is held in front of a pink snout pushed through a clean hole in a clean fence, and if the cherry just once lightly touches the snout, it steers successfully to the cherry, plucks it neatly off its stem, and the traveler, even if skin path had not much to do with it, will find years later that he has this occasion safely stored among his recollections of the fjords and mountains of Scandinavia. The laboratories of Scandinavia have made some of the shrewdest of modern studies of the senses.

·Our skin, our sometimes tired yet usually willing skin, that barricade with its sentinels, is with us early and late and mostly we pay no attention to it. A warm bath on a Sunday morning while the

church bells are ringing through the March air caresses the mind
with the smooth water and the friendly temperature, skin paths
busily delivering coded messages to headquarters, the human mind
not bothering to question whether what is happening fits into any
classification of touch, pressure, et cetera, when without warning
there is an absent-minded turning of the wrong faucet and the soul
is slapped with an arctic blast. What dozing skin receptors, dozing
skin paths, dozing analyzers, flew into action at that instant. Then
perhaps one nevertheless realized again the pedestrian order of
everything. A warm bed cover lies innocently upon us, our skin
receptors adapted, hardly reporting the touch and pressure, we
soothed toward sleep, when a crumb of last night's snack uncouthly
awakens a crowd of the sleeping sentinels, instantly whole analyzers
in action. Or there is an itch. Or there is a tickle that does not make
us laugh. Tickle is touch plus something slightly destructive. Everywhere coded messages are traveling from the outposts of a bel-
ligerent who is defensive when he is forced to be, offensive the rest
of the time. Despite all the accumulated knowledge, all the tracing
of skin paths, all the mapping of sensory cortexes, who in hell or
heaven could account for the feeling of velvet rubbed against him in
the dark?

# PAIN

## Complaints

Why do we have pain? Why do we have this sense forever waiting?
Someone, too indiscriminating, called it music. Minor-key then.
*Doloroso.* The whole orchestra, all the violins especially, first, second, viola, cello, bass, now and then a complaint from the oboe, a
crash from the brass, a rumble from the drums. Why? Is it only
what everybody believes and tells everybody: Nature's determination for her own purposes to protect the living creature, that is
tough and delicate, that can take only so much damage if it is not to
go back to the chemical beginnings again? The creature must be
forewarned. Pain is taking thought. Pain is preventing destruction

going too far too fast by making a quick announcement whenever anything has begun threatening the flesh, and something almost always is. But Nature—she is only maintaining those actuary tables. Pain's concern is not for the single ones, Mary, Michael, but for the population, for the species. How heartless—from the human point of view. Needless to add that no teleology is meant, none implied. Anyone is invited to think of these things according to his temperament, and, while he is thinking, pain will be preventing his getting careless about any absent-minded pin pricks or automobile crashes.

We inherit the capacity for pain. Now and then someone does not, is doomed for his life long never to feel it, not in his depths, not in his skin. We wonder about that man's mind. We try to picture it. He bleeds into a joint, tears off a fingernail, steps casually down from a chair only to find later that a part of his body does not operate as it should. The X ray shows that he has hurt himself grossly. They splint his leg. That should hurt. It does not. A strange world he is living in, we think. He does not. He leans against a hot stove and smells burnt flesh somewhere in the room. Surely that mind is blighted, is it not? When he gets older he may join a circus and be paid for letting people see him push needles through his cheek. He might—one such was—be prevented driving nails through his feet because he thought he was Jesus Christ.

Usually there is a lessening of the pain-sense rather than such a complete loss of it. It might sometimes be, instead of the inheritance of painlessness, the inheritance of a temperament for extreme self-discipline. When he was young he was a good child; when he got older he was a courageous man. In Sparta the sensation of pain was experienced, but the national rule was, deny it. An Irishman, a father, would not pay his pain so much as the courtesy of recognition. His doctor believed the pain frightful, shook his head. "A patient Irishman." His wife objected to that. "Why shouldn't he be patient when he has brought the whole thing on himself?" She was German. Her mother, his mother-in-law, endured the pain of her cancer in a silence that accused all those around her of their good health. She was German too.

Pain might begin in the fetus in the womb. It seems to sometimes. Any psychology, any sensation occurring in the womb, would have been a theme for comedy to Laurence Sterne, for mystery to Samuel

Taylor Coleridge. That in the womb was Act One. The newborn on the first day of life had a cramp in the belly. Act Two, Scene One. The eleven-month-old slipped off the baby-sitter's lap. Act Two, Scene Twenty-three. Thus on, the pain of this life always having somewhat the quality of this particular life. A distinguished American regularly fell asleep in the dental chair, bragged about it, but that falling asleep was part of him, of his nervous-system, his lethargy, his philosophy, but also his dampened pain-sense. Besides this pervading quality of person, pain does depend on other sensations, other facts, on the current state of his world, is more severe or less severe yesterday than today. The sharpness comes and goes, for an hour is larger than the universe, then is nothing. It would be sure to be chalked up as hysteria by the neighbors, and it may be, or partly be.

Pain is forgotten once it is over. That often strikes us as remarkable in ourselves, though commonly it does not strike us at all. We just forget. A mother screamed while the head of her baby was pushing open the exit from her birth canal, which women say is a special pain, yet the next day she remembered clearly only that she had had a bad time. Some kind of thrifty economy of our mind requires that. A man or a woman may, we all know, on the contrary nag at the memory of a pain, and then be caught having forgotten it. One must be careful in interpreting. Some edge of the damaged flesh may after all be nagging right now.

A tall woman with syphilis of the spinal cord had the pain of her disease for twenty years. Drugs did not relieve her nor surgery. "They operated on me four separate operations to take away my pain, and look at me." Poor skin-and-bones, pain had used her up, and she groaned, moaned, squirmed, feebly. She would have been worse off had the pain been present for twenty years and she not moaned, been muted by it, as happens. A man cried from his open window into the street when a kidney stone passed through his ureter, and though his ureter was the same as anyone's, and the cause of the pain the same as anyone's, namely drastic distension of a living tube, yet even that pain, that choice variety, as every doctor knows, will in some definable respect be that man's own. It has taken on his character. A big-bodied man let out all the vowel sounds in the English language while a gallstone crept through his gall duct, but even to an ordinary ear his vowels were his vowels. Each of us does think of his pain as his own, nurses its privacy, may exaggerate in speaking of it, may minimize, but he never lets the pain become a public affair.

All over the planet we and other creatures crouch in all gradations of silent or verbalized pain. It is a painful planet. We have so much practice in the verbalizing. We do it well. Nevertheless, every doctor accepts how roundabout he may have to be to find the facts of a pain. It is part of his routine. It may be talent. Does the pain throb? Does it burn? Does aspirin stop it? Where is it? Is it? Is it pain or is it only unpleasantness? Is it nothing at all? Is it not on some other subject? What is the subject? Why the vagueness? At a moment there may be no vagueness at all. The pain may be as exact as a zero morning, unlike anything on earth or in heaven, "and shall I couple hell?" Even the bitterest pain keeps its fringe of other things, its mixture of accompanying emotions, of sometimes unrelated thoughts, sometimes the mellowest of a man's life. One man in the clinic nodded his head as if there were someone in space. "Good old pain," he said. A woman writhed when a paroxysm knifed through a corner of her. "Have had that happen ever since I was married." She had got that stated. She immediately quieted. She pointed to that corner in her as if the withdrawing spectre must be visible to anybody, and in her face the haunted look transformed to weariness, she appearing to be meditating on the meaning of things, or perhaps she was just daydreaming of thirty years ago.

Headache that old hag is in the head, somewhere up, has done its part to dampen joy, discourage Pollyannas, deflate administrators, defeat psychiatrists, dent sobriety, make the confessional and the tax collector unimportant, reduce to size gossip-column morality. It may start from teeth, sinuses, eyes, muscles of the neck, membranes of the brain, hemorrhage in or around the brain, harassed blood-vessels inside or outside the skull, many flesh causes. Then there are the mind causes, the mind as elsewhere employing anatomy and physiology. Should there have been tension in one's mind, and one's blood-vessels kept in prolonged spasm, then the tension relaxed, and one's blood-vessels relaxed, over-relaxed, pulled on, dragged on, that is one headache. The mechanism has been described so. But apparently either the spasm or the relaxing or both can be the cause. A man all week has suffered anxiety and has been constricting the blood-vessels inside or outside his skull, feels a gnawing but not much more, then comes that Monday, his vessels suddenly give up, dilate, over-dilate, pay him back for what he did to them.

Trouble in a man's life may be the cause of a specific headache, migraine, that is nauseous, usually quick in onset, usually accompanied by what briefly might seem serious neurological signs, usually on one side of the head, attacks recurring over years, then he gets bored, it appears, lays down regular money for a novena, visits an understanding doctor, visits a psychoanalyst, marries an artist, his trouble lessened or anesthetized, or he has gotten old and everything has lessened, and he may not even know when the migraine finally departed. A few physicians still insist that all migraine is invariably an allergy.

How curious despite their commonness are some of the facts of headache. An occasional headache has a bull's-eye precision and a thrilling quality, and a man allows the thrilling because he knows there is no danger, as when something as agreeable as cold ice cream in the mouth activates something as disagreeable as that icy fire high in his nose. A surgeon while operating on a brain under local anesthesia drags on a blood-vessel, the patient instantly experiences a headache that also has bull's-eye precision, may also have the thrilling quality. The man was not expecting a headache at all.

One could write the entire history of body and head, and not once leave this evasive elusive intransigeant subject of pain, as one could write the entire history and not once leave the subject of the blood. Indeed, one might write the whole history of mankind, and animal-kind, and cover the ages, and not leave the subject of pain; then choose not to publish the book.

Why, one asks again, did the blind spinner spin this sensation? The dying Claude Bernard would have been willing to accept pain were there any use. What he meant—use to him. He was dying! What use? But pain grinds on when there can be no imaginable use. It must do that. It must be indifferent that this single wretched man, Claude Bernard or fool, pays for the good of the rest. When someone says "good of the species" it cannot but be a bitter realization to that single wretched man. He is stricken with untreatable cancer accompanied by unbearable pain. He is being alerted to nothing that he does not already know. But Nature reasons these matters in her own way. Pain alerts the flesh, all the flesh of earth, and so saves every possible one from annihilation. Claude Bernard must accept. There are good people who would insist that Claude Bernard's pain was of use to him, the highest in him, strengthened something more important than his body, his soul. Generations have said that. The truck-driver husband of a pregnant woman wanted to know why she made such a fuss, her pangs could not possibly be that bad; women had borne babies before her. Had that excellent male that instant then been kicked by that woman in the scrotum he might have reevaluated the situation. Claude Bernard's racked body does for the two-thousandth time make us reflect that we are all damned in the end. But why dwell on Claude Bernard? Because he was a great man. What happens to a great man gets the emphasis of his greatness. It should. We are prepared by his experience for our experience. We all must probably face pain, all must inescapably face death. Claude Bernard's pain would of course in fact not have been different from that of an obscure woman. The blind spinner would lightly add agony of flesh to agony of spirit in a dying woman in an empty room at two in the morning, then let her die, no one to benefit, the experience lost, buried, if you like, shortly before her body. The blind spinner is blind. And is stupid, if intelligence is measured by any of our measures of intelligence. Whatever the discrepancies, quick warning, and quick exciting of defending re-flexes, is in 1969 still the entire known reason for pain.

# PAIN PATH

## *Bad News Gets Priority*

From the skin inward to headquarters the path for pain is much like other skin paths. Two questions arise. Does the pain path of itself tell us the location of the pain, and does the pain path explain referred pain, where the cause of the pain is in one place and the sensation in another?

Beaded naked nerve endings, no insulation of fat, are the pain receptors in the skin. Soon the nerves lead into treacherous-looking nerve nets that overlap and underlap. There are two sizes of nerve-fibers. Large fibers carry the bad news swiftly, are responsible for the brilliant burning of pain. Small fibers carry the news slowly, are responsible for keeping the pain hot. Together they sometimes make a Dante's Hell. The slow news reaches the top of the nervous-system instants later than the swift, the two sensations quite separable in the mind, as can be roughly tested by giving a sharp crack to some faraway spot like one's ankle.

When the cause of a pain is below the head, the path leads toward the spinal cord, when in the head, leads into the brain. The paths cross, right to left, left to right, then continue centrally into the nervous-system: medulla, pons, midbrain, thalamus.

The thalamus is the great way-station for the senses. From it paths go directly or indirectly to the top of the brain. Along the course of this total routing, the pain path, particularly along the higher levels of it, somewhere, somehow, by a chemistry, by a physics, possibly by both, possibly by neither, there occurs that miracle (miracle if one could be objective about it), usually disagreeable, sometimes unbearable, the sensation of pain.

The location of the pain? How do we know where we are suffering? Some specialists have emphasized the path, the anatomy, where the path started, where it went. Some have emphasized learning, what the creature taught itself, how it saw the location of the damage, where the pain was caused, saw that location again and again and presently knew the location, put down a mark on an uncharted map, kept that up and so produced a charted map. Infants and children hurt themselves hundreds of times, and that would be the period for rapid learning. The other senses would have helped, touch especially. And as with any learning some would have been acquired from the experiences of others. Then one day when the left foot was damaged, the brain now having its map knew that the pain was in the left foot. Had the damage been in the left ovary the map would have been less useful, but useful. Those persons born without the capacity for pain, the "painless people," still meet a destructive world successfully, else their bodies would not have survived.

Parts of our bodies are less equipped to experience pain than other parts, fingertips less, but fingertips more equipped to experience touch. Arms and legs at their junction with the trunk have four times more pain points than touch points, and, whatever the explanation, the creature would then acutely know when destruction was starting in the direction of its vital organs. The surface of the eye, the cornea in the middle, as every person has experienced, is highly sensitive to pain, and in that middle are found the naked nerve endings. Recently, by testing the cornea with nylon threads of different coarseness, it has been possible to excite touch separately from pain, so there must be touch receptors there. Jets of air of varying temperatures directed toward this area have shown it sensitive to temperature.

So, the cornea senses pain, touch, heat, cold, but has only naked nerve endings. Naked nerve endings must respond to several kinds of stimuli. Furthermore, in the skin pain-points and touch-points and other points appear to change their localities from day to day. If these facts and some more like them are correct one would say unarguably that we do not yet understand all of pain's mechanisms.

Pain paths exist that will not be called upon once in a lifetime, nevertheless stay vigilant.

What explains referred pain? We all experience it sooner or later. The heart is ill but the left arm aches. Physicians every day make diagnoses with the help of pain that has been referred to the surface from organs inside chest or abdomen. A theory is that the paths from the damaged organ and from the healthy skin reach the same junction in the nervous-system, this junction barraged by stimuli from the organ, so that the mildest stimulus from the skin, that ordinarily would cause no sensation, now does. Brain and mind have had much experience with the skin. The species gained by that. An animal was guided by pain in the skin to pluck out a thorn from exactly where the thorn was. The species would probably not have gained had the animal been able with equal exactness to locate pain in an inner organ; there would have been little use in distinguishing an abscess of the liver from an abscess of the pancreas. Vague pain might have reduced the animal's activity, and that might have been useful, whereas a geographic locating of pain in a deep-lying organ might only have inspired the animal to lick and gnaw and bite and increase the damage. We know how animals bite at themselves with the greatest enthusiasm. Sex often has excited them to a wild activity in spite of painful destruction of the flesh. As for the top of the top, the cortex, possibly it is not as much concerned with pain as has sometimes been thought. Assuredly, however, it has been concerned with teaching the human being how to alleviate pain, how to write a prescription, how to extract and concoct and administer morphine. The extracting of roots, seeds, leaves, flowers must have been among man's earliest skills and calling on his highest nervous-system capacities.

# SMELL

## Gossip Column

On Easter morning moving in the direction of St. Patrick's Cathedral on the other side of Fifth Avenue there went a small bouquet of violets. When the smell particles had crossed the avenue they were so diluted that not even the micro-methods of the modern chemist-physiologist could have proven that anything noseworthy had walked that way to the Easter service, but it had. The blind man at the corner knew perfectly.

Methyl mercaptan when diluted in the proportion of one molecule in fifty thousand of breathed air, registers garlic. From that it follows that if anyone dislikes garlic it is always spoiling his life, whereas an admirer can tell whether it was three times or only twice that the chef waved the clove over the salad bowl. A delicate sense, the sense of olfaction, truly. Artificial musk registers when the concentration is four one-hundred-thousandths of one gram in a quart of air. That is a smell-fact repeated book-to-book. Skatole, putrefaction molecule dancing in human residue, reports at $0.4 \times 10$ to the $-6$ milligram liters. Oh, well. Animals often have saved their lives, successfully battled through brute history, because they were specialists in the manufacture, the recognition, the study, of excrement. So you and I will simply have to bear up. They are dedicated students. They and we share the one world together, sty and laboratory.

Yet a pig's smell picture, even a dog's smell picture, is confined in a frame, and insofar as a dog's thoughts derive from olfaction they can begin only on one side of that frame and go no farther than the other. We are things framed. We are creatures hung in a gallery, and what the gallery is we never know, but we operate within its thresholds—within thresholds.

To the smells that are ours, many, lumped this way by this specialist, lumped that way by that specialist, add the four primary, the conventional tastes—sweet, salt, sour, bitter—to give the flavors, and the possible sum of that comes to such a richness that it is a complete mystery why cooking can be so bad. In themselves the tastes are dull things. One would check them off as mere background were it not that they had other importances, help our automatic selection, lure us to substances our body requires, table-salt, calcium, vitamins.

Taste buds are the receptors, odd-looking microscopic crockery distributed over mouth, throat, tongue. As we get older they atrophy, but they do it kindly, so gradually that an old man has to be taken aside and told quietly that, really, he is not tasting much of anything any more. He thinks about that for two minutes. Then he agrees. It might hurt his feelings if he were told too suddenly that he also was not smelling much of anything any more. (That last has advantages.) On the tongue the taste buds make a pattern. Salt and sweet are mostly at the tip, bitter at the back, sour at the sides. The taste-pattern travels inward over taste-paths, meanwhile the smell-patterns are traveling over the smell-paths with a way-station where there are opportunities for modification, increase or decrease, excitations or inhibitions, modifications, known and unknown, and earnestly studied, in the electrical characteristics of the neurons and their junctions, as well as all possible characteristics of the receptors, complications and combinations in the flesh that anticipate if they do not explain the complications and combinations in the mind where chill mechanism somehow becomes warm experience.

Ammonia draws in a different sense. They call it common chemical sense, a burning. For us it is a burning, and what it is for an insect we do not really know, but it strikes us, with evidence, that it is especially useful for insects.

Insects and vertebrates were evolved over widely different routes yet have taste talents and odor talents that sometimes are quite near, or apparently. Their sense-world is forever closing in upon them too. The minimum of sugar that can be tasted is nearly the same for bee, butterfly, man. The mind of that last one is always trying to play above the minds of all of them. Minds? Call it what flatters you. Bees have been used by man as experimental animals—funny—to investigate the so-called stereochemical theory of odor, and have been claimed able to recognize the same primary shapes of molecules as we, meaning, not flippantly, that they should be able to enjoy the smells of our flowers, that bouquet on Fifth Avenue on the way to St. Patrick's.

Everything here is easy to give words to, difficult to give satisfaction to scent-pursuing scientists. Some of these believe there are seven primary shapes of molecules for odorous substances, these, imaginably, fitting into shapes of smell-receptors, or some more chemically-adapting ingenuity. Those receptors are in the olfactory membrane, that is a sheet spread out, about a square inch in size, in the high attic of each nostril, protected there from bacteria and dirt and cold. A few odorous particles would rise up, warm as they rise, dissolve in the fluid that bathes the receptors, a quick transmission, the mind get a nudge, the nose give a sniff, many particles accordingly rise, the mind unmistakably informed.

That olfactory membrane is yellow, and color appears of some significance. Negroes have a dark membrane and great sensitivity to smells. Albino animals sometimes are poisoned because their smell is poor and they eat what they should not.

For the rabbit smell substances that are soluble in oil are said to stimulate the back of the olfactory membrane, substances soluble in water, the front, some degree of solution in oil or water always deemed necessary, the rabbit frequently used for experiments in olfaction because its smell-system is easily approachable surgically, the overlying bone chipped off the system, and the parts of it more easily separable than in other animals. All questions can be studied more singly. Also, much learned from rabbits appears true for man.

Blunt naked nerve-ends equipped with five or six hairs poke

through the olfactory membrane. Each nerve has two parts, two poles, one directed toward the cavity of the nose, one toward the brain. Thousands and thousands of such make a bundle, the bundle passes through a perforation in the skull, entering a longish bulb, named olfactory bulb. In smell-talented animals this is enormous, has two principal types of cells. Is a bloody area. Is actually a small brain. Cajal studied it microscopically and in late years it has been studied electronically and with other stimulating energies. A rhythm comes off it. This was pointed out by another winner of a Nobel Prize. He called this intrinsic rhythm. (We could always think we were in a fairytale except that for the time we are scientists and maintaining ourselves in the coldest daylight state.)

Up into the nostril sneaks a smell, and that starts something electrical. One rhythm might consequently be playing into another rhythm, and smell-pattern after smell-pattern advance into the brain, and a man be seen to look around him and say he has a haunting thought and does not know where he got it. The thought departs. The smell adapts. The adaptation could occur either in the receptors in the nose or in the depths of the brain. All that happens in the brain, the known, though much no doubt happens, could no doubt be shrunk into a not overlong English sentence. In the nose, then, an extrinsic something induced by a smell-substance meets a receptor and an intrinsic something generated by the bulb, the intrinsic yielding, slowing, stopping, the mind getting the message, then the intrinsic returning, the mind dropping the message.

Dulled for skatole the sense of smell still would be fresh for Chanel No. 5, then would adapt, would dull for Chanel No. 5. But one does not need to be a physiologist because who has not started into a roomful of friends, been knocked back, continued on in because he was polite, soon found himself happily panting with everyone else.

Far far back in time and in the old old world these senses were evolved in the seas and rivers where the one-celled as now was lured by food. When this lure acted from a distance, that was probably the ancestor of smell, and when from nearby was probably the ancestor of taste. Their dinner floated past those hungry ones and they partook. How long living things have been hungry. Were one to add up all the hunger since hunger began, how awful.

It is stated that if a wad soaked with meat juice is held to one side
of a catfish it snaps to that side, and if the pertinent nerves are cut it
no longer snaps, nevertheless continues to distinguish salt from acid,
and this would suggest that the catfish employs, let us hope also
enjoys, both smell and taste, therefore enjoys flavors, as we do. Nice.
Some insects walk in their food, have receptors on their feet, that is
to say smell with their feet, or taste with their feet. Not nice. Cock-
roaches in the kitchen were annoying enough as it was. A fly has
taste receptors on its feet, and if a left foot were not to pick up a
strong enough impression, the right foot would reinforce the impression. Some insects have taste receptors on feet, on antennae, in
the mouth, and their world should be rich. Again though, and
unfortunately, we do not know about the sensation side of the
matter, having in fact confusion enough with our own. To a male
eggar moth a female eggar moth is recognizable at two miles. He
presumably thinks of her in smells. A male dog communicates with
a female dog in the same, valentine greetings. To a hound and a
bitch sitting together in the sun on the porch the fresh earth of
spring is a Haydn trio, quartet, symphony, whatever their word for
it is. She slightly turns her head, he turns his, both heads in the best
position for the smell music.

A human being has a nose too, minimized when we think of a dog,
for ours surely is not as acquisitive as a dog's, or apparently a
moth's, a shark's, nevertheless quite adequate to convey the quality
of a steak, a spray of jasmine. Also adequate to lure a small boy to a
peanut stand, his father to a house he meant not again to visit, his
grandfather at the beginning of the century to a midnight bakery,
coal-hot ovens from a cellar that filled with odors that city street,
right by the side of a stagnant canal. Like Amsterdam, Holland, it
was, but plus the hot bakery. "Rose is a rose is a rose is a rose." Not
true, but could be true if anyone knew what Gertrude Stein meant.
She said it that way because Gertrude Stein could say it arrestingly
no other way, and she tried to be arresting, and she looked as if her
olfactory sense might be scant.

Some psychiatrists say that some odors are acceptable and some
not, that we never can be sure which, if shoved under and denied,
will entangle our mental life. Furthermore, they say that odors from
the depths usually are bad, suggesting that life at its roots is bad,

rotten. Neurologists are concerned with odors because they lead sometimes to diagnosis, a frontal-lobe tumor sometimes detected because of a changed acuity of smell. Brain surgeons say that the odors awakened by stimulating the brain usually are bad too. The family doctor is concerned with odors. The garbage man is. Artists are. Maupassant enriched many a page with an odor. Baudelaire did. Possibly the air around those Frenchmen was olfactorily more enticing than that around us but likelier that their noses were more sensitive, born more sensitive, genetically more sensitive, but more disciplined also, though neither as sensitive nor as disciplined as the noses of their dogs digging in the dank French soil of April.

# WAVE MOTION

## News Flash—Tree Fell in the Forest

"Hear ye! Hear ye!"

That now is the Town Crier. He is in love with his cry. His cat—his dog always sleeps on the opposite side of the fireplace from the cat—is not. She has turned him off, has switches at several points in her telephone system, in her brain, to do it. If all switches were open and she chose to give her mind to his cry she could hear it over its entire range. Hearing brings to cat and dog and Town Crier that part of the heard world that each needs to remain cat, dog, Town Crier. For us, when our body does not at the moment need some part of the heard world we still have a type of mind that easily builds up a curiosity about it. There was the quarrel in the next booth. First it was all in whispers. Then suddenly the girl screamed. The boy shouted. She sobbed. In unison they continued decrescendo, adagio, molto adagio, molto molto adagio, and the waiter turned out the lights. He has been a waiter a long time. In the dark one could not see his face but one knew him by his tones, Spanish, almost song.

The stimulus for hearing?

A stir in the air, a wave motion, a complex wave usually, a wave that repeats, and for any animal-society to pay attention to it, it must fall within that society's hearing span. So there are the limiting walls. For all the senses there are the limiting walls. Where the walls stand varies sense to sense, individual to individual, and, inescapably, species to species. We live our lives in the sense-hedged ghetto or garden where we were born.

As for the wave motion, it is molecules of air swept together by some force, the shout, the sob, then immediately because of their elasticity springing apart. Swept together. Springing apart. Hence a wave. Textbooks speak of condensation and rarefaction of air, speak of increase and decrease of atmospheric pressure, so-and-so-many times per second. That is to say, a frequency. This then sets up a corresponding frequency in the appropriate machine-parts of some living body, and after various handling within those machine-parts, not all of them understood but investigated, a mind hears, and that item is not understood at all. The source of the frequency may be inside the body itself. Everyone remembers a humiliating gurgle at some intimate silence when he was young, the gurgle transmitted from inside him to outside, to air, to ear, to mind, someone else's mind, that gurgle a genuine misery. Most of the time the source is outside the body, the call of a bird, the rasp of a locust, the wail of a police patrol. Empedocles, the Greek, with his earth, air, fire, water, did already recognize that the nothingness around us is not nothing, may even have recognized that it is full of tremblings, but hardly could have recognized how complicated these are.

A musician strikes a tuning fork. The prong of the fork moves one direction, compresses the air ahead of it, leaves an emptiness behind, air rushing into the emptiness, is pressed in, is sucked in. Then the prong moves the other direction and everything is reversed, that back-and-forth repeating until the energy put into the fork has been spent and this locale is at tonal peace once more. The back-and-forth traveled at the rate of 1,180 feet per second if through air, four times that fast through water, fifteen times through steel. Air waves are not like water waves, cannot be seen, but the movement of the tuning fork can easily be recorded. One back-and-forth is a cycle and the number of cycles per second is the fork's frequency. Air's elasticity allows frequencies up to 100,000. Middle C on the piano is 256. Orchestras tune to A, a high A because that makes for brilliance. The lowest frequency to which the human ear can respond is about 16, lower than the low of a bass violin and heard more as a beat than a sound. Highest is 20,000. In children 40,000. The total span is about ten octaves. Music falls mostly within seven and one-half octaves. Supersonic is any frequency above the creature's range. Dogs hear higher than men. Insects higher than dogs. Bats guide their flying by frequencies between 50,000 and 100,000, first produce the frequency, then listen for it as it is reflected from objects around, a radar that the bat uses with an efficiency comparable to our vision. A housefly hears best when the frequencies are close to the hum of its wings, and the hum of other houseflies' wings.

One day man will know one thousand million such facts and this to the dedicated student is no doubt what he means by the millennium.

Male canaries have singing contests. Male grasshoppers do. It has been reported that if a male grasshopper goes to a man-built telephone and a female grasshopper thirty miles away hears his voice she sits straight up, or whatever a grasshopper does, anyway stays at the other end of the line to harken to his wave motions for as long as he is in the mood. A fish's machinery is evolved for two cycles per second, which would enable a five-star fish to eavesdrop on the promenades of a worm. Parrots hear less intricately than we but have excellent analytic capacity inside our speech range, and that is partly why a parrot can be so clever about imitating his Madam. She thinks him clever too. "That bird listens." She probably has only an anecdotal idea of the intricacy of that listening. The ordinary voices of her regular customers the parrot distinguishes as correctly as she. In many an impasse the survival of Madam or parrot or worm or fish or grasshopper or canary or housefly or bat is owing to the fact that the frequencies important to their daily affairs fall within their auditory range. Each has selected for itself—its ancestors selected for themselves—a spot of universe where friend or enemy will be neither supersonic nor subsonic. (Friend is he who is eaten, enemy he who eats.) For any individual the arrangements may not be perfect and an occasional mouse because of its squeak is the victim of a bat that drops between wires that it does not see and pounces on a squeak. The bat depends on overtones, and that merely adds oddity to all the oddities under heaven.

Where man differs from the others is that he can if he chooses stand off from frequencies, despite that he is always a prisoner within them, make observations upon them, draw science and philosophy and often error from them. When that tree fell in the forest one thousand million years ago there was the stir, the cycles per second, the spread, the dying out, but no living instrument to receive them. How careless of somebody.

# EAR

## *Receiving the News*

The instrument that picked up the pressure variations in the air, and did it with high efficiency, was of course the ear. Its receptors, in the case of our human ear, are set in a cavern in bone, this communicating with the outside via complicated corridors, and the bone is the hardest of the skull. Two ears are customary—two silent pianos waiting to be played on.

It took a long time to evolve an ear, many a hot afternoon of primal labor, the earth younger then.

Three divisions: external ear, middle ear, internal ear.

*External ear.* The pinna is that part of it that we see, that sticks out from the side of the head, has a canal that runs down in, and this we ordinarily do not see. A man cleans his a bit when no one is looking and a dog cleans his when everybody is looking. A dog has only a blunt paw to do it, a few fingernails, no effective fingers. It is the monkey and his relatives that have the effective fingers. Add soap and washcloth to the effective fingers, and that is culture.

Cartilage gives the pinna rigidity and cartilage blocks X rays. A medical student's ear in the dark against the ocean green of the fluoroscope is as diaphanous as the wing of an insect against a lamp, but tacked to a nightblack chiaroscuro skull that may appear wood cut out with a rough saw. Yet skull and pinna usually appear nicely proportioned in the daylight. Pinnas differ person to person, may be small, may be large, may be rolled, unrolled, objects of charm, gargoyles. Every woman who uncovers hers knows why, and knows

why if she doesn't, also why she hangs on pearls or seventy-five
blinding rhinestones; alongside gross ignorance of the body runs
cunning comprehension. A person whose pinnas stick out too far has
an early seriousness thrust on him. If he lacks one, was born without
it, the chance is he will develop into a sentimental altruist, a super-
man, or some other kind of psychological cripple. Efficient pinnas
may spell the difference between life and death in the antelope, one
graceful pinna lifting appraisingly in the breeze, changing its angle,
changing its angle again, the second pinna coming in to assist, the
grace of the two no less because they have important work to do.

Man does not especially need his pinnas for hearing, but he needs
his ears and he is better off when he has two. Two help in establishing the direction of a sound. If the impact of the air-waves is greater
on the left or the crest or some other part of the waves arrives earlier
on the left, the mind locates the commotion as on the left. If the
length of the waves is shorter than the side of the head and they
approach from the left, they are stopped by the head, must travel
around it to get to the right ear, the mind recognizes the delay, correctly locates the commotion as left. By this acoustics and others,
direction is established and even if a man is sitting in a bare room
where reflected waves rebound everywhere off the walls he establishes direction with surprising accuracy. Two ears are useful, too,
because they enable him to sleep on one and have the other free to
bring him bad news.

*Middle ear.* None of it do we see from the outside. It is part of
that cavern hollowed out of the bone, is filled with air, has three
ossicles inside the hollow, three small bones, and it is separated from
the external ear by the eardrum. Two hundred and fifty eardrums
stacked one on top of the other would make a stack an inch high, so
each must be thin, one-tenth of a millimeter, tough, presents a large
surface for the air-waves to pound against or lap against, is convex
toward the middle of the head. It is sensitive to anything vibrating
in its neighborhood, vibrates with it, provided the vibrations are in
its range, a soprano's thrilling tongue that has set into motion all the
air between her house and this, and all the eardrums in this, that
have had quite enough soprano but are forced to swing; also the
eardrums of the cat in the kitchen, and of the kittens on the porch in
the sun. At some frequencies an eardrum can respond to a vibration

that is no wider than one-tenth of the diameter of a hydrogen molecule. (The textbooks say that.) Students of the ear have wondered how our mind is spared the annoyance of hearing the flow of blood in our head. That might result from some mechanical block in the flow, some block in the nervous-system, some block in the mind, and might also be merely a matter of cues, because only the cues that are cues are cues, to keep up the doggerel. Our mind responds to what it wishes, as does, with some redefinition of the word *wishes,* the nervous-system under it. What we can be sure of is that no more than a fraction of the life-related energies of world and universe reaches our nervous-system, and only a fraction of that fraction, our mind, with no doubt an enormous unknown beyond, a fact that should be remembered by priest or philosopher or scientist, anyone feeling smug, anyone convinced that contemporary man is on the heels of final meaning.

The pressure of the air pocketed in that cavern of the middle ear is kept close to that of the outside air by a tube connecting with the throat, accordingly with sea level, accordingly with Pike's Peak. Old anatomists called that tube Eustachian tube; young ones speak of auditory tube. When an airplane rises or a train dips down under Hudson River, passengers swallow, open the throat-end of the tube, so equalize the pressures on the two sides of the eardrum, making those newspaper-reading heads more comfortable again. The heads only vaguely knew they were uncomfortable. Every mortal at intervals swallows for the same unimmortal reason. At intervals we spit, at intervals cough, at intervals pick our noses, yawn, stretch, scratch our backsides, and each of these performances has physiological or psychological purpose.

The three ossicles, bones, partly because of their names, partly because they caught our imaginations in the eighth grade, are more companionable than is most anatomy—a hammer, an anvil, a stirrup. Ossicles. Smallest of the body. The handle of the hammer leans against the inside of the drum and when the drum moves the hammer moves, and the anvil and the stirrup move because they are latched to the hammer by joints. The stirrup has what is called a footplate. That fits into an oval window in the wall between the middle ear and the inner ear. (We are being escorted through the house.) Those three ossicles act like a lever, and when the mechani-

cal advantages accruing from lever and drum et cetera are calculated and added, the force of the wave motion that arrived at the
drum will have been multiplied thirty times at the footplate, or,
subtracting for losses, eighteen times, or, by the calculation of the
most noted living student of the ear, twenty-two times.

Some say that the footplate moves like a door on a hinge, some
say like a piston, fast in either case. For the comparatively low
frequency of middle C the piston moves in and out 256 times in a
second. Two tiny muscles are attached to the ossicles, and these by
their graded contraction reduce the blow delivered by too-loud
tones, especially by low loud tones, giving quick, reflex protection to
the system. If the tones are explosive the muscles may not be quick
enough and the ear is damaged. When everything praiseworthy has
been said, we know that it has left only a crude description of what
occurs in the two ears, the two silent pianos, when the word
Scheherazade shakes the nothingness.

*Internal ear.* It also as said is a cavern in bone but filled not with
air, with fluid. It is called cochlea. Cochlea = snail shell. A conical
snail-shell spirals two and one-half turns around a column of bone.
It is intricate. It is a jewel. Each turn is divided into an upper and a
lower gallery by a spiraling membrane, the upper gallery again
divided by a spiraling membrane; so it is three galleries filled by
fluid placed deep in the bone on each side of our head. Fluid is in
the internal ear, air in the middle ear, and that means air is pushing
against fluid. That is not mechanically the most effective. Fluid is
incompressible, and when the footplate pushes from the middle ear
toward the internal ear something must yield. Something does. A
membrane. It fits into another window, the round window, that
yields in the direction of the middle ear.

Thus has the scene been set for action. If now a tree falls in the
forest, a firecracker fires, a two-ton truck snorts, a schoolbell rings,
or my beloved blows her nose, two instruments are waiting to
dispatch the variety of the news to my mind.

# BASILAR MEMBRANE

## *Sifting the News*

The membrane that divided the two and one-half turns of the cochlea into an upper and lower gallery was the basilar membrane. Highly sensitive. Highly specialized. Hermann Helmholtz studied it. Also studied ears. Studied sound. Studied music. Almost a century has elapsed since the Steinway Company presented him with a piano. So he was known beyond the academic world. He developed a theory—the harp theory—that has stood up against ninety years. The modifications of it have been largely to bring it in accord with sequential advances in the general field of physics, and possibly that would have appeased the ghost of Helmholtz, who was one of the greatest physicists who ever lived, and one of the noblest of men.

Impossible to think of Hermann Helmholtz as the diapered infant in a crib, lying there the usual months, then in bed seven years, a sickly boy, worried his aunt, not his mother, she saying that whenever he lifted his face he smiled and she saw intelligence and spirit and knew therefore that he would come out all right. He did. He did not disappoint her. He was born in 1821, at fifteen had chosen science for his lifework, by twenty-seven, independently of Robert Mayer, formulated the law of the conservation of energy (the energy of the universe constant), gave it mathematical statement, and it was one of the gigantic achievements of the human mind.

In his harp theory, successive groups of fibers of the basilar membrane corresponded to successive strings of a harp or piano. The living strings ran side-to-side, each had a frequency, as do the dead strings, and vibrated sympathetically whenever that frequency was set up in the neighborhood, by the schoolbell, the two-ton truck. Later the basilar membrane was shown not to operate just that way. It operated more complicatedly. At low frequencies it did respond frequency for frequency. At higher frequencies groups of impulses went into the brain. At still higher there was something quite different. A fluid-wave moved along the two and one-half turns of the internal ear, and this put a shearing force on that other nearby membrane, exciting several kinds of electrical play, nerve-endings stimulated, and impulses went into the brain. It was the crest of that fluid-wave that had excited the dominant frequency. Brilliant studies had established this later observation and interpretation, and for it Georg von Békésy won the Nobel Prize of 1961, deeply deserved but pleasantly unexpected. Helmholtz's overall idea meanwhile continued to stand much where it stood.

In a human ear the basilar membrane is somewhat more than an inch long, mounted on flesh pillars, tiny of course, two rows of them, leaning toward each other, and on either side of the pillars are cells with hairs, these the receptors, and above the hairs is the membrane. As the fluid-wave moves along, the hairs are alternately squeezed and released, squeezed hardest in the neighborhood of the crest of the wave, electricity generated, nerve-impulses traveling. An important fact had not been anticipated—to either side of the crest other nerve-impulses also traveled, inhibitory impulses, that dampened, pedaled the tones to either side of the crest, leaving the dominant tone less interfered with, more isolated. It is one of those neat arrangements that when one heard it first and had not yet got dulled to it caused one to find oneself rather approving of Nature, for the hundredth time, or the thousandth.

The nerve-impulses that result travel inward over the great auditory nerve. It has twenty-five thousand fibers.

At its one end, at the peak of the snail shell, the basilar membrane, with another membrane above it and sensitive cells between, is broad; at its other end, at the base of the snail shell, it is narrow. The narrow end, as the harp, vibrates to high frequencies, the mind hearing high tones, the broad end to low frequencies. Boilermaker's disease produces victims deaf to high tones. At post-mortem the damage is where expected. All of the facts fit. All is mechanical. All is inevitable. Another strait jacket is by this flesh-and-bone structure, immediately at the receiver-end, where the machinery for this special sense meets the world, put around our minds. Fatality always is leaning in upon us. If the fatality had a mind one would say it was both annoying and humorless, because unrelenting, because solemn. The very beauty of the orderliness affects our philosophy, our morals, drives some temperaments to cocktails, to avarice, to sitting growling, other temperaments to sunny optimism, and a small fraction between living waveringly in that get-up-every-morning space.

Helmholtz had decided on medicine as a career because he could embark upon it without having money in the bank, the government paying the expense of his schooling. It did mean barracks. His premedicine he took in Potsdam, and the premedicine as he understood it was to include the study of Latin, Greek, French, English, Italian, Hebrew, Arabic. He wanted a piano in his room. He insisted. He got it. But he selected the wrong roommate and there was the trouble of finding another. His words about the new one were: "Lanky, unskillful in mundane affairs, plays the flute with no sense of timing, but has ideals!" Much of our knowledge of acoustics and of the physical basis of music began, accordingly, in a barracks. One thinks of how times have changed, thinks of all the shiny wonderful electronic equipment everywhere in the world, thinks of the candle-lit dining rooms for students in Rockefeller University.

Nowadays, motion pictures can be taken of the basilar membrane and the structures attached to it. A glossy falseness may lie over those pictures, as over most motion pictures, but safer to have the pictures than to leave everything to the imagination. And here the imagination might be too modest. We might for instance imagine the basilar membrane as a stiff prim parchment on each side of our head. It is the uttermost reverse. With proper stroboscopic lighting and a magnification of a hundred times or more that membrane is mobile past imagining, is never at rest (like the fimbria in the uterus embracing the egg), its surface shimmering in the lighting, and when the biggest fluid-waves rise because some crash has beaten the air one's mind sees the surface of the sea. If that last strikes someone as clumsy exaggeration, he is wrong, but let him think instead of Symphony Hall, the strings of the string instruments and the drum-heads of the drums and the air columns of the woodwinds and the organ, all dutifully vibrating in their totals and their fractions, setting the loose molecules in the space of Symphony Hall into a corresponding vibrating, while in front of the orchestra, built into his civilized anthropoid head, are the two highly cultivated analyzers of the conductor, periodically beguiling themselves, periodically annoying themselves with picking out a single sour note in a single viola, while behind him in the dark glow of the hall two to four thousand pairs of basilar membranes are swinging in a sinuous harmonic motion, the energy of the swinging resulting in electrical messages that travel inward of his sideburns to a final destination in the hearing cortex in the temporal lobe of his brain. Each ear connects via way-stations with both temporal lobes, and that is why serious damage may be done to one side of the brain and there be no hearing loss in either ear.

There are maps of that hearing cortex in the temporal lobe. Each point on the map corresponds to a frequency. A cat or dog has a temporal cortex that resembles that of other cats and dogs, and there are reasons for thinking that ours resembles theirs but resembles more that of other men and anthropoids. In each of our skulls there is a strip of cortex that has a point-for-point correspondence with the basilar membrane. The high frequencies are forward, the low backward, and every two millimeters it means another octave. Upon that final keyboard of cortex the final news is played. Each side of the brain has, in fact, two keyboards. Even three. A second and a third have been mapped in animals and the order of frequencies of the second found reversed, low forward, high backward. Recently as with the other senses it has been shown that the nervous-system responds effectively only if the animal (cat) is paying attention, not if she is thinking other thoughts. It has been shown, furthermore, that the brain itself can to some degree determine whether a sound shall be more powerful or less powerful.

No one understands all of the mechanics but in general understands well enough until the attempt is made to pass from mechanics to mind, when everything is suddenly blurred, and thin theory takes the place of hard fact.

Late in life Helmholtz traveled to America. His government had sent him to what was called the Electrical Congress of 1893, in Chicago. By that time he was known far beyond the academic world. He was, besides much else, the leader of all scientists in the physics of sound. Joachim the great violinist was his friend. During the return trip from America he fell down the ship's stairs, may have had a stroke and therefore fell, did definitely have a stroke the next year, July 1894, deteriorated until September 12th, when surrounded by his large family he died. Lenbach's famous portrait is probably a good likeness. The face is kind but distant, gentle eyes, fire shining through them from somewhere not near.


# SOUND

## Newscast

And in this roundabout way we come to sound itself.

Sound is frequencies heard.

If sound is an evolution, as to the twentieth century everything is (Darwin's evolution, Lamarck's evolution, cultural evolution, the evolution of the pretzel industry), and if mind is an evolution, then sound would have been one of the sculptors of mind. A woodnymph, mind, inexplicably born, would ever since that birth have been trying to disengage herself from the primordial tree-trunk where she was trapped and remained captive, frequencies in there with her. Somewhere sometime sound had begun to exist, sight had begun to exist, smell, touch. So, mind had begun. What a strange epoch had with that been introduced into the Genesis of the physical world.

But, go back, before mind, before sound, just the frequencies. They had been beating since Chaos, which is to say, before the seven days of that famous week. Then came the moment—faintly faintly faintly something was heard, faintly faintly faintly mind was appearing. (Was a master mind listening? Nonsense!) Then, both were more than appearing, both were positively emerging, and, after that, the two continued to work with each other, against each other, friends now, adversaries now, enriching each other, till, one twilight, sitting alone in his house in Stratford, Shakespeare heard Sonnet XXXV read aloud by himself to himself.

Textbooks say sound has three qualities: pitch, intensity, timbre.

*Pitch* places a sound in a scale. E flat. Pitch depends on frequency, but, there is the ear, that instrument, and there is the mind, forever intruding on everything. Not surprising, therefore, reasons enough, that sound should be wrenched somewhat from the place allotted to it by its frequency. High sounds for instance, are heard higher, low lower. High sounds are not as masked by a streetful of high as by an engine-room of low. *Intensity* is volume, loudness, depends on wave-size, and that depends on the energy put into the wave to enable it to journey, to reach to the farthest seats in the new Met, there to modify minds. *Timbre* particularizes sound, separates a human voice singing on the beach from the foghorn moaning at the end of the pier, from a cricket who is on scene as an interloper, a 1969 cricket, which takes one not back to Chaos but part way back, starts one wondering whether this cricket is the same cricket as that of the late Paleozoic, whether the Paleozoic cricket was really a cricket, whether his shrill music has been sounding across the land ever since, and making mind. Timbre depends on wave-form, complicated often, repeats for as long as the source produces. Our ears break up the wave-form, but fuse it also, and our mind bats back and forth with all of it. A mathematician reminds us that mathematics breaks up the wave-form too, but only an intellectual simpleton would think he could by what is called Fourier analysis account for the enveloping glory in a tone of Casal. Timbre lets us distinguish a clarinet from a flute, a single flute from six flutes, each flute belonging to the species flute but obligated also to its flutish self to be true, else it goes straight to Hell, as whoever to himself is not true.

Pitch, intensity, timbre, these allow us for the sake of the defense of our body and for the sake of the slow evolution of our mind to isolate the peculiarities inside a sound. They allow the physician with his stethoscope to know if the fetus in the womb has its pump still pumping, that therefore it probably will arrive alive, add that one more to the directory. Allow the physician also to hear inside a sick man's skull a dangerous rush of blood that will shortly subtract one from the directory, cancel out for him who is arriving. Then comes 7 P.M. and the physician dons cutaway and drives to Boston Symphony with a mind too tired to hear anything at all.

A plucked violin string vibrates through its length and produces the fundamental, the base tone. Any ordinary ear without training hears that. The halves of the string vibrate independently and produce the octave. Any ordinary ear hears that too, hears fundamental and octave as individual tones, at the same time hears their fusion. Then there are the halves of the halves, and the thirds, and so on, all the overtones that the ordinary ear has long since stopped hearing individually, but does hear more or less of the richness that comes of their fusion, the mind in consequence continuing its evolving, perhaps exceedingly slowly now.

The ear itself, the instrument, sifts frequencies and intensities with an efficiency that no one stops to consider as he reads the cold notes off the page of piano-music and bangs the keys. From these siftings, whatever gets past the ear and on into the nervous-system, more is lost there, and that which finally is left for the mind is the sound. For the cat-mind similarly. A cat-ear picks up what it is built to pick up, the cat nervous-system shutting off what at the moment need not have the cat's attention, the cat-mind then no doubt shutting off what it pleases. Our mind too. Cat-ear and cat-mind, our ear and our mind, are interesting. (And the word *interesting* is not interesting.)

A tone that repeats rhythmically and is agreeable to the mind is music. Bring together legions of such. Bring in changing frequencies, changing intensities, and different wave-forms. Bring in, that is, trumpets and the rest. Bring in accelerandos, retardandos. Bring in quavers and semiquavers and arpeggios and minor scales and major scales, and modulations from one key to another. Bring in a new theme, or a rearrangement of an old theme, let it fade, emerge, fade, emerge. Bring in knowledge of the sonata as a form. Bring in a composer—Beethoven would do. Bring in, if you prefer, one of the uninhibited that Stokowski plays and aids, or the still more uninhibited, but then in the second half of the program bring back Beethoven. Let him fit the small into the large, the large into the majestic, the majestic sweep toward the horizons, toward the sky, toward the bottom of the sea, become the last movement of the Ninth Symphony, and our mind, if we can get it free enough to truly hear all that is singing in the air, may through sound, as Beethoven claimed his did, talk with God, or whatever a scientific determinist or a clergyman permits us to say may be God.

Noise is different. Noise also is sound. Noise also is a passage from frequencies through a physiological machine to the parallel-lying mind. (How embarrassingly clean that statement was.) Noise is disagreeable to the mind usually, unwelcome to the old music if not to the new, excited by tones that do not repeat rhythmically, or that leap from here to there with inconsiderate speed, or slow to the pace of a croon, or a salad of frequencies flopping from the LSD-soaked brain of a hippy-composer, or an angry brash slash crashing through a unison, or a silence that one affectation hears as music though a person close by is hearing nothing, or, if the silence slash-crashed abruptly, hears that as noise, or a waxing and waning of intensities that produce a beat that scratches the soul. The sources? A wagon spring. A pencil against slate. A falling dishpan. A village orchestra stumbled into misfortune. A violin whose string snapped. A bad line in a bad poem that stayed bad because among other reasons the poet did not have the patient ear of Keats. The out-of-rhythm bogus bass of the boss. Gum-chewing Lucy's hi-fi through the hotel wall at 2 A.M.

Mind is warned by noises. Informed by noises. Inflamed by noises. Enraged by noises. Driven to despair by noises. There are noise-meters that give numerical values to noise, and satisfaction to meter-readers. A psychiatrist was of the opinion that noise sometimes has been the straw that broke the psyche's back. Even when not that destructive it still is morally destructive to have music fed fitfully between clicking typewriters into a bank in order to sweat more typing out of mind-deprived Lucy.

# AUDIOMETER

## *Applause Meter*

It was the frequency of the waves that gave the tone its pitch. The size gave intensity, loudness. Speech therapists and some doctors use an instrument that relates pitch and intensity. The audiometer. It produces pure tones electrically. The therapist needs only to turn one knob for pitch, another for intensity. Right and left ears are tested separately. If an ear does not hear a certain frequency at a certain intensity the therapist turns up the intensity knob. If then the ear hears, a mark is made on a printed chart, the successive

marks connected by a line, and the chart examined and filed. It is a
rough but useful description of that human ear's capacity. Normally, the two ears of a person test much the same, and audiograms
of members of a family are apt to be similar.

Alexander Graham Bell invented the telephone and in his honor
the intensity unit is called the *bel*. One-tenth of a bel is a *decibel*.
The pressure of the vibrating air molecules (everything in the
universe is vibrating) might itself be measured but the measurement technically difficult. Instead, the tones are graded. A tone
close to the lower limit of hearing is the reference tone. If another
tone is ten times the reference tone, that is 10 decibels. If a hundred
times, 20 decibels. Ten million million times, 130 decibels. (It is the
logarithmic scale, 1 for 10, 2 for 100, 3 for 1,000.) A 130-decibel
tone is almost more than the ear can bear even though the difference
in actual impact on the eardrum between 10 and 130 is not great.
Our breathing sounds are around 10 decibels.

The ear has an indescribable sensitivity. Much study has gone
into finding how mechanically it can be that sensitive. It is far more
sensitive than the eye, possibly because in the primeval forest night
was more dangerous than day. At any rate, it is. A good ear can
eavesdrop on anyone's gossip.

The rustle of leaves in a gentle breeze also is 10 decibels. A
whisper at four feet, 20 decibels. A suburban London street, 30. The
center of New York at quietest night, 40. Steel plate hammered at a
thousand feet, 110. So also a loud motorcycle. A jet plane with
afterburner, 160. Conversation is apt to be around 60. In human
speech there are shifts of intensity and frequency that can be
fantastically quick. Around speaker and listener there is always,
except in an experiment, some level of noise. (Much is always
occurring at the same time.) The ear selects, enhances this, suppresses that, as is known, and as has been recorded and interpreted
by laboratories all over the world. In our country there are the
laboratories of the Bell Telephone Company and numerous university acoustic laboratories. We are steadily receiving a fuller accounting of our capacity and animals' capacities, but it is still not all
explicable that we can intelligently hear a small Parisian boy speak
a quiet *Comme ça va,* but, the next moment, the grotesque assailing

of ear and mind by the clipped instructions of a gay Parisian gendarme. The frequencies of human speech stay much around and above one thousand cycles per second; our hearing is acutest there. Strange meanwhile that a bird can be singing as if the whole sky were listening but not be hearing half of a noisy child's range. Strange, too, a man orating and his cat paying absolutely no attention, able to hear but not concerned, at that moment concerned with vibrations from a cat world quite beyond the man's range. A complicated telephone switching, selective suppressing, selective inhibiting, in nerve, in brain, in mind, alone makes possible that cat's deliberate indifference. It would not reduce the complication if someone were to say that the indifference was simply dents put by air molecules into a polymer of flesh macromolecules.

However that, each creature in this sense also stays in its own reserve, stays within its walls, may try to reach over, can't. Each creature begins its auditory comprehension of the universe from that reserve, the detail of the comprehension different between any two cats, any two men, and the universe of cat and man also at points overlapping.

Suddenly the cat has stopped washing her face. She is as stirless as a metal Egyptian cat. Peculiar. Cats are peculiar. Evidently she is listening, but what she is hearing has not decibels enough for us to keep up with her.

# DEAF

## *News Flash—Void Revisited*

A human being may have lost his hearing for no reason that anyone knows, or for tumor, or for infection. He may also have had some abnormality of genes, or something went wrong in utero, either way a silent creature there in that silent womb. He then arrived in the outer world, deaf.

As many as ten million deaf have been credited to this nation, and a more enthusiastic figure would put it at 80 percent of the population, the figure to be brushed aside because it surely ignores that each of us is rushing toward old age and getting there, and getting deaf. We may manage both tantalizingly gradually.

An eccentric said he was pleased he could no longer hear the world, but more commonly a man insists rather that he is not deaf, not very, that what you are noticing is absent-mindedness. He has other things than your voice to listen to. His liberal neighbors make a praiseworthy effort not to hold it against him that his body has a flaw; and he looks right through his neighbors. That small button in his ear, they say, no one would ever notice if he were not always calling attention to it, if he were natural about it; it might even give a fillip to his personality as pince-nez did in an earlier day. The button would give nothing to his hearing if the damage was in his hearing nerves. It must be somewhere on the way to the nerves, somewhere in the transmitting system; then a hearing-aid helps. His neighbors are right, his aid has been so cleverly placed that no one would see it except for that self-consciousness of his. One kind of aid amplifies all frequencies, puts practically a loud-speaker into the ear, another kind amplifies only the impaired frequencies, then the deaf person fidgets all day to keep the instrument adjusted.

That word—*deaf*—is soaked with mood. It is gray or black. The limiting walls have moved in closer.

A giddy girl asked another giddy girl whether she would prefer to be blind or deaf, a harmless pastime, and neither really cared. They paid no attention when you reminded them of a woman they know who since she went deaf dashes about less, reflects on things more, has one sense fewer to nag at her, and she who used to be ill-humored is good-humored. The giddy girls immediately decided that you yourself were giddy. One of them pointed across the street to the deaf old man who hates everybody. She is right. People think it is all original sin, his mean disposition, but that came mostly in the years he was getting deaf, when his whole point of view changed. He still had his eyes, you might say, but that in one respect was a disadvantage, because eyes are in every instant bringing in something, attacking our minds, so that the deaf man in a way has not even his earned isolation. He is worse off because he sees— to say it that way is of course exaggeration. Also, this would differ person to person, and difference in disposition is brought out by deafness. This man behaves as if he were constantly thinking of his lost antennae, thinking how his power to grab has been reduced, everybody else's power increased. So he just sits, throws into the wastebasket the good-advice letters from his sister, who hears. Deafness may produce a wasp, or a pillar of sadness, or a gusher, or an honest philosopher, and which depends on what the persons were to begin with.

Rarely, but it does happen, a person is born both deaf and blind, from the beginning heard no evil, saw no evil, thought no evil. It is as apt to be, heard nothing, saw nothing, thought nothing. It is quiet inside his cave. One might conclude that he had been left far back on the road of the living. But it is not so, or may not be. The person was born with the intelligence and the force to circumvent anything. The latter was stepped up by his loss. The force may be excessive, as if the mind correcting for a loss did not know how far to correct, when to stop correcting. He may have persuaded another to join her fate to his. That may have helped. The community may have made a community project of him. Better to have that happen to a dog, go from house to house for a pat and some meat. On the contrary, he may have made a project of society, he reach farther into it than it into him. Society might instead have been scrupulously honest with him, but that is too much to ask. Even in the presence of affliction society is interested in society, prefers the afflicted to behave as it does, have its mentalities, its prides, live in its world, accordingly flatters the afflicted to go about as if its senses were his, as if he actually saw a Cézanne, heard Debussy, thus still further unsettling for him the boundary between fancy and fact, and so depriving his mind and its own mind too. This man could give it aspects of the world that belong uniquely to him, that reach him through senses superbly sensitized, powerfully disciplined. Touch. Pressure. Pain. Taste. Smell. Even temperature. However, we are all too tired to break through our own numbness so as to assist him to assist us. The deaf are too tired, unless they are exuberantly young and aggressive, as happens.

So, here we have a human being whose two dominant senses communicate nothing and yet his mind remains a mind. It is a most sane mind. It has brightness. It has comparative selflessness. Instead, it might have had maniacal selfishness, maniacal imperious-ness. The point is that this mind has not died, has not grown dull, and never may.

Each of us to keep awake and alert and thinking requires a different hammering from his senses. There were those college students a decade ago, everybody then talking about them, who were paid twenty-five dollars a day for allowing a maximal reduction of the hammering, their vision cut off by goggles, hearing by soundproofing and deadening the experimental room, touch by wearing cardboard cuffs extending beyond the fingertips. These students lost concentration, lost the capacity to do simple arithmetic, lost contact with what is called reality, soon showed a tendency to distort reality, became hyperirritable, abusive, finally had hallucinations. The facts are facts. They were reported abundantly. The interpretation is not that easy. Excitatory stimuli were indeed prevented reaching some of the principal receptors of these students, but also a massive total inhibitory stimulation was imposed, a sense-heaviness that might operate like hypnosis. Those nervous-systems were affected positively instead of negatively, or so it could be reasoned, the experiments striking one less as evidence of our need for sense-stimulation as for the danger of abnormality when our senses are drugged. The mind fighting off such weight goes mad. Something like that. The deaf of our everyday world are not mad. They are cheated. Were they mad, and were there as many as ten million, the planet would stagger in its orbit more than it does.

It is said of two deaf-mute children that they were always nudging each other by way of stepping up their jokes. This was not painful to see. Rather it left upon the persons who saw it an astonished amused admiration.

# LIGHT

## *"And There Was Light"*

Vision brings us the evening star. Hearing could not do that. No other sense could. The word *vision* has as its principal meaning that common one, sight. He has vision—he sees. She lacks it—she is blind.

Light excites vision, and light is two: (1) dead radiations, (2) a living receiver. "And there was light." This could not have been written had there not been the two, what arrived from outside, what waited inside.

Of the universe's vast series of radiations only a comparatively narrow band, about one-twenty-fifth, can be sensed by the eye as light. Toward one end of the vast series of radiations are the short X rays and the still shorter rays, for which we have no sense organs, no receptors, and toward the other end is the large domain of heat waves, of long radio waves, of broadcast bands, of electric waves. About midway are those, then, that make man and monkey and dog and cat and many creatures so vision-bound. The rest of the series also plays through space and slips into the dark closet where a man goes to think or to hang his hat, but the man cannot know that anything is there, except by his intellect. He reasons his way to the rest, does not sense them. He may do experiments with them—intellect still. But his everyday life has not much to do with them—rarely, for instance, would he have gotten an X-ray burn.

Pressure on the eyeball excites vision in the mind. A shock sent through the skull does the same, but neither of these has much to do with that which lets a body make use of an energy that has traveled ninety-three million miles from the sun or three feet from an oil lamp, the oil lamp being the sun in a disguise, a burning taper another disguise.

Even a creature of one cell, the entire creature one cell, may live and be directed by light. The sun's radiations arrive, cause a change in the creature's chemistry, it turns toward the light, or turns away. A beetle flies out of the night toward a street lamp, crashes, has used those radiations to die. Some invertebrates are light-sensitive over half their bodies. One nocturnal species buries itself in sand, has a tube to siphon in air, shoves out that tube, and if the radiations of the sun strike it, back in it goes, has received the message, buries itself safely again. Life has obeyed light. Other species defend themselves or are offensive, withdraw or attack, because in dim illumination they are able to concentrate light, to see, though better not say see because some scientific purist may object, better say receive. Other species have the shape of a hollow sphere with an inside light-sensitive lining, and a pinhole opening to the outside.

Others have light-sensitive units fitted one against the next, as the dragonfly. Many inventions. A spider's compound visual system produces a mosaic of blotches, which becomes a signal, a language, and the spider says in translation, "There approacheth a fly." What the mind of the spider is seeing we cannot know, or whether it is seeing, because a spider is a spider and we are we, but the way it runs from a moving shadow, which is a warning, assures us that the spider has something that makes use of dark and light, and that is worth its having. Two human beings may have a surge of human feeling because both can see, then a surge of melancholy because what the one is seeing the other is not. Rays come from taper or oil lamp or sun and flash all directions and somewhere someone has the proper receptors to receive them, to add them to its chemistry, to be notified ahead of time about something that is still at a distance.

By their vision all human beings except the blind have been tuned in with the sun, and the suns beyond that sun, and any ephemeral puffs produced by the one fire-producing creature when he is in a humor to light and heat his foolish way. By their vision, too, they are aided in their sculpture, their architecture, their painting, their pen drawing, their etching, and in the devious pursuit of their quenchless psychology.

# EYE

## *A Camera*

The eye is the instrument.

The eye is the camera. That has been said a thousand times and is correct enough, on the surface. Both eye and camera have a talent for light. (And for dark.) We creatures were born with the talent and its carrying-out depends upon facts and rules that have been and are and will be studied.

The foremost student of the chemistry of the retina was one of three recipients of a 1967 Nobel Prize; all three recipients were discoverers in vision. Those living test tubes there in the retina are two kinds of cells known to all intelligent grade-school students: the rods and the cones. Their shapes give them their names. They are set in the hollow on the inside of the back of the eyeball. Whole books have been written about that sensitive many-layered inside, the retina. Man's has 125,000,000 rods and 4,000,000 to 7,000,000 cones. At the fovea, which is a pit in the macula, which is a circular patch near the center of the hollow, there are 120,000 cones to every square millimeter. So, at the fovea, cones; at the macula, cones and rods; at the outer edge of the hollow, rods. To risk another figure, the eye is a silent lantern hung at the front of the brain waiting to be lit from beyond itself. The two eyes, two lanterns paralleling each other in construction and in work.

The comparison of eye and camera does help children to understand, also helps to deceive them, or could, might start them thinking that scientists know more than they know, that experts who experiment with eyes of frog and cat and man surely would not be confused, that the universe is a reasonable universe put together by a man-type mechanics, and tomorrow all things will fall within our grasp. Had that comparison of eye with camera been with a motion-picture camera it might come closer, still not very close. To be honest we ought to say that the nature-built has possibilities that could not be dreamed in the man-built.

Why is the white face of the clown, that is brought by the eye's lens to the retina as an upside-down tiny image, seen in the mind right-side-up? Why? And seen actual size? Actual is of course only a word. But the facts can be bewildering. We ought also to remind the children, if it is children we are trying to help understand, that they are not looking at that upside-down image, which is only so-and-so-many light-excited points (or light and dark receptor areas) at the back of the hollow sending messages via several way-stations to the back of the brain. Yet we would need to be prepared to have an exasperatingly logical child ask the same exasperating questions for the back of the brain. What happens in the visual centers in the depths of the brain? And it would be no answer to lay everything to experience, because this would be to make the whole problem one of learning and that problem has similar ambiguities, and, anyway, we are born with our visual system and the evidence is that it begins to work at once. To be honest with the logical child we ought, furthermore, to explain, if we can, why the white face stays out there in the circus ring. The image is in our own heads, is it not? Why does it stay out there then? In fact, why do we see at all?

An old and famous and simple experiment encouraged support for the comparison of eye with camera. College students perform it, or used to. They took an eyeball from a freshly slaughtered cow, thinned the rear wall of the eyeball with a knife, fitted it, front-side frontward, into one end of a mailing tube. Then they pointed this rude instrument toward, say, a tree across the street, and on the thinned rear wall as on the ground glass of a camera there immediately appeared an upside-down picture of the tree. Tiny picture. Delightful! Surely we must be near to the solution of the mystery! Those that look for comparisons of eye with camera would snatch onto that ground-glass picture. They would say that the retina there was the camera's film. Future historians may say that the invention of the camera was the Adam's curse with which the fall of art began. Instead of coming home from a vacation with mood-renewing inaccurate memories of Luxor, we come with a box of accurate color-exaggerated Kodachromes to bore our friends who next February must sit in the dark and lock up their intellects and look. In the dark they can nod, of course.

Another experiment, old too, discouraged that famous comparison. The experimenter put a patch over one of his eyes and over the other fitted an inversion lens, so that images on the retina now were not upside-down. Whatever traveled back to the receiving parts of the brain should be the reverse of normal. Notwithstanding, in eight days the experimenter's mind, or whatever, had completely made the adjustment, and if one were to reason from the way the images on his retina previous to the patch and inversion lens had instructed his body he would now be walking with feet toward the ceiling and head toward the floor. Of course, he walked as he always walked. Facts are facts and explanations are explanations but at least we cannot help being astonished at how adjustable our sensations are, how pliable our learning, how detachable and reattachable our mind to our brain.

The front of the eye has its transparent living glass with the pain receptors and the touch receptors. Rays of light reach this curved glass, are bent, and focusing begins. The glass is bathed behind and in front by a salt water that keeps it clean and keeps it shrunk, the salt water only slightly salty, steadily manufactured, steadily drained away, manufacture and draining steadily balanced. That is necessary. Any sharp rise of pressure inside an eyeball could cause sudden blindness. From the cornea the rays continue inward, strike other transparencies, the most important being those of the lens. If a lens is lifted out of a cow's eye it still bends rays as a glass lens would. Helmholtz is reported as saying that had his lens-grinder ground a lens as poor as Nature's he would have discharged the man. Helmholtz probably was speaking only of the ray-bending power of the living lens and of its optical errors, not of that miracle that grows in the embryo from apparently nothing, soon is expert at its performance, serves in all illuminations from high noon to midnight, is glassy only at death. Being living, it consumes oxygen, gives off carbon dioxide, must receive nourishment, must have its wastes removed, and all of this depends on flowing blood, yet blood-vessels are not permitted to cross a lens and cast shadows, so the chemical transactions must all be via that salty fluid.

The curtain that we all know, the one in front of our own lens, has a round hole in it. It is the iris. An iris may be a heavenly blue in the young but usually grows darker. Its inside surface is continuous with the retina but has no rods or cones. The iris as we look from outside is dark dark-red in brunettes, light dark-red in blonds, chocolate in the Negro. The tint is a chemical, melanin, and the corresponding tint in the camera is black paint. In both cameras, accordingly, reflections are reduced and blurring is reduced. As for the total eyeball, that rolls harmoniously with its partner in a fat-protected orbit, the harmony faltering in sleep when one eyeball may wander, swing upward and outward, make grandpa dozing in his chair even older and more foreign than he already is to his five-year-old grandson. An eyeball is about an inch in diameter. From it the optic nerve-fibers, that carry the messages, pass diagonally backward along the middle-bottom of the brain, meet the nerve-fibers of the opposite nerve, the inner fibers of both nerves crossing and continuing via the way-stations to the back of the brain. The right half of each eye does thus connect with the right half of the brain (that takes a moment of thought), and if the right half of the brain is destroyed in an apoplectic stroke the opposite half of the world is still there but is not reporting. In spite of this, brain and mind will soon have adjusted, each of the two eyeballs will henceforth swing somewhat farther and somewhat faster. The subtlety of response of any spot of the receiving retina to light-turned-on, to light-turned-off, to direction of movement of light and dark, probably the rate of movement, and the responses of corresponding parts of the brain, and of right and left eye to each other, all is widely studied by physiologist, ophthalmologist, engineer, MIT, Harvard, Los Angeles, London.

That person with the stroke may be quite convinced that he is seeing all there is to see. Every one of us is willing to be convinced that he is seeing all there is to see, visually and otherwise. A quite blind person may insistently claim that he sees.

# PHOTOPIC

## Day Editor

Sir Isaac Newton—Shakespeare's contemporary and countryman with his so different quality of mind—directed a beam of white light toward a glass prism and out the other side of the prism as out of the spray of a garden hose there spread a band of color. (The tough mind of Sir Isaac would probably not have expressed it that unmathematically even the first time he saw it.) White light had the stimuli of color in it, the glass prism had broken the beam into its component radiations, these were bent by the cornea and lens and other media of the eye, were focused on the retina, caused changes in the chemistry there, impulses went over stretches of nerves, crossed junction-points, and the mind saw red, orange, yellow, green, blue, indigo, with the hues between. It saw a rainbow, and no rain, red at one end, violet at the other. Recombine all of these and there would be white again. Various other combinings, as red and green, any so-called complementaries, also gave white. Amazing forever to the two naïve minds left in the western hemisphere that a red light and a green light in right proportions will give a white light.

So, color too signals to the mind from the void, from sky and earth, from the glittering blue of yonder double stars, from the green of yonder traffic light. Color informs. Color rouses. Color puts stained-glass windows into our churches and our poetry. Socrates thought color led the way to geometric form. He was smart but maybe he was not a biophysical scientist.

Black is color. That point was debated to apparent conclusion through a century. We see black. We require, that is, a retina for there to be black. Where there is no retina, nothing. Black is not nothing. But when we try to carry any radiation to the brain, and then on to the sensation, we still are in trouble. Black is the apparent result of all color stimulation being withdrawn from the retina. The no-stimulus becomes the stimulus, this transmitting in and out of the communication pools of the brain, the nerves and synapses for vision, and, again somehow somewhere, there is the alchemy and a mind is in possession of black. Not long ago it was experimentally shown that when an image was fading from a tiny patch of retina, that patch in the corresponding part of the mind became an intense black. The increase and decrease of stimulation was the stimulus. This might be a so-called psycho-physiological experiment. Vision is raging with experimentation, on the sensation side, on the chemistry side.

Most intense of colors would be that black against white, unless you object to calling black a color, and white a color. Each of those two does mix with other colors to give still other colors. Olive results from black and green, and that is one of those facts that may seem a mere hop-skip-and-jump to the psychologist and to the chemist, but still might confuse a truck driver and a physiologist.

Your dog sitting quietly at the edge of the universe and looking out sees black and white and shades of gray, nothing else, it is believed. Whether the experiments done on him should or should not be called psycho-physiological makes no difference, but within that poverty of color, within that limitation your dog as fatally as you within your limitation senses the visible world, and has the restrictions on his thoughts that follow from the limitation. A parrot sees color. A lizard sees color. Bony fishes do. Crabs do. Turtles do. Rats do not. The cat does not. We have proven this for the cat, we believe, but no cat will commit herself in writing, and it is possible that she cheats during the experiments, but the fact that impresses us is the chill casualness with which Nature has dropped so useful a

458 S E N S E : *Message from the World*

talent as color-vision here and there in the ranks of the living. Some
birds see almost the same colors as we and have notches chipped
out of their rainbow as we.

A human eye discriminates a hundred distinct colors, fewer or
more, and if all mixtures are taken into account, a thousand. Ultraviolet the human eye does not see, unless the person has cataracts
and his lenses have been surgically removed. Ultraviolet might give
a lovely tinge to parts of the earth, we believe. The bee does see
ultraviolet, turns it into honey, money, does not esthetically evalu-
ate it, we believe. If our daylight included ultraviolet, excited by
wave lengths beyond the shorter end of our spectrum, our rainbow,
that small strip of that vast stretch of radiations that exist in the
universe, we might better understand the bee's point of view, better
know how to confuse the bee, make it self-conscious, take the
accuracy out of its sting, and bees would worry us no more than
flies, though a minimum of worry, everybody in the learned world
has told us, is good for us lest things worse than sting of bees
happen to us, like insomnia and insanity. (Human photography, as
one would expect, since it is radiations that are filtered into the film
emulsions, can photograph beyond what the human eye can see,
beyond our spectrum, numerous wave lengths, and though we
cannot see what it is in the photograph, we know it is there because
it can be put to practical use.) Various insects employ the longer
wave lengths, the direction of the spectrum where our eyes see red.
Everything in this bewildering world remains bewildering for
everyone who has the capacity to remain bewildered. The sun
strikes a sweater, and the dye in the sweater absorbs some wave
lengths, reflects others, and what is absorbed the eye does not see,
and what is reflected the eye sees as the color of the sweater. If the
wave lengths for the greens and blues have been absorbed, then the
sixteen-year-old next door, with her head held sharply up and her
hairdo on top of that, has come out of her house this morning in her
prettiest red sweater.

Color is the glory of day vision, photopic vision. Color requires the
cones, though there is evidence that, for blue, day vision gets assistance from the rods, and similarly the rods get assistance from the
cones, and there is obviously much to learn, much even that might
throw current ideas out the window. Cone chemistry has been

studied and many facts are known but not as many as for the rods. And there has been much study, and much has been written about the psychology of color, yet, when one reads an explanation of what goes on in the retina, in the various levels of the brain, in the mind, we think the author has meant to befuddle us. Possibly he is be- fuddled himself. The subject just is difficult. One of the greatest of living authorities says that we still know nearly nothing about the physiology let alone the psychology of color in mammals. Nature tucks away her ideas and keeps us searching. A chicken is not a mammal. A chicken's eye has frank drops of vivid oil in its cones, and these drops are color filters (acting much as in the photographic film), set the limitations on that chicken's color vision, color life, color thoughts, a hen's meditations on a bright-tailed rooster. Man knows about color filters, manufactures them, is pleased when evolution does what he does, and confidently goes on speaking of the chicken as a stupid bird.

In the time-honored Young-Helmholtz theory there were three primary colors, red, green, violet; three colors, three cones, three chemicals. When the first chemical was decomposed, red was ex- cited in our minds, green for the second, violet for the third. A pickup electrode has been placed on one fiber of the optic nerve—in an animal—then one color after another flashed into that eye to find which color caused a response in that fiber. In another experiment a beam of colored light was flashed on one spot of the retina, then on another spot, till a spot was found that responded to that beam, established itself as belonging to that beam. The scientist who originally did this charming experiment found seven primaries in- stead of the Young-Helmholtz three, but also found the seven to fall each near one of the three, further supporting the old theory, as he himself pointed out, though he also pointed to the trickery that remains. At the center of the retina each cone owns a private nerve line. As that line travels deeper there are interconnecting lines, which immediately complicates any explanation, and everyone sus- pects that there are complications we know nothing about.

The color-blind are classified on the three-color theory, are blind or enfeebled for one or two or three of the primary colors, therefore need to experience their world in what is left. Color-blindness is inherited, women ten to twenty times less subject to it than men. A completely color-blind person would be rare, one who saw his world in shades of black and white, and very likely he would not be as discriminating for blacks and whites and grays as his dog. A color-blind painter might not necessarily know he had a deficiency, would paint a gray sea with gray waves and gray clouds, his painting to the rest of us look either wrong or piquant or highly original.

The purpose of color vision?

Please do not say it evolved for survival. It seemed to. Do not say either that it was to get the best mate, discover the best prey, play dead in one's own color against a background of concealing color. It seemed to. If you could not resist putting it so, then at least admit that the human mind was driven also by a wish to reach out, that it had some of the passion of Prometheus, some of the patience of Sisyphus, sometimes climbed to the highest state of wonderment that it could, and color sometimes played a part in this. A face or a countryside had not only planes and lines but had color to start dream and cogitation. The mind thought in color. It used color-words in its writing, *mauve* and *sepia*. One musician saw colored pictures when he composed, another saw them when he conducted, a blood-red sunset, blue bluebirds in flight, the fresh pink of a young figure of Renoir. In disease of the brain there might be periodic flashes of color. The electrocuted ought to see crimson, then see a nothing that is nothing. Is that how it would end?

# SCOTOPIC

## Night Editor

Night changes all of this. Night changes what is seen and changes the eye that sees. Night prepares the eye for the affairs of night. An Indian walks in the moonless forest and never stubs his toe or bumps into a tree, Boy Scouts tell us. An infantryman can kill in a fox hole in Okinawa or Vietnam, or a citizen murder a citizen in the springtime in Paris and wish only that the Parisian night had been less embarrassingly bright.

The murderer would have been fitter for his business had he first spent some time in the dark, gotten his eyes ready, because eyes do get ready. (Minds get ready too.) That getting-ready of the eyes has been studied on and off for three-quarters of a century. Many of us have had the experience of the X-ray specialist waiting a time in the dark before he squinted at our stomach. If he had worn red goggles he could have come straight from daylight and squinted, because only the radiations that excite red pass through red goggles, and the night-vision chemical does not absorb red and a ray must be absorbed if there is to be chemical action, so there is no action, the night chemical is not used up, therefore the X-ray specialist can squint at once. He need not wait for his eyes to dark-adapt. This they do rapidly the first five minutes, slowly the next thirty when dark-adaptation is near to complete. Thirty minutes is also the time from sunset to night, suggesting that our eyes were evolved to get us ready for night in the same time it takes night to draw its curtain across the sky. The process continues triflingly for another hour, vanishingly triflingly for twenty-four. The adaptation in those five minutes was in the cones, the daylight receivers of the retina, because they too must adapt; the remainder was in the rods, the night receptors.

Some persons are night-blind, cannot adapt to night, some day-blind. The chicken's eye, that has that excellent scheme for color vision, has no rods, hence no night vision, and poor twilight vision, which is why a chicken decides wisely to go to roost early. It ought to be a satisfaction to hen and rooster as they doze away to dwell on their excellences and not their deficiencies, as we try to, not always succeed.

The stimulus for night vision as for day vision comes finally from sun and stars, from light energy packeted in bundles, quanta. The size of the quanta varies with wave length. One quantum decomposes one molecule of the eye's photochemical. Two or three quanta are enough for a sensation. Two or three molecules! By some later evidence it is less, one molecule. A single decomposed molecule starts effective messages running back into the brain; that is an amazing fact even if one does not quite understand it or possibly only because one does not.

"And there was light."

When we go from broad daylight into the movie we see nothing, bruise our shins, or somebody's, then after a while see surprisingly everything. Later we come out, and the long immersion in the darkness has allowed large amounts of photochemical to rebuild, the daylight attacks it, a fierce decomposing, a fierce rushing of messages, a fierce brilliance, not the happiest sensation, almost pain. People in the far north exposed to the cutting reflections of sun on ice or fishermen exposed to the sunlit ocean may seriously injure their night-vision machinery and it is a habit among them to shade their eyes whenever their hard life allows. Then comes that other part of the year, the months and months of dark. "It has been a long night."

Visual purple is the photochemical in the rods. Light bleaches it there in the rods much as if it were in actual bottles. The human vision machinery with its 125,000,000 rods could be fancied as having visual purple stocked in 125,000,000 microscopic thin-tall bottles huddled together in the retina waiting on those radiations that excite toward green or blue-green. (The red blood-cells were a different kind of bottle.) Green or blue-green in dim illumination are indeed more easily seen than the other colors. On a summer evening as day dies down and one color after another is turned off for the night the greens of evening grow strong and have to an unusual degree the power to cast a spell. That last, no doubt, is because of past summer evenings associated with contentment.

The visual-purple molecule is large, a protein with a pigmented something attached, light breaking this attachment, a number of compounds formed, visual yellow and visual white among them. Vitamin A is required for rebuilding the visual-purple. If one's vitamin A intake is insufficient one can eat liver, which has it. Oils of the liver of fish long have been employed in the treatment of night-blindness. Liver gives us so many good things that we ought really to enjoy it and when we do not it must mean that Nature left something out of us, or had reasons that you and I cannot yet fathom.

An unpleasant dramatic experiment dates back to the last century. A rabbit was kept in the dark, then for an instant one of its eyes was exposed to an old-fashioned daylighted laboratory window with bars, the next instant the rabbit killed, an eye removed, its retina immersed in a solution of alum, as a photographer immerses a film in a developer, and there on the retina was the window, dark bars and the bright blotches, the unbleached and the bleached. Almost the same experiment at almost the same time was performed secretly upon the retina of a guillotined criminal. A Frenchman did that one.

What scenes are spread on our retina and quickly wiped out! One thinks of the night pilot the last moments before he crashes into the mountain, how exasperatingly calm the words coming from the control tower two hundred miles away, how calm the electric lamps over his instrument board, how calm the chemistry in his retinas. The large molecules are methodically dividing, the coded messages marching over nerves into the depths of the brain, the kaleidoscopic record rolling, then the final frame, the camera smashed, and the universe keeps piling up light-years. In the future a chemist will figure out how to reach chemically into and beyond the eye of the mangled for data on the crash. A detective story written at almost the time of the rabbit and the guillotined criminal did have a detective getting evidence about a murder from the retina of the victim. All murderers since then have practiced how to work from behind.

At night, one recognizes especially furtive movements, which are a succession of flashes of dim light of short duration, and the rods are sensitive to dim light of short duration, the report to the mind dim but authentic. Since the number of rods increases as the microscope moves outward from the center of the retina, it is to be expected that out there we would pick up the prowler prowling in our bedroom, but do not yet see whether he is richman, poorman, beggarman, thief, drag him into the bright light of the hallway, converge our eyeballs on him, and with our uncompromising cones nail the theft of our wife's diamonds to her bosom friend.

Sometimes four hundred rods connect with one nerve-fiber, and that is to say that an area instead of a point, four hundred rods instead of one rod, report over one wire. This also inclines night reports to be vague. Night reports are like night as our mind knows night. Night sights are night sights. Night opinions are night opinions. Night fears are night fears. The beauty of night and the mystery have the vague besides the dark. It is all different from day and the glory of the difference is due to that chemical, visual purple, and to those four hundred messages over one wire, the clarity blurred, softened, even before the messages start into the depths of the brain, where somehow night interpretations attach themselves to night sense data. The glory is somehow still something else. *Scoto* means darkness and scotopic vision is night vision. As for the light source, though heaven's great candle is out, and its small candles out too, somewhere some wax candle is going up a stairs past a window to bed, somewhere some shimmer slips from beyond the black shutters or the black clouds. The human eye at its minimum in the night responds to one ten-billionth of its maximum in the day.

Cats in all respects are interesting animals. "All cats are gray in the dark." Thomas Lodge (1558–1625) said that. "At night all cats are grey." Cervantes (1547–1616) said that. "In the night, sure ev'ry cat is grey." Lillo (1693–1739). "*Quand il fait sombre les plus beaux chats sont gris.*" Beaumarchais (1732–1799). "When all candles be out, all cats be grey." John Heywood (1497–1580). These eminent men were impressed by cats. But, are cats gray at night? Interesting animals, cats.

# FORM

## *Layout*

Shapes of night, shapes of black against black, black against gray, against white, shapes of shrill daylight, of shadow, of color, of straight lines, of curved lines, of crooked lines, these registered in eye and brain give to the mind the form of the visible world. Give glimpses also of the world of others, of other eyes, how these see form, how Georgia O'Keeffe places unforgettably a feather and a steer horn and a screaming splash of departing western light, all within a cold steel-ribbon frame.

Already a century ago it was understood that two stars would be seen as two (sense-registered, brain-handled, mind-interpreted) provided the lines that came from them made an angle of one minute at the eye. The eye will be able then, as one says, to resolve them. Even a century earlier it was recognized that "a spot of the moon's body" must make an angle of at least one-half minute for it to be seen on earth. Some eyes do better, some fantastically better.

If, instead of two stars, a thin thread is rightly lit, rightly set in a contrasting background, has a right reflecting texture, the angle need not be one minute, or one-half minute, but, learned from an extraordinary experiment, one-half second. At the retina that would be less than the diameter of a single cone, one of those six or seven million daylight receptor-cells. That cone is able to "see" that thread. This explains why the tailor at his bench is free to think of other things, of his new wife, while he works. The thread gives his eyes no trouble. Some birds of prey do better than the tailor.

When an eye is not able to resolve a sequence of points they fuse into a line. A line has length and direction. Add genius, and the line may produce the famous one-line cat in the drawing of Hokusai. Hokusai wetted his brush in black Japanese ink and by a single sweep left us the cat. He placed the cat in his mind. He pointed the brush. He held its bamboo handle vertical to the rice-paper. Then the sweep. The cat was there forever, or a thousand years.

When an eye fails to resolve close-lying lines they fuse into a surface, and a surface may have many qualities. A black-and-white reproduction of Hokusai's self-portrait would have both the sequences of points to fuse into lines, and then the close-lying lines to fuse into surfaces, and the resulting play of blacknesses against white passes on to us how that greatest of Japanese artists saw himself. In any wire photo the eye only partially resolves the points, hence the grain. The eyeball, because it is a roving thing, perpetually in movement, contributes to the fusing of the picture. Even in the womb the fetus's eyeball is known to move. It could be training for the coming life. It could also be, if some dream researchers are right, evidence of the fetus's dreaming. Of what would it be dreaming? Where is the clod would not wonder about that? Other researchers have other explanations.

Recall the eye chart in the doctor's office. There are rows of letters of successive decrease of size. The breadth in any stroke of any letter in any row or the space between strokes is such that the eye can resolve stroke or space at a distance from the chart that depends on the letter's size. The person being fitted for eyeglasses sits 20 feet from the chart. An average eye reads the 20 type at 20 feet, has 20/20 vision. Stroke or space there is making an angle of one minute at the retina. When the eye can read only the 60 type at 20 feet it has 20/60 vision. The fitting for eyeglasses becomes an investigation into this human being's two-dimensional vision.

Form may be three-dimensional. Form may have depth besides length and breadth. Depth can be seen with one eye, better with two. So there is monocular and binocular depth vision. Illumination expectedly helps. A spherical surface looked at head-on—with one eye—is spherical because the light reflected from it diminishes progressively toward the edges of that spherical surface. Shadow helps. A soft corduroy sleeve has a thousand depths, or a hundred, but there is not any doubt of what values this fabric, corduroy, owes to those depths. Contrast helps. Whoever has picked black fuzz from black cloth knows it would have been easier had it been white cloth. Parallax helps. Parallax is the apparent movement of objects relative to one another. When looking out the window of a taxiing airplane one recognizes that the moving faraway distance, the middle distance, the near distance, have different speeds and directions. Our mind, long before it jerked itself up and paid attention to such speeds and directions, knew about them, and from them learned something more about depth.

When as children we saw and understood that the mountain beyond our town was big, and the robin in our back yard small, we were collecting data for all mountains and all robins, and again were learning to see depth. However, Lashley's rats confined in the dark from birth did leap perfectly across a gap the first time in their lives that the light was turned on, warning us not to forget how much of everything needs also not be taught, what capacities or parts of capacities are built-in and ready in the nervous-system. By night out the south windows of the College of Medicine are two lit clock faces in two church steeples, and a valley of ghostly houses between, and the eyes of those who work in the College by night directly measure the depths in that ghostliness, and see those steeples as three-fifths of a mile away. Perspective helps of course. Two lines approaching each other in a drawing give us a street running off into the distance, and if those lines are felicitous the drawing may require not another detail. Tanyu, the Japanese painter, painted an arc on a rectangular *kakemono*, just the arc, placed it low on the *kakemono*, did it with perfect knowledge so that to anyone's vision that arc is the evening sun sinking into the sea. Interference helps. When an oak gets in the way of a maple the eye sees the maple as farther. Blue haze helps. We have so often seen the blue over the faraway that the yellow trees are near.

Two-eyed, binocular, vision is better than one-eyed. The right and the left eyes are not looking at exactly the same scene, the right reaching farther around to the right, the left to the left. Textbooks instruct us to lift a finger in front of our noses, then alternately to open and close right and left eyes, the finger then alternately jumping side-to-side, proving that each eye singly was not looking at exactly the same view of that finger.

Strain in the muscle inside the eyeballs may help, does in birds. In birds this muscle is not involuntary, smooth muscle, as in us, but voluntary. Its tone, its tension, is registered somewhere in the bird, let us say its consciousness, that appraises the tone, and the appraisal assists its sense of depth. In stereoscopic pictures and three-dimensional movies every action has been photographed from different positions, and the eyes of the viewer draw together the photographs, the views, and that gives to the mind, whether we completely understand or not, sometimes an even unnaturally deep depth.

The developmental psychologist has said to us often that the various senses help each other. We rub our fingers along the baby's crib at the same time that we see the crib. We touch the baby's face. The baby touches our face. Hand teaches eye. Eye teaches hand. Whoever has watched children bending over their drawings knows how the entire body helps, how then after a time the formless takes on form, the two-dimensional becomes the three-dimensional, and it is always striking how this occurs suddenly, overnight, one thinks. Latterly a Puerto Rican has been teaching the blind of his country to draw. In human beings deprived of their dominant sense a teacher has taught the hands without benefit of vision to put form into minds. Our artists teach us. Our painters and sculptors see shadow where the ordinary eye sees none—may avoid seeing in deference to an earlier tradition of painting and sculpture. The face of a Japanese geisha in a Japanese print has little or no shadow because through generations she has been painted so; then one day we come into the presence of the flesh-and-bone geisha and her powdered white face appears round and shadowless, though her anatomic face is of course not round and not shadowless.

Shadow and substance signal from the void, and signal from an ink drawing, from a water color, from an oil, signal from a mouse that flees along the side of a restaurant kitchen and into a black hole next a pipe. That mouse is a light source, stimulates a human retina, and a human mind thinks mouse thoughts. "Wee sleekit, cowrin' tim'rous beastie. . . ." And a mind wishes it were not in this restaurant.

# BODY-IMAGE
## *News Flash—Man Loses Limb*

Each of us knows he has a body. When did he find that out? Usually the question never occurs to him. It may at that horrible moment when he is introduced to someone and reaches to shake hands, and the person has no arm. He may be sure at that moment that his senses teach him everything. In any case he knows that he has a body and the woman has no arm. He knows that he is in possession of an infinitude of masses and shapes and lines and textures and moistures, all apparently reporting to one another, that he is suspended in radiations and vibrations that produce for him sunlight and moonlight and color and heat and cold and sound, with the perpetual possibility of pain. These apparently do somehow give him his image of himself. The image has two legs, two arms (usually), a trunk, a neck, a head.

Until a few decades ago no one realized there was anything remarkable in this. Then it was discovered that there were and must always have been persons who did not with any of their senses sense a body, so did not have it, or did not have some part of it. A case was reported of finger agnosia, no knowledge of fingers, these lost to the mind, the person's own fingers and everybody else's. The person could not name a finger, not point to the correct one if named, had a blank spot in the mind. Nevertheless, the sensations when tested proved normal. Skin receptors were normal. Vision normal. Smell normal. News came from the world as it should, but no news of fingers. No fingers in the world. And that person, as it turned out, was not unique. One case having been recognized, others were.

It appeared on study that the damage might be in the flesh of the nervous-system, or it might be in the mind. Nervous-system damage. Mind damage. The person might be totally unaware that anything was wrong, or he might be suspicious, might be tormented.

A man washed the right side of his body, laughed miserably when it was pointed out what he was doing, and he proceeded to wash the left side too. He had overlooked the left side because it did not exist, to him. This case was reported as nervous-system damage, demonstrable blighting of mind because of demonstrable blighting of flesh. The case was brain damage.

A young girl lost her body, the whole of it. This occurred suddenly. Earlier in her life she had had a hopeless reduction of force and feeling in one leg, some reduction elsewhere. This was indeed nervous-system damage. She had gone along well with her trouble, was brave, as one says, but that had been because she had an imaginative family, and had a close friend her own age. The friend fell in love—the patient lost her body. She could talk calmly about both losses, the presumed loss of her friend, the loss of her own body. The psychiatrist gave her a Rorschach test. Everybody has heard of the Rorschach. The subject tells what he or she sees in some carefully worked-out ink blotches. This subject saw only animals. Blotches that nearly everyone sees as the whole or the parts of a human body or face she saw as animal body or face. She had wiped off the human body. There were to be no more human bodies in the world—because she had such an unhappy one. It struck one immediately as being her mind's way of protecting itself from dwelling on a disability with which she was doomed to live. The case was mind damage.

When persons who have losses in body-image make a drawing of a body they produce what could be mistaken for the work of a child, though of course the psychiatrist does not make that mistake. To him every detail of that drawing has adult meaning. If one listened to the psychiatrist talk (or the psychologist) one could soon think that every line in any drawing had meaning, any detail, that everything has mind-meaning. Everything has, though not necessarily worth deciphering, and not necessarily possible to decipher.

When then do we find out? How do we come originally to know we have a body? Perhaps we learn it. Perhaps the developmental psychologist is correct when he suggests that our senses are our first and last and only teachers. I touch my hand. I hear my one hand strike my other hand. I can tell in the dark where my hand is. And those facts are connected in nervous-system and mind. That teaches me that I have a hand, and similarly that I have a body. The knowledge keeps accumulating in every hour of every day. I blow my nose and learn about noses. I sense the weight of my feet and learn about feet. I listen to my voice or without listening hear it beat in the rafters of my skull, and learn about a most complicated sense. To the knowledge of my own body and its parts I add knowledge of other bodies and their parts. The baby looks, looks, looks at another baby, also looks, looks, looks at himself. Somewhere in the course of this he comes into possession of that other baby, that embodiment into a single of what are already many parts, and possibly with his eyes closed he now can see that other baby. I also press her cheek with my hand, her cheek with my cheek, learn some of the subtle sculptural facts of cheeks, suspect (if I were old enough to suspect) that the Lord Almighty had practiced on many a cheek before He could have made her cheek.

Yet the whole of my knowledge of body-image cannot be a mere addition of the teachings of my senses. It had to have been built on something I was born with, the sense-machinery that could some-how sense, could bring scattered pieces of various senses together, make totals. There is excellent evidence for an inherited location in the brain for such uniting into totals. At post-mortem in persons with loss of body-image, damage repeatedly has been found in a definite location, parietal lobe, one side, toward the back. But admittedly everything about this perplexing subject is perplexing, including the examination of the patient and the interpreting of the post-mortem.

A patient was asked to look at her left arm that was absent to her mind, then to put her right hand on her left shoulder, even though that might be difficult because that arm was absent, then move her right hand from that shoulder down that arm. She saw clearly what she did but adhered to the logical opinion that a person must believe her intuitions and not her eyes, then presently admitted that there was an illogic in that also. Another patient. He had his arm hanging off the side of a hospital bed. The doctor asked: "Whose arm?" The man answered: "Yours." The doctor lifted the man's arm in front of the man's face, asked again, whereupon the man suggested, yes, it was conceivable that that arm was his own arm.

For centuries there was the notion, less common in this less superstitious century, that each of us at a time and place enters his body. His soul enters—or his mind, if that is how his temperament, tutelage, era inclines him. Each of us might therefore under unusual circumstances, or even under usual ones, leave his body again, and not necessarily Biblically speaking. Indeed, if a mind exists that does not know it has a body, or anyone has, would not that make the disembodied a fact? One gets a new respect for such a simple after-dinner performance as taking one's body out for a walk.

# XVII

# VOLUNTARY

## Threading a Needle or Felling an Oak

# MOTOR SIDE

## *Juggler*

There was Grandfather—one certainly remembered him. He undoubtedly was dead. They said so. He was lying in the parlor. They
had taken off the chair-covers. He was very still. Every white hair of
him was still, in that black box, like a King. We had heard about the
dead but we had never seen one. Grandfather did not move. Grandfather was dead.

Life. Death. Movement. Mind.

An American physiologist years ago pulled these bluntly together.
He said: mind exists for movement. That often is implied when not
said. It could be true. It could be questioned. Originally when one
heard what that physiologist said, it was objectionable, that mind
should have arisen on the planet merely to move the skeleton
around. Mind was for perfecting movement.

It would be logical then to investigate creature-mind in creature-
movement.

In the laboratory the physiologist cuts up the brain and observes
the effects on movement. In the street we all interpret movement.
Children watch the juggler. We watch the children watch the juggler.

Notwithstanding, mind exist for movement? Exist for the sake of
that small boy's melting head-to-toes in tears, thinks he has lost his
mama, moves erratically among the people? Mind exist to make that
melting, that erratic movement so convincing that the crowd would
go to some trouble in his behalf?

We speak of *voluntary* movement. This introduces a further difficulty. Voluntary?

"Eyes, do this," we say, under our breath, or something says, and in consequence the proper eye-muscles pull the eyes around, yet the voluntary, our will, as ordinarily we mean will, knows nothing of those single muscles. This fact has been pointed out many times. True, the will referred to was the old-fashioned that man through the centuries has been trying to understand, not this peculiar reasoned will, this mind, that exists for the highest perfecting of movement. "Foot, do this," we say, and the proper foot-muscles obey a nervous-system that obeys—we do not know quite what.

A high-jumper might have more courage than a juggler, or maybe only less imagination.

It is nighttime and we are in Madison Square Garden. The high-jumper just has risked leaping seven feet and one-quarter inch straight off the earth toward the dome of the sky in the general direction of the island universe of Andromeda. He has accomplished it, too. Has cleared the crossbar. Has posted a new world record. That was a minor miracle. Any motor act is, to the performer also, if he can get himself into the right observing frame of mind, separate himself from his own performance.

He started diagonally from the left, a leisurely run-up, had learned it must be leisurely, so held back his body while he pictured it in the air above the crossbar, allowed a margin of safety up there, twelve inches. *Saw* his body up there. Momentarily it was a still body, but that stillness was of the living, not the stillness in the black box. Step on step he advanced toward the launching step when his left leg must give its powerful thrust. The left leg bent. Suddenly the left leg straightened. A flea's skinny leg straightens in the same way. A flea's mind?

High off the earth toward Andromeda shot the high-jumper, the force of his contracting muscles overcoming gravity, the right leg leading. Immediately the right arm, smooth-moving, like all his machine-parts, joined his right leg. Then both legs were horizontal, tremorless, floating. (Must be an agreeable sensation.) He had become that picture in the air. Abruptly, the left leg shot the other direction. Cleared the bar. The weight-center of his body shifted. The body rotated. One hundred and sixty pounds of muscle and bone dropped into the dirt below. That drop was almost gentle. Pride took over. There is movement assuredly in pride. The high-jumper rose to his full height. He rose erect, a model of self-conscious grace, every muscle except the muscles around his mouth variously contracting, shouting: "I am the world champion!"

As to the island universe of Andromeda, it is quite near, only 2,000,000 light-years, a good opera glass able to find it, and a light-year, the distance light travels in a year, only 6,000,000,000,000 miles. As to the flea, for its weight it is the best high-jumper on earth, accomplishes its powerful thrust by an exceedingly rapid contraction of muscles.

A seamstress does not high-jump, but a seamstress's movements are no less extraordinary than that flea's. This morning she threaded a needle, will many times before her day is done, without an instant's hesitation put the thread through the steel eye, and each fine part of that when expertly fitted into each other fine part would tell a story as dramatic as the juggler's, as the high-jumper's, more dramatic because we never thought of her except as patching holes in pants. Rembrandt could have painted that seamstress threading that needle, could have made us feel the thread go forward, the needle come halfway to meet it, the one movement guiding the other, sense of touch helping, skin against thread, skin against needle.

Albrecht Dürer could have painted her too. More likely Dürer would have sketched her. Leonardo would have painted. Any one of those three could have made us realize that a motor act may be worthy of God. A friar in his cell mumbles that all acts take place in the mind of God. Whether this is or is not fully accurate, every artist knows that every fine part in the perfecting of that seamstress's movements pleased her mind, moved in her mind. But does that in any way prove that mind exists for movement? That that is why she has mind? Why mind came to her originally? Why mind came into the world?

# INFANT MOVEMENT

## *Earliest Routines*

Before high-jumping and threading of needles there was creeping, and before creeping, at two months or so, there was the sucking of the thumb, and, before that, the more and more expert sucking of the nipple, and the beginnings of the movements that enter into the production of a smile, then at four or five months the grasping of something seen. "At first the infant, mewling and puking in the nurse's arms." Shakespeare was a student of infant and neonate and everything else.

Long long before that human-infant's creeping there was creeping; something living crept. If the creeping had mind in it to assist it that mind would have been the wispiest wisp of mind. So it was late, actually, that baby crept. His mother followed with her eyes and heart, that infant, that dear boy shaped like a platter. If there was a psychologist anywhere about, he followed too. Those earliest routines interest him. He would have consulted his timetable. Any first movement in any new citizen occurs by the timetable and the timetable does not much change despite anyone's eagerness.

At three weeks, baby's head bobbed under its great weight; at four weeks, still bobbed; at five weeks, bobbed but bobbed less; at six weeks, eight weeks, twelve weeks, approached the adult state which it attained at sixteen weeks, and at twenty weeks no more bobbing till old age. "Can't you make Grandpa stop doing that when we're having company?" That was the older sister complaining. She will be doing it herself sooner than she anticipates.

By the second week the sucking was indeed visibly expert in the control of all the relevant movements, promptly recovered the nipple when lost, unambiguously separated nipple from the neighborhood of nipple. From then on one after another the perfected motor behaviors made their appearances. The stance of the body was more secure, which took movement, and control of movement.

"Dear boy, he isn't falling out of bed on his head any more." (That was his mother.) Many times he had to fall out of bed to learn. Learning is mind, or can be, and he will go right on falling out of bed, all kinds of beds, and learning. "What a good mind he has." (Again his mother.) The perfecting movement was easy to see. The perfecting mind was not quite as easy to see, and not at all easy to see what that physiologist believed, that mind was being evolved for movement, though movement was definitely creating a world that was definitely outside of this body.

From birth his head was more to one side, right side in his case. At sixteen weeks, the head was in the midline. Movements produced by contracting muscles accomplished that, pulled the head there, kept pulling it there, it not just staying without this continuous reflex effort, nerve-impulses racing over nerves for the continuous control of the muscles that controlled the head. A head is a heavy object. It is easily imbalanced. He, this one, may have had something to do with it, but difficult to weigh that against the huge amount that was sheer species. At thirty weeks he sat, briefly, then sat longer. The dullest in the family could recognize what an achievement this was. Nature from the start had been trying to get his body off the earth, get it upright, give it the status of Homo sapiens, free its ancient motor patterns to express themselves, and at the same time add that private bit of its own, add the voluntary to the ancient, a little and a little and a little until at last his body, this person, could hustle through city streets, be a lover, be a murderer, be a convert who with genuflexions hoped to make peace between flesh and soul, and with a few final fading facial grimaces expected to bring even his deathbed to an oratorical conclusion.

Studies have sought to relate the first appearance of this or that movement to the first appearance of this or that change in the brain.

The fetus in the womb at eleven weeks—fetal timetable—clenched its fingers into a fist, called grasp reflex, inherited from ancestors who lived in trees and grasped branches, but grasping is useful also for creatures who do not have branches around, and not surprising that the power to grasp should express itself early. However, a fist to be useful in daily affairs must be able not only to close the grasp but open it. The infant out of the womb at twelve weeks—infant timetable now—could do this, could release the grasp. Some change in the brain, as found in dead brains, did suggest that at the twelve-week stage a new type of motor work might be done. That work might be inhibition. The fetal nervous-system could then be imagined (this is too free and easy) to excite the closing of the grasp, the infant nervous-system to inhibit the closing, open the grasp, thus a completed act be chipped from this marble that is only less old and hard than the marble of the hills. At twenty weeks, the smart boy did efficiently grasp a spoon, and at thirty weeks, not only grasped it but moved it to the thumb side, thumb and index finger being the most capable digits. "He does not drop one drop of prune juice into his lap any more." (His mother again.)

An infant were it left on its belly or its back would die like a bird dropped out of the nest. Its extremities would be useless. Ever so often the weight of its body would pin down an arm. Its hand would bat its nose. Its legs when its body was held in the air would carry out stepping movements, but machine stepping movements.

Wait.

Allow the preordained time for the preordained to have freed itself, and the hour would come when this infant, this boy, not of his own will, as his fond mother was thinking, but by the dogged will of the ages, pushed his chest off the bed. Pushed. Pushed. Triumphed. The chest was off. He settled on his arms. Was down on his belly again. Rolled to his back. Rolled the other way. Rolled. Rolled. Rolled. Rolled as if this were the day of a great voluntary emancipation, but the rolling too was by the clock that ticked impartially for all. Where was the mind? Who can be sure?

The first creeping came at thirty-five weeks. "Today, baby crept." Creeping was locomotion by alternate bends and thrusts of arms and legs. The first standing came in stages. At three weeks, the legs could not take the body's weight even when supported. At twelve weeks, the legs momentarily could. At thirty-two weeks, about eight months, a breathless quarter minute when the small male stood if supported. He could at this point be imagined (it looks that way) to be getting some assistance from that mind that should at this point have advanced a good distance from its downy beginnings. At forty weeks, he unmistakably stood if supported, his legs apart like a sailor's, and at fifty weeks, stood alone. As to walking, that might be somewhat ahead of the timetable, somewhat behind, but at one year he was sure to stamp, stamp, stamp, holding tight to his mother's hand, or his grandfather's if his mother was cooking the carrots, and, a few months more, he had become at last the independent light traveler.

Every item of this was only on the average, and only if when he was put together out of so many parts a single part was not forgotten. If a part was, then he probably fizzled out, whether he dragged on, as we say, or passed on, as we also say. Movement now seemed everywhere in his life. The voluntary seemed everywhere— but so difficult to be sure of what is voluntary. Mind could appear to be in and around every item. How to interpret that mind? How to think of what that physiologist thought of movement and mind? How to think of what others, not physiologists, so vocally often, have thought? How honestly to interpret the interpretations? Free movement could be close to free will, close to the voluntary, and free will has always been difficult to think about.

# FLEXION

## *Jerks Away When Hurt*

A movement among movements, an every-hour movement, is to draw away, draw some part of the body toward the core of the body, usually. May be an isolated movement, may be a complex pattern, may be a total behavior. The total creature draws away. It can. It can fold away. Jerk away. Flex away. Flex in degrees. Flex reflexly. Flex deliberately. Flexion may be lifesaving. Often it has military snap. Often it is a sequence of flexions, a number building up to an act. The infant, the seamstress, the juggler, the juggler's cat that sleeps with the juggler, each has the capacity. White muscle-fibers carry it out. They are swift. It may be no more than the withdrawing of a toe from the bottom of a cold bed, and then that reflex, as if unsatisfied, travels up, involves knee, hip, till the body appears to have but one intent, escape.

A sleeping dog may show a luxurious array, four legs toward abdomen, head toward chest, tail slinking between legs, the smart dog giving gravity the least possible trouble. Such a wide-spreading flexion can be so excellent it looks contrived.

A wide-spreading flexion doubles up the body in an attack of appendicitis, the movement rightly termed withdrawal. In rheumatism, flexion-reflexes withdraw a part, diminish the drag on it, and the distress. Flexion-reflexes may overdo the job, muscles become useless because continuous overuse has ruined them, their opponent muscles useless because never used. Deformities ensue that could fit the miserable man or woman into a Greek tragedy.

The scratch-reflex would fit him into a Greek farce. Flexion is part of scratching, and on this itchy planet scratching often flows over into warm emotion, into contentment, into mind. Aristophanes might have introduced the scratch-reflex into his theatre business. Molière might have when he was director besides playwright because even the politest Frenchman scratches some impolite place when he thinks nobody is looking, and if an entire audience suddenly was looking, discovered him, as is said, that would be farce with belly laugh in Molière's day as in ours.

A flea employs the same technique as the Frenchman. The flea lifts a limb, brings its outermost joint exactly where something itches, scratches. Jonathan Swift was so interested in that flea that "Hath smaller fleas that on him prey." A kitten first flexes her hind paw to the place that itches, within millimeters, the paw poised there, then follows through with the rhythm of the scratch, far from simple, thumps enough to shake a house. If an artificial gnat is placed on that kitten's back her paw goes where the gnat goes, not where the gnat does not go. The destination is precise, and, unless interrupted, the rhythm is as regular as a pendulum, four beats per second. If the kitten has been scratching on her right side and the artificial gnat is pushed across her spine to the left, at the instant of the crossing, down comes the right paw, up comes the left paw. All the movements are perfect. Nevertheless while one contemplates the perfection a chill may run over one, the too machine-like in the living exciting something close to horror. Or what is it? Is it the evolutionarily old intruding upon the new? Is it the impersonal upon this highly personal cat?

If an experimental animal with the topmost parts of its nervous-system removed lies helpless on its back, and the researcher pinches its ear, a hind limb lifts, executes a few scratches that may look hesitant but plainly are scratches. Medical students sit in a classroom and see the limb lift, see the scratch, nervously laugh. No student knows why he laughs. The professor does not know. It would satisfy something to know.

Flexion is part of locomotion, of walking, of running, and each time either leg does its forward swing it must flex so as to avoid scraping the ground. Motion pictures of the expert runner reveal the flexion as occurring with sharp mechanic precision. How beautiful is the runner's almost naked body. How beautiful is every fraction of his movement. How beautiful is one single contracting muscle when for an instant it becomes visible under the skin. Do beautiful women select athletes for mates? Does the all-American marry the seamstress? Years later the all-American dies, too young, and he so careful of his health, regular meals, regular sleep, regular exercise, a quiet walk at midnight along the river—isn't it lucky for those five children that she knew how to sew?

The runner is off—a mile in four minutes. Light is off from the sun—186,000 miles in one second. That light keeps coming, keeps coming, keeps coming.

# TONE

## *Never Slumps*

Gravity reaches us through weaning, wedding, bedding, bearing, dying, pulls us down. Our body must automatically push us back up, has the nerve-muscle machinery to hoist. Each of us is like Uncle Remus forever sinking toward the earth, victim of his weight, in the nick of time saved.

Flexor-reflexes help a man to jerk away. Extension is the opposite of flexion. Extensor-reflexes do the opposite. Instead of folding him like an accordion, they straighten him.

Stretch is the stimulus. Stretch means stretch. When the body tends to sink, to flex, extensor muscles are stretched. A report goes into the nervous-system, the stretch-stimulated muscle contracts, pulls on its tendon, that readjusts the joint, the muscle is back where it was before the stretch. That is the story told over and over and over and over.

The receptor that picks up the report of the stretch is called stretch-receptor. It is microscopic. It is in the muscle. A scientist who brilliantly studied it states that there is no more extraordinary receptor anywhere in the body. The stretch-receptors are specialized muscle-fibers locked among the regular muscle-fibers. They accurately register the muscle's changing length throughout the time it is stretched. There are also stretch-receptors in the tendons. The two sets are vigilantes on guard day and night watching the tiltings of the body, telegraphing, readjusting, keeping us upright.

The sloth is held in its normal state by flexors. But the sloth's normal is to hang upside down on the branch of a tree.

Our body may slump any direction.

Imagine it to have been rocked forward on the ankles. As it rocks, that calf muscle at the back of the lower leg is stretched. The tendon of the calf muscle, that is tacked to the heel bone, is stretched. If the rocking was forward and to the left, only those fibers are reflexly stimulated to contract that will rock it backward and to the right. Any other movement would be uneconomical. The body would only be saved from one imbalance to be pitched into another.

Our lower jaw escapes slumping because each time it starts down because of the pull of gravity, stretch-receptors jack it up, and a crowd of human beings looks no worse than it looks. If human jaws all slumped, would the Associated Press stop dispatching around the country photographs of the new brides in Washington and Northampton, or would we just revise our eyesight and our esthetics and admire slumping jaws?

What, then, is tone?

Everybody uses the word. The ancient Greeks used it. They compared the taut vibrancy of our muscles to the tuned string of a musical instrument, but they would not have had the data with which to understand the tuning. The family doctor uses the word freely, speaks of the tone of a man's liver as excellent, another's as bad, prescribes tone in a tonic, knows that the tone of a man's stomach and gut can be no better than the tone of the smooth muscle in it. Then, he too thinks, not necessarily deliberatedly, of the vivacity or lack of it in his patient's voluntary muscle-ma-chinery.

Tone does not visibly move the body. Tone lies behind movement. Tone prepares for movement.

All the muscles that cover our skeleton have tone, flexors as well as extensors, but extensors more. Our limbs are heavy. They create ceaseless pull, ceaseless stretch, and the body is making use of its own weight. Stretch is operating in all directions in all parts exciting this minimal contraction, a fiber here, a fiber there. If one puts one's finger on the skin of an awake body one feels the tone, feels it less but still feels it on the skin of the sleeping, not the corpse. Our great gobs of muscles never are at rest, not at noon, not in the small hours, not on the Sabbath.

# LEVELS

## *First Is First and Second Is Second*

Consider now the earthworm. Consider its rings, its segments. Con-sider how evolution added segment to segment till there was as-sembled a self-respecting worm moving above or below the surface of the ground. Each segment of that worm has some independence of the others, a group toward the tail-end much independence of a group toward the head-end, and that is why a schoolboy can cut an earthworm in half and coolly observe the halves go their separate ways, the head-end with a fling, as if she had kicked her petticoats behind her. Yet even in an earthworm the head-end has great au-thority over the tail-end, and the higher in the scale of creatures the more is a tail-end cut from its head-end a hapless piece.

John Hughlings Jackson, Englishman, was the first to see here a special idea, a principle. Earthworm, juggler, high-jumper, seam-stress, ladies and gentlemen hustling along Madison Avenue, all have nervous-systems built in segments, with dependences and independences, and Jackson believed it correct to conceive a hier-archy, conceive that in the course of evolution one segment was lifted above another, the highest and most recent with the fewest automatisms settled into it from the past, therefore the greatest freedom from the past. The lowest must slavishly repeat the past.

Jackson's idea was inspired. It has everywhere affected thinking about the movements of the creature. Has affected experimentation. Has seemed to avenge itself when neglected, and a discovery made later that might have been earlier, as some of the delayed understanding of the controls on breathing.

Today it is thought that levels were evolved not only one above another, but front to back and almost any direction, and that they need not be levels so much as areas of influence.

Hughlings Jackson died in 1911. He had been a practicing neurologist. Before him there had been no such specialty. While he was creating it he was increasingly a Londoner of Londoners, got his diploma far from London, in Yorkshire, a medical college of eight or nine students. He had been unsure at the time whether he ought to be a doctor, feared, as he expressed it, that medicine might not offer enough room for his kind of expanding intellect. Hippocrates thought the opposite. Several persons of Jackson's day recognized his talent and kept him on the line of it: namely, physician, neurologist, physiologist, writer, man-of-the-world. He was a great book-buyer, would rip pages out of a book, send them to someone he thought would be interested, left a useless library. Had an impatient temperament. Laid that to heredity, an uncle who would poke him in church and say, "Come on, John, we've had enough." As neurologist, he drove to all corners of London in a carriage and pair, visited patients who had strokes, and from them came his idea of levels. Immediately after the stroke the victim might be unconscious and when he regained consciousness body-movement on one side might be gone, a left leg slide off the bed, hang there. Later this paralyzed leg took on a rigidity, the foot twisted inward, heel lifted, and it was possible to fancy that if this victim were stood up on that leg he would have the stance of a toe-dancer. (For ten seconds.) Not only had an active tone been poured into that leg but it was a tone that would imbalance the body. Such a leg was spoken of as spastic.

Jackson reasoned that the hemorrhage of the stroke, if hemorrhage, had cut the low parts of the nervous-system from the high.

Normally the high restrained the low. The low now was freed. Jackson called this release, as we do, summarized his views as follows:

> Roughly, we say that there is a gradual "adding on" of the more and more special, a continual adding on of new organizations. [He was speaking of the evolution of the system.] But this "adding on" is at the same time a "keeping down." The higher nervous arrangements evolved out of the lower keep down those lower, just as a government evolved out of nation controls as well as directs the nation. If this be the process of evolution, then the reverse process of dissolution is not only "a taking off" of the higher, but is at the very same time "a letting go" of the lower. If the governing body of this country were destroyed suddenly, we should have two causes for lamentation: (1) the loss of services of eminent men; and (2) the anarchy of the now uncontrolled people.

This is vivid writing, and pompous, "eminent men," "uncontrolled people," belongs to the day of Thomas Carlyle, Herbert Spencer, Queen Victoria. At least, though, we understand it. We have no trouble with the idea. One result of the idea has been that all around the curving earth for three-quarters of a century researchers have produced experimental slicings of the nervous-systems of cats, dogs, monkeys. They have been as destructive as Nature in a foul humor, have studied how levels influence, how they are influenced, and every year we have in consequence learned more about body-movement and posture, but at the same time have grown less dogmatic about what is a level. Recently, also, the idea of levels has been pushed not only into and up through the brain but into and up through the mind.

Because of this we can follow in greater detail the harmony that threads a needle, that fells an oak, the harmony in the movement of frog, flea, juggler, high-jumper, Hughlings Jackson in his carriage and pair riding through the streets of London on October 26, 1890, glancing this way and that, the poise in his mind conveyed in the tone of his excellent nerve-muscle machine. He had a hairy face, like Charles Darwin, or one of Darwin's apes, more groomed. Fancy him, Neurologist Hughlings Jackson, driving his carriage and meditating the injustice done a lady on the part of a scoundrel in a

French novel that lies open in his one hand while with the other he rides along the Strand to the house of the next apoplectic. He himself suffered migraine.

## SPINAL MAN
### *Lost His Head*

A woman caught her husband in a place he ought not to have been, caught him, as people say, in the act, therefore aimed a revolver at him, fired, a single bullet, and it went straight through his spinal cord. Did him no special harm, except that it separated two sets of Hughlings Jackson's levels and disconnected the brain. The bullet traveled sufficiently below the neck, else the transection might have been equivalent of the hangman's. The gentleman lived. The cord below the bullet was in excellent state, but the legs, because the lines to and from the high were cut, could neither receive nor carry out any orders for voluntary action, and could send up no reports of what had been accomplished, so the man did not get on his feet and move about, indeed stayed hour after hour in one place, giving him at least the leisure to meditate on his life. Had such a problem interested him he might have collected observations as to what a fragment of spinal cord can accomplish of itself, regarding it as so-and-so-many Jackson levels, or regarding the entire topless piece as one level, spinal level. Any animal that for any reason is in this state is called a spinal animal. Spinal dog. Spinal monkey. Spinal man. The unfaithful husband was a spinal man.

Man is always flattered when he can place Nature's productions by the side of his own, and that was the pleasure of the neurophysiologist who thought of the spinal segments as Swiss watches, one stacked on top of another, segment, segment, segment, finer Swiss watches than any legendary Swiss watchmaker ever saw shimmering in his Rhine-wine dream, each adjusting its tickings to the tickings of the others, the final synchronizer being that master clock stowed up there in the skull. It should not be disconnected. Each watch of itself is a miraculous Swiss watch.

A physician testing spinal segments taps routinely beneath the knee-cap. There the tendon of the large muscle that lies on the front of the thigh bridges across from thigh-bone to shin-bone. The tap kinks that tendon, which stretches the muscle, a stretch reflex results, and the lower leg kicks forward, a kick that the gentleman did not intend. Syphilis may destroy the ingoing path of that reflex. A tumor may destroy the whole neighborhood. Poliomyelitis may destroy the outgoing path. In each case the kick does not occur. The spinal segment and the wires to and from it must be intact. Usually the physician is tapping to test whether the total nervous-system is keyed up or keyed down, which the knee-jerk may reveal. One physician is more expert at this, and at everything else, than another, a difference between physicians that the patient need not know, because the patient, Park Avenue or Grover's Corners, had better be allowed to think he has selected for himself the best of possible physicians—once he has selected him.

That husband had by his irate wife been subjected to a somewhat unique surgery. Less unique is to insert the blade of a scissors into the mouth of a frog, close the scissors, and off into the slop jar go skull, brain, face, eyes in their sockets. Lucrezia Borgia says, Ah! Instead, the brain by itself might have been pithed, as thousands of students pith frogs' brains every year, squash them with a needle against the inside of the cranium. (Or they used to. Maybe they don't now—thousands.)

The frog as a result lies a small green heap utterly collapsed, a state known as spinal shock. The frog continues to lie there, seconds or minutes, then appears to wake. If it was the pithed one, lost its brain but not its head, it might to some observer seem now quite an excellent frog. Spinal shock, the transaction farther down, in dog or cat lasts not minutes but hours, in monkey not hours but days, in man or great ape not days but weeks. The brain in these was not pithed, just separated from part of the spinal cord, as by that bullet. The man's mind is working well. Soldiers in battle have been shot through the spinal cord and thought their bodies cut in two, were able to think that. Of a monkey in a laboratory it was related by a famous observer that after the cord was transected the head-end amused itself catching flies. The head-end was alive and useful to itself, the tail-end too, but no old-time instructions could be going either way. The monkey just went on catching flies.

Gradually after any catastrophe some movement returns. Gradually reflexes return. Gradually the cord goes to work again on its own, but blighted. Any movement of the limbs appears to lack good sense, lack the sheen that mind throws over limbs, the muscles without proper tone, the losses greater in one species than another. The return of the reflexes is unpredictable. In the highest species everything below the transection may seem a sorry blunder. A spinal dog has tone enough to stand if set on its feet and even spinal man may stand a brief shaky interval before his legs give way, experts speaking of it as spinal standing. A monkey's losses are greater than a dog's, a chimpanzee's greater still, man's greatest. Spinal man is not rare. We who live outside Veterans Administration hospitals may forget that the wars that made us safe—a phrase difficult to get out of one's throat—have left behind them fractions of bodies with blood circulating as does water through dead fountains. Immediately after the disaster the victim's limbs lie like the frog's. A rigidity later develops, not of much use, and one mutilated veteran lying in his bed may display the same detached interest in his own legs that he displays in those of the mutilated veteran in the next bed, the two gossiping, passing away the tedium of a midwest August afternoon. At least, the hospitals are air-conditioned.

# DECEREBRATE

## *Lost Less*

If the slice—revolver, researcher, apoplectic stroke—had cut one level higher, two inches inside the skull, two inches of brain-stem left attached to the spinal cord, everything above guillotined off, essentially, that would have presented a different motor picture. A literal guillotine has indeed been used in some laboratories, efficient and excellent because quick.

Poor small laboratory dog, poor bejeweled beribboned silky Marie Antoinette—on some days of one's life, on some days of history, it does seem that nobody ever could help anything.

The result of that guillotining inside the skull was the decerebrate.

The laboratory decerebrate, because of the state of tension running through it, because of the widespread observing and thinking of that inventor of the guillotine, and the thinking of some who came after, enlarged the meaning of posture and movement. It never had been suspected that there was so much meaning. Now it was suspected that the enlargement might go on and on, the experimental ingenuity go on and on. An experimental genius had found a method and he knew what to do with it. Were the preparation still more thoroughly understood, what it lacked, what it had had but no longer possessed, it might have taught something, too, about mind in either sense—the extraordinary, that mind existed for motor output, the ordinary, that mind existed for that bundle of enthusiasms and worries that each of us is, that most of the time does not so much as remember it has a muscle-covered muscle-driven skeleton attached.

Not a pleasant sight, the decerebrate. If the guillotine had slipped, had cut somewhat lower, the routine laboratory preparation would have been, once more, the spinal animal—the spinal animal that the researcher could carry with that miserable limpness over his arm. It could be a dog, a cat, other animals. But—the decerebrate would be as far as possible from limp.

Imagine it to have been operated on at 10:30 A.M. In operating-room parlance the anesthetic had blown off and the clock now stood at 11.

Anyone would immediately have noticed that dog's legs. All four were extended. In fact, they were over-extended, starkly, if the experiment went well. (From the experimenter's point of view.) An excess of unbalanced tone appeared funneled into the muscles, the resulting rigidity sometimes so great that the decerebrate could be picked up like a carpenter's horse and stood on its four legs, four props, stay there till tipped. If tipped, it must be stood up again, could not stand of itself, imagining enough of itself left to decide for itself—could not automatically recover from a tipped position. If it was laid on its side the legs continued over-extended. They stuck out. The head was drawn back. The jaw was in lockjaw. The back was flat, not arched as normally, and the tail might point—Hitchcock horror—toward the stars. The breathing was one or another type of gasping, that fit the picture.

The animal knew nothing of this. Its knowing regions had been sliced off. One university professor called another university professor a decerebrate nightingale, but he was being figurative. A caricature of standing was how the scientist who originally produced the state described it. High tone was in it, high force, giving an idea of the energy that must be ordered by the nervous-system into the muscles if they are to hold the body up off the earth.

Nothing in this world is terrific if the tension in the decerebrate is not. The lower parts of the nervous-system have been released in Hughlings Jackson's sense. The normal inhibition exerted by the high has been lifted away from the low and the tone there is eerie besides excessive. In unmutilated life tone is adjusted to inconspicuousness, but the force is present. The force is keeping the body off the earth and toward the stars, though, in the decerebrate, no seeing eyes to see the stars. (Possibly eyes should be regarded as a cheap commodity, two to a head, billions of heads.) The force, the tension, the tone must be great because the body is heavy and without that force would collapse. It is the crafty modulation of that force in normal life that seems such a wonder, checked, as that force is, in various degrees in various places and from various points, enhanced, suppressed, to the end that all dogs, cats, men stand, walk, lie, swim, float in the soft manner they do. The nervous-system produced the tone, the excitation, and produced the inhibition that balanced it.

The researcher took hold of the head of the rigid decerebrate (decerebrate cat) and bent the head toward the chest, such a change sweeping over the cat that it would have startled a hangman. All rigidity melted from the forelegs, the animal settled down on them as if looking under a sofa. That phrase—looking under a sofa—was first employed by the researcher who first discovered the maneuver that brought about this astonishing distribution of tone. For the animal in its everyday life, when it had the impulse to look under a sofa, or under anything, the patterns to carry out the action were ready in its nervous-system. Mind gave the orders. Ancient automatisms carried them out.

Again the researcher took hold of the head, this time bent it backward toward the spine. The rigidity melted from the hindlegs and the animal settled on them as if looking up on a shelf. That phrase— looking up on a shelf—also was first employed by the researcher who discovered the maneuver.

A moving crane struck a workman and produced a two-legged decerebrate. He lived nine years. Where during all that time that workman's mind had gone was anyone's guess, or whether there was mind (Schrödinger and his East Indian would have been sure there was), old mind or new mind, actual mind or theoretical mind. The surgeon who had the responsibility for that workman sometimes took hold of his head as he lay there in the hospital bed, rotated the head so as to cause the chin to point to the right shoulder, when an increase of tone flowed down the right side of the body, at the same time a decrease down the left. In meningitis in children physicians have seen similar broken-doll effects. Tuberculous meningitis used to be common, produced human preparations that neither physician nor surgeon could do anything about, just let lie and die, and often in those damaged machines one saw the exaggerated operational design. The antique patterns were present but useless. Descartes's junkyard was just over the hill.

# POSTURE

## *An Act Frozen*

Plainly there can be more tone or less tone. There can be more in one part of the body, less in another, the quantities quickly shift, the body quickly shape and reshape. One wonder of life is the postures that the animal can assume, the leisurely roll-up of the cat, the stance of soldiers leaning into the wind on the flight-deck of an aircraft carrier. In the leopard, as in any four-footed, tone will appear simply to be moving from front legs to back, or more so. One posture is erased, another takes its place, tone flowing from front to back, or right to left, or left to right. There is infinite possibility of rebalancing—no other word but infinite is adequate. A sea lion is able because of that rebalancing to reveal in its movements an edgelessness that harmonizes with the edgelessness of its anatomy, with the apparent softness of the texture of its surface, with the softness of the water through which it glides for so much of its life. Posture follows movement. A posture, then a movement, then a posture, continuous and forever, and in that manner can a living body, our human body too, yield to its always-changing world, changing even in sleep, changing only more ostensibly at other times, as when a man steadies his opera glass to find the moons of Jupiter or to keep up with the horses at the Derby or see around the curtain at the Folies Bergère. His manifest life, or the batter's waiting for the ninety-mile-per-hour baseball, is a succession of postures, and that is not metaphorically speaking. We exist in postures. Years ago a famed neurologist stated this with a simplicity one never forgets. "Posture follows movement like a shadow." That is what he said. A movement, its shadow, then another movement, its shadow, and one could think of mist, or of moonlight or clouds seen from above from a jet, or of utter noiselessness, but then again of a living body, of the back-and-forth between the behavior of the creature and the world around it. Someone says ethology and adds not much. A silent progression of postures is our motor life. The single posture is the single frame in the motion-picture. Possibly it is that silent progression that becomes (to someone) the presumed mind presumed to exist for movement, which is fact or not fact, imaginable or not, comprehensible or not, an idea to play with, or just play, like croquet, meanwhile watching movement. Something freezes momentarily, melts, solid then liquid, never really solid, and this for as long as this machine of Hamlet lasts. Posture is movement stilled.

# LABYRINTH

## *His Head Must Know Where His Head Is*

All our lives our head like the rest of our body is regularly tipped over, re-established, tipped, re-established. We do not often think of the tipping, need not, ought not, only cannot help but think when two balancing instruments in our head get out of order. Those two guide, monitor, that perpetual re-establishment.

They are built into the bone of the skull, our skull, or the pigeon's. They register tipping, the nervous-system is signaled, it signals the right muscles, these contract the right amount, and the tipping corrected. Muscles in the neck correct the head, and the body follows. Even while we are quietly lounging there are small tippings, small changes of direction, small gains of speed (accelerations), small losses (decelerations), and these may be up-and-down (linear), or round-and-round (angular). The last in an exaggerated form provides the pleasure of the carousel, of crack-the-whip, of the nose dive of the pilot, of the dive of the nighthawk, of the astronaut's swinging into orbit, if that is a pleasure and not a gilded misery.

Usually before one has advanced far in college, in some biology course one has dug a set of those instruments from the skull of a dogfish. In the dogfish they lie in soft cartilage. In us they lie in the hardest bone of the body. Non-hearing labyrinths is what they are. Hearing labyrinths were those earlier-mentioned snail shells. Three parts to the non-hearing: (1) Semicircular canals. (2) Utricle. (3) Saccule. How labor is distributed between them is uncertain.

The semicircular canals look like their work. They also look like jeweler's items. Two of three stand upright, vertical, when the head is vertical, the third horizontal. Within each bone-canal is a membrane-canal, fluid inside and outside of it. When the head speeds up, accelerates, at the beginning of the round-and-round in the carousel, the fluid inside that canal which is at right angles to the movement lags, as fluid would. This causes pressure on a kind of gelatin with hairs embedded in it and shaped like a cupola,

and in response to the pressure the hairs bend, nerve-impulses telegraph to the central system, and promptly instructions for increase
or decrease of tone go to the muscles of neck, eyes, arms, trunk, legs.
Not only is the head helped to maintain a correct relation to the
earth despite carousel, nose dive, smaller violences, but the entire
body helped. It comes closer to holding on to its self-respect while
the mob at Penn Station shoves. Because her labyrinths were not as
good as they used to be, but the bouncing of the rickety bus on the
bad road outside Leningrad was exactly what it had been in the
time of Catherine the Great, an old Russian woman giggled when
she got home that night, having without benefit of dentist extracted
another tooth.

# NYSTAGMUS

## *Eyes Assist*

Let us fancy a young girl, having swept to the center of the ballroom, had decided she would pirouette. No, let us rather inquire
into what happens in a humbler species, say, an alive shark.

A scientist is holding a shark's face in front of his own face. It is a
small shark—this one is perfectly safe. The scientist rotates its head
clockwise, its right side dips down, and instantly and automatically
that eye, the right, looks up, the left looks down, the shark looks
coquettish. Then the scientist rotates the head the other way and
the eyes do the other thing. Which probably inspires him to change
the technique of his impudence. He takes hold of the shark's tail,
bends it side-to-side, the eyes now rolling as when the shark is
swimming in the sea, eye movements and tail movements being an
interlocking mechanism that helps keep the shark's movements in
balance in its world, the sea. During the side-to-side bending of the
tail the fish, one felt, was maintaining its gravitational decorum
despite the scientist's indecorum. Even the fins can be shown
hooked with eyes and tail in robot inevitability. Fish and men all are
moving in amidst the movements in and around them, that are
mostly incalculable.

A squab (has not yet got its feathers) placed on a turntable and spun, instantly begins to execute corrective movements of head, trunk, eyes, age-old movements evolved to re-establish for that squab the world whenever it took to going giddy. "He that is giddy thinks the world goes round." A grown pigeon, with one or more of its semicircular canals experimentally destroyed (the pigeon is a common candidate for that), when forced to move, or moving of its own accord, is in a most unhappy state, most wretchedly disoriented, so much so that the experimenter may feel himself disoriented too. Squabs and pigeons and other living experimental material ought at least not to stray into laboratories.

A gentleman with his eyes open, and swiveled respectfully in a chair, reveals an automatic side-to-side swinging of his eyeballs called nystagmus, his vision wanting to keep in balance with the world, his eyes moving correctingly, or what on a basal level would be mechanically correctingly. Nystagmus looks artificial. It may occur spontaneously in brain damage, or with some kinds of bad eyesight, or with poor illumination, as miner's nystagmus, the eyes, finding it difficult to hold a point, grope, and the groping assumes a rhythm. Everything in the universe is chockful of movement and chockful of rhythm, and a pair of Japanese chess players sit silently in a rut of street and move their chessmen, and their bodies rhythmically sway, and every passerby moves quietly and respectfully undisturbingly around them. It must be granted that on all sides the movements of the living are telling stories.

Two components were in the nystagmus of that swiveling gentleman. While his body revolved toward the right, his eyes fixed a point that seemed to be running off toward the left, held that point to the left as long as they could, then snapped back to the right, fixed another point, held, snapped, held. The swing while the eyes held was slow, the snap fast. The explanation is debated.

A dancer states that in pirouetting she deliberately "spots" some exit light, or some bloated male face, keeps her head as stationary as she can while she spots, lets her body revolve under her head until at an instant by a swift sweep she makes her head catch up with it. That means that during the pirouetting she is always giving her labyrinths a rhythmic rest. Mei Lan-Fang, the great Chinese actor, placed the sharp blade of a sword at his neck, or the blade appeared sharp, kept his head stationary, pirouetted under his head, and he too made his head catch up, his controls so marvelous that one never saw when his head made the swift sweeps. Soon he had cut off his head. His face through the whole time looked straight at the audience, and just before the decapitation he added a touch, put his eyes through the movement of a swoon, and just after the decapitation let his head roll limply and then drop like a dead head down on one shoulder. Champion ice skaters who pirouette with tremendous speed seem to have been able to suppress their labyrinths, throw them out of gear, out of the system, incite the suppression with their minds and no need for their eyes to spot. When beginning their training these skaters show as would be expected a violent nystagmus, have lost it entirely when the training is complete.

Had the swiveling gentleman kept his eyes closed during the swiveling there still would have been a nystagmus behind his closed eyelids. For this nystagmus, vision obviously could not have been responsible, but the labyrinths were responsible, especially the semicircular-canal portions. In life (outside of the laboratory) the visual and the labyrinthine systems usually are operating together. If that swivel had been suddenly stopped and the gentleman opened his eyes, during the next twenty seconds there would have been a nystagmus with slow and fast swings reversed. The explanation given for this is that when the swivel stopped, the fluid in the semicircular canals, that had lagged one direction, now lagged the other, like water in a pail swung and stopped. Hairs again were bent. Nerve-impulses again traveled over nerves.

Complicated head and body movements like these would be expected to have one region of supreme control. This is in the midbrain.

An entire head may swing and that is head-nystagmus. An abnormally sensitive labyrinth may twist a body until it has the wound-up pose of an athlete ready to throw the discus, called discobolus position. If still more sensitive, as in some brain tumors, the body falls a vomiting lump to the floor. Nature gave us labyrinths not to produce nystagmus nor wound-up poses nor vomiting, nor was it for the education of scientists, nor to annoy fishes and squabs, but to keep the parts of the bodies of creatures in motor harmony during the twirls that earth-life subjects them to. Labyrinths assist posture.

Labyrinths assist movement. Dogs vomit when placed in a swing and passengers vomit on a happy Caribbean cruise when the ship pitches and rolls and corkscrews, each of these movements having in it accelerations and decelerations, up-and-down, round-and-round, changing in direction, the passenger's labyrinths and eyes going mad and shortly joined by his stomach. Sometimes even before the ship sets sail a man's stomach thinks ahead, and so does the man's next to him, who talks to him. "How do you feel?" "Not well."

# RIGHTING

## *When Dignity Slips*

The study of the righting reactions began with a cat. It was lifted with its feet toward heaven and let fall and landed nimbly and economically on its toes, as everybody knew it would, even if blindfolded. A small girl regularly threw her cat out the second-story window. It landed. The maneuvers that the cat body goes through to make that landing have been elaborately analyzed. Motion pictures have helped. Football coaches take motion pictures of their cats, show the cats the pictures afterward, of course do not talk about righting reactions.

First, the regular cat rights her head, or the head rights itself, there being righting reactions to right the head, get it in a proper plane with the horizon. The muscles accomplishing this are in the neck. They contract, pull the head into place, and, once it is in place, the upper body is drawn into line with it, the rump with the upper body. Usually the reactions can be shown to operate tail-toward-head as well as head-toward-tail.

The machine-parts concerned have been pried apart for experimental study. The parts are labyrinths, eyes, neck muscles, trunk muscles, rump muscles. Each part is surgically isolated. The researcher gets his preparation, cat, dog, monkey, into what he calls zero. That is, all parts from which righting reactions could start are destroyed or made motionless, except one. The cat is allowed to keep one. In successive experiments (and successive cats) she keeps a different one; in that fashion the story is pieced together.

To find, for example, what the labyrinths contribute the animal's eyes are blindfolded, or they are enucleated. If such a cat is held in the air by its pelvis, no matter what direction its body is twisted the head does not hang like a bag of money, but compensates, rights to the horizon. Next, to find what the eyes contribute, the labyrinths are destroyed. Next, eyes and labyrinths both are destroyed. That lets the researcher study what one might not think of, how the cat's body acts on the cat's head. Such a cat, no eyes, no labyrinths, if laid on its side, automatically lifts its head. That head appears to insist on keeping gravitationally proper. The reaction is set off, oddly, by the imbalance between the down-side of the mutilated animal, and the up-side. Even the rump may assume responsibility. If the cat is on its side, and pressure put on the up-shoulder, immediately the rump looks as if it had heard the alarm clock, rights itself, and the rest of the body follows the rump.

The cat is interestingly made.

One researcher dedicated his book to the house cat. Another to the house dog. That ought to settle scores. A dog might agree. A cat would not, would scratch out the researcher's eyes if she could get at them.

As a result of years of this type of research we understand better than did our forebears the behavior of that cat that was lifted in the air and let fall. The head was righted first. The falling body was rotating. The legs were flexed toward the body. Then, just before the landing, front legs and hind legs were straightened out, extended, and the final stance of the cat was as triumphant as the high-jumper's.

# WORK OF CEREBELLUM
## *A Glass Too Many*

It has been said of the cerebellum: three parts, three origins, three tasks. The tasks concern (1) equilibrium, (2) posture, (3) willed movement. That suggests that the way the cerebellum operates is well understood. It is not. It has been compared to a computer, but everything is.

*Equilibrium.* Harvey Cushing, great American brain surgeon, claimed he could make a diagnosis of tumor of the cerebellum if he saw the gait. He needed no more. He would ask the patient to walk, often a child, and it would walk like a sailor, legs apart, give itself a broader base, a gross unhappy clumsiness, no paralysis. For the sailor the difficulty is that the outside world is heaving, for the child, the inside. A staggering cerebellar child is a pitiable sight and might cause anyone to think afresh how astonishing healthy movement is, to wonder how it is accomplished, to think therefore of what the cerebellum contributes to normal movement. For Harvey Cushing the child's walk made the diagnosis. It was there before any other sign of tumor (vomiting, headache, failing vision), and in a few weeks or even one week the child might be dead. It might not be tumor; could be abscess; could be the knife of the vivisector; frightful signs always. Sensation is not involved. True, the cerebellar patient feels dizzy, but with the body so thrown about, the other parts of the brain are reporting the sensation of dizziness. A boy with a cerebellar tumor was asked to get out of bed, then stand, then walk, which a month ago he could have done easily and naturally, now without warning was thrown violently backwards onto the hospital floor.

*Posture.* If that ghastly stiff decerebrate creature, whether rendered so by the researcher's tools or God Almighty (we should credit God with our agonies as we credit Him with our beatitudes), has the cerebellum electrically stimulated in its forward part, the stiffness melts. The posture softens. In the decerebrate dog lying on a surgical table the stiffness is wiped off before our eyes. Posture and the forward part of the cerebellum were recognized as related years ago.

*The Willed.* The willed is associated with the back part of the cerebellum, the recent part. Since man is recent, and since he performs more willed movements than dog or pigeon, we are not surprised that in him the back part of his cerebellum should be highly developed. In the ape too. A man for his pleasure wills to pick a berry off a bush and an ape to pick an ant off a stone, complex performances, impossible after damage to the back of the cerebellum. The tone in the muscles may be tragically less. With the cerebellum experimentally destroyed, a monkey may be so weak that the drooping of its head can seem one of the major sadnesses of the world, the monkey's intelligence remaining sharp, alas.

In cerebellar disease there may be a sidewise flinging of limbs, noisy tapping of the feet, other signs that made some neurologist think of a drunkard, even postulate that the drunkard's motor behavior might be cerebellar. Not convincing, but in a negative way tells again how a normal cerebellum assists in movement. A lady or gentleman whose cerebellum has been damaged by chronic alcoholism might, like the child with the tumor, fall backward when meaning to walk forward. The god of the grape is not sober, yet in paintings and statues of him he comes down to us well-coordinated, his righting reactions operating, tone admirably distributed, also ruddy, fat, utterly capable of the voluntary, the graceful, in an acted sotted way.

# STANDING

## *Getting Up in the Morning*

At thirty-two weeks of age there was that breathless quarter-minute when the male stood if supported. He had triumphed. His will (to all appearances) had triumphed. Movement and posture had confronted the world with success, and he was up, had pushed himself up off the crust of the planet, though it was more as if he had pushed the crust down under him. Earlier it had been the same with the wall of his nursery; he had stuck the flat of his foot out between the slats of his bed and pushed the wall off from him. Pushed his life free—or freer. Anyhow, now he stood. He was a fool ever to have done it because, once he had, he would need to keep doing it for nobody knows what number of years, stand, stand, stand, get up in the morning, stand, stand, stand.

Other animals too have pushed themselves off the crust and found two-footed standing a problem. Birds can be said to have accomplished it. Any sleeping city pigeon is comfortable (though always swaying) on one foot. Bears when in the humor stand on their hindfeet. An unusual dog may stand on hers to satisfy her curiosity about that small additional stretch of earth made available to her when she stands, spies a rabbit that other dogs can only ferret out vulgarly with their noses. The primates are up and some apes come close to achieving the one and only uprightness, ours. We are the chosen.

The troubles that go with standing date far back. Far far back there were the fishes. Some of these of an afternoon would drop themselves to the bottom of the river, poke out their breast fins and their pelvic fins, like the four legs of a folding table, support themselves on the four legs, get a short rest from their swimming lives. The lung-fishes notably. A resting fish may rest so without movement that someone else down there in the water could think it just an inedible stone. A fish mistaking a fish is a cartoon. Later, in evolution, a bone, to appearances, oozed from each shoulder out into each breast fin, and from each hip into each pelvic fin, and two more bones beyond those bones, with joints between, and some more bones, to make an ankle and a wrist, and some more, to make a hand and a foot. (This is Walt Disney, in heaven, producing a film and sound track for the amusement of Charles Darwin, smoking a cigar.) Yet even after a creature has pushed itself up to its full height it is still down on the ground, only a few inches up, or a few feet.

Add self-consciousness to the creature, high self-consciousness, and you have a man. You have a just-graduated doctor or a just-graduated anything, standing. It is that glorious moment when he accepts his diploma, stands, thinks his academic gown is hiding his quaking knees.

Contemplate the lizard. Contemplate the rhinoceros. Contemplate the tallest of them all, the giraffe. Contemplate how the giraffe's parts must be balanced in order that standing may put the giraffe's nerve-muscle machinery to the least possible expense. Many have studied standing. Some of the greatest of physiologists have.

We are told that where the weight center of an animal is, whether forward or backward, determines which foot that animal may lift for recreation without falling on his face. The horse, for instance, lifts a hind foot. The kangaroo's weight center allows it to settle well back as zoo visitors know. The chimpanzee can not only settle back but be absent-minded about it. And man has done it so efficiently that he is truly upright.

Standing is not stillness. Standing is perpetual movement. Standing employs neck muscles, trunk muscles, rump muscles, extremity muscles, usually the eyes. Employs touch. Employs the powerful upward push of the entire antigravity system. Employs the cerebellum. Employs the basal ganglia. All of it costs the nerve-muscle machinery a minimum because of the careful contracting everywhere of opposing muscles, reducing the need of perpetual restoration, giving also that softness and grace to the standing body's slightly changing postures, that readiness to shift from one to the other. Each of us knows this when he is tired and the shifts are exaggerated and his body in every second wants to be thrown out of balance and in every next second must return itself to balance. Standing differs person to person. Neurologists watch it. Physical education directors do. Orthopedists. Psychiatrists. Pediatricians. Ninety-five medical students are instructed to watch a tottering tot. In their freshman laboratory a student may have fastened one end of a string to a fellow student's head, the other end to a writing lever, and thus the swayings of the standing body recorded on smoked paper. A time-honored clinical test has the patient close his eyes, put his feet together, the physician prepared to catch him if he begins to fall, which he rarely does, because he ignominiously opens his eyes.

Should anyone still underestimate the achievement in the minute-long day-long life-long act of standing, let him at the next opportunity help a corpse to his feet. Irishmen do it at a wake, as is widely known, stand the leading man in the corner and offer him a drink, yet even they and even in the excellent spirits they are must have that corner. He is dead if he does not accept the drink.

# WALKING

## *Entrance Self-Conscious*

If it is a risk to stand on a globe traveling one thousand and thirty-eight miles an hour at its equator, the moon its companion, both at the same time doing a once-a-year around the sun, what a risk to walk! And to run?

The leech draws itself over the earth by its suckers; the turtle paddles; the shark swims; the swallow flies; milady sweeps into the drawing room on two interchangeable I. Miller heels, and milady's cat slinks in on no-heels. The top half of milady has been freed of the burdens of locomotion in order that both halves of her may have the burdens of anxiety—no need ever to set *her* forefeet on dirty pavements.

A ten-month-old human, knowing nothing of the risk, one morning starts out across the wide plains of the nursery. He walks. He follows left foot with right, right with left, has within his nervous-system the mechanisms to do it. Has the mind to do it. The will to do it. At the moment he is so busy with his feet that he cannot be bothered with thoughts of nervous-system, mind, will, et cetera. The motor patterns of the eons have become expressed, appear to have the concentrated passion of what has waited so long. Foot touches nursery floor, weight bears down on foot, toes spread, leg stiffens. That leg looks like a pillar and was spoken of as a pillar years ago. A pillar. It is as if something had been poured into the mold of a leg, solidified there. The something is tone. The other leg meanwhile is bending in all its joints, flexing, and since flexing at the toed or fingered end of an extremity is necessary for needlework or digging in the garden, it was spoken of as an instrument, also years ago. In walking pillar alternates with instrument. Pillar—instrument, pillar—instrument, pillar—instrument: that is a gentleman in black, walking beside his lady in black, toward the Cathedral in Seville, Spain. Her strides are shorter but her advance is as fast as his, and by actual measurement she expends less energy than he, which doubtless is why all ladies whatsoever regard all gentleman whatsoever as uneconomically put together. In galloping, left and right legs instead of operating alternately operate simultaneously. The kangaroo on a hurried errand across the Australian countryside gallops, but it looks more like flying, soft and swift. A gallop frozen when the two legs are bent, is sitting. Arms help walking. In Seville toward the Cathedral if the lady's and the gentleman's right legs swung forward, so did their left arms, and vice versa, the arms that way preventing the body lurching toward the foot that is set down. Eyes help, fix a line, keep the body on that line. Head helps, wavers just the right amount to add its pennyweight to the body's balancing. A pigeon's head overdoes this, we think, pumps forward-backward, forward-backward, subtracts from the pigeon's stateliness when it goes out walking, we think.

Neurologists watch walking even more than standing. We all do. The elite police, the Chief of Police says, watch a man's gait without ever directly looking, may suspect what he is about. A repertoire of gaits parades through any clinic any day. A leg swings in a half-circle outward from the hip because that patient cannot bend his knee and if his leg did not swing outward his foot would scrape the ground—the man has had a stroke. A leg swings outward because the foot itself is paralyzed and if the leg did not swing outward that foot also would scrape the ground. Another walks on his toes. Another on his heels. Another on flat feet. Another sways dangerously. Another advances slowly with mincing steps. Another with scissors-gait. Everybody in the clinic has seen a woman walk like a duck. One arm of a man swings normally, the other does not swing at all, and that difference makes one think instantly of something wrong. But outside the clinic any day we know our friends and our enemies and everybody else by the way they walk. There is a woman has tenpins for legs but walks like the pigeon. There is a woman loops her right foot all the way over to the left side of her left foot, then reverses that, seems impossible, but she does, and one has the feeling that she has given herself dancing lessons to accomplish it.

Between two-legged and four-legged walking there is not much to choose, the advantage in two-legged being in that other fact of the top-half freed for stealing from one's neighbor and breaking the other commandments. The chimpanzee walks on four legs, or three legs, or two, according to the social situation.

# MOTOR CORTEX
## *Top of the Top of the Top*

In the year 1691 a young man had a skull fracture that would take a place in the history of our knowledge of body-movement. A fragment of splintered bone pressed on his brain, the part that later would be thought the highest level of motor control, which should be near the most concentrated mind-exists-for-movement region, if that idea has by now accumulated any truth, if mind does not exist for something, let us say, nobler. Robert Boyle, the natural philosopher, reported the case, said the young man suffered a dead palsy of arm and leg, and that when the splinter was lifted, the palsy was lifted. A miracle. It happened in the last year of Robert Boyle's life.

One hundred and seventy-nine years later, 1870, time of the Franco-Prussian War, two Germans, Fritsch and Hitzig, stimulated electrically in that highest level. The University of Berlin thought it did not have laboratory space for the two, so they used the dressing table in the house of one of them. What Fritsch and Hitzig (sounds like a comedy team) discovered was that when the cortex of a cerebral hemisphere was stimulated there was movement on the other side of the body. This was in a dog. The spot stimulated and the movement were as responsive to each other as a button and a bell. Contrariwise, if the spot was destroyed, that was again Robert Boyle's palsy, and an animal's limb that did not look different from its other limbs collapsed under the body's weight. Today this surprises no first-year medical student. He knows the spot is in the strip called motor cortex, and the strip is on the outer surface of the hemisphere, top of the world of living tissues, has been scrupulously plotted, photographs in the textbooks, plotted later, also, for man during surgery. All the loftier beasts have a motor cortex. Our decision to walk to the grocery, drive to the theatre, sail to the South Seas, implicates it, though we may not be able to say convincingly what the word *implicates* means, nor the word *decision*.

The stimulations of Fritsch and Hitzig were throughout the next one hundred years extended to many animals, macaque, gibbon, baboon, chimpanzee, orang, gorilla.

Whoever has had the opportunity finds unforgettable the first time he witnessed the stimulation of the motor cortex in the monkey. A laboratory can be a quiet place. The monkey lies under light anesthesia. The experimenter knows where the motor strip is. He surgically exposes it. He places his electrode on one spot of it, expects and gets a lifelike movement, say, in the monkey's face, on another spot a movement of a group of muscles, on another the contraction of a single muscle. All occurs sufficiently predictably if the point stimulated is small enough and if it is explored carefully enough. Not too easy to do. There is a fluidity, an instability, the same spot serving different functions as any of us would know of parts of his face. Forward of the motor strip is the premotor strip, and there he expects bigger and less precise movements, that might be a background for the smaller and more precise. For years it was claimed that between the strips the electrode could put a brake on movement. Successive parts of the body are represented in that strip, the term, represented, indicating not an explanation of what occurs but largely where. Toes and ankles are in the trench between the two hemispheres, knees at the crest of each hemisphere, and as the experimenter proceeds downward over the outer curvature it is hip, shoulder, elbow, wrist. Still farther down are neck and face. The face has large representation, especially if jaw and tongue and the machinery for chewing, swallowing, talking are included. We animals must chew expertly and talk well enough, or the equivalent of talk.

So, the experimenter identifies the neighborhood, narrows the identification, narrows to a small area finally, the size depending on what he is seeking. He is keeping his attention for the moment on the monkey's hand, an exquisite piece of sculpture, hair over the back, yellow-brown palm. He stimulates. The monkey's thumb lifts.

It appears to lift by the monkey's will. It may even appear that the monkey is preparing to express annoyance. But no, poor monkey, he is a slave to an electrode. It might also be that these movements are not lifelike except to an enthusiastic scientist; at least they are startling, especially if the laboratory happens to be that quiet place. The lifted thumb settles again. Lifts again. Shakes several times convulsively. Settles. One always remembers that first time. One thinks of it now and then. One fails to understand it, beyond the obvious. Since the time of Fritsch and Hitzig the electrodes have gotten smaller, finer, tougher metal, the controls on them more precise, but the gains to science in the long run possibly not proportional.

# HUMAN EXPERIMENT

## *Dr. Bartholow of Cincinnati*

Mary Rafferty when an infant in Ireland fell into a fire and burnt off her hair, later wore a wig held in place by a whalebone, that eroded her skull, produced an ulcer, and at the bottom of the ulcer one could see the pulsations of Mary's brain. She presented herself at Good Samaritan Hospital, Cincinnati, Ohio, in 1874, and was attended by a Dr. Bartholow. The good doctor had read of the stimulations of the dog brain by Fritsch and Hitzig, and thought, as he stated it, that it was desirable to elucidate the functions of the human brain. He prepared electrodes of various lengths, reckoned they could be introduced into Mary's brain without material injury, as he stated it, and introduced one of them into the left hemisphere, and stimulated. Mary felt no pain at first but later besides pain described an unpleasant tingling of her extremities. The good doctor listened and watched. He wrote as follows: "Notwithstanding the very evident pain from which she suffered, she smiled as if much amused." Mary began to weep. Perhaps after all she was feeble-minded, as the good doctor suggested. Both galvanic and faradic currents were used. He wrote: "The needle was now withdrawn from the left lobe and passed in the same way into the substance of the right." There were movements. They were on the opposite side of Mary's body. He wrote: "Very soon the left hand was extended, as if in the act of taking hold of some object in front of her." There were sudden single movements. A leg shot forward. An arm was thrown. The head was deflected. Following one stimulation Mary had a convulsion, her eyes became fixed, lips blue, breathing disturbed, a frothing at the mouth, a loss of consciousness that lasted twenty minutes. Dr. Bartholow postponed the "experiment" to two days later.

Mary arrived. She was admitted to the "electrical room." She was pale and numb and had difficulty walking. He wrote: "The proposed experiment was abandoned." The following day Mary was worse, and in a brief time she died of what must have been a septic meningitis. Dr. Bartholow, Professor of Medicine, an honest scientist in this context, reported in a journal of medical science what transpired. "It has seemed to me most desirable to present the facts as I observed them, without comment." Altogether he had made six observations. He called them that. The title of his article also was factual. "Experimental Investigation into the Functions of the Human Brain." The article appeared in April, 1874. That year a new comet was seen in the heavens. Coggia's. So many interesting things happen every year, do they not? The episode in Cincinnati gave proof of the vitality of the times, of the vitality of the clinician, of the vitality of man, no doubt, and gave proof that there was a motor cortex in the human brain as in the dog's. Movement could be stimulated from there. It was the top of the motor system. If mind did exist for the ultimate refinement of movement, clouds of mind should hang around that cortex.

Poor Mary, one does hope that her personal affair worried someone, some night nurse maybe. Dr. Bartholow's "observations" would have interested those two Germans, Fritsch and Hitzig, and that canny Britisher, Hughlings Jackson, and the even cannier natural philosopher, Robert Boyle, and the mathematician-philosopher Frenchman, Descartes, and the ruthless Greco-Roman, Galen, and those two Greeks attracted by all living things, Aristotle and Hippocrates. A British medical journal was sharp in criticism and quoted Dr. Bartholow's article at length. He defended himself but seemed in general to concur. He wrote: "To repeat such experiments with the knowledge we now have that injury will be done by them—although they did not cause the fatal result in my own case—would be in the highest degree criminal." Dr. Bartholow was a doctor who when his own time came to die would have had the courage to meet that episode with becoming deportment too, we trust.

# XVIII
# INVOLUNTARY

---

## Night Watch

# ANALOGY

## A Ship

*Soma* means body. Somatic refers to the outer wall of the body, the body-wall as against what is inside, the viscera. Somatic muscles keep the manifest body, the covered skeleton, on its lifelong march, hold it upright when standing, when sitting, do not permit it quite to collapse even when lying. Somatic nerves supply somatic muscles. Somatic nerves make up one of the two large divisions of the nervous-system, somatic nervous-system.

Autonomic nerves make up the other large division, autonomic nervous-system. Autonomic means self-governing. Autonomic emphasizes that of our body's life that is less manifest, has less to do with the perambulating skeleton, less with the voluntary, more with the involuntary, with the underneath, with the below-deck.

The below-deck suggests an analogy. It suggests a likeness between a living creature and a ship. People of the sea think of their ships as creatures, as alive. They think of them as headstrong and free some of the time, as traveling at high speed some of the time, as leisurely on a cruise, as a yacht sailing in a race, a tug, a liner, a destroyer, a freighter, a tanker, a scow. One sees his friends. A motley crowd, yet when in their natural habitat, water, graceful enough, adapted to yield like birds to the physical forces around them, a complicated adapting. In the beginning on their maiden voyage they have their beauty of line. After that they go on plowing the seas, rough, mild, winter, June, and whenever anyone has the chance to look at them disappearing in the offing he falls in love with them all over again. It is not too different when they are tied to a dock, they are always floating some, and with their captain and pilot they look voluntary, and have what one can choose to think the involuntary inside them.

Where is the pilot? One sometimes wonders about the pilot. Rarely does one have the privilege of a visit with the pilot. And the captain? Him one sees only at the captain's dinner. And the night watch? If one sees him at all it is dimly on the bridge. The night watch is a romantic figure. Less so is the chief engineer. He works below-deck.

The engine room is below-deck. The stokehole is. Down there are fuel and fumes. Pipelines start from there and go everywhere, also to the outermost outside of the ship, to just under the surface. In the creature they go, for instance, from the heart to the muscles of the body-wall and to the painted skin. A vast below-deck machinery makes possible the sight we see, a ship alive and sailing. To that end everything below-deck is important. The pump. The ventilators. Supply lines. Outflow. Bulkheads. And peculiarly important are the controls on the underneath itself. In the creature all these controls are exercised via autonomic nerves, autonomic ganglia, autonomic central masses in the brain.

Too soon a ship gets old. It is the destiny of ships. Seems yesterday she left the ways. Or she has a mishap. A noise rises from the below-deck to the above-deck, a knock, a squeak. A part begins to rattle and when one part rattles it will not be long till other parts rattle, till the whole affair rattles on the calmest day. Then comes the day when she will have run her last run, and one nostalgic satisfaction we do have on that day, we know her all the better because we knew her when she rattled.


# EXPERIMENT

## *Supply Line*

This now has to do not with a ship but a rabbit. Claude Bernard, the Frenchman, he who gave us our idea of the constancy of the internal environment, cut a nerve on one side of a rabbit's neck.

Promptly the ear on that side blushed. A rabbit has a good-sized ear and this, plus the coloration of her diapery white posterior as she runs away, makes her such a pathetically easy hunter's target. The blood vessels of the ear stood out. Some that had been invisible became visible. More blood was flowing in them. The ear was warm.

Here was a fresh fact about nervous-system of rabbit and man.

What had been discovered was that a nervous influence plays upon blood vessels, closes them, holds them closed in different degrees, lets them open passively, just relax their tone, but also open actively. The eye of science henceforth saw the nervous-system as two. Structurally this had been seen earlier, but now it was seen in an action, seen dynamically. Claude Bernard had supplied that.

The following year Brown-Séquard, half French, half American, again cut that nerve in some rabbit's neck. The ear blushed. The cutting had left two raw ends. He stimulated with electricity the raw end toward the head. The ear paled. A dramatic sight it must have been that first time. It still is in 1969.

So, there was one experiment one year, another the next, and together they argued that in any rabbit as it hopped along, or in any man, an autonomic influence, an autonomic tone was always playing upon vessels to keep them closed, and if the autonomic nerve was cut the tone was cut, the vessels opened. The ear blushed. Stimulating the raw end brought back the tone. The ear paled. Blushing and paling were signs of emotion, signs that ran along with other signs flashing from the mechanic underneath of the body.

The two experiments suggested further what today seems inevitable, that blood can be delivered where it is needed by squeezing shut vessels where it is not needed. During exercise blood is needed in the muscle of the heart and in the muscles that move the skeleton over the earth, and digestion and the like can wait. After supper blood is needed for digestion, this blood might in part be shunted from the brain, and an hour after supper a person might begin not quite to see the TV, not quite hear it, not quite possess the tact to remember that legs stretched out across the room are ugly. Those legs are a sign of mind too.

# HEART-RATE

## *Knots*

Imagine yourself on the ship's morning inspection.

You have descended the steel ladder to the below-deck. The busy organ you see over there is of course the heart. It shakes everything, ancient pump, tough, sensitively governed from inside itself, from outside itself, force and rhythm modified a trifle at every beat, governed occasionally by thought, occasionally governing thought, late evidence indicating that when the heart rate goes down, the attention of the creature picks up.

The degree of stretching of the entering blood modifies the beat. Endocrines modify. The two divisions of the nervous-system modify.

The two divisions are the voluntary and the involuntary, somatic and autonomic.

And the autonomic itself is divided again, sympathetic, parasympathetic. Everywhere in body and head the two—sympathetic, parasympathetic—exercise their control, sometimes in opposite directions, sometimes same direction. Frequently the sympathetic appears to operate alone.

Both play upon heart-rate. Both play upon the force of the beat. Since both play also upon the blood-vessels, the circulation is everywhere under autonomic regulation.

In 1845, the Webers, two brothers, found they could apply electricity inside the nervous-system of a frog and stop its heart. There was a center there. In the nervous-system a center must usually be allowed some leeway, and often is plural, centers.

The center of the Webers was low in the brain-stem, not far above the spinal cord. They also discovered the nerve that goes from the center to the organ to deliver the orders. They could not yet have spoken of the parasympathetic division of the autonomic, would have said vagus. It was a cable. For centuries men had seen that cable traveling from head and neck into the underneath of chest and abdomen. Galen saw the vagus.

Today the freshman medical student and biology college students see it with no difficulty.

The student is taught that the vagus goes to many places besides the heart. He exposes the heart of a turtle. Locates the vagus. Touches it with electricity and in his own chest feels a queerness that lasts as long as the turtle's heart stands still, seconds by the clock. What the parasympathetic does in the life of frog and turtle and little Jason is to hold the beat on the slow side, and the weak side. Sympathetic fibers control toward the fast side, and the forceful side. Thus the pump is taking nervous advice from two sides, balanced between the sides.

At birth an infant's heart rate may be 130, the parasympathetic not yet doing the job of slowing. At puberty the rate may be 90. The adult rate in a woman is 84, average. In a man 78, but before breakfast at that black hour when men drink black coffee to wake themselves from the gloom left from last night's brilliance, the parasympathetic is dominant, 60 perhaps. Athletes develop high parasymphathetic tone, less than 50, the force strong, and then when the moment for exertion comes with everybody and his girl cheering, the controlled machine-part quietly and rapidly increases force and rate, sometimes 180, even 200. Faster would be wasteful because the heart's chambers would not have time to fill between beats. Horses have slow rates, strong parasympathetic tone, and horses like athletes are capable of prolonged exertion. The tone lessens in the aged, and a ninety-year-old's heart may run skittishly.

# TOTAL NERVOUS-SYSTEM

## *Deck and Below-deck*

Control of the speed of the heart, control of its force, control of the bore of blood-vessels, control therefore of the entire machinery for the supplying of blood, and the shifting from where not needed to where needed, hence the sustaining of life in every nook and cranny, all depends in some respect or largely on the autonomic. The bore of the bronchial tubes depends in some respect or largely on the autonomic, hence air, hence again life in every nook and cranny. Also, tension in stomach, intestines, rectum, bladder. Flow of sweat. Movement of hair. Amount of light that enters the eyes. Much of the machinery of sex. Much of the control of the heat, water, utilities generally.

The other division of the nervous-system is the somatic, and somatic and autonomic join work at many moments and in many places, and the endocrines join both. We know better than ever, in 1969, what a fascinatingly complex integration this is, during health.

To think of it another way: what lies beyond the will, what lies underneath, is nearer the autonomic. What lies grossly and comfortably on top is somatic. Anyone meanwhile is free to define the two words, *will* and *mind*, according to his philosophy and his experience.

In the figure of speech of the ship the autonomic controls the below-deck, the somatic the above-deck. The latter is what most of the time the passenger considers the ship. Everything that a child would think was the other child playing there next him in the sand—somatic. Everything that even a bright child would hardly put into the other child's body—autonomic.

The latter is the system that is divided again, sympathetic, parasympathetic.

Like the somatic the autonomic has a hierarchy of Hughlings Jackson levels. They may not be quite so manifest. They begin low, add level to level, each with more nerve-cell masses, until the top where the amassing is enormous, and where we are in the habit of thinking is the all-weather-sensitive self. A ship rides over the planet and with the ship rides the self, always by itself.

# DIGESTION CONTROL
## *Fuel and Ash*

The morning inspection is piecemeal. It tries to make sure that everybody sees everything there is to see of the below-deck, not understand it all, but it should assist them in always remembering, afterward, that the whole ship is one.

Digestion control.

Saliva, gastric juice, bile, pancreatic juice, more, are the digestive juices, and mucus flows along everywhere. Food taken in from outside is processed by the juices, and the resulting products, finest of submicroscopic particles, are delivered to the cells of the body, the leftover expelled. All of that requires control and part of the control is autonomic. Immediate questions are: which juice, how much, when is it to be released, when to stop being released? Important questions. The autonomic provides answers. What is the situation under natural as against experimental conditions is sometimes difficult to establish.

Saliva is the first juice. Saliva is controlled exclusively by nerves, not at all by hormones, apparently. Pavlov delighted the small world interested in such matters by his demonstration of the nicety with which saliva meets its obligations. He studied it for half a century. Drops of saliva in right quantity are practically counted into our mouths. Quality of saliva also is controlled. If something must be spit out, if the body must reject something, rotten fish, the saliva is watery. If something must be pushed down the gullet, the body needing it, steak, the saliva is thick and lubricating. Three pairs of salivary glands do the job. One has a duct that all but invites the experimenter to insert a glass tube, in a dog, and when he does, and when he then stimulates the parasympathetic nerve, a watery saliva rises in the tube, when the sympathetic, a lubricating saliva. The blood for the manufacture is at the same time rushed in, that also under autonomic control. The gland swells. It gets warm and red. Some workers claimed they could trace autonomic nerve-fibers directly into the gland cells, others deny this.

What is true of one pair of salivary glands is probably true of the others, and of the digestive glands farther along, the millions of microscopic gastric glands, small-bowel glands, the digestive part of the pancreas. An observer might get the impression that the juices were poured out with casualness, the advancing food cracked up, the submicroscopic particles brought into the cells. The casualness is illusion. There is bank-teller accuracy. All the evidence is that the animal must accept, deliver, expel, precisely. Through-lines do lead from the highest parts of the brain, helping us to understand geographically why when a lady's gastric juice behaves quixotically it usually means garbled instructions coming down from garbled thoughts.

While the food is chemically handled by gland action there is chopper-conveyer muscle-action. Chemical and mechanical operate together. While the juices are disintegrating the dinner the chopper-conveyer action is helping the juices get at it, meanwhile pushing everything along, gullet to anus. The overall controls are: (1) local nerve-nets in the bowel wall, (2) autonomic nerves from beyond the wall, (3) chemicals, (4) the temper of the temperamental gut. If an inch-long strip of bowel-muscle from that gentleman hanged at San Quentin (they do not hang them today but kill them in the spirit of the day, with chemistry, a pellet of cyanide) was stimulated by electricity it responded only weakly, might not respond at all, might whimsically, not like skeletal muscle, though more apt to shorten if the tone was low and lengthen if high. It is a fact that both parasympathetic and sympathetic arrive from beyond the bowel, and there must be a reason, but it is also a fact that even large nerves can be cut and the bowel look disinterested.

Ordinarily our mind is not paying attention to our bowel, a pinch of pain in the abdomen on the evening of the 4th of July or a twinge of nausea on any Sunday morning, but the existence of the through-lines from the highest parts of the brain helps us to understand the nervous-system side of the matter when some internist gives it as his impression that a silent ache in a memory, a silent gnawing, want of love, might produce a chronic groaning in the pit of a poor woman's stomach. Also why our own worry about next week may make our gut seem an annoyed person, kick up a diarrhea or a constipation. For one poor woman it was suggested—it shows how far medical fancy may go—that something in her mind produced an operable something not in her bowel but in the bowel of her nine-year-old girl child. What if it were true?

# BRONCHIAL CONTROL

## *Draft*

Still the morning inspection—attention on the ventilating system.

The conduits for air—bronchial tubes—make a passageway between the envelope of gases around the earth and the alveoli of the lungs. As the conduits advance into the lungs they get smaller and more numerous. Fresh air enters. Used air exits. Since the composition of the gases in the alveoli, that exchange with the blood, must keep reasonably steady, the total volume of the in-out varying, the bore of those conduits gave evolution problems. Not only are they getting anatomically smaller as they advance into the lungs, there is a mechanical getting larger and smaller that goes with each breath. Then, there can be an active muscle-controlled change of size. Indeed, the bore is changing and is changed every moment of every breath. The active muscle and the nerves are in the walls of the tubes. A long stream, a flow of muscle, it appears, contracts and relaxes. But the words give no idea. This one knows when one sees an X-ray motion picture. *Flow* was the word one used spontaneously. The fibers of the muscle run obliquely not circularly in the tube, therefore their action, which appears so exceedingly soft, alters the total tube not only in diameter but in length. It is a rhythm of bronchial-tube inside the rhythm of breathing. First the tube is longer and wider, then shorter and narrower, and it is nerve-muscle control both directions.

Both divisions of the autonomic operate. It may not be finally understood, but for the breathing-in, the accepting from the gases of the air, the sympathetic relaxes its tone and this allows the widening and lengthening of the tubes. Vice versa for the breathing-out, the rejecting of the gases from the lungs. Meanwhile the parasympathetic also is involved. It may even be dominantly involved. (This expresses the current uncertainty.) Billions of gas molecules are always moving one direction, billions the other direction.

Should anything foreign, food, drink, dust, threaten to slip into the lung, or something simulate such threat, a plug of mucus, a nodule of cancer leaning somewhere into the conduit, a spasm will be excited in the muscle of the wall and there will be a valve action, a sphincter action, this occurring in the smallest tubes just before the entrance into the alveoli. Should that threat become chronic, the spasm chronic, no matter what its cause, that would be asthma. Other mechanisms are described. The breathing-out of the lung-trapped air gives the trouble. Asthma is a common illness, its signs and symptoms known to anyone who has lived long in any street. They may be frightful.

The top of the brain, or rather that ghost the mind, can supply the spasm-exciting nudge. If it does (and it does), again the through-lines to carry the reports and the instructions are built-in. They travel through whatever junction-points from the top of the brain to the muscle in the wall of the tube. Feelings and thoughts have a roadway for shutting off a poor devil's air. The feelings, the thoughts, as has been known since the Old Testament and before, let alone since Freud, have their history, more curious or less curious but never banal, and reaching back to something dated or undated. Physicians skilled in their art all can tell stories.

Asthma comes in attacks. The muscle-fibers are thrown suddenly into spasm. Since the breathing-out is giving the trouble, the chest in consequence of the effort of pushing out the air, month-long, year-long, is slowly made barrel-shaped, not as drastically reshaped but not differently from those chests in high mountains where the pressure of the gases of the air was insufficient. Despite this blind wish of the body to correct, the victim may not get in the required fresh air, not get out the vitiated, his ventilation inadequate. He may be half-asphyxiated most of the time. He may be plum-colored. Acid may accumulate in him, appear in his urine. The dammed-up carbon dioxide may act like a drug to his mind. There can be such unremitting nagging at the person that the nagged asks himself why he was born, then shivers and goes to confession, or shoves reality out the door and invites in what he has found often has helped, a memory, a radiance when he was twenty-two, or forty-two, or seventy-two. Today there are physicians who believe asthma is invariably explained by some action, some habit, that came with feelings and thoughts, as crying held back. Sexual anxieties may be close. Disappointment. Hysteria. Other physicians believe otherwise, require pollen, or horse hair, some allergen.

In health the two divisions of the autonomic play in harmony. All day and night it is like the rise and fall of a wave on a calm day. Then, once, a man is startled. His mind gives an order to his brain (whatever way that is accomplished), which sends curt orders to the autonomic, the sympathetic lessening the tone in the tubes, the parasympathetic opening them still wider, oxygen given an easier entrance, carbon dioxide an easier exit, the body thus better prepared to meet whatever it was that startled the mind.

# EYELID

## *Dust and Glare*

Plato called the human neck an isthmus. It was a bridge between head and the rest, between high and low. Imagine yourself crossing the neck in the direction of the head. Behind lie the entrails, the gluttony, the below-deck, the mortal, in Plato's opinion. All around in the head is the divinity of the high, chips of it, delicate *objet d'art*.

An eyelid has of course use besides being an *objet d'art*. It is the first defense of the vision machinery. Regularly it drops a lampshade down over an eye. Decides the amount of eye other eyes are to be permitted to see. Hides an eye completely for a night or a wink. A wink is voluntary blink. A blink is a rapid wink—let us say that it is involuntary. Some underprivileged persons cannot wink, cannot shut one eyelid by itself, do not have the skill. Children begin early, try, fail, look funny.

Eyelids contain voluntary and involuntary muscle, have nerve-lines from both of the great divisions of the total nervous-system. Both join forces to keep hoisted this combination of skin, cartilage, mucous membrane that an eyelid is. If the involuntary begins to desert a man as sleep advances on him he manages to keep at the dinner table with the voluntary, with the *levator* of the upper lids, may also bring in the eyebrows, wrinkle his forehead. It is like raising the roof of the theatre to pull up the curtain. He may go farther, throw back his head, peer down his nose through the slit of eyes that still are open, or one eye open, bite his tongue, clench his fingernails, all the tricks we all know against determined guests and determined lecturers.

That involuntary muscle is only a few shreds, but deprive those shreds of their motor power and a human face is drastically altered.

The nerve-line to the shreds starts from the vertebral column toward the top of the inside of the chest, which is where it has left the spinal cord, travels up the neck, reaches a junction point under the skull. There it makes a coupling. The new nerve-line enters the skull and goes on to the eyelid. A tumor could cut the communication line in a number of places, in the chest, in the neck, in the skull, would cause the lid to droop, produce a sleepy-lid. One sees such a sleepy-lid now and then in a city street. Other disfiguring alterations occur at the same time because nerve-lines to other places are wrapped in the same bundle of nerve-fibers. A blush lies over that side of the face—meaning more blood flowing in the paralyzed vessels. The skin is dry—sweat glands are paralyzed. The iris of the eye has undergone a gradual loss in color. What once may have been a beautiful face is now a face beguiling but damaged. The physician sees this combination of signs, makes a quick diagnosis, and the victim sees them, especially if the victim is a woman, because the signs are in her face and even the busiest woman looks regularly into a mirror. The abnormal has reached from the underneath to alter the outside of Mary Jones and for that reason Mary's mind and our mind toward Mary. The abnormality may be inherited. In that case it will probably be two sleepy-lids, which is better. Such may characterize an entire family, the family known over an entire suburb, look oriental. In one such all the children appear sleepy, never are. The autonomic is working normally elsewhere in the bodies of the children of that family. An odd game for a mutant gene to play.

# PUPIL

## *Illumination*

Through the pupil—the hole in the iris—the owner of the eye looks out morning and evening to see who is there, see what she is up to. The size of the pupil is controlled by both divisions of the autonomic, many parts of the brain brought into the play. The operations of the two divisions are here reciprocal: when the one is thrown into action, the other is thrown out, must be, must get out of the way, and accordingly quits.

Each of us knows what delicate work his pupil does and rarely and usually healthily pays no attention to the machinery.

A pupil can be small—three millimeters. Can be large—nine millimeters. A trained physician looks at a pupil and estimates its diameter within half a millimeter. He with comparative ease looks straight in through the pupil. That is because Hermann Helmholtz did so that first time more than a hundred years ago with his newly invented ophthalmoscope.

Light enters through the pupil from the sun, enters from an oil lamp, from a wax taper, from footlights in the evening, each of these illuminating the insides of the two black boxes. Billions of human beings have been born with two such boxes darkened on the inside, have lived with them, died with them, and nobody knows how many cats. The cat's pupil has a crazy shape, as has struck anyone who drives the highways behind automobile lights by night. Man's has the crazy shape, thinks the cat, so far as she bothers to think. There are many species with pupils, and many shapes. The fox's is elliptical when contracted.

If a person takes his eye from his soft brown study into his garden that is assailed by the daylight and reflections from the sands of Tunis that are both fierce (Tunis because Tunis is a pleasant place to be), the appropriate parts of the nervous-system go into action. The receptors, the receivers for vision at the back of the eye, are excited, electrical potentials rise, nerve-impulses increase their speed nevertheless retain their methodic click, click, click, advance over sense wires to a traffic center in the deep middle of the brain, reroute, return over motor wires to the eyeball, into the eyeball, proceed onward to the frail muscle-fibers that control the size of the pupil. In consequence the pupil is brought down. That is a neurologist phrase. The pupil is made smaller by exactly the right amount. All day, all night, the pupil admits the right amount of light, not more, not less, and the nervous-system makes the arrangements.

The pupils get smaller also if the eyes roll inward, converge to look ruefully at a broken fingernail. The getting smaller has reduced the entering light, has reduced glare, has confined the rays to the center of the lens, optically the best of it, so has improved the depth of focus. We see better. If the eye were now to forsake that fingernail, look off at a lonely cloud in the blue sky, the pupil would have gotten larger somewhat. But, call it back. Bring it into the brown study (the cat must stay outside) and the pupil now is moderately large, until there is a sharp knock at the door, the police! The pupil has gotten very large. Emotion. At the middle bottom of the brain there has been quick action. Plainly, a pupil is sensitive to physical and to psychological influence. Police are psychological.

The time-span for these events? For the getting smaller, that took fractions of a second, the getting larger, minutes. When a pupil is large, the iris, the flimsy curtain floating in a watery fluid, hangs free in the space in front of the lens. When the pupil is small the iris hugs close against the lens. The muscle-fibers that made the pupil small run in a circle around the hole. They are a band of muscle-fibers about a millimeter broad, one twenty-fifth of an inch. As muscle-fibers they are said to produce more force proportional to their size than any other muscle-fibers anywhere in the body. While the circular fibers are contracting, their antagonists, that radiate from the hole like spokes, relax. When they contract they pull away from the hole, enlarge it. Of those two nervous-system controls, that on the circular fibers is the more powerful, and it follows that if the nerve to them is cut, by a scissors in an animal or by disease in a man, the pupil, freed of that powerful influence, and having only the lesser influence of the radiating fibers, gets large only part-way. On the contrary, if the radiating are cut the pupil gets pinpoint.

A neurologist directs his flashlight toward the left eye, watches the left pupil, which should come down. Flashes again toward the left eye but watches the right pupil, and that should come down also, because the nerves from either side of the brain go to both pupils and both respond to a light flashed toward one. Whether a pupil responds as rapidly and completely as its partner is important to neurologist, ophthalmologist, horse-and-buggy doctor. They speak of a pupil as prompt, sluggish, fixed. Fixed means that it does not react at all. Now and then however they draw an unexpected conclusion. "A brisk reaction, but I think the woman is in a bad state and will die." Thus do the pupils give diagnostic help. Both may be small, not respond to light, nevertheless respond when the eyes converge on the broken fingernail. The pupils may get large when they ought to get small and then they are called paradoxical pupils. Many combinations, not understood, or well understood or refinedly understood, nail down what and where in the system the trouble may be. The pupils of the blind will or will not respond depending on the location of the disease.

By using drugs the physician may override one of the two divisions. Atropine dripped into an eye, to examine for eyeglasses, blocks one division and the pupil, freed of that, gets large, stays large for hours, the physician able to look as long as he likes, see in on the living blood-vessels, the optic nerve, see into a workshop that is a sample of the underneath of the entire body.

Spanish ladies formerly made their pupils large with belladonna and by that cosmetic trickery put into their dark faces two black spots, gave their eyes a blacker brilliance, were willing to accept the incidental blurred scene, every lamp with a halo of fireworks. The atropine of the ophthalmologist is an extract of the same belladonna. Possibly the Spanish ladies enjoyed staying behind their blurred vision, staying alone with their warm thoughts. Morphine makes the pupil exceedingly small. At the moment of death the pupil opens wide, in animals too. Is that a last look at this or a first look at that? Is anyone looking? Is anyone there? Has the host gone out to a party?

# LENS

## *Binoculars*

In some fishes the lens is a sphere. That sphere cannot change its shape. However, the entire eyeball can move backward or forward, and this for those fishes is the mechanism of accommodation—the way that this eye adjusts so that it can see clearly at different distances. In some shellfish the eyeball itself, instead, gets longer or shorter, and that is accommodation. In fishes generally the outer glass of the eye, that living glass, the cornea, is not able to help bend the beams of light to bring them to a focus because its substance is almost as watery as the water into which the poor fish peers. We should stop saying "poor fish" because the sea is no more ruthless than the land. In birds, the cornea does change its curvature to keep the changing world clear, the sparrow at one instant able to see a seed in front of its bill, the next see the hawk descending. In the human eye, the cornea does not change shape, the eyeball does not get longer or shorter, does not move backward or forward, but the front surface of the lens under guidance of the autonomic alters its bulge, rounder for reading print, flatter for gazing across the prairie.

That alteration of bulge is brought about as follows.

A ringlike ligament rings the rim of the lens. Muscle-fibers, a circular set and a radiating set, control the ligament, and the autonomic controls the muscle-fibers. The eyeball itself being a taut structure, and that ligament being within the eyeball, the pull of the muscle-fibers upon it is from a secure anchorage. If the radiating fibers contract they draw outward on the ligament, which draws outward on the rim, which draws outward on the lens, and the lens flattens. At the same time any contraction in the circular fibers is gotten out of the way. Those fibers relax. This is for far vision. For near vision the circular fibers contract, the radiating relax, the lens is freed, bulges, is rounder. Near vision. Between far and near there are as many stages as anyone cares to imagine. Throughout the action, the eyeball remains the small hard structure that anyone who touches his own knows it is, kept hard by those fluids inside.

The hardness, the tautness, besides making possible the attachment of the delicate muscles of accommodation on the inside of the eyeball, allows for attachment of relatively gigantic muscles on the outside, six of them, to move the eyeball smoothly all directions within the eye's range. These are the fastest and most precise muscles of the entire body.

The nearest point that can be brought into focus by a maximum bulge of the lens, called near-point, is very near when we are young. At ten years, print can be read at four inches. Double the age, double the inches, and it can be a sadness or a satisfaction (depending on how our life has gone) to recognize that even at ten years the lens and we are getting old.

Someone will insist that to swing his eyeballs to look into the next room is a voluntary act, controlled by the voluntary system. So it is. However, in the course of swinging, the lenses are steadily changing shape, and by the time the two eyeballs do look into the next room, the objects there are in focus, and this has been unconscious, involuntary. (*Voluntary* and *involuntary* again are words and their meaning no less elastic than the meaning of most words.) Since eyeballs and lenses and pupils work in harmony, it should not surprise us to be told that there is a super-control center, a fabulous small area of brain from where the orders come, and to which orders go either from the outside world or from the mind.

# LACRIMATION

## *Washed Windowpanes*

The eyeball, the globe, is mobile, and the surface exposed to the outside is kept moist in a special way. We can weep.

Weeping is salt-water plus, flooding an eye. It may be lubricating and somewhat thick, may be thin, no substance wasted that the body should save. An inturned eyelash can start it. A harsh light. A harsh word. It is always flowing some. The word would operate via the mind, and weeping for a mind cause, psychogenic weeping, should perhaps not be accounted an affair of eyes though they are the instruments that express the state, a state honest sometimes, dishonest sometimes, the onlooker always embarrassed. Wet eyes are executing orders of a nervous-system that is executing higher orders, or however that is, and this mechanism as elsewhere not well understood.

How is a tear produced?

The report, say it comes from that inturned eyelash, goes to low regions of the brain, a pain report. It travels over ordinary pain wires, in code, nervous-system code. In the brain there is a rerouting, the path then straightforward or complicated but reaching the nose. In the nose there is a ganglion, a junction point, a new connection established, then the path continuing to the secreting cells in the tear gland, lacrimal gland. A tear begins to form. Other glands will have been contributing some of the fluid. And orders will at the same time have gone to other places—to the arteries that must deliver the fluid to produce the tear, and these on receiving the orders actively enlarge or their walls merely lose tension and they passively enlarge, more blood flowing, more fluid available.

The tear gland—the manufactory—is located at the upper outer corner of the eye socket under a fold of eyelid. In this gland are cells that chemically and physically can create their portion of the tear. Before it can be called a tear enough of it must have accumulated for it to form a drop. Ducts deliver that drop to the surface of the eye and its windowpane is washed though it may not seem particularly in need of washing.

Each time we wink, each time our eyelid closes to shut out dust or glare or wind or cold, each time the eyeball moves, that special salt water with its small amount of protein is spread over the windowpane and the efficiency of vision aided. We wink voluntarily. We wink involuntarily. We wink with regularity. We do a good deal of winking. The fraction of tear that is unevaporated drains off at the inner angle of the eye through two drainpipes—it may be stretching metaphor to say drainpipes because unlike man's plumbing these have the superlative anatomic grace of the living. In this same vicinity is a small sac of flesh believed by some to put suction on the unevaporated and the excess, the sac too included in any description of the lacrimal assembly. As also are included the glands that in man make the oily layer of the tear. It has been calculated that of every thirteen tears six are unevaporated. No tears overflow the lower lid in usual unemotional states because grease glands edge it. All in all the tear system is one where admiration for the architect, who seems to have thought of everything and fitted everything into so compact a space, makes the medieval monk that is somewhere in most of us recall with a shock to which we get accustomed that the entire body, that mammoth compact system, is nowadays generally considered to have had no architect. Charles Darwin thrives healthily among us and Girolamo Savonarola is not so much as dust.

Those two drainpipes are a half-inch long and they drain into the nose, which is why the father of the bride blows his nose while his wife wipes her eyes. The cornea, the outside surface of the eye, is exposed to trespassing bacteria, but that fact also has not been overlooked, tears having in them an enzyme that decomposes sugar, therefore gets rid of the sugary creature that each bacterium partly is, destroys it. Nothing on earth or in heaven or in hell is left to chance, not even a tear.

Tears are emotional and protective—psychogenic and reflex. Students of tears give us still other classes. No animal but man has emotional tears. And in man they do not form in the newborn, the newborn probably not yet having had time to unpack its emotions. The same Charles Darwin who observed tears and everything else stated that the earliest he knew emotional tears was at forty-two days, before which date the cuff of his sleeve had brushed the eye of his infant, that eye pouring protective tears because it was bruised, but the other eye, though the infant screamed, remaining dry. To Darwin this meant that emotional tears were not yet available. All of us know scientists who would by-pass such an informal experiment if a less eminent sleeve had performed it. Recently some fifteen thousand infants were examined and found to begin weeping between thirty-five minutes after birth and eighty-eight days, and Darwin would perhaps not have bothered to find where on the curve his own infant belonged. One does wonder whether so commendable and voluminous a study might somehow prove Darwin, detailedly, incorrect, as Einstein in a different kind of context proved Newton, detailedly, incorrect. Besides emotional tears and protective tears there are what are called continuous tears, and these flow day and night.

Tears express us. If a person weeps every afternoon or twice a month, if the weeping is profuse or scanty, may be an important part of that person's history, may be only a dramatic or a melodramatic part. Actors create tears at will and all of us are sufficiently actor to know one way he could do that: could imagine a humiliating scene or a dreadful situation, whip it up with dialogue, wait till the dialogue had got the right reality, be convinced himself, then weep, naturally, emotionally. The talent of an actor might be the possession of a strong imagination plus a low threshold for feeling. He imagines vividly, feels easily. Some actors assure us that their weeping on the stage is cold, that they can coldly call for tears, and tears fall. This could be the voluntary nervous-system calling on the involuntary, as often in our lives. We may get more than we call for, may will an act and much embarrassingly unwilled may come along. Women are said to weep more convincingly than men, more copiously, and this might be because they use weeping as part of their defensive-offensive gesture machinery. Women frequently are more vivid than men, especially to men. The chimpanzee also, high animal though he is, does no emotional weeping, cannot, it is said, though there has been some dispute. The monkey cannot. Only at the top of the scale does one find the most delicate employment of both sides of the nervous-system, the most elaborate or subtle stage properties for expression "either for tragedy, comedy, history, pastoral, pastoral-comical, historical-pastoral, tragical-historical, tragical-comical-historical."

# PARASYMPATHETIC CHEMICALS

## *First-Officer Duty*

Far and wide over the body there is a parasympathetic chemical. It is manufactured, stored, released, destroyed.

Name of the chemical?

Acetyl-choline.

To the chemist that is one little innocent word.

Acetyl-choline oozes from electron-microscope-visible blebs that are at or near the ends of nerve-fibers. At that point are the synapses. Acetyl-choline is responsible for sending the messages across those junctions, hence the orders for the gland-cells to secrete, the smooth-muscle-cells to contract, any muscle-cell to contract. Therefore it is found top-to-bottom in the body. Is found in the heart.

A suspicion as to the heart was in the air for years. Then one night came the clinching experiment.

It was 3 A.M. It was 1921. Otto Loewi had a dream. Seventeen years earlier he had had an idea. On the night before the 3 A.M. night, which was the night before Easter Sunday, he waked with something in his still murky mind, wrote words on a scrap of paper, next morning could make no sense of what he had written. However, the following night, the 3 A.M. night, he waked not only with the dream of the night before but with the idea of seventeen years before. The dream was a completed experiment. It was instructions printed in his mind as in a laboratory manual. He got up. He went to the laboratory, performed the experiment on the hearts of two frogs. Because of that dream science has a fresh understanding of the heart, of the nervous system, of the total body, of a miserable illness, of the further subtlety of the living, of another linking of our species to other species.

The experiment was a sequel to one of equal or greater pith performed three-quarters of a century earlier, 1845, by the Webers, those two brothers who stimulated with electricity the vagus nerve of a frog and its heart stopped. Without that earlier experiment there would not have been the 3 A.M. dream. It might therefore be right to say that in the sleeping mind of a century there had lain in dim shape what appeared now in sharp shape in a dream, led to an experiment, led to a principle, the principle standing behind discoveries that keep accumulating up to today. It was a famous and possibly the most famous instance of a scientist working in his sleep.

Two frogs were used for the 3 A.M. experiment. Their hearts were cut out. The vagus nerve was left attached to the first. Both hearts were suspended. Both were beating. A glass tube was stuck into each. A salt solution of proper concentration was run via its tube into the first, then its vagus stimulated for about one minute. The heart stopped—as expected. But when the solution was siphoned from this first and dripped into the second it stopped too—not as expected.

Here was one of this century's fatal experiments.

Why did those stopped hearts after a while start again? The answer to this was easier because now there was an idea, a direction in which to go to seek the answer. The hearts started because of another chemical, another molecule able to alter the acetyl-choline molecule, able to destroy it. That chemical too was found top-to-bottom.

Name?

Choline-esterase.

Acetyl-choline produced the effect, choline-esterase stopped the effect. Acetyl-choline. Choline-esterase. To the chemist—two little innocent words.

Finally, another chemical, a third chemical, eserine its female name, made experimenting more practical because it blocked choline-esterase. The choline-esterase could not reach the acetyl-choline. The eserine prevented it. So the discoveries had been one, two, three. Eserine saved experimenters from having to race as they had in the past. Later, eserine was prescribed for a disease whose chief symptom was a terrifying weakness—too much choline-esterase was removing too much acetyl–choline and the muscles lacked what kept them stimulated. They were weak. The ailing person had lost strength everywhere, eyelids drooped, head nodded, jaw was too tired to chew, death descended because the chest-muscles, as in Claude Bernard's rabbit with the curare, were unable to contract enough to bring in air enough.

Acetyl-choline, choline-esterase, eserine. To the chemist—three little innocent words.

A triple discovery foreshadowed in a dream, a dream that re-peated, that insisted, that produced a wave of experimentation that spread outward to faraway shores. How fussless and firm is science when one views only the end-results, does not live through rainy mornings and sleepy afternoons and watch professors grow old.

# SYMPATHETIC CHEMICALS
## *Second-Officer Duty*

Adrenalin and noradrenalin are sibling molecules, same parentage, similar but also a difference in chemical structure, similar but also a difference in the effects on body actions. Both are manufactured moment to moment. Both come off with the vein-blood of the adrenal medulla in amounts depending upon the body's needs. Noradrenalin, especially, comes off sympathetic nerve-fibers. That is, sympathetic nerve-fibers when stimulated give off noradrenalin, possibly some adrenalin.

In general, noradrenalin works in a direction opposite to acetylcholine. Nerve-impulses travel down a sympathetic nerve-fiber and noradrenalin is released all along its length but particularly toward its endings. Depending then upon the tissue or the organ, noradrenalin excites, or it inhibits. The organ is spoken of as a target organ. This is the same word, target, that is used with the same meaning for the endocrine glands.

What happens when the molecule of the noradrenalin meets the molecule of the effector organ is not known exactly yet. It will be. And it is already, somewhat.

There is a phenomenon called sensitization, strange.

The sympathetic nerve-supply to the heart of a dog was cut and the medullas of both adrenal glands were removed. That was two weeks ago, Monday, 2nd. Happy dog, clean cage, fresh water, chemically correct food. Today is Monday, 16th. An attendant brings the dog. If an attendant wears a high-style white laboratory coat he may be the researcher himself, so be careful. If he speaks of the "design" of an experiment, he is the researcher.

He now excites the dog in a way that causes it to struggle and to try to get away, bark. Under any circumstances this emotion would cause an oozing from the dog's sympathetic nerve endings, and from its adrenal glands, and the oozing would travel in the blood and though small in quantity would speed the dog's heart. However, this dog had the sympathetic nerves to its heart cut, and the adrenal medullas removed, so there ought to be no speeding, but there is. Noradrenalin caused it. Noradrenalin entered the dog's blood from the ends of sympathetic nerves in many places, but the quantity too small to have any effect usually, but here the cutting of those nerves two weeks ago has sensitized the heart. One way to look at that heart is as itself a mammoth complex of chemicals, with many receptors, chemicals, to be affected by noradrenalin, a chemical, that stays right near those receptors, accumulates, is not carried away because the nerves to the blood-vessels are also cut, so it is there in sufficient quantity to speed the heart despite the cut nerves. There are other explanations. At least, the sensitization is fact. It is made use of. It gives another picture of body and head.

# EMERGENCY
## *All Hands Join In*

Walter Cannon thought that in emergency the sympathetic division went into action top-to-bottom. That includes the outflow from two chains of nerve-ganglia along the vertebral column. Also, he thought that the medulla of the two adrenal glands went into action.

By emergency Cannon meant what we all mean, something that threatens, that we must meet, early in life, late in life, every day, any day, big threat, small threat, imagined threat. The top-to-bottom action he seemed to think was mostly defense against the sudden, the suddenly dangerous, the body mustering its scattered forces, uniting them to resist attack. Total sympathetic action was required for that. The adrenalins were at the same time oozing into the blood to give endocrine support, as Cannon understood it, and he doubtless understood some of it wrong, and some was wrong, and some right, and he was the first to gather it all together.

Cannon's overall idea was sweeping. Sweeping ideas were his forte. Maybe he was America's greatest physiologist. His idea became the drive behind many investigations throughout a third of a century. Cannon called his formulation a theory but like any healthy theorist believed it was more. Not much in the interval since his death has been brought forth to upset the large lines. During crisis the maintenance of the constancy of the *milieu intérieur* can seem to depend upon total sympathetic action plus adrenalin.

What categorically were emergencies?

Pain. Hunger. Suffocation. Anesthesia. Hemorrhage. Severe cold. Severe exercise. Severe lack of sugar. Strong emotion. Severe body abnormality. These were, or, if Cannon had been asked, would probably have been placed on his list. Any outer or inner environmental violence that the body is thrust into would have been. Storm overtakes the sailboat and all hands join in. Even a great ship gets to be a sailboat when the sea swells around it.

Heart, brain, lungs, skeletal muscles must go into action, must have more blood. This is shifted from stomach, bowel, and other places that can for the time wait. Heart rate rises. Bronchial tubes widen. The liver releases sugar. Cannon thought the spleen released red cells. (It seems it does in some animals.) A cat's hair stands up. (Excellent to depress the spirit of the cat's enemy.) A man sweats, is thus self-lubricated, wherefore his assailant when he grabs slides off—certainly an extravagant notion but it reveals the theory's capaciousness, and, who knows? Anyway, a sheet of water does lie over the man's skin and does its part to keep down the heat of his contracting muscles. In the cat there is an increase of saliva, for spit. The lenses of the eyes flatten, which improves vision for the distance, whence the enemy might be approaching. (Excellent to keep up the spirit of the cat.) The eyelids lift, the pupils dilate, more light enters, both actions dominantly sympathetic, and into the face of the highest creature because of the lifting of the eyelids by those shreds of smooth muscle there comes an expression of astonishment. The eyeballs protrude. This also is sympathetic.

For his cats Cannon used a natural stimulus, a barking dog. Samples of blood drawn from these cats he would label either "quiet" blood or "excited" blood.

He invented the phrase "fight, flight, or fright" usually just "fight, flight," and the phrase, like his word *homeostasis* has permeated not only scientific literature but general literature. (The opposite defense would be to play dead.) A sophisticated coed drops "fight, flight" casually into a description of the state of her socio-biological affairs in the automobile on the road home last Friday night after the dance. The indignant coed, the aroused cat, fights or takes flight, or just settles quietly into a pleasant fright, all the battalions of the sympathetic rushing in to help her. She still may not win, and still not be sorry, because emergency besides its own excitement can have sharpened her thoughts as to important decisions.

# SEX CONTROL

## *Modus Operandi*

Both divisions of the autonomic work at this, the somatic too, this guiding and consummating of the act of acts, the whole length of the neuro-anatomy, spinal cord, brain-stem, cerebral hemispheres, dovetailing flawlessly, harmoniously, in the mood of a regatta, sometimes.

In the male the chief consideration is a fantastic tenfold shifting of blood into the male organ, a detaining of the shifted, erection being an active opening of the sluices, a vaso-dilation, an inrush of blood, its impounding, the engorged vessels and the blood spaces blocking the thin veins that otherwise would immediately let that blood slosh off somewhere else. In short, bright red blood from the arterial side is trapped and fluid is incompressible, hence the rigidity, the organ's ability to push its way in, the beginning of the mating assured.

That opening of the sluices, the erection, is primarily controlled by the parasympathetic, though in cat and dog it seems proved that the sympathetic assists. The release of the semen—ejaculation, emission—is sympathetic and somatic.

In male cats under deep anesthesia the stages of the sex act, erection, ejaculation, emission, subsidence, all have been demonstrated by the sequential electrical stimulation of single nerves. The researcher touches a nerve, touches another nerve, touches one-two-three, the stages of the act unfold one-two-three. Any era but ours would have thought that an odd research. In female cats parallel stages ensue upon parallel stimulation. Also, upon stimulation deeper in the nervous-system. Ancient mechanisms like these are probably the same for men and women as for cats. Everywhere sympathetic joins parasympathetic and reports throughout go to the highest centers.

For impregnation, for maintenance of pregnancy, for delivery of the child, the highest centers are not necessary. A human spinal cord has been snapped in the middle of the back and the classic scenes have notwithstanding played off first to last, brain and mind left out, could look coolly on from a distance. That is, a man with a snapped cord has mated, produced a child. It appears a mere display of skill, Nature's, is ironic, appears to toss procreation unflatteringly off to the aft of the ship.

For the total *modus operandi* of the sex act there has been postulated a solitary center in the cord. Destruction of the sympathetic nerves coming from the center renders the human male impotent usually, not the female, who despite such destruction has conceived and carried through to pregnancy and delivery. Nature never gives up. Nature adjusts every aspect. Nature also adjusts the human mind, perhaps, in every unsubtle and every subtle way, creates romance, makes physiological use of the moon, brings out the stars, whips up a whispering warm wind.

At the culmination of the act an excitement spreads over the entire ship, over the entire organization, beyond the ostensibly relevant, involves autonomic and somatic and physics and chemistry and hormones and endocrines, other considerations for the time in eclipse. The experience should all be pleasantly remembered, for the next time.

# HYPOTHALAMUS

## Pilot

The controls on heart-rate, digestion, sex, et cetera, were all low in the brain-stem, though for each it was evident there were higher. Two of those higher—hypothalamus and autonomic cortex—have a voluminous history and literature.

Hypothalamus and cortex.

Pilot and captain.

The hypothalamus has a concentration of controls. Every year adds a few. One thinks it must really be big, so much found in it, an alluring puzzling region, and actually still small, still lies flat on the middle bottom of the brainpan. Shakespeare used that word brainpan. All our lives Shakespeare turns up in all kinds of places, here in the below-deck, dissecting theatre in Padua, city barroom, county morgue, and of course a Denmark graveyard.

The hypothalamus has overall autonomic controls and small discrete controls. In "emergency" it is overall. Among the discrete: regulation of body water, body temperature, satiety and appetite, partial regulation of sleep, and, if somewhat more neighboring brain is included, patterns of sex, signs of the emotions, possibly the emotions, though here as with all things of mind one immediately moves carefully. The satiety-and-appetite control is fascinating. Two centers. When our stomach is empty, our blood sugar low, and we hungry, the electrical activity of the appetite center can be shown demonstrably to rise, that of the satiety center to fall. As we eat, the electrical activity of the two centers reverses.

There is a history. In the nineteenth century a pathologist who did thirty thousand post-mortems, personally inspected the insides of thirty thousand corpses, noticed that infection at the middle bottom of the brain was apt to be accompanied by peptic ulcer. This was a fencing-in. It was the relating of a place in the brain to a disturbed performance of body: middle bottom of the brain and peptic ulcer, hypothalamus and digestive tract. Later a clinician thought he saw a connection with the pituitary gland in a disease where there was obesity with loss of libido. After more years the obesity was shown associated not with the pituitary but with the hypothalamus above. This again was a fencing-in. The obesity resulted from imbalance between those two centers, for satiety and for appetite. Other facts. During epidemics of sleeping sickness it was noted that the brains of the corpses showed damage around the middle bottom, associating it somehow with sleep, and during surgery slight pressure around that middle bottom brought unconsciousness. In this zigzag fashion did place and performance become more and more related, a small place, a wide range of performance.

After more years it was evident there had been too much fencing-in. Fences were torn down. The locality was less restricted than thought. It extended basally and medially into what long had been considered smell brain but in the highest creatures had nothing to do with smell. Plainly that middle bottom extended into transitional brain, septal brain, visceral brain, limbic brain, extended tailward and headward, was a crowded concourse where much of the life of the nervous-system came and went, much of the life of the endocrine system, much of the life of the creature.

By weight the hypothalamus remained only one three-hundredth of the brain. When electrodes were planted in it, the animal allowed to recover, shocks delivered, out came autonomic signs, one or another or several from apparently the same point, clear-cut often, confusing often. A drop in blood-pressure. A racing heart. Sweat from footpads. A terrific-though-frail change inside an eye. Decrease in gastrointestinal activity. Increase in gastrointestinal activity. Release of hormones. A spot when stimulated (in a goat) caused the animal to drink long after it was sated, even to drink acid. The Commander of Oceans and Ships seemed to have bespoken for this region of the brain the role of head-end of the autonomic. Influences scattered all directions, to the mind too, whatever direction that is, and contrariwise from the mind to the hypothalamus, to the body, to parts of the body, destructive influences sometimes.

# BLADDER CONTROL
## *Blissful Moment*

Bladders have different capacities. A gentleman is able to be inattentive to his for several hours, an occasional lady all day unless she is perturbed or pregnant or has nothing better to do, and a contented person of either sex sleeps through the night and then in the morning liquidates the entire obligation in thirty seconds, though not that fast for an elderly gentleman.

If a corpse has water run into its bladder till full, then more run in, the bladder bursts. This world is a place to get out of. If not in a corpse, if alive, and drip-urine run in conventionally, the bladder yields. It objects if two hundred cubic centimeters are run in all at once (through a catheter), contracts annoyedly several times and then yields, has adjusted its capacity. A definite report will have gone up to the owner. If that two hundred cubic centimeters had been drip-urine, the report will or will not have gone up. If four hundred cubic centimeters it will have. Emphatically if five hundred. Urgently if six hundred. Imperatively if seven hundred. Yet as much as one thousand have been known to accumulate before the stubbornness of the individual was overridden by the common sense of the machine. The infant's blissful moment will have arrived earlier because the nerve-lines for delaying the act were not yet connected, a reason for rubber pants and wet fathers.

In the control of expulsion the autonomic is joined by the somatic. Urination centers are at successive points along the nervous axis. Some self-rule exists in the bladder wall itself. There is a urination center in the spinal cord, and centers at three or more levels of the brain, also at the highest. Nerve-lines run from there through brainstem, through cord, to bladder; the reverse also, bladder to intellect.

A road map is in the mind of every neurologist or neurosurgeon. He sees the events. Urine enters the tank, rises in the tank, a report goes to the owner, who reaches a decision, telegraphs initial orders from mind to brain, that frees the spinal-cord center of its duty to stem the tide, and bladder wall and its double-valve outlet go into action. The bladder contracts, the internal outlet valve relaxes (autonomic), the urine keeps pressing, the external relaxes (somatic), the floodgates open, and all is well. Assistance is given by the abdomen, whose muscles are holding firm (somatic) while everything below is relaxing (somatic plus autonomic). "A machine made by the hand of God." Relaxing could have been changed to contracting if the gentleman had, for instance, decided to finish his crossword puzzle.

Broken backs have added knowledge. Immediately after the catastrophe urine runs out unrestrainedly, but soon there is a tight contraction of the double-valve. The bladder fills. It overfills. No report goes to the victim's mind because the wires to carry the report are down. But a paralytic bladder must not be allowed to overfill, not once, and every alert physician understands this. In World War I a man with his spinal cord cut across died. In World War II he did not die, is living all around us and there is reason to think he has a chance of living his normal span. Better drugs, better hospitals, better knowledge of the autonomic, better recognition of the danger of an overfilled bladder have given him that chance. Some weeks after the catastrophe the center in the cord begins to operate reflexly, is quite effective. Pressure rises in the bladder, the center gauges the situation without benefit of mind, the floodgates open, urine runs out, the gates close. Some urine might stagnate in such a bladder, become foul, and for that the physician has his sulfa drugs. This reflex urination can be initiated by the patient himself. With his hand, the unparalyzed one, he strokes the inside of his thigh or the bottom of his foot, in that way stimulates the spinal-cord urination center and the emptying starts. The rectum might start at the same time, the controls not as discriminating as they were. The mind looks on. A mind often has had to look on at a bizarre unhappy situation sadistically put upon its body.

Everything indeed leads toward mind. Even a crushed spinal cord might get its greatest interest from the mind that looks on.

As for the autonomic, the sympathetic division carried impulses to inhibit the tone in the bladder wall and to excite tone in the inner valve, the urine thus kept stored in that cellar; the parasympathetic did the reverse, assisted expulsion. The outer valve was meanwhile controlled by the somatic, by the voluntary, by you. How much attention now and then but how little usually we bestow upon our bladder. It gives us a nudge, lets us go on day-dreaming, both situations for our good, then after a polite interval nudges again. True, the orders that come from above may get garbled up there, wherefore a troubled mind wets the bed.

A chronic hormone-induced overproduction of urine does occur. The bladder also overworks. The afflicted is thirsty, drinks, urinates, is thirsty, drinks, drinks, drinks, urinates, urinates, urinates. Anyone would think immediately of that pretty bambino of the fountain, water running in at its mouth, running out below. A daily output of urine for a normal human being might be a quart and half. With the chronic overproduction the daily output might be five quarts. Twenty quarts would be astounding. Forty have been claimed. You spend the day there. A camel would scorn these figures because a camel deprived of water, according to two talented scientists who worked in the Sahara, after losing water until its hide hung in folds, drank without stopping one hundred gallons, its hide filling out like an inflating tire.

# TEMPERATURE CONTROL

## Thermostat

There are persons in the hills of India who never have had a thermometer under their tongues. We, most of us, began early, rolled one a bit, wished we could cheat, at least strongly hoped it would come out Fahrenheit 98.6°. The human machine performs best there, or near there. Eons of evolution have established for the body a temperature that fits its chemistry, and possibly the other way. A dog's best temperature is higher. A pig's still higher. A bird's higher than a pig's. One man's differs from another man's by only a degree or two, whether Eskimo or Zulu, whether hod carrier climbing up a ladder or fop slumping in a contour chair, whether December 25th or July 4th. But each man's does fluctuate through the hours of the day, regular fluctuations, also personal fluctuations. During the racing of a hard race though the temperature dramatically rises it as dramatically falls, quickly. The astonishing fact however remains the constancy. (Claude Bernard's constancy.) This all refers to the temperature of the inside of the body. The outside, the skin, the exposed parts have not the same temperature as the unexposed, vary easily and widely. When the air is Fahrenheit 70°, the skin of the forehead is apt to be around 90°. That core, that inside of the body, the organs, have the steady state. In recent years to get a human temperature as near as possible to, for instance, the brain, the registering instrument has been pushed high into the nose, or deep into the canal of the external ear. Small differences there are. The heart is hotter by half a degree on one side than the other. Man's steadiness despite environmental unsteadiness makes him a member of a class, the homoiotherms. A frog is a poikilotherm; its temperature goes up and down with the weather, its chemistry up and down, its behavior, sluggish in winter, waking up in spring, and in danger of its life if in August it cannot hop out of the sun.

The living body has first of all the heat characteristics of any physical structure. It cools by conduction, convection, radiation, evaporation. Cooling by conduction is when something touches something and heat flows from hot to cold, the efficiency depending upon the material, silver excellent, wood poor. Hottest organ may be the liver, from whose hot cells heat flows outward through blood, muscles, fat, skin. The organs of the core—thorax, abdomen, pelvis —besides being in general hotter are indeed steadier, the important chemistry of their cells steadier, not disturbed because of heat fluctuations.

Cooling by convection occurs when a layer of cool air lies next to a surface, is warmed, rises, another layer takes its place. Cooling by radiation is energy transferred without anything touching anything, a wave motion. The small human body radiates. The large sun radiates. The distant quasars radiate. The eventual source of our heat, as of our light, is the sun, and because of the sun the moon radiates, and the walls of our houses radiate, the pavements, the ground, usually to the body's disadvantage, toward it on a hot day, away from it on a cold. Ice skating is comfortable in winter because radiation is for the time in the body's favor, the skaters warm in the cold, and happy. Cooling by evaporation occurs because for every gram of water turned to steam more than half a calorie is drawn from whatever that water touches, asphalt or flesh. The delivery of fluids toward or away from our depths is accordingly a heat-regulating device. Slow total redistribution of fluids is part of our or any animal's acclimatization to hot and cold parts of the world. Visible sweat falling off the body in drops is obviously not evaporation, cools nothing, or little, and true too for drops dripping from the tongue of a panting dog. Air currents assist in cooling. They continually carry away the wet warm air. A panting dog produces its own air currents. Even in the Sahara light work is possible if there is a wind. Already in the eighteenth century it was known that a man exposed to dry air at 250° would come off unharmed, because dry air would accept the water vapor that was coming off his body and cooling him. A beefsteak placed in a pan next to him was broiled, as would have been the man if he had not had his heat-dissipating machinery. He wore shoes so as not to singe the soles of his feet.

To sum up: on a hot day a man exerts himself only if the humidity is low, gets out of the way of radiations, wears clothing that has the fewest layers to trap poorly conducting air between the layers, turns on an electric fan to blow away the layers of body-wetted air, turns on the air-conditioner to cool and dry the ambient air, arranges his bedding so that the least possible heat may flow from it to him, eats non-stimulating foods, eats less so that his chemistry is less, swallows ice floating in something else, contracts his muscles only enough to reach for another glass, or only enough to throw a wave of animation into his eyes, which cools his mind, which cools his body. Both divisions of his nervous-system, autonomic and somatic, join in the work.

Now, a cold day. Closing of skin blood-vessels—vasoconstriction autonomically controlled—keeps in heat. Reducing the area of uncovered skin keeps in heat. The person's will is in the affair now. A woman on a winter night waiting for a bus huddled up to the man next her who himself has no more than his teeth and spectacles exposed, are two poor cold people presenting the least they can to God's cruel night. The night gets crueler. The two begin to pace, slap their arms, stamp their feet, like the man who sells chestnuts the Saturday before Christmas, the pacing and slapping generating heat by contracting muscles. Adrenalin at the same time oozes into the blood-stream, travels to the cells everywhere, speeds general chemistry. Suddenly, the machine automatically throws in an oh-so-valuable shiver. Shivering is contraction one after another of small muscle-groups, may double basal heat, comes about either because body temperature actually is dropping and the cooled blood excites the hypothalamus where the heat-control center is, or because reflexes are excited from the body's surface, as when a chill stone operates through thin pants.

All levels of the nervous-system take part in these regulations. The hypothalamus houses the thermostat. It has been located exactly. It employs both nervous-system divisions, autonomic and somatic. It assuredly is no simple device. Every system of the body except possibly the reproductive may be involved.

The forward part of this thermostat operates particularly against rise of temperature, the backward against fall. Rigid regulation occurs through the twenty-four hours, the daily, perhaps always somewhat personal pattern, whatever it is, unvarying over years. A metal grill has been surgically fitted into the forward hypothalamus, the animal allowed to recover, then placed in an electrical field, which warmed the grill, deceived the thermostat, which mistook the body as being too hot, swung it toward heat loss, caused it to throw away heat that it could not afford; the animal might lose five degrees. Disease can destroy either forward or backward parts—anyone is now in possession of enough information to calculate the consequences. They are not as obvious as one's calculation, but would be if one knew all the facts, which one never does, quite. Abrupt shifts of temperature occur in the mentally ill, the temperature up, the patient up in bed, looking wild, then in a few minutes the temperature down, the patient depressed. Abrupt shifts occur in fracture at the middle bottom of the skull. Brain surgeons operating there may inadvertently bring about shifts so great that they go on to death, but it is a neighborhood where it is easy to mistake what is going on to death. A temperature soaring to 106°, or 108°, or even 110° if sustained cooks the nervous-system and its cells do not recover. Fever relates to this neighborhood of the brain. Bacteria, viruses, toxins tinker with the thermostat and for a week it is set at 103.3° instead of the routine 98.6°. How it is set, or reset, is not understood. Infants have not been about long enough to have their thermostats stabilized and their temperatures dance around and would horrify the young mother except that she has read *Redbook*.


# PILOMOTION

## *Windbreak*

Hair—much more in many animals than in us—is part of the equipment of temperature regulation.

Each hair is a lever. When the muscle hooked to it contracts it stands up, lies down again when the muscle relaxes, stands up faster than it lies down, has character. Experimentally, that standing up and that lying down have been stimulated from many points in the brain, till one sometimes thinks that the totality of the brain is involved in every autonomic decision. (But what a funny man an experimenter is, pokes about in a brain to make a hair stand up.) Goose flesh is the bumpiness produced by the contraction of hair muscles, but no hair, and goose flesh does look futile though it may not be.

Human hair provides no great windbreak, nevertheless a man is apt to suggest to his barber to be a bit conservative in winter. Animal hair provides more. Feathers do. Both feathers and hair trap air, that we know conducts heat poorly, thus feathers and hair are part of the bird's and the beast's heat-regulatory equipment. Hairs are useful too in an animal's self-expression, belong therefore in what one might say are the higher echelons of its nervous-system. Hairs stand erect in the aggressive, make the lion look bigger, but are supine in the frightened, make the victim look smaller, help him to play dead. Useful also as sex panoply, man or beast. Man with his large brain consciously recognizes the situation, exploits it, regular haircuts on Wednesdays, a beard in some countries and some periods, a mustache and sideburns added for other countries and other periods, beard or mustache or sideburns on one kind of man in any country in any period, or the reverse, every blade shorn off to make the head more masculine, hopefully. Human beings, that is, have deliberately reassigned their hair to decoration and are convinced they put it to better use than the wolf that merely keeps itself warm. Does the wolf merely keep itself warm? One doubts it. The wolf's hair top-to-bottom rises in emergency. Polly's feathers do. Sex and defense and temperature regulation all employ pilomotion, hair-movement, in Polly feather-movement.

The sympathetic controls the action of hair, the hypothalamus controls the sympathetic. If one special spot in the hypothalamus is touched with electricity, hair rises, another spot, not only does it rise but rises with alacrity. There are higher controls than the hypothalamus, even places in the highest brain, the high operating through the low.

The rise and fall of hair is involuntary, but it has been claimed on occasion to be voluntary, like some other involuntary acts. A person says to his hair, lie down, and it obeys. A small Negro girl saw the hair rise on the back of a hairy white man's forearm, felt tenderness, said "Ah!" then proceeded to pet down the hair with her hand, as if she expected him to purr. Probably he felt white-inferior. Probably he felt understanding and affection.

There has been much debate, to no conclusion, as to when and how and why man lost his hair, and why he kept some of it in some places of him, whether the advent of clothing had anything to do with it, or a greater freedom from ticks and lice, or a dislike of hairy women and therefore less productivity from hairy women. It could seem a simple thing to remain unsettled.

# AUTONOMIC CORTEX

## *Captain*

On a ship, pilot's and captain's quarters are separated. Captain's are high—autonomic cortex is high and it is spacious.

William Beaumont, celebrated American who studied digestion in Alexis St. Martin's stomach through the peep-hole left by the gunshot wound, reported that the red lining of Alexis' stomach turned pale every time that gentleman-drunkard flew into a rage. To turn pale means constriction of blood-vessels by muscles controlled by autonomic nerves. From the region of Alexis' thoughts (high presumably) and the region of Alexis' feelings (lower but still high) nerve-impulses traveled down to the region of Alexis' stomach (low, as Plato said) and pinched off the vessels, and the red lining turned pale. Thoughts and feelings were expressing themselves in a man's stomach. The divine (Plato) was expressing itself in the gluttonous.

Other clues accumulated. It was more and more evident how parts of the cortex and parts of the autonomic were linked in an up-and-down back-and-forth manner. It was a further weaving of the nervous-system into one to produce a human being who was one. Our subtlety, capability, ambivalence, treacherousness, dependability, undependability, happiness, sadness are one. We are one.

Long ago physicians had noted that in an apoplectic stroke, when hemorrhage separated the top of the nervous-system from the bottom, there were, besides the signs of paralysis and numbness on the opposite side of the body, abnormal flushing and later abnormal pallor and sweating. The paralysis and numbness meant the cortex cut off from its lower somatic controls, the flushing and sweating from its lower autonomic controls. Brain-surgery meanwhile was being performed more frequently and under local anesthesia, and it was noted by an occasional surgeon that when he stimulated electrically high-up, where he was working, he might see autonomic effects. The effects were fact, the explanations by no means fact, somewhat wobbly still today. This surgery was in human beings. Experimental workers were noting the same in cats and dogs and monkeys. There has been an accumulation of data over the years. Blood-pressure effects were commonest, sometimes a rise of blood-pressure, sometimes a fall. It was never possible to be sure that the effects during experiment and surgery were the same as those appearing over the body when an excited mind read, say, that strange ballad of a ship, *The Rime of the Ancient Mariner*.

Heart-rate might slow when one spot of cortex was stimulated, speed when another. Saliva might drool. Sweat might roll. Human brain-surgeons obtained fewer effects than the researchers. Gastric juices might ooze. Bowel muscle might churn. Bladder muscle contract. All of these were autonomic evidences and obtained by stimulating of the human cerebral cortex during surgery, but just what they indicated seemed still to require more data. Other spots stimulated gave joint autonomic and somatic effects. A leg might lift and at the same time increase in size, the lifting somatic, the increase in size autonomic; the leg's blood-vessels apparently had been opened to let in increased blood to make possible the work of lifting. That is, a somatic performance appeared to include an autonomic performance, but it was again not easy to be sure that the same mechanism was operating in the experiment and under natural conditions, because though the lifting did result from stimulating that spot of cortex, it might be that not the stimulus was drawing in the autonomic but rather something in the act of lifting. Or, a leg might increase in size and the kidneys decrease, blood apparently shunted from kidneys to leg, urine-making put off temporarily, this double autonomic effect brought about by an electrode touching one exact spot of cortex. Many such spots could be identified. Autonomic joined somatic, or somatic joined autonomic, and the important end-behavior, however much of it might still go unexplained, was lifelike. A left and a right eyeball moved in unison (somatic), the movement accompanied by tears (autonomic). Chewing movements (somatic) or tongue movements (somatic) were accompanied by a flow of saliva (autonomic). Movements of fingers (somatic) were accompanied by sweating (autonomic), the sweating to improve, it was surmised, the delicacy of touch, and our touch was more and more showing itself a sense to reckon with. All along the motor cortex, autonomic and somatic appeared to trot in the same yoke. One would expect them to. They do.

It is difficult in the light of the entire evidence not to be convinced that joint somatic and autonomic actions are not operating in the same way through the same single spot of cortex. Everything that is needed to get done a single job has been ordered from a single spot. But the requirement of cautious appraisement remains. Communication lines for both systems do start from one such small area at the top, and go from there for whatever distance, and via whatever junctions, and lead to the bottom. The hypothalamus is often a way station for some action (reflex) that started at the surface of the body, went to the top of the nervous-system, came back to execute the action. The hypothalamus itself might be the top. Because these communication lines could be traveled, in fact because they existed, an informed father might explain to his unhappily-married son how it can happen that if the top of a woman's brain, where her mind seems to cavort, is adjusted, the bottom is apt to be, and, if not, her ship too somewhere begins to rattle, and that he ought to take that into consideration and behave accordingly. People nowadays talk of such matters in a bus, in powder rooms, at cocktail bars, inspired to it often by unrest in their own instead of someone else's body, and using their own language.

In our living ship much regulation is undoubtedly autonomic. The pilot's directives come from the pilot's quarters. The captain's come from the captain's. Telephone lines join the two quarters, and may be busy. Water regulation, heat regulation, circulation regulation, sex regulation, others, each has a place or places in the pilot house. The captain leaves much to the pilot but can and does interrupt. Sometimes the pilot nags at the captain. So, the nagging may go both ways. If it is too frequent the all-important passenger may get his mind unsettled, get his body unsettled, get seasick. In his stateroom at night he sometimes is awakened by a stab inside him, coming from somewhere below, and this sharpens his attention, may make him thoughtful, philosophical. Then the stab withdraws. He falls asleep. Next morning he forgets. Or he thinks it was a dream. Before noon he probably will be leaning on the rail, admiring the shiny keel of his ship, proud of her, as if he owned her, as if he had built her. He does not even often reflect on the labor that is keeping her sailing. Best during a sea voyage to forget machinery. The ship's doctor sensibly says: "Forget." The passenger obeys. Forgets the doctor too. At least, though, it would be sensible for passenger and doctor both now and then to remember that the visible of the ship has the invisible always underneath it.

# XIX
# THE HEAD

———

## The Rest Exists for This

# MIND'S BIRTH

## *From the Wings*

Imagine a father in a distant village. Let it be Ecclefechan, Scotland. It might be New York, United States. He in Scotland is looking with God-fearing fervor into a crib, into the face of his newborn son, Thomas. He is looking and seeing there Thomas's mind, right in the pink flesh, or somewhere near, and he is not in the least distracted by the still unformed putty of Thomas's nose or by the two arms that pump and pump as mechanically as oars, or the legs that equally mechanically kick the bottom of the crib. To that father this newborn son might still be having some trouble breaking into the world, might be held back somehow by the flesh, appear shy, but the person exists. The mind exists. If he were asked to say something more about that, where for instance it was, he would not say anything about genes, does not know about genes, but might unperturbedly answer, "Oh, somewhere around that small head." At the least he is mistaken about the size of the head. It is not small. It is one-quarter as long as Thomas's body.

Imagine another place. Imagine an infant developmental diagnostic laboratory. Yale for many years had the famous one. Imagine a psychological tester and Thomas in an interview. A secretary has been writing down everything the tester said. A cinema camera has been photographing everything Thomas did, and tape has been recording the coos and screams. To the tester Thomas has exactly five days' worth of mind—because he is five days old. Five days since birth—five days' worth of personality. Mind is building, has been building from the beginning, womb to now, nothing to riches, and will go on building by the method of the building-association, week after week, month after month, year after year.

Plainly, there might be two ways of imagining the birth of mind, and no doubt others.

Of the two the first is closer to the church, any church, any religion, all religion, the second is closer to science, biological science, psychological science, any science, all science. Mind to that psychological tester sprouted and grew from no-mind, personality from no-personality. There had been nothing. Now there was something. It branched. It was gene-directed, but it emerged always as a product of the back-and-forth between this machine and a world, between a small lump of molecules and a large lump of universe. It followed: change the world, change the universe, change the mind, change the personality, and that is the only way mind and personality can change, in fact could be born in the first place, could appear. There is nothing strange, mystical, inexplicable, even if some items are not yet explicable.

Quite the contrary, to Thomas's father, had he troubled to think about it, had he been able to think about it, mind was—well, it was there in that crib. It not only already existed, it always existed. As the astronomer said of the astronomical, it always was, it always would be. The back-and-forth that in the psychologist's view was producing the first dawn of it, in Thomas's father's view was setting it free, was allowing it for a while to go about on the earth. It arrived at 4:20 A.M., with an I.Q.—no matter what the I.Q., the mind was busy, if tentatively, establishing its contacts with mother, father, siblings, neighbors. The first wisps of a capacity to give attention had occurred in the first hours, by the kind of evidence the scientist properly accepts, the most eminent of child psychologists having produced it.

Later, after a thousand yawns, Thomas will puff himself out for his present round-the-world tour, at some port where the ship docks, the plane puts down, will buy and stick a chrysanthemum into his button-hole, then politely walk over, pin an orchid on her dress, the two continue in duo, to Paris, to Patagonia, to the North Pole.

Thomas's father's view was the view of the centuries before Copernicus, before Darwin, before Freud. Those three changed the view.

Copernicus. Darwin. Freud. In Thomas's father's view mind lost its body at the end of a life, or some distillation of it did, possibly even lost its body again and again, as Socrates suggested, became disembodied, at each transfiguration waited for its cue in the dark of the back-stage and when the voltage was turned up one could see the body, chubby or skinny, shortly see the mind too, perfect or not-yet-so-perfect, and anyone who had an interest would be invited to the baptism, the confirmation, the wedding, the funeral. Thomas's father's view must strike a computer age, any late age, our age, to be as distant as Ecclefechan, but it is just possible that one ought not to forget that Ecclefechan still is right there in Scotland. Also, though you are a robust citizen of one of the colossal clans of the last of the twentieth century, United States, Soviet Union, China (add any other), it might be wise, or rather it would be courteous not to dismiss Thomas's father's view too entirely and too easily. Call it the senile view, or, if you are feeling charitable, call it the old-man's view, and, after you have called it that, show it historic respect, show it the respect sometimes shown the old.

# MIND'S FIRST STEPS

## *Downstage Center*

The turning up of the voltage continued. The emerging of the mind continued.

Often the small face looked illumined, the light not cast on it but somehow coming from it. Healthy infants glow that way. A pale doomed one may. However the newborn seemed on arrival, now there was mind unmistakably. The turning up of the voltage continued. Even at its pudgiest this baby never was what one pediatrician said a baby is, a bag to stuff with canned applesauce, canned meat, correct milk formula. The turning up continued. The infant turned its head at its mother's voice, was discovering her to be more than something to suckle, and she was discovering it more than something that suckles. They stroked each other. The behaviorists say that that stroking is first love, as restraint, any restraining, is first hate, or first rage.

Days went by. Weeks went by.

The infant tumbled less awkwardly. Its parts looked less those of a toy put together, or blubber. Its actions looked less automatic. True, suddenly the eyes still might stare as if mind were somewhere else or nowhere, stare for minutes. Then as suddenly mind was in the eyes again. Then there was that special moment, if anyone was around to observe, when the eyes discovered other eyes. Fledgling birds do that too—bird eyes discover human eyes. In the infant the effect of this was astonishing because intelligence appeared to flash from eyes that had none. Moving objects were watched more continuously. The restless was being separated from the quiet. Persons were the most restless. The world had been all one with this one—there was no outside. But now something was being pushed off, gradually, gradually. Something was being separated off. Psychologists speak of "subject" separating from "object." The world was lost, found again, lost, found. At moments one thought that one could actually observe how sensations from different senses were floating together, sight and smell and touch and hearing. At moments they seemed mixed, two of them, three of them. Then there came a time when one knew that something knew that a wall of persons had closed in around its crib. A person in there was recognizing persons out there, then forgetting. A black face was selected from a row of white faces. A mouth was looked at with what appeared fascination, then blanked. That wall would remain, happily, unhappily, for the balance of this creature's life. Over weeks and months the eyes were succeeding (and it appeared with some recognition of succeeding) in breaking that wall into mother, father, sister, brother, the child next door, the police, the Junior League, faculty of the university.

Abruptly the infant heart-breakingly wept, stopped. Some weep their way out of the womb and go on weeping through life and into old age. Weeping in an infant seems not much touched with mind. The cramp in the belly to judge by the face and by the tears was not a pain in the mind or a pain in the belly but a pain in space. The cramp faded, the weeping faded, but so slight an event nevertheless seemed an event in a universe.

Not so with laughter. When no one was expecting it, a smile, meaning attached, slipped shyly out, slipped shyly back, next time might stay out longer. Mind was emerging with hesitation but emerging. A laugh on the right side of the face pulled the left side along, then was all over the face, then died. A harsh noise from the street had killed it. More and more something generally comprehensible attached itself to laughter, and those around the crib were watching and were pleased to see how the solitary island was beginning to join the mainland. Mother, father, sister, brother, each at a time was a piece with the laughter. Weeping stays off alone.

At six to eight weeks it was evident to anybody that the infant was seeing increasingly with something more than eyes. Call it mind. Call it a forerunner of mind. If your definition of mind requires language your definition is wrong; there is mind without language.

Anyway, the infant was noticing. Parent and pediatrician and expert of the infant-developmental-diagnostic laboratory noticed the noticing, and, to those who notice, noticing is credited with mind. (Pavlov would have credited his dogs with no more than the exploratory reflex.) Whether this infant arrived on earth with mind or with no mind, now there was mind, and now no need of a definition. Nevertheless, without warning the mind still would blank into the nowhere. A pup may do that, concentrate on you, blank, concentrate. A macaw may. Until yesterday the eyes appeared always to swing in an unnatural parallel like two beacons, today they rolled inward, focused, converged, or converged more often, and not yet upon a letter of the alphabet or a decimal point but upon an artificial butterfly that was attached by a string to the crib. To the watching mother that converging was a sign of an intelligence so high that it simply could not be baby intelligence. (Periodically he still clung to her like a wild beast.) The mother mused. Perhaps "baby" is only a mother's hope to keep a creature all for herself? In an enlightened vacancy he sat now in her lap—and saw.

How tinged with mind that converging of the eyes made that face.

Who would not wish to fathom what the butterfly was reporting? What was going on in him, in his personality, definitely a personality now, when through his eyes he was seeing butterfly? Was butterfly attaching itself to his own plump hand, to which his eyes also shifted, on which they also converged for long periods. Is this what a mind does, brings one thing into itself, and having brought one, brings another, just adds, adds, adds?

The infant began to play. The mother called it play. What is play? Why is there play? Is it the machine oiling its operations in order to be ready for the living of its life? It may be. But is he—in there—not living his life now? He is of course. And maybe he will not be much less an infant—in there—when he is forty. What we can say of this immediate act is that that infant was doing something that undoubtedly looked like play, that his mother joined in, and that both laughed.

The pediatrician throughout has had the stages of that developing mind as clocked as the incubation periods for chickenpox and measles, would think abnormality if ever the gains did not fit the timetable, though he does admit that any good clock may be somewhat slow.

Faster and faster the two eyes went whatever direction the world called. At twelve weeks, the pediatrician could demonstrate to onlookers that the eyes focused on him. At six to eight months, if there were two like the pediatrician, the eyes selected one, deserted the one, went to the other, came back to the one. Mind was rampantly on its course. At ten months, the eyes selected one from three. At a year, one from a roomful. Another half year, if the eyes were seeing one, and another got in the way, that other was bodily pushed aside, hands joining eyes to accomplish it. The infant now could decide to push. It was deciding this or that all day long, mind willfully on its course, and an old man could stand at the play-pen and watch the sleeping child as it waked with its nightmare upon it, watch the pool of tears from its broken heart, shake his head, know that the ancient story told and retold was being retold once more.

Ears, too, were emerging from the cotton, and these ears from the start had been sensitive ears. A heard world was needling and weaving into the fabric of a seen world. A child's psychology still was taking its first steps. Earlier the child had wailed each time it was startled by a sound, now listened instead. That was a large advance. To tones of lower and lower intensity it listened. The face was more and more a listening face. Mind was receiving water from many sources, many rivers. Arms and hands were continuing to help in what way they could. They reached, groped, clutched, let go. Sometimes the hand reached where the eyes had reached previously. Hand had joined eyes. Sometimes the left arm and hand operated independently of the right arm and hand, a further independence of body-parts that went along with a further independence of mind. Skin-sense and muscle-sense both were woven into the fabric. Sounds in the laughter also were changing, had gross variety, were like canary sounds sometimes, like monkey sounds sometimes, and at sixteen weeks were unmistakably helping this male, this person, to express what he had to express, which was much. Much more had been going on—in there—than was indicated in any movement. But movement did indicate. For another year he remained unpredictable and impoverished as to sounds, but every month reduced the poverty, and by the time he was five years old all infant characteristics would have been erased, parents and the neighborhood willing. Mind was enveloping the person as the ocean a reef.

Unexpected gains accrued. It was natural that a bright young mother should aim to have her bright young son housebroken ahead of the Joneses'. It would convince the Joneses of the value of good stock. Triumphant days were followed by days when the young mother despaired of anything ever bringing order into the ghetto habits of her genius. Then he fooled her and performed expertly. Lapsed. Performed. Observations like these were bound to inspire research. One of two identical twins was placed on the chamber every hour and the responses to the "chamber situation" scientifically recorded, the other poor twin not given the advantages of his brother. He was scientifically neglected. Notwithstanding, without benefit of training—imagine everybody's surprise—he performed as expertly as his brother. Training before the nervous-system mechanisms are matured is useless, perhaps may even be harmful, may take away a freshness from the mechanisms when they do mature. It was a Saturday morning after many mornings when the young man, no, the two young men approached their mother spontaneously and well ahead of the physiological act, whispered each into one of her ears, conveyed to her their thought, accompanied her down the hall, disappeared, reappeared, had chalked up two more neat victories.

## DOG'S

### *In an Ounce*

In 1892 Friedrich Leopold Goltz, Strassburg, Germany, reported on three dogs whose cerebral hemispheres he had removed, both hemispheres, a formidable surgery, especially at the time. It was the removal of the latest highest portion of the nervous-system, with which is associated, or so it was and mostly is thought, the latest highest portion of the animal, its mind.

In a dog, the cerebral hemispheres are a large fraction of its brain. In us, a much larger fraction. Hemorrhage was Goltz's problem. His dogs bled to death on the operating table. Again and again he failed before his historic successes.

The first of the three survivors lived fifty-seven days, the second ninety-two days, the third eighteen months and would have longer except that Goltz wanted a post-mortem to make sure of what he had removed. He operated on that third in stages: June 27, 1889, a portion of the right hemisphere; November 13, the rest of the right; June 17, the following year, at a swoop, the left.

So, June 17 much or most of the presumed organ of that dog's mind had been erased. It would have, should have made anyone stop, perhaps look up, perhaps reflect—on mind.

Another dog, a bitch, Goltz allowed to keep her hemispheres, removed nothing of her brain, but she had to give up her spinal cord. The consequence was a dog paralyzed from her neck down, a spinal animal. She had her brain but what could she do with it? Except think. Perhaps think. Goltz arranged for her to conceive and to carry a litter to birth. A macabre triumph. It would do its part to affect our understanding of both spinal cord and brain.

*Der Hund Ohne Grosshirn* (*The Dog Without Cerebral Hemispheres*) was the title of Goltz's classical monograph. He reported those three "brainless dogs" in language a tired market woman could have understood, their behavior, as we say today, and the fact that he does not mention mind somehow calls attention to it. One of those dogs might stand up, start walking, was apt to walk in a circle like a numbed prisoner or a numbed inmate of an asylum. On a rough floor it walked securely enough, on a smooth slipped. If it had not been fed for a time its pace increased, or it might stop the circling, rise to its hind feet, put its forefeet on the bank that surrounded the cage, stare out. (Mind in that?) Meat had to be placed right under the dog's nose if it was to eat, which it did like a stoker. Drank the same way. Emptied its bowels with that put-on, it always appears, absent-mindedness of a normal dog, began with the ritualistic spiraling that Goltz described as having been "very lively and rapid."

Sleep descended on the mutilated as on the unmutilated, and Goltz reported the quiet breathing, the closed eyes, the motionlessness. (Mind gone deeper in?) The dog slept too much. (Bored?) Loud noises were required to rouse it, the surgery having blunted its hearing or blunted its attention, probably both. Most effective, in Goltz's words, was "an instrument that is recommended to bicyclers to warn innocent wanderers that are coming along the road."

At the assault on its hearing the dog "twitched with its ears, as if it wished to be rid of something unpleasant, then finally stood up." It would push a paw toward the ear nearer the sound, and that action, again for some reason, not clear, suggests mind, a worried mind. In spite of the fact that those dogs of Goltz had almost normal voices they never joined in the howling that swells from laboratory kennels the cursed world over, particularly on a Sunday. Hunger waked the dog. Stroking might. Rough handling did. If the dog was lifted from its cage a spell of rage took possession of it, it growled, bit the air, had lost aim. Lifted back into the cage, the rage dropped off like a cloak. It seemed an act. The curtain was raised. The act was played. The curtain was dropped. It left upon one an impression of a de-individualized animal, a member of a species, René Descartes's machine, no longer a dog with the dignities and privacies that to a dog's degree add mind to the world. Goltz's own cerebral hemispheres were substituting for the dog's, brought the necessities to the dog, protected it against injury, gave defense to a body that would otherwise have stumbled into death.

Experimental brain-surgery might be anatomically more exact after this, might be technically cleaner, parallel human brain-surgery, might be more revealing neuro-physiologically, but never again was it as ground-breaking, or more haunting. It was bound to excite the continuing effort to understand the back-and-forth between brain and mind. Experimentally, it was an achievement of the first magnitude, and Goltz shines through it not only as an unusual man but one who in spite of appearances was concerned about his dogs. He knew them. He recorded that Moog UU was a "stupid dog." He knew their characters. Between him and them there was concern. "The human being," Goltz wrote, "can pretend to have feelings that he does not have. The unarticulated, the unlearned speech of animals may be a much more dependable expression. If I step on the foot of a dog I will not doubt that I have hurt him, especially if he cries and howls." Simple words that one may make too much of; one remembers them.

# MAN'S

## *In Three Pounds*

Thousands of years ago men already believed that the human body and the human mind met inside the skull. That idea fascinated them. That idea fascinates us.

It is understandable that Nature built such a stockade around that precious flesh, bone that is light but strong and shaped into a form that has elicited the admiration of architect and engineer. Neolithic people bored holes into that bone, sawed out windows. The Incas did. With them it was religious rite. It might have been to shoo out evil spirits. It might have been sheer bravado—one hopes it sometimes may have been. It was taking a risk with the risky; at least, the human head broken into that far back does excite feelings. Sometimes this could not have been bravado, was a physician trying to save a life after some fool had got a skull fracture in a brawl. Fracture lines can be distinctly seen in an occasional ancient skull. Often it has been perplexing that even with large penetrations there were no signs of infection, meaning either that the victim died promptly or we are not understanding something. Hippocrates in patients with sudden blindness bored holes to relieve pressure.

Galen six hundred years later bored holes. Physicians in the Middle Ages did in cases of head injury. We do.

Editions of *Gulliver's Travels* formerly had a picture, may yet, of a Secretary of State and one other, two heads with their tops sawed off, so as to interchange the half-brain of a leader of one party with the half-brain of a leader of another party, to achieve peace. Swift explained.

> The method is this: you take a hundred leaders of each party, you dispose them into couples of such whose heads are nearest of a size. . . . It seems, indeed, to be a work that requireth some exactness, but the professor assured us that, if it were dexterously performed, the cure would be infallible . . . the two half-brains being left to debate the matter between themselves within the space of one skull, would soon . . . produce that moderation . . . so much wished for in the heads of those who imagine they come into the world only to watch and govern.

This was Jonathan Swift's cleansing irony, was not meant to give any view of mind, had not anything to do with the problem of brain-mind, and yet the old picture did kindle the thoughts of a child. It was a quite satisfactory picture. It was the one of all remembered years afterward, the first inkling, it could be, of the awesome intimacy of brain and mind, of body and head.

If people in the future dig up our bones, will they think of how we painted paintings, wrote books, used electron-microscopes, built cyclotrons, explored atoms, explored space? Will they suspect what sophisticated Martini-drinkers aware-of-our-fleeting-life depressed-elated neurotics most of us are? Will they consider the top-end of us to have been the residence of a young earnest President of the United States, the penthouse of a Miss America? Aristotle, quench-less boy, argued that the mind was farther down inside the body, in the beating heart.

# THE OPERATION

## *Exploring for It*

No modern surgeon enters a skull to find mind. If he is entering a baby's he is not trying to find the beginning of a personality. It is to cure. We may not be interested in his opinion of mind—depends on the surgeon—but we may still be interested in the vantage point at which he stands. There is glamour there. It may be only the glamour surrounding the mountain climber, for someone below, but even to work one's way from station to station around that conical mountain, see up toward its peak, never seeing clearly, has some of the daring and doubt of exploration.

A brain surgeon's operating room should vibrate differently from other operating rooms, dental, rectal. Like any operating room it was built to be kept clean. Whoever enters wears a sterilized costume, a laundry for this on the 10th floor. (A New York hospital.) All knives, scissors, chisels, suturing needles, catgut, silk, gauze, cotton, towels, are sterilized. Because human beings exhale bacteria with their stinking breaths, mouths and noses are covered by gauze masks, double gauze if anyone has a cold-in-the-head, in which event he also avoids talking expulsively. Instead of gauze it might be copper screen. A cap covers every head. Shoes are grounded so that no spark of electricity may set off some tank of inflammable gas. Each person in the room has scrubbed his hands and scrubbed his hands and scrubbed his hands, then plunged them into sterilized gloves. That is not enough. Even an informed hand may stray and spread infection, on which account the chief surgeon, and the assistant surgeon, and the assistant assistant surgeon, and all the others of the hierarchy make a routine of keeping their gloved hands lifted in front of them until the work begins. They look, those green-gowned green-capped hand-lifted ladies and gentleman of various heights and girths, as if about to be initiated into a secret order. It is. It has as a mostly unrecognized incidental project the searching in that bony container for one of the secrets of the world.

Whoever watches TV knows most of this.

For one operation the surgeon works with his patient sitting, for the next has him stretched out, or lying on one side, head lifted, or head dropped, or face down. In that last position a fixture spans the forehead. The table itself may in the still-voiced middle of an operation be tilted to some other angle from that at the start, orders gone out from the chief surgeon. A few of the pioneers, Harvey Cushing notably, were responsible for much of the spirit of this surgery, and Cushing was responsible also for much technique.

The chief surgeon has precisely in his mind the steps he is planning to take. So has his assistant. So his assistant assistant. So the nurses. Each is skilled at his own task but the best always informs himself of the total task.

All equipment is now ready on stands, and not only the items scrupulously anticipated, but every item is in its military place, because, at the moment needed, life or death could conceivably follow from whether it was or was not in that place. An eminent surgeon discharged a nurse whose eyes wandered toward the distances out the operating-room window, got an unearned bad reputation for doing it. An ultraviolet machine may be spraying the air in expectation of killing bacteria, the ultraviolet directed away from any cut tissues, and, whatever the ultraviolet may or may not be accomplishing, it does add to the mood, though there is mood enough.

The patient is wheeled in. Drapes have been thrown over him, a space left in the drapes where the green-gowned chief surgeon will with his tools penetrate that presumed sanctuary of mind. Quickly he stitches into the skin the towels that edge the space, to prevent their slipping. So there is now a neat frame around what is spoken of as the operative field, an area of naked skull. The entire skull, indeed, has been shaved and scoured and the scalp sterilized. Often nowadays the operation is under a local anesthetic that must be injected by a succession of fine punctures like mosquito pricks, not really painful in skilled hands, even possibly half-useful to the patient's mind, because he senses some mingling of touch and pressure with whatever pain, this taking hold of his mind that might otherwise slip off toward frightened thoughts. He is of course groggy, has been made so. A nurse sits under the drapes in a kind of tent, and there she can undisturbedly check pulse, breathing, blood-pressure, writes everything on a chart, leaves a record of the patient's shifting body-states. Should the patient at a point seem not to be doing well she promptly but with professional quiet notifies the chief above. He may regularly ask her. A formal dignified dialogue may run between them, yet everybody at the same time, more in one operating room than another, tries to be as light-hearted as the circumstances allow.

Bottled blood is visibly ready for transfusion, and a needle to deliver the blood is in an arm vein, the needle held in place by adhesive, the bottle raised on a stand so that the blood can drip from high to low. The total setting, including the ultraviolet, including the general theatre of an operating room, cannot but have helped to create the emotional pitch that this surgery at least appears to require.

Mind hovers around that place. One thinks of it. One thinks of how it is in every worker because no doubt one is thinking of how it is inside that shaven skull. Here one does not need to be an East Indian to feel mind as a unifying cloud enveloping all. Possibly the individual mind does float in and out of that unifying enveloping cloud. There would be the chief surgeon's mind. The assistant's. The assistant assistant's. And, again and always, the patient's, that would be somehow somewhere in the vicinity of that substance where the work now begins.

By quick knife-cuts the surgeon establishes definitely the boundaries of the operative field, a line of scarlet following his knife and serving as a guide for deepening of the cuts. Down through the scalp to the bone he goes. The patient feels the cuts but no pain, the anesthesia successfully blotting pain. Rules for establishing those boundaries and indeed for all entrances and exits at different levels of the nervous-system have been laid down by a long experience, though any everyday surgeon on any morning may add some unexpected datum to the accumulating files. The assistant surgeon has kept just behind the chief surgeon's cuts, mopped the blood, pinched off and tied the bleeding vessels. The patient feels the mopping but no pain. The scalp is stripped back, the bone along three sides of the stanched four-sided field scraped. Procedures differ, but burr-holes now are drilled into the bone at the four corners of the field. This today is apt to be a motor-driven trephine but if hand-driven it would look like any auger that was cutting out buttons of bone, the touch of the surgeon keeping him informed of the depth his auger is going, the substance of the brain also mechanically safeguarded by the construction of the trephine. So, four buttons. It takes some prying to get them out. A flexible strip of grooved metal with the groove facing upward is carefully fed into one buttonhole and brought out the next, a wire saw slid along the groove, handles fitted to the two ends of the saw, and the surgeon saws, up and down, up and down, up and down. The bystander holds his breath thinking that the saw may snap. It sometimes does, but no harm because the substance of the brain is now protected by the metal strip. Three sides of the four-sided field are thus sawed, the fourth side left unsawed, and the surgeon slowly and tenderly pries and levers and lifts at the slab of bone, then, this appearing sudden and rough, cracks through that fourth side. He folds the slab back along that side and away from the brain. The result? A door has swung open. Long forceps are clamped across the cracked fourth side so as to squeeze off leaking blood, and the door, with the attached scalp and wrapped in gauze, turned out of the field.

If a man enjoys the sheer carpentry, if he is dead to any mystery and to all wonder, or forever is shy of any show of sentimentality, he may be a better surgeon, but he may not be. He must employ whatever bulk of body he has because this surgery often goes on hour after hour. True, an occasional skinny man does brilliantly, as with surgery elsewhere in the body, makes up for his lack of bulk with his nervousness and his fervor.

What one is now seeing before one is deeper into that lair of the mind but not yet seeing the brain because the membranes still cover it. The outermost is the dura. Should the pressure inside the skull have risen abnormally high, as happens, and the dura bulge through the doorway, the surgeon pushes in a trocar, a sharp-pointed hollow tube that he advances into a ventricle where the cerebrospinal fluid is, drains off some of it, the bulge receding. Next he cuts a doorway into the dura corresponding to the doorway in the bone, and now at last there is the brain itself. It is a human brain. We may think of it as a formidable habitat, but now it is heaving gently each time this patient takes a breath. A hallowed moment for an onlooker but for the surgeon who has seen this sight often it is bound to be repetitious.

The character of the disease, which often has mind signs, dictates what happens next. If a tumor, the surgeon tries to determine the tumor's extent, whether there will be more loss than gain to the patient in removing it, or removing part, and this depends too upon the kind of tumor, the kind of cells, the area of the brain, how critical the area, whether it might be better not to do anything, just relieve pressure, close up. Instead, he may be needing only to trim out an old scar. Open an abscess. Explore for hemorrhage. Tie or clip an abnormal and dangerous blister on a blood vessel, an aneurysm. He works when he is talented with what could well be described as courage, may make one think of the sculptor who must strike boldly if he is not to damage the delicate living thing inside, because any faltering may forever silence that mind, or worse, keep a maimed mind roaming and alive. A brain surgeon moving with his fingers in that so-called organ of the mind can strike one as an archeologist digging in a buried city, small but ancient city, asking questions of the stones. Most days most surgeons are asking no such questions, are thinking neither as sculptors nor as archeologists but as doctors maneuvering in a brain and not at all embarrassed by the fact that a mind may be potentially in that neighborhood.

Who has not wished he could just for one morning be a brain surgeon, for one morning get his own mind inside some living skull, be the principal guest at the four-hour or ten-hour visitation in this house associated with human thought, feeling, personality, observe the patient before the surgery, perhaps talk to him throughout, talk to him afterward if he is in a state to talk?

The critical part of the surgery complete, blood-vessels are tied where necessary, the operative field flooded with salt solution that is syphoned off, the field scrutinized, flooded again, syphoned again, scrutinized again. Is any small vessel oozing? It must not. All bleeding must be stopped, this according to principles that studies in blood clotting have made routine. The dura is stitched back into place, the bone wired back, the scalp sewed back, this according to principles that studies in wound healing have made routine. Then, he or she, head wrapped round and round with gauze fashioned into the shape of a great white turban, is wheeled back to his or her private room, H219, looking rather jaunty. His mind went out of the operating room with him, body and head went out together, and a person standing by may well be disposed to look up and away.

# HEMISPHERECTOMY
## *Throwing Away Half of It*

An American surgeon was the first to remove an entire cerebral hemisphere from *Homo sapiens,* half a brain. The other half stayed where it belonged in the skull, else there would have been no mind. An eager reductionist might argue that there would have been, that the future will look for and find some in any fragment of nervous structure. But common sense finds none. The hemisphere that is removed must be the right in the right-handed because if it is the left the man will have become what people for some reason like to call a vegetable. (Recently there have been several cases reported where the left in the right-handed was nevertheless removed, with some recovery of mind, some recovery of speech, the risk taken because of malignancy that could only go on to death.) This species of surgery is not experimentation, though at first thought one might think it is, but then one remembers again the patient, the person, how everything living, including man, wants to live, keep on living, however absurd. It is forty years since one-half of a brain was first removed from a human skull and hemispherectomy had become an accomplished fact. Having performed one, that surgeon performed a series. In the case of an elderly colored man the total flesh cut out including the tumor weighed 584 grams. A healthy adult human brain weighs on the average 1,350 grams. The elderly gentleman had been brought to the Johns Hopkins Hospital in coma, was operated on, lived three and one-half years and died because of a recurrence. Three and one-half years is not a negligible span in which to observe and reflect.

Human hemispherectomies were adding and in the future might be anticipated to add to our understanding of mind, not much, though that might depend on who and how, and they would keep awake the controversy around the brain-mind.

That is, it might, just might, add another view, half view, to our views of Fuji.

Hemispherectomies began to be reported both in this country and in Europe, oftenest performed on children born with a damaged hemisphere, the damaged interfering with the healthy, so, better the sick be removed. Another elderly patient who survived a hemispherectomy for more than a decade was reported puttering and hobbling and losing his balance and enjoying his garden and leading a generally contented if bodily enfeebled life. A decade would be a still less negligible span in which to observe and reflect.

Walter Dandy, the American who first risked the operation, left us snatches of dialogue with a patient whose right hemisphere he had removed, a man fated to live only a few days, and who had a fever while talking.

| | |
|---|---|
| PATIENT: | "They could get a lot of things around here to build!" |
| QUESTIONER: | "What things?" |
| PATIENT: | "Grass and things." |
| QUESTIONER: | "What is this place?" |
| PATIENT: | "Johns Hopkins Hospital." |
| QUESTIONER: | "What would they use the grass for?" |
| PATIENT: | "To build a nest for the young ones." |
| QUESTIONER: | "Who would?" |
| PATIENT: | "The mice." |
| QUESTIONER: | "I didn't know that mice used grass for nests." |
| PATIENT: | "Yes ma'am, and feathers and cord and things." |
| QUESTIONER: | "What made you think of mice?" |
| PATIENT: | "I don't know, I just thought of them." |

A human mind associated with a half-brain, or dwelling in a half-brain, or even being a half-brain, as some scientist would consider it, was still able, judging from this dialogue, to poke about in what interested it, a half-brain apparently space enough for that, or whatever way to regard it. Presently the conversation went elsewhere. Then $3.00 was mentioned without provocation.

| | |
|---|---|
| QUESTIONER: | "What do you want $3.00 for?" |
| PATIENT: | "To get to Cambridge." |
| QUESTIONER: | "Why do you want to go there?" |

| PATIENT: | "My wife wants to go." |
|---|---|
| QUESTIONER: | "Why?" |
| PATIENT: | "Because her mother is there." |
| QUESTIONER: | "Does it take $3.00 to go there?" |
| PATIENT: | "No, $1.65." |
| QUESTIONER: | "Don't you have that much?" |
| PATIENT: | "Oh, yes'm." |
| QUESTIONER: | "If it is only $1.65, why do you need $3.00?" |
| PATIENT: | "You see my wife's got to go and I've got to go." |

That last was spoken, it sounds in the last line, with some annoyance. Realize, it was a dialogue between a patient and a nurse, or, to be needlessly literal, between one cerebral hemisphere attached to a brain-stem in one skull and two cerebral hemispheres attached to another brain-stem in another skull. (Jonathan Swift again would have ruminated.) Lately, animal brains have been surgically sliced through much of their length, right side cut from left, and, after recovery, each side taught to do something independently of the other, in the experimental psychologist's sense of taught, and what the one side had learned the other side did not know. Guesses can be made as to the meaning of this but not firm conclusions quickly drawn. (Jonathan Swift would have ruminated some more.) The old large question did insistently recur: could the objective methods of science be applied to the subjective questions of mind?

# SHERRINGTON

## *Not One But Two*

Instead of a whole brain there might be the scant nervous flesh involved in a reflex. It might be of the skeletal system. It might be of the visceral system. That might be another station from which to view the problem of brain-mind.

The scientist who first saw the reflex in its full variety was Charles Scott Sherrington. He came to it slowly, came to it with great talent, soon was completely absorbed by it, and not for weeks, for two generations; then late in his life he reached a conclusion about mind.

There was a cholera epidemic in Spain and Sherrington, a young Englishman, went to help, no doubt to help himself, was striking out to find himself, as young men before and since. He was a physician inclined to research, the direction of it not settled. He met Cajal. (Cajal's autobiography does not mention the meeting.) Years later Sherrington would say of the Spaniard that he had "introduced a new conception of the nervous system as a whole, and no man who ever lived did this to his degree." It was an accurate statement, and it was praise from one who by that time had proven himself. The other had more than proven himself. They were different.

Sherrington's research, once he had settled the direction, was the scrupulous study of the simple reflex. Dust blows into your eye and you wink. The doctor taps under your kneecap and your leg kicks out like a jack-in-the-box. But the reflex was not simple, most of it, in fact, unknown, unsuspected, the intricacy in that unit of our machine. Sherrington over the years had demonstrated the detail of how it was not simple. He derived concepts from it. He pursued it with persistence, with intensity, with some combination of artist and artisan, may have dreamed of it at night; would have been an odd thing to dream of stripped as he stripped it. Usually his experimentation kept away from the terrifying complexity of the top of the nervous-system, the brain, but by no means always, was fascinated by and worked with that complexity; but usually he stayed down in the spine where the technical problems were more manageable. He did stimulate the motor cortex in primates and his mapping of it was regarded as correct and standard for years; it was in all of the textbooks still at a much later date. Once the primate was a gorilla. On the day of the gorilla an assistant had a revolver ready because a gorilla with a point of view has the stamina to carry it out. For the spinal studies he used mostly cats.

In the course of the prolonged, only outwardly unexciting experimentation, though exciting to his quality of mind, a roamingly intense mind, he gathered the data that with his and later interpreting gave a more fundamental comprehension than any before of how the body stands and moves, and, that firmly set, that truly granite base, he was able in English-gentleman fashion to worry the scientific community with his carefully thought-out statement of what was the nature of mind, or rather what was *not* its nature.

So, he can be said to have made the reflex, somewhat indirectly, one station around Fuji.

Sherrington's career branched originally (there would be several branchings in a strict history) from that of Friedrich Leopold Goltz, the German who in 1881, which was four years before the cholera epidemic, had made the trip from Strassburg to London to the International Medical Congress to report on the dogs whose cerebral hemispheres he had removed. Goltz had one of the dogs with him. A glittering Congress. Charcot was in the audience, French neurologist and psychiatrist at the peak of his power. Ferrier was, British physiologist, demonstrated a monkey paralyzed in one limb, consequence of a sliver trimmed with jeweler accuracy off the outer surface of its brain. Charcot saw the monkey, burst out: "C'est un patient!" Whether Sherrington heard that remark is not recorded, but he also was there, might be considered the younger generation that comes knocking at the door. Later he would be Sir Charles. Later still he would be Sir Charles Scott Sherrington, O.M. For him the importance of the Congress was the opportunity to join in doing a post-mortem on the nervous-system of that dog "which was exhibited by Prof. Goltz at the International Medical Congress of 1881." Those were Sherrington's words. Goltz had killed the dog in front of the audience, and taken out the brain and spinal cord, and Sherrington's teacher, Langley, got that material, and together teacher and pupil studied it microscopically. What were the changes in the low parts of the system when the high had previously been destroyed? This low affected by the high was to be important in his later conceptualizing of the nervous-system, as was his talent for histologically studying the detail of the cord. Sherrington's first article dealt with that, a young man's first published article. The importance of the Congress for the world was that it turned Sherrington finally into a field that he would plow and plant and grow and weed and from it glean not only a new neuro-physiology but also, for whoever cared, another view of brain-mind. The latter he would state neither with apology for presenting the subject at all, nor with romantic pitter-patter.

Like Cajal, Sherrington shared a Nobel Prize, Cajal getting his early, Sherrington late. Graduate students began coming to him from near and far, many Americans. Looked at historically his effect on neuro-physiology because of his students became even more world-wide. A generation moved by, another, another. He had acquired the toothless look, peered through pince-nez, but his thinking remained clear. Arthritis knotted his hands but his penmanship was so incisive that some psychiatrist would be sure to have thought it needed explaining. People began saying that he had always been a kind man, and he doubtless was, but what we on the other side of the Atlantic also may feel is that he had always been the Englishman. He experienced lapses of memory, told another man's story to the man who had told it to him, and in 1952 at ninety-four years of age he died.

He left a heritage that was like Cajal's in this: it would never cease being drawn on. He had worked with the reflex as methodically as a general of the army who isolates one regiment of the enemy, decimates it, isolates the next. Ever since Descartes, three hundred years earlier, the reflex had been a unit mechanism of the machine-of-the-body. What Sherrington did with his imaginative thinking, with his careful experimentation, with the techniques of one and two generations ago, was to make that mechanism quantitative. (Modern man feels he is scaling off one more layer of guilt each time he says "quantitative.") For his recordings he used the old heavy muscle-lever, essentially that Helmholtz used, later to be replaced by light fast electronic recordings.

Since the body in action could be regarded as an unrelenting hierarchy of reflexes Sherrington did add his part to the century's conviction of fatality. In the laboratory he was cool. He built that guillotine that chopped off the top of an anesthetized animal's head together with a consistently same amount of its brain, the animal thrown into that state that Sherrington labeled decerebrate rigidity, ghastly to look at, painless to the animal, which last fact Sherrington always took seriously. He vigorously supported legislation against laboratory cruelty or indifference. Most physiologists at one time or another feel an uneasiness around this point, an uneasiness toward society, an uneasiness toward themselves.

The decerebrate state was seen sometimes in human beings after an automobile crash, after an industrial accident (as in that work-man who was struck by the moving crane), also seen as a transient phase in the fading of motor controls during the closing hours of any life, but never before Sherrington satisfactorily explained, though with Hughlings Jackson it had begun to be, and other men had seen it and thought some about it. At the half-point of his career Sherring-ton wrote *The Integrative Action of the Nervous System,* a book which sixty years later still was worth reading, still appeared assured of its kind of immortality. It made the machine-of-the-body even more machine than Descartes had intended, and Sherrington had not intended that either; it screamed from his work.

Liverpool had him as a citizen for his eighteen happiest years, but probably had little notion of the distinction. Even the medical school may not have been sure that what he was at was worth being at. (It was sure subsequently.) Liverpool was one of the world's great ports, hence rats, hence cats, and whenever Sherrington needed a cat he had only to send a boy down to the docks. Crazy world, Professor Charles Scott Sherrington preying on cats, cats on rats, rats on whatever rats prey on—on cobra eggs, they say, and the cobra preys on the rats. Unusual man though he was he apparently let himself be caught in the usual feuds of higher institutions of learning. Between his department, Physiology, and the neighboring department, Anatomy, there was a door, and the two professors each kept that door bolted on his side. Funny for anyone not mixed up in the situation: two bolts, one door. A poor lecturer, he would have fits of absent-mindedness, turned away from the class, made some calculation of his own at a corner of the blackboard, turned again and continued the lecture, on another subject.

His work with the reflex completed, as he saw it, he was ready to turn his mind elsewhere in it, toward aspects of it, and to turn generally elsewhere, his fame truly international, truly de-served. Oxford offered him its Chair of Physiology. Honorary de-grees began to cover the walls. Invitations to memberships in foreign societies arrived like the first of the month.

Meanwhile he kept producing more and more non-physiological writing, wanted, needed, to be a writer, steeped himself in litera-ture, like Schrödinger sought (hardly so simply and hardly so effectively) to clarify for himself and for the reader his view of life and the world, attained a vexatious style, and his ruminating book *Man on His Nature* has exasperated many a reader in and out of science. It attempts so much, too much, but it is Sherrington, goes back-and-forth to a physiology four hundred years earlier, compares the physiology of that day to the physiology of his own, drops in vivid descriptions, but one gets so tired because a brilliant book is not really one book.

First to last he was composing poems that were praised by physiologists and by, there could be a difference of opinion about this, sentimentalists.

> I think I love old people most,
>     they seem so staunchly brave;
> their eyes say little of the ghost
>     that they perforce all have.
> And they go full of things that were,
>     and carry them about;
> and where we do not care they care,
>     yet doubt not where we doubt.

In contrast, his biography of the sixteenth-century physician, Jean Fernel, who defined physiology and pathology, had the scholarliness of the trained historian. It is flatter. It is not so dripping with color. It kept at a minimum Sherringtonian mannerism, showed him capable of gathering data from books without wringing the vitality out of the data, capable of thinking of historic relations, putting them in right order, and then, every now and then, also spreading his wings. To accomplish both sides of that is difficult. *The Endeavour of Jean Fernel.* That was the title. Sherrington's literary command was self-conscious but it was literary, and labored over.

Anyone who has intelligently and scrupulously and year after year worked with the nervous-system, let alone worked with genius as this man did, must inevitably and frequently have reflected on how mind relates to brain. Of the gross functioning of that unique flesh he said: "As to its mechanism, perhaps the point of chief import for us here is that those who are closest students of it still regard it as a mechanism." (A Sherringtonian sentence.) What was mind? What was our be-all and end-all? What was it? Where? How? In many a twisted phrase he argued that the nervous-system was everywhere the same, bottom to top, everywhere input-output, only tremendously more input at the head-end because there were the eyes, the ears, the nose, yet not anywhere in that tier-on-tier could he find mind. And he did not find it among Cajal's neurons, not in the mammoth assemblage of them at the head-end, not in any multiplication of his long-studied reflex. Someone might comment that he was unwise to look in those places, but the someone would have to remember how sustainedly this absent-minded present-minded gifted man looked, how shrewdly, how he had disciplined himself, how prepared he was by temperament to find his way through that orderly jungle the brain. What he said *must* be listened to. The brain he described as an organ of liaison that made connection with mind, was associated with it, but in what manner he disavowed understanding. To him we were two. We were brain and mind. "As for me, the little I know of the how of the one, does not, speaking personally, even begin to help me toward the how of the other." The dualism never worried him. "That our being should consist of two fundamental elements offers, I suppose, no greater inherent improbability than that it should rest on the one only." (Not an especially Sherringtonian sentence, and a clear thought.) Was mind energy? A twentieth-century scientist might be surprised that a twentieth-century scientist should ask that? Sherrington thought it unanswered, and unanswerable.

# COMPUTER

## In Metal

Having accepted the figure of speech of Mind-Is-Mountain, Mind-Is-Fuji, with stations along its base on the road between Yedo and Kyoto, pilgrims moving from station to station, stopping, looking up, trying to see, the other vulgar figure, computer-is-a-brain, becomes a matter to take or leave.

Sherrington would have rejected as nonsense any computer-is-a-brain analogy.

Two decades ago Sherrington—a young man of ninety in a

wheelchair in a nursing home, his long study of the reflex praised throughout the world—labeled the reflex an exhausted gadget. Two decades ago, Cajal's neuron and the uses to which it could be put had been available for a third of a century. By that time the computer had comfortably settled itself among us. It was everywhere, up with the astronauts, below with their earth-bound controllers. It was in medicine. In physiology. So-called special-purpose machines were part of nervous-system experimentation, and general-purpose machines were serving the computer-is-a-brain analogy. Computer models were replacing one another like automobile models in the springtime.

At the computer's earliest appearances someone implied it was a mind. "In principle . . . I do not see why it should not . . ." From mathematicians and psychologists and other scientists came the phrases. "No reason a computer should not be as original as a mind, just a matter of space enough and money enough; with space enough and money enough . . ."

Computers can undoubtedly compute. They can help a scholar. They can help a librarian find for him what is in a particular book on a particular bookshelf in a particular city. Some disillusion on the brain-mind-side—which seems some realism about what can and what cannot be accomplished—was bound to come.

Everyone agrees that a computer can receive data touching the living if the data are mathematically expressible, that the computer can handle such data, that the data can be symbols. In our society what cannot serve as a symbol can serve as a model, and a treachery of models is that they may start as an aid to thinking and end by seeming the thought. For example, the computer began as a model for a brain, became a model for a mind, then computer was mind.

The 1935–45 war, statistically called World War II, gave a push to such ideas, was in general a spree for the engineers, encouraged them inventively to kill. Three developments touched the life sciences. One, the atomic bomb. Two, the computer. Three, any device that employed feedback.

A thermostat was at the time Example #1 of feedback, as Norbert Wiener explained in his *Cybernetics*, that was responsible for the first great thrust of the mind-computer analogy. The temperature in our house rises, the thermostat receives the information, sends a report to the furnace, that turns itself down; temperature falls, furnace turns itself up.

An anti-aircraft gun was the happy-warrior example during World War II. Example #2. The gun swings toward its target, swings too far, corrects, over-corrects, corrects back, keeps that up, at the same time corrects for target-speed, wind, rotation of the earth, the corrections smaller and smaller until missile and aircraft meet at X with the desired dead soldiers. Our muscles by innumerable corrections attain the smoothness of the dance floor, and if driven by genius attain the manual skills of Michelangelo. The human ear corrects the human voice, the voice the ear, back-and-forth, till Pagliacci hurls his tragic perfectly-pitched lament.

A second engineering development of the war was the computer. Though it hobbed-and-nobbed with physiologists and psychologists it still was merely the great grandson of the adding machine. It matriculated at MIT, matriculated at Harvard, entered government service. It entered wherever there was budget, and in return gave status. Will computers escalate until there bursts over the planet a savage war of computers? A question like that is lightly raised, but earnest underneath. The clerics have come in too. Several years ago a Church-of-Scotland clergyman using his computer had a bitter quarrel with a Massachusetts Episcopal clergyman using his computer as to the authorship of the epistles of St. Paul. Was St. Paul six authors? The conclusion seemed to be that he was one. In Italy five computers commandeered by a cleric, if those press releases could be depended upon, interpreted the shades of meaning in the thirteen million words of St. Thomas Aquinas, wherefore thanks to five computers each of us might hope to arrive on schedule in the House of God.

Throughout, the computer goes on doggedly solving problems in arithmetic. These must be presented as the human mind presents: input, order, output. *Program* is the term for this. Everybody half-knows it. A possibility: those persons who do not possess the capacity to be programmed may after a time be quietly discarded from our society, the uncomputerizable simply not honored in the land.

In 1969 the computer is a station at the base of the mountain, undoubtedly.

To dub a human mind a computer is good or it is bad. Good where it encourages anyone in any field to use this tool to minimize human drudgery. Good where some special-purpose computer is expediting some special job, as exploring the activity of any living cell. Good wherever human life and the human mind have produced hills of arithmetic in places where there were only anthills and the hills needed to be leveled to the plains. Good—but now we must be careful—where the idea computer-is-brain-is-mind pushes the thinking of someone who honestly wants to perceive, even more than to understand, the ways that brain and mind may conceivably relate. Say that the someone is a researcher, capable of keeping separate fact from fiction, talented in the building of computers. He collects data. Constructs curves. Goes to a meeting. Returns to his drawing board. Builds a better computer. Compares this to the brain. It compares well but not well enough. Builds another. Compares. God Almighty might begin to be uneasy if He happened to have accomplished His job at the start by trial and error.

Bad to dub a human mind a computer when it introduces more machine thoughts into a society already bulging with them, and the computer is a suffocating machine thought. Is there first thing in the morning in the telephone. In the fingertips of the girl at the airline office when she informs you in forty-five seconds of the reservation you may or may not have between Fairbanks and Nome next January 30. Every week our lives and minds are more enmeshed. Time-sharing has come in. Dartmouth has a distinguished time-sharing system. A computer there does computer-work for four colleges and twenty-three secondary schools. There are anticipations of the entire country computerized from Washington or Arlington, a Computer Pentagon, teletype serving the computer, machine-minds getting machine-answers to machine-questions.

Should someone feel a twinge of disappointment that he may after all not be a computer as quickly as he hoped, a 176-pound suitcase-type computer, let him call up the university to find what computer-courses are available. To cheer him while he hurries his supper to get to his night-course, he will receive in his mail want-ads for the computer-trained. He will be told: "The study of the disorders of these metal brains will help the human brain meet its own disorders." And lives there the man with soul so dead he will not give his dime for etc., etc., etc. At some lunch hour, after a rapid physical examination over the telephone, a person should be able to press a button and know his illness, its outcome, his statistical future. "Why not . . . In principle . . . With space enough . . . With money enough . . ."

Should that other person, who from long back dreaded the computer that he thought he was destined to be, sense now a computer breathing down his neck, let him not despair either. A noted cyberneticist-psychologist-specialist a few years ago told us that the metal brains up to then had attained no more than the developmental stage of the head ganglion of an earthworm. Ought to cheer the earthworm too.

In New York City at 590 Madison Avenue, the then world headquarters of International Business Machines, IBM, there dwelt for four years, 1948 to 1952, a computer visited by thousands. It was second in a line of great computers. The first, Mark I, was at Harvard. That at 590 Madison appeared to be a room that one walked into and looked around. Rows of vacuum tubes. They covered walls, everything busy counting, averaging, comparing, gathering information, storing information, all this information releasing other information. Toward the back, punched tape kept moving along. In cabinets, cards were being punched, being read. Confessedly, one did not immediately think of a brain. "Of course, of course, but . . . In principle . . ." The 590 was fabulously fast, its descendants fabulously faster, the upcoming faster. "And what will he do when he gets there?" said the Chinaman watching the racer.

The 590 had judgment. When a crossroad in a thought-process had been reached and the computation must continue north or south, the machine decided which. This judgment was self-determined by the arithmetic, whether so-many billions or so-many. The 590 had sex, female, and when a group of professors from Harvard visited, up she perked, performed phenomenally, said the engineer.

Name any aspect of mind, the 590 had it. Norbert Wiener years ago had already developed a psycho-pathology, a lexicon of mind diseases, for computers. A computer could have a neurosis. Could have a psychosis. Norbert Wiener should be considered the Hippocrates of computer-physicians.

But could a computer have the capacity of the living to learn?

The engineer thought it could. On one occasion the 590 solved an old problem faster than a new, so it must have learned. Someone suggested that the machine-parts for the new might not have been used for a time, that dust might have accumulated on unused contacts. Which would make absence-of-dust learning. Presence-of-dust often has been.

René Descartes's machine-of-the-body has then in our time been pushed up into the skull. A cocktail party where machines walk about or sit on stools. Ingenious machines. They dance in the Bolshoi Ballet. Talk of Sartre and Ionesco and Beckett. Perpetuate themselves in small replicas that grow. One wears a summer tuxedo. One wears bifocals. One wears contact lenses that he takes out and loses. One has a tonsured pate. One invented the lever. One wrote the Song of Solomon. All consume vegetable, convert vegetable to machine, return the unconvertible to the earth as nourishing chemical that by the eternal recrudescence of everything converts to fresh vegetable. In Atlantic City at a national meeting each wears a white ticket so that everybody may know which machine he is. A hundred fiddle in the symphony. Art, science, equations issue from that vent each has at its head-end. They call theirs the machine age. Not entirely to agree with this, or something equivalent, marks a man as an unsocial machine. One died on a cross.

# CONDITIONING

## *It Learns*

Returning to flesh, getting away from metal and plastic, getting away also from the simple reflex that Sherrington studied, would bring us to the only less simple reflex that Pavlov studied, the conditioned.

This is known around the world. This is added to day after day right into our day.

This would be a large station, many pilgrims, more pilgrims convinced that what they see there is not only glimpses of the nature of mind, but mind. Someone says the stations are too separated, mind is not. True, but that is a difficulty we cannot avoid. We still can only do what we are doing, superimpose views, keep walking, consider this, consider that, tarry, walk, look, annoyed or tempted by the mist that forever hangs around that cone-shaped mountain.

The conditioned reflex.

A dog stands alone in a room, a trained dog, a small room. It could be that the dog is tired of these daily experiments but if ever he happened to think that, and went farther and thought, for instance, that the experiments, dog-speaking, were ridiculous, he quickly would cross himself because—is it not a human being performing them? The experiments must be sublime. Any vivisectionist who took advantage of that trust in the dog would be a low beast, wouldn't he be? Pavlov said of the dog: "This friendly and faithful representative of the animal world." Pavlov spread that feeling through his physiological laboratories; gentle places they were, and physiologically lively.

The dog stands facing a wall of the room. A dark room. No sight. No sound. No smell, so far as a man's nose knows. No stimulus. No change of energy. The intention is that for this dog the world will change only when and where the experimenter chooses to change it. For the duration of the experiment the room will be, sense-speaking, unvarying except as something pleasant or unpleasant or neutral is introduced by this meddling member of another species, the researcher. Russia saw the first such room almost three-quarters of a century ago, but it has since been duplicated in every country where science thrives, modified of course.

The following is a typical and also the original conditioned-reflex experiment.

A button is pressed. That rings a bell, or sounds a buzzer, or lights a geometric figure in the wall the dog is facing. Almost immediately another button is pressed and the dog is served an exact portion of meat-powder in a cup, the dog needing only to bend its head to eat, like breakfast in bed. The portion is small. A hungry dog stays more alert, has better intelligence, will do a better experiment; we are the same, a bit of starvation exciting our intellects. Loose straps are slung around the dog's legs to prevent it wandering, prevent it investigating, exploring, as is said in the laboratory.

With the eating of that meat-powder the first experiment ended. It will be repeated any number of times on this day and on subsequent days, each repetition will be spoken of as a reinforcement, meaning that what has happened will have been etched deeper into that dog's brain.

The experimenter throughout the experiment has let the dog be alone in the small room, the dog therefore not see him, and be aroused, and wag his tail, have his mind stimulated. That last the researcher would never have said. Nervous-system stimulated, yes.

The researcher is watching the dog via periscope or one-way glass.

How can he be sure that in the course of the repetitions of the experiment something is building up inside that dog? How can he prove it? How can he know that something is being etched deeper into the flesh of that brain? (Or, possibly, that some path that was from the beginning in that brain has been made wider, opener?) Simple. He counts the drops of saliva that drool from one of the dog's salivary glands.

Historically and philosophically interesting that so much of the whole indifferent world has heard of those drops of saliva.

In the classic experiment a minor surgical operation was performed in advance, a slit made through the dog's cheek, and the duct of the parotid gland, which is a conveniently close salivary gland, brought out through the slit, saliva from then on trickling not inside the dog's mouth but outside on its cheek. During the classic experiment a funnel was sealed over the slit to catch the drops. (Some other reflex whose end-point was not saliva could also be used.) The first time the button is pressed and the geometric figure lighted there are no drops because the figure does not yet speak meat to that dog's brain. But from then on the figure does speak meat.

The effect of the stimulus is first generalized in that dog's brain. Generalized is the inventor's term. This means that the stimulus reaches anywhere, and, possibly, from what we today know of the interconnections in the brain, everywhere, signifying nothing specific to that brain, but with the repetition the stimulus does signify something specific. It has become particularized, concentrated, means meat. It reaches one exact point in that brain, the point isolated from the rest. This was the inventor's explanation.

To emphasize another side of the matter, a more everyday-body side of it, when the stimulus is particularized, reaches the one point of brain, the dog's physiology now neither wastes saliva nor does it fail to provide it. Saliva flows for meat or the promise of meat and the amount of saliva is the right amount for the job. That brief lighting of the figure was the stimulus. It sent out notice that the meat was coming, and the brain now responds accordingly.

To emphasize still another aspect, a more obscure aspect, more difficult to establish, there is no emotion when there is no assurance of meat, but when there is assurance there is emotion. The geometric figure in its overall effect, emotional and otherwise, is like the dinner-gong that starts saliva, and on a rolling ship starts nausea.

When the experimenter on a first occasion places meat in a pup's mouth, saliva flows. This saliva does not depend on previous conditions in that pup's life, is unconditioned. It is an unconditioned stimulus and the reflex that provides the saliva is an unconditioned reflex. On the contrary, when the geometric figure is lighted and saliva flows, that saliva does depend on previous conditions, the figure is a conditioned stimulus, the reflex a conditioned reflex. The stimulus, as suggested, might have been anything else, anything that alerted the dog's brain because important to the dog's life. That brain has learned.

A second type of experiment goes still farther into the dog's thoughts, if one can say thoughts, a word that would have thrown the inventor of the reflex into a fury, a pink flush over his pink-white Russian face. In this second type the lighted figure again is presented numerous times, with meat, the dog's brain learning, then presented numerous times with no-meat. And what the brain does now is unlearn. It sends no-meat signals to its salivary glands. These secrete less and less. For this the inventor's word was *extinction*.

Curious what did in fact happen. As the repetitions with no-meat continued, the spot of brain concerned was pushed down not merely to zero but below zero. It now had negative force. It had the power to resist excitation—had the power of inhibition. And this inhibition could accomplish things. It could lessen force as a brake can. It could, as the inventor understood it, reduce something taking place in its neighborhood, could separate off and make exact some spot of excitation by digging around it a moat of inhibition, which to the inventor was how the brain separates: to take an example from our day, one blue dress from all the other blue dresses in *Harper's Bazaar*. The separation can be exceedingly fine. Thus did the eons build precision into our senses. For the dog it is not shades of blue but shades of gray. All shades except that which signifies meat are inhibited. All pitches of sound except that which signifies meat are inhibited, and a dog thus develops a hearing as sharp as Toscanini's, but not for the glory of God, for meat. All smells also, and indescribably acute is a dog's sense of smell, until he gets senile, as when we get senile. The inventor's phrase for this inhibition was *internal inhibition*. He worked with it widely. He got everybody in his laboratories to work with it. Each year the techniques were modified to take advantage of developments in other sciences. Many scientists at the same time were denying its existence, or denying the interpretation, or scorning both, saying both were nonsense, all a lie—in the polite terms scientists call one another liars.

Later, electrodes were implanted in the brain along the path of the conditioned reflex, and attempts made to follow millimeter by millimeter what occurred. This was indeed the beginning of the busy implanting of electrodes for any and many purposes in any and many parts of the nervous-systems of any and many animals, and in man. It is going on somewhere at this hour, and every day, and in many laboratories or surgeries.

Meanwhile the discoverer of the conditioned reflex often is doubted, often reinterpreted, often misinterpreted, something like hated by some who borrowed from him. The pivotal position of his reflex in experimental physiology may be forgotten or buried under respectful acknowledgment, but mostly all over the world his reflex is not forgotten and in the Soviet Union where there are no gods he has become a god. His forward-leaping imagination saw his techniques as applicable to mathematics, politics, language, literature, all aspects of mind, psychology, psychiatry. He was full of plans, but he died.

# PAVLOV

## *Not Two But One*

He was even more Russian than the Ivan Petrovich by which every-body in Leningrad knew him.

Pavlov himself not only never used the word *mind*, he forbade his students to. What could they know of an animal's mind? They had trouble enough knowing their own. Instead of mind they should say higher-nervous-activities. If they didn't they were fined, real money, kopecks.

Pavlov had his opinions. He had his convictions. He knew black from white. He had his vehemence. He had his passion for work. He had his scientific imagination. And he kept something of the child. Aristotle did too. Genius always does probably. Whatever happened to man or beast Pavlov squeezed into his conception of the condi-tioned. He spoke of it to the world—and for everything in life he spoke with his earthy statement. Even dedicated salesmen have second thoughts. Not he. During the thirty-five years that followed his first description of his reflex he used it to study hearing, vision, neurosis, sleep.

He was born four hours by train from Moscow in the out-of-the-way city of Ryazan, had a church-school education, as a young man left for St. Petersburg, continued his education in the capital. First it was general science. Then it was medicine. Then it was research, initially on the control of the heart, later on digestion, finally on those higher-nervous-activities. He was in St. Petersburg when it was renamed Leningrad, lived through the violent events associated with that turnover, went on living in Leningrad for the rest of his life. Married a schoolteacher, a warmly successful marriage, and more and more people began to know him. Soon everybody every-where in the capital knew him, knew his birthday, and the entire country and much of the world knew his deathday. That was thirty-four years ago, February, 1936, he approaching his eighty-seventh year. He had said he would live to be one hundred and fifteen, arrived at that figure by mathematics, calculated the number of years it would take to complete the experiments he was at. Notwithstanding, he gave up the ghost in February, and probably it was because his son had given up his the previous November. Pavlov had invested heavily in that son, his youngest son, one of four children, three sons and a daughter. That son was planning—actually it was planned for him—that he should carry on the work after his father's death. Pavlov would that way be getting two lives, cheating Homer's gods, but the son died, and the doctors said cancer, and the father died two months later, and the doctors said pneumonia. Men sometimes die when their principal motive is thwarted, or their arrangements thwarted. Possibly Pavlov did not die of pneumonia. Possibly they misdiagnosed his illness—should have been death-of-a-son.

Every morning and every evening he walked the same Leningrad streets, flat streets by the side of the River Neva. People saw him, were instructed that he was a great man; if he looked as if he were reflecting on something they must not disturb him. The energy in that walk could be seen at a city block, exaggerated by a limp from an imperfectly aligned fracture ten years earlier. No life-is-settled was in that walk. Youth flashed through old age. A free body. A free head. Fire in his eyes though their color pale. Love of physiology was love. Some in Leningrad said his was the only free life in the Soviet Union of Stalin's day. Pavlov had described a reflex-of-freedom. He had described a reflex-of-slavery. Dogs revealed both. Every day in those years Pavlov added to the territory of the conditioned. Every day he was looking, and every day he found or thought he found something new. There were scientists who at the same time were thinking that he distorted the only and true and everlasting meaning of reflex. Cajal would not have thought so. Sherrington not.

Pavlov had a white mustache, a white beard, white thatch for eyebrows, reminded people of George Bernard Shaw, and this may explain why Shaw said that the Russian was a mere vivisectionist who had uncovered nothing about the dog-mind that was not self-evident to anyone who owned a dog. Viewed from the moon it is absurd that each of us should insist on looking so like himself despite that he was built by a plan so monotonously like everyone else, but not absurd the distance from London to Leningrad.

Pavlov's father had been a priest of burials, a low priest therefore, his grandfather almost a mujik, a serf. The grandson stepped off from all of that, as afterward from tsarist St. Petersburg, communist Leningrad, took a world place in digestion physiology, in nervous-system physiology, first place in Russian science. Everything went fast. He won his Nobel Prize two years before Cajal, in 1904, when Nobel Prizes were hard to win because the committees still had long lists to draw from, when unmistakable achievement could be the sole criterion, when committees did not have to have meeting after meeting after meeting. Their eyes could simply sweep the earth and see the peaks. Pavlov's "Nobel," as the debonair say and as Marie Curie for instance would not have said, was for the work on digestion. Winners of Nobel Prizes are apt after the winning not to gamble with what they have won, risk only variations on the theme. Not Pavlov. He was well into his fifties. He dropped digestion, never returned to it, never gazed back over his shoulder, year after year struck at those higher-nervous-activities, struck at the nature of mind, as he would see it.

In digestion he was impressed, and impressed the world, with the orderliness. The successive juices did not flow willy-nilly from countless glands, but were precise in composition, precise in amount, precise in timing, flowed for what was there to do, and when it was time, or just before, for the smell of the steak, the sounds from the kitchen, the sight of the cook. So he came to use saliva to study those higher-nervous-activities. Since something in the mind could determine something in digestion, to wit this saliva, might that be turned around? Might drops of saliva, counted, tell what was transpiring in a mind? Could spit do that? Pavlov liked to recall how he asked himself this question for four years, by that time had convinced himself that the answer was yes, plunged. Before he was finished he had the whole world paying attention, physiologists, behaviorists, psychiatrists, psychologists. To condition became as vulgar as to repress. In America there was an avoidance conditioning, an instrumental conditioning, an operant conditioning. Classical conditioning versus operant conditioning became Type I versus Type II conditioning. It was all that box within box within box that is the way of science, and must be. Rarely there is the old explosion and expansion and out of a smaller box or even an entirely new box there comes a grand big box and everything starts over again. During the same time there was developing a behavioral approach to psychology and psychiatry and, at least it is arguable, a winding road to an existentialist philosophy. Overnight, one realizes in retrospect, he had excited international experimentation and had himself achieved global size. Also had the global self-confidence. All of that was written of course in his genes. On the Soviet Union his hand kept its grip, even tightened the grip after his death, may be tighter today than ever.

He believed that the conditioned reflex, his idea and his method, could be applied at every level of nervous activity, and had anyone doubted that mind was a level, doubted that it belonged in the physiologist's domain, Pavlov would have blasted him with blunt Asiatic scorn. He did not deny the subjective but fused it with the objective, psychology with physiology. The two were one. "What could have happened to Sherrington?" he burst out angrily concerning the Englishman; an intellect like Sherrington's accept dualism? As for himself, it pleased him to have done his part to place mind where it was, *in* the flesh. Indeed it was flesh. It reminds one of Einstein saying that a phenomenon, a mind, was in time and space, or it was nothing. For Pavlov it was flesh. His was a satisfied monism. What right had we to impute to dog or man anything that could not be measured with a yardstick? Dog-mind and dog-brain were the same and man was an evolutionarily later dog. No worker in the Leningrad laboratories would have dared let slip a phrase suggesting the psychic—the dog suffers, the dog is bringing back a memory, the dog is having a dream—on that February day when Ivan Petrovich died, when eighty trained workers came at 9 A.M., left at 5 P.M., as punctually as workers in a factory. This was 1936. In 1969 Pavlov still hovers over those laboratories, his ghost, that also goes dancing around the world, which its body never was pleased to do.

A poet's fervor was in him. He was sparing with it. In his research he as ruthlessly ruled out the emotional as he did the reporting of it. This was not like Sherrington. This scientist had no secret ambition to be a writer, put off even the writing of a scientific article, was always late for the publishers. He was late nowhere else.

Despite his priest-of-burials father, Ivan Petrovich admitted nothing left over for the other side of death, as who would expect him to; never anything of mind floating up to the skies. The idea of a soul was absurd. (That idea early in the century was by no means as widely absurd as now.) He laughed at it. When he fell ill and cold-in-the-head became pneumonia he called a neurologist, wanted to discuss with an expert his own deteriorating mental processes until these would no longer discuss. Somewhere in the course of the discussion he commented that mortification was setting in, his brain was mortifying, his mind shortly would vanish. So it did of course. He seemed like those flatlands that run off from the River Neva where he took his walks, the horizons near, the distances measurable, but the Neva on a wild wintry evening. Sturdy, direct, stubborn as a peasant, sure of himself, sure of science, he anticipated a day when mind and body would be expressed in a single equation. That gave him satisfaction. Like the physicist he would like a mathematical equation to take care of everything, body, mind, planet. One equation. His talent was so powerful that it was bound to reach into the future, was bound even to affect the cold war, be part of the missile. Confronted with that last credit, it would have disgusted him. He expected others to speak as literally as he, had neither capacity nor patience for the indirect. To make his physiology a plinth under Soviet ideology, as has been done, was as dishonest as it may have been politically useful. The two were uncompromisingly materialistic but otherwise merely happened to grow up together, in time and space. Pavlov never thought of equating them. That he helped Russian science and Russians to believe in Russia is fact, and good for Russia.

Once he visited America, gave lectures in his rough tongue, a translator translating. The translator was his eldest son, who interested him less, the Professor of Physics at the University of Leningrad. Many Americans remember that. While his train waited in a railroad station someone stole his wallet. He was glad when that visit came to an end, relieved to see the shores of America recede, which is the way any of us may feel when we leave the other country; but there was the wallet.

# ELECTRO-ENCEPHALOGRAM

## *Spirit-Writing*

As long as human brains have been around on the planet brain-waves have been coming off them, and until recently no human mind gave any heed. That most sensitive instrument in the world has been ceaselessly speaking in a language of electrical fluctuations and nobody listening. Now brain-waves are written, from normal brains, abnormal brains, help to tell where and often what the abnormality. Particularly from the skull of epileptics brain-waves can make a diagnosis. How does the brain code? Many experts believe that these electrical rhythms give us our best chance of creeping in upon the continuing obscure relations of brain to mind.

So, the fact of brain-waves provides a station from which to view the mountain, worth tarrying there.

Up-and-down, up-and-down go the waves of this sea of the brain, our mind sitting on the shore, and wondering.

In 1875 Richard Caton spoke of the "feeble currents of the brain," but not until 1929 came the dramatic news that electrical waves had been recorded from the outside of a human skull. Hans Berger had done it and he gave what for the time was an extraordinary summary. Two principal electrical rhythms registered, one when the mind was attentive, the other when inattentive. There were other rhythms. They changed with changing states of consciousness.

Berger's studies did not arouse the excitement one now thinks they should have. Whether or not these electrical evidences were incidental, or were indeed a way of creeping in upon the essence of mind could not be said by Berger and cannot be said today.

Berger was a psychiatrist in Jena, Germany, had crude equipment but nevertheless made this startling discovery, that if electrodes were properly placed on the human scalp brain-waves would be written. They had a rate that varied, so-and-so-many per second, and a voltage that varied. Berger worked in secrecy for five years. His electrodes were metal plates fastened one on the forehead and the other on the back of the head. Subsequently he stuck needle-electrodes into the scalp. He found that if his subject's eyes were closed, his mind relaxed, nothing wrong with his brain, the average rate was ten per second. Berger called this the alpha rhythm. It was best obtained from the back of the skull over the vision areas. When at last he did publish he was doubted. The most noted doubter was an Englishman (still vigorous and working), had the best equipment anywhere, and soon had not only verified Berger's discovery but wanted the rhythm to be called Berger rhythm. It seldom is. Men's names do not attach themselves to discoveries as they used to.

Another rhythm is faster than the alpha, eighteen to thirty. Another is slower, four to seven. The rate from the cerebellum may be three hundred per second. However, this simple kind of rate-amplitude analysis could not be satisfactory for long in our computer age. Furthermore, the brain appears not to operate in that simple way. The neuro-physiologist in late years seeks with his computer to relate the background noise, from the depths of the brain, to the conventional brain-waves.

Berger named the recording instrument the electro-encephalograph, and the record, the electro-encephalogram. This later was alphabetized in our clinics and laboratories as EEG.

Radio tubes amplify by ten million the feeble currents of Caton, and a wave-writer, an oscillograph, faithfully follows the instant-to-instant rise and fall. These can instead be displayed on the face of a cathode tube, and photographed, but usually are written on moving paper by a magnet-driven ink-pen. By convention, the moving paper is ruled, voltages and rates read off directly, as are the reports from the heart.

Sixteen pens reporting the brain at the same time became common, even thirty-two. From sixteen scattered points the pens were bringing together one story, one view.

Even Berger's ordinary alpha rhythm remains largely a mystery.

Each brain-cell generates its small electricity, this can be made to discharge, therefore it would be natural to conclude that some summing of the electricities of groups of such cells might account for the brain-waves. However, the impulse over the frog nerve is many times one up-and-down alpha, a tremendous difference, so brain-waves could not be produced in that direct way. A wave as understood today might relate to a combination of factors: to the membranes of the nerve-cells, to their so-called impedance, to the chemistry inside and outside them with resulting hyperpolarization and depolarization, to some electrical spread between cells, to local circuits that might reverberate among cells, to some intermittent bombardment of a huge population of cells. Great numbers of cells would need to be drawn into unison, and that is why pacemakers were proposed, but they were not found.

Sheets of dendrites of neurons are directed toward the curved surface of the hemispheres. Myriads of those microscopic threads are close together and parallel and all directed outward. That represents an enormous membrane. When the axon of a nerve discharges, is depolarized, the depolarization invades those sheets of dendrites, is followed by a slow hyperpolarization, so the subsequent depolarization could not occur quickly. This alternation may pace the waves of the electro-encephalogram.

When a man or his dog is idle, is not paying attention, is in a state possibly for loosely remembering, for floatingly thinking, for freely hanging on some brink of creating, the ten-per-second, the alpha, is apt to be dominant over a large part of the skull, but particularly over the back. The ten-per-second is slow. One can keep up with that fluctuation on the oscilloscope face.

The ten-per-second is far-reaching among animals. Many besides man have it. Whatever one day may be the final accounting, brains for some reason have required the ten-per-second, the alpha, have required that tone, if that is what it is, behind all their hell-heaven-earth activity. A water beetle has a vision ganglion useful for experimentation, and when this beetle's eye is illuminated, all of its light-sensitive cells evenly stimulated, after they have discharged once, one-tenth of a second must elapse before they discharge again. It takes that long for the rebuilding of that particular chemistry, and something similar probably explains the alpha.

One student surgically undercut a thin slab of the surface of a cat's brain without disturbing its circulation, dug a moat around that slab but left an isthmus connecting it to the mainland of surrounding cortex. The electro-encephalograph proved that that island and the mainland were beating in unison. They were connected. If now the isthmus was cut through and only the island remained they no longer beat in unison. If however the slab had been thick instead of thin, and the isthmus cut through, that island would continue to beat. This suggests that something beneath the surface is feeding upward, that in your brain and mine the ten-per-second may be paced from beneath, from the thalamus possibly. Further experimentation made this doubtful. Another student thought that the thalamus excited the cortex, which then excited the thalamus, a back-and-forth. Further experimentation made this doubtful too.

In the last years the computer is increasingly brought into the study of brain-waves, to compute averages, the brain operating on averages, operating probabilistically, it being more and more evident that with as many units as the brain has it could not possibly operate in any other way. For this there are special-purpose computers. The operator is able on the oscillograph-face to watch the constantly-changing reports from the constantly-changing brain, can stop the instrument at any moment and dwell on that moment. We are steadily if slowly knowing more what the brain is. Via the computer it is revealing an electrical flexibility that we know it evidently must have, its present always having to meet its past.

What haunts us and what mainly interests us is the simple fact of the beat, the ebb and flow, the relation to its background noise, our brain being electrically an exceedingly noisy place.

The electro-encephalogram has long been recognized as varying with our thinking, our feeling, anesthesia, action, age, open eyes, closed eyes; an encephalogram is apparently unchanging if circumstances are unchanging. It speaks. We do not know well from where it speaks, or what it speaks, but it speaks. Into a rhythm of our brain the affairs of our lives throw another rhythm. Those affairs may be a story. May be a sight. A soft memory. A sharp hate. An impersonal hitting the mind through an ear.

A man's character has been thought readable in his electro-encephalogram, has a pattern recognizable if the writing is examined carefully enough, a pattern that consistently stays with him Monday through Sunday.

Even before birth, activity can be picked up from the forward part of the fetal skull. Four months after birth definite activity begins at the back of the skull, may have been there earlier but masked. The rate is slow. Gradually the rate increases. By the time the child is thirteen the characteristic adult rhythm is settled and remains so until illness or old age wears it down and death wipes it out. Hans Berger took his original recordings from the brain of his young son. Berger believed us to be as individual in our brain-waves as in our fingerprints. If so, it would make us ask once more, and not expect an answer: what does all the individualizing mean, the high cheekbones, the winking nose, the straight gait, the blood literal, the blood figurative, and these computerized electro-encephalograms? Identical twins have similar electro-encephalograms. Is it all just an incidental mathematics? Is it all psychologically irrelevant? It surely is not philosophically irrelevant. But is it philosophically discouraging? Is it just replicating genes? Is it just creature? In the lowliest that have a nervous-system some dominant rhythm pounds in some part of that system.

A healthy medical student, eyes closed, lies on a couch with electrodes pasted to his scalp. They look like curlers for the night. Wires lead from them to the electro-encephalograph, where yards of ink-written paper roll off. The alpha repeats. The student opens his eyes and daylight strikes his retinas, impulses travel through his brain, and even an inexperienced observer sees that the writing has changed. Is it faster? It is irregular. The student closes his eyes; the alpha returns. He starts talking with eyes closed; the alpha goes. The observer is half amused that not only has the ten-per-second gone from this brain but some equivalent change has occurred in the brains of all those watching. Our brain is more sensitive than the surface of water. So are these fluctuating electrical signs of our minds. The student stops talking. His alpha returns. He is asked not to keep his eyes closed but to think hard of a dachshund. He does. The alpha goes. He is seeing in his mind that geometric design, dachshund. A lapse. The alpha returns. He opens his eyes. He is asked to look at a uniformly lighted surface that flickers. The brain-waves follow the flicker, up to twenty-five per second, the brain appearing a stupid instrument, to follow a flicker. The student's eyes now are uniformly illuminated, and there is the alpha, which stays when he closes his eyes. He is relaxed now. Begins to doze. Keeps dozing. Possibly a classmate slipped a sleeping pill into his snack. His eyes have been closed for fifteen minutes but not until now could it be said confidently that he has left the waking world—high voltage and slow rate in the brain-waves, every now and then a spurt of fast waves called spindles. He snores—large high-voltage waves that look random. The watchers nudge each other. If the student claims afterward that he heard everything that was said, he lies. He opens his eyes. The alpha goes. He closes his eyes. It returns. Someone hints at an escapade last Saturday night. Peremptorily, the alpha goes. Everybody laughs and the record skips and hops. The student will soon be an M.D., so he recovers his poise but not his alpha. Plainly it is appropriate to call the alpha the rhythm of inattention, and about last Saturday night this student is helplessly attentive, so has lost his alpha.

Has any of this anything to do with what interests us more than anything else during our long-short sojourn under the moon? Were any of the hints indicated in electro-encephalograms coming from patterns of cells and sheets of cells and lumps of cells at the surface or at the depths of the brain adding anything? Does any of this help us to understand the mind, that rests like a filmier film on the film of life that hugs the planet?

# EPILEPSY

## *Divinity Hints at It*

Every day in hospitals and clinics and even in basic-science laboratories there may be talk between electro-encephalographers and epileptologists. (The terms make one shiver.) Experts consult each other. It is their problem to understand the disturbance in body, brain, mind of persons suffering seizures, fits, those unwelcome visitations that have been adding to human knowledge from far back in history, medical history, just history. Nature has been conducting these experiments in streets and houses, and no doubt before there were in our sense streets and houses. The Gospel according to Mark described the fits. Hippocrates objected to their being called divine.

The first epileptic attack one ever witnessed one did not forget. A playmate in school. An old man living up the street—one had previously thought him merely disreputable. An over-proper woman who sat straight-up in the bus and the next minute was squirming on the floor, and the bus rolled on. (A distinguished neurologist related a case where sun flickered through trees into a bus and started a first seizure.) One had been told about fits but being told was not enough, the woman shook so, her face was so twisted, and what must be going on in her mind one could not dismiss from one's own mind. Maybe nothing was. But one decided that her shame at the spectacle she was creating would have made anyone hate the whole world. How miserable everything suddenly seemed. Also how ghoulish. The bottomless appeared to open up. It all occurred quickly, but it felt slow: a frightful caricature had been there, faded, and the woman was back again. She had lost consciousness. Though she did afterward sit straight-up, it was in apathy. Everybody in the bus knew. She knew. Her friends knew. They all always had known.

Epilepsy has put footnotes into the biographies of some of the greatest, Napoleon, Baudelaire, Mozart, Cellini, Caesar. They each had experienced this one of the more dramatic unpredictablenesses of human body and mind. If it was a facet of mind it was a blemished facet, an ugliness.

Drugs might have helped that woman. Drugs have made epileptic attacks rarer in schoolyard, polling booth, town meeting. Most seizures are controlled by drugs, but the sufferer is of course drugged, more, less. Physicians the world over have learned about the drugs and how to administer them. Physicians have explained the disease as far as they could, and could quite far. They have added to our compassion and to our realism, have lessened the lurid, lessened the mysterious. It would be wrong to draw the conclusion that because the mysterious had been removed from fits, it had been removed from the universe. The universe keeps its frame of mystery, but we do with urbanity accept what in Caesar's Rome would not have been possible to accept, that the epileptic has a flaw in his physiology, in his chemistry, his anatomy, damage directly or indirectly in the flesh of his brain.

A fit may start always in the same part of the brain, a part in some respect different, the person born with the difference, or got it after birth, a visible difference sometimes, a molecule difference some-

times, a still utterly unaccountable difference sometimes. It may all
be on the outer surface of one hemisphere or it may be deep. Electrical energy piles up at a point, and when enough has, there is the
discharge, and with it the agitation of body or mind, or of both.
That is how the attack has been explained. The cause might also
have been some control cut off, so that that which was controlled
was released, hence the wildness of the muscles. That rising electrical energy has been likened to a gathering storm, abruptly the
thunder and the lightning, and the body shaking and the mind
getting queer.

If such a brain comes to surgery there may be a scar from an old
head injury, may be a benign or a malignant tumor, a clot of blood,
a wasting. In these cases the epilepsy is spoken of as symptomatic,
the person showing the symptoms of brain irritation. Probably all
epilepsy should be spoken of as symptomatic, and the causes of the
irritation not necessarily so gross, merely a rearrangement of molecules, a metabolism altered. It might, also, somehow be a mind's
effect on molecules. A mind that is molecules, if anyone insists,
touches a brain that is molecules. One can be as playful with one's
ideas as one likes. The troubled mind usually is not playful. If no
flesh evidence can be found but the probable neighborhood of the
source of the symptoms is surgically cut out, the attacks may cease.
When this happens it must be believed that the tissue had something hidden, something secret (cryptogenic), some abnormal pattern of nerve-fibers, lack of enzyme, focus of trouble. When the
focus is definite, is found at surgery or post-mortem or otherwise can
be diagnostically pinned down, there is still another designation,
focal epilepsy. Designations have come and gone.

The electrical storm may be small, brew a time, a rumble, occur
most infrequently, one attack a year. The storm may be large. A
hundred attacks may pass over body and mind in a single day, the
misery so unrelenting as to make another moonlight, another dawn,
another glorious sunset not worth waiting for. Sometimes the
gathering storm is not quite adequate to induce an attack. Sleep
may so have depressed the high region of brain, the cortex, or disconnected it from the low so that the high does not get through to
the low, so that the body does not display the signs, and the man
never knows. Had he been awake and his nervous-system not de-

pressed, the same storm might have pushed through and there would have been the convulsive movements in face, trunk, arms, legs. (Actually, attacks are commonest in sleep.) A man may have been asleep, but his elbow pushed into a cramped pain-stimulating position, or a flea bit him, agitated his cortex, reconnected what had been disconnected, and, bang, a full-blown fit.

In all of this the inexorableness of brain-mind is upon us. The inexorableness of light-years is upon us too, but we have not the incisive imagination to maintain the thought of that, and it is not as personal, does not as harshly shine through our private Venetian blinds or our New England shutters.

A shaking thumb may have been the first evidence of an attack, the shaking advance from thumb to hand, that shakes, to arm and shoulder, that shake, to the whole body. That focus was in the highest area for the thumb, the attack triggered from there, marched from there. (*Marched* is the verb used for the advance in the brain of the sweep that began in the focus.) The attack may have begun and ended in the focus. The thumb shook, no march, and the mind watched the thumb as it shook, could do nothing about it, might have been something like fascinated by it, a dull fascination. Custom dulls our agonies. An attack may begin at the angle of the mouth. In an eyeball. In a toe. Anything sudden may be the provoking stimulus, a door opened, slammed shut. Through the attack one may make out the lit-up architecture of the nervous-system, some area of it. One sees the anatomy. One sees how mind-tinctured our mind has made our brain, sees what the physiology, if it can for the moment be considered physiology, has done for anatomy.

It is another view of Fuji, that peak, from another stop on that road.

Epilepsy has in this manner often cast streaks of light into the architecture. A lamp is carried through the temple by night and out of the shadows leaps what one knows is there but never sees this arrestingly. Even a casual observer may think he is learning in furtive flashes something additional and important about the living. Even the specialist may pick up qualities that no drawing, no dissection, no photograph, no word could give him. The word would be best. Writers on epilepsy have sometimes had strong literary talent, and the epileptic has sometimes been excellent literary subject, has beckoned some of the shrewdest pens in medicine, Hippocrates and Hughlings Jackson, to mention two 2,000 years apart. But a quite ordinary physician might listen with helpless interest to his epileptic patient, keep on the heels of his story, the anatomic story also, this possibly interrupted by the clear-cut statement of the misery of this person's daily life, which might have meant most of all to the physician, and been most important for the patient to have revealed.

Instead of that shaking thumb there could have been a delicate trembling of one side of the face, the stricken put his hand there as if to wipe away the trembling and appear to do it. There could have been a mechanic turning of the head to the right, the eyes straining that direction, the turning and straining continuing as far as the joints allowed, then the arm on that side lifting, as a machine-arm might, or the arm of the cadaver be lifted passively, and out at the end of the arm the fingers rhythmically contracted and relaxed, then were quiet, then the arm dropped, then the head returned where it should be, then the eyes. The stark sequential character of such a motor complex alongside of the always pliant character of mind as we know it—the two underscore each other. One begins to ask old questions, but the epileptic's face, that seemed the front of a vacated house, seems occupied, and one lets the questions lie.

An event outside the person may trigger the attack. A street scene. The smell of a barber-shop. Music. Musicogenic epilepsy is a kind. In a well-known case it was the music of one composer, Tschaikovsky. The mind of the dead composer reached into the mind of the living citizen, mind to mind, or, for whomever prefers the other emphasis, brain to brain, flesh to flesh, touched a spot, started an electric turmoil, this probably increasing, the body then thrown into a convulsion that frightened the small boy who was watching, did not frighten the citizen himself, who had got used to this rough handling of his body, no longer rebelled, despite that it disfigured him so. To the small boy this might have been his introduction to the unnatural, made Grimms' fairytales more grim, more real. Recently in a widely publicized case of musicogenic epilepsy rock 'n' roll and jazz had no effect but soft music brought on an attack, and a tedious long-term treatment with the techniques of Pavlov's conditioning cured the patient, the extinction one after another of the complex of conditioned reflexes that were excited by the soft music.

The autonomic nervous-system apparently might from some one of its parts trigger an attack, excite some motor stir in the digestive system, or a burst of fever, a bout of sweating, a dizziness. Emotion might whip up any type of attack that was latent. Emotion, some believed, produced all attacks that were not explained by gross brain damage.

The somatic nervous-system apparently, too, might from some one of its parts trigger an attack. A color might appear in nothingness. It would be a hallucination of color. Beginning to end the sense-experience would be all inside a brain and mind. There would be nothing there where the epileptic's eyes were directed. Instead of a color it might have been a colored object, a Georgia peach, no peach there of course. From the vision region of the brain, which is at the back, the electrical activity might have moved toward the hearing region, which is at the side, with a hallucination of sound, the electrical activity continuing to move, burning out finally in a hallucination of smell, at the front. Subsequent attacks would have followed the same order, vision, sound, smell. Two Americans in a study pursued over years showed epilepsy originating in one part of the brain during infancy, invading another during childhood, another during adolescence, and there burning out. It had migrated, left one place, left another, finally vacating the house altogether. Pain might accompany an attack. This was rare.

An attack might include furor, the breaking of furniture, the leaping through a glass window, the intent of that mad mind not known to anyone, but the whereabouts of the disturbance in the brain pinned down by the expert with justified confidence. The violence might have been such as a murderer would act on the stage, but this was no theatre, the murder was real. A thirteen-year-old boy shot his mother, reloaded his gun, shot her several more times, and though he did not forget having shot her, had no amnesia for that, he did have for all that led up to it. Later under questioning it emerged that his mother for years had been an annoyance to him, this an aside in the report. It deserved amplification. Wherever there has been, as one says, a life situation, it deserves inquiry by the physician; even when the electro-encephalogram shows a definite peculiarity in the brain, as in this boy's so-called fourteen-and-six per second.

Epilepsies for decades have been considered three kinds. There were other dividings. Some believe that dividings have always been justified and are inevitable, others believe that when the smoke has blown away there will be no dividing and each epileptic will be a single patient and a single person with a single mind. The three classic kinds are *grand mal, petit mal, psychomotor seizure.* In grand mal there is a stiffening of the entire body, an intense tone that spreads over it, twists the head, this phase called tonic, and grafted upon that is the violent rhythmic shaking, this phase called clonic. A tonic-clonic convulsion would be grand mal. A frothing at the mouth might go with the attack, an epileptic cry, a biting of the tongue, a blueing of the skin, a helpless urinating, and to conclude this drama that Egypt and Greece and the Bible and the Middle Ages knew, stupor and sleep. The episode might come and go in seconds. It probably is ushered in by a warning, called aura, and the aura probably mind-tinctured, a wisp of sensation, a smell, a bad smell frequently, voices, a face, a crowd, an action, a strong memory, a story.

In petit mal there is no convulsion, though convulsion may occasionally be added on and change the character of the epilepsy. The person briefly loses consciousness. That may be the whole of it. The mind goes off. Petit mal has been called an absence. The French have called it that. A blank gaze, a blinking, a nodding of the head, then in a moment the eyes again see the actual world around and the young girl is without any recollection of what has occurred. If her head did nod, it lifts. The eyelids that twitched are quiet. The work that she halted she takes up again. Petit mal is an illness of the young, commonest between the sixth and twelfth years. The mind usually is not damaged. Furthermore, month after month this illness may reach into this person's life, and he never know it, never notice this fleeting trickery, and, more astonishing, his family never notice. (Eddies in the stream of mind, whirls, may flow right past us and we pay no attention.) As to the span of the attack, that can have been five or six seconds. Dozens of attacks can have occurred during a single day.

In psychomotor seizure the person performs some unusual act.

There is some abnormal behavior. It is a trancelike state that may come and go in seconds. A bout of chewing. A bout of disrobing. But the seizure also may be more prolonged, suggest less of mind, less of body, or more of mind, more of body. A New Yorker finds himself in San Francisco, does not recognize the name on the driver's license in his own pocket, and the newspaper reports that an amnesic was picked up last night at the docks. (This degree of drama would be rare of course.) The New Yorker's mind and the San Franciscan's mind were the same, presumably, but what separate compartments there sometimes can seem to be in a mind! Here each of the two compartments was unknown to the other, though both were associated with one brain and one body. The mind frequently is damaged, gross deteriorations. It is the commonest of focal epilepsies. And the focus, the scene of the brain sickness, is the temporal lobe or its neighborhoods, the hippocampus, the diencephalon.

Helping to characterize the three classic kinds are their three electro-encephalograms. Berger already knew that abnormal electro-encephalograms were produced by epileptic heads. (Berger was of course the first to record any electro-encephalogram from any human head.) The three epilepsies and the three electro-encephalograms often correlate so closely that an expert may appear to believe that the record rolling off his instrument is itself shooting off these Roman candles. He could for a moment forget what he perfectly knows, that no electro-encephalogram is yet fully explained, that it is only some kind of sign of brain-mind, may not at all fundamentally explain. Brain-waves do however let us look (not see but look) into the pit of the brain while it is perpetrating these deeds.

Nevertheless, many researchers continue to think that the electrical approach to the mind is the best that science has.

Each of the three epilepsies, and some less categoric but related disturbances of brain-mind do have strikingly individual electric characteristics. The pens may write a sharp spike. The pens may write slow waves that quickly wax larger and faster, get irregular, get unpleasantly regular, get unnaturally slow, because the man who was in tonic-clonic convulsion now is in stupor, finally write sleep waves, because he is asleep. The pens may write a sharp spike followed by a rounded dome, a monotony of that, spike-and-dome, spike-and-dome, three per second, but even in this repetitious electro-encephalogram of petit mal there may somewhere be individuality. The focus is in the depths of the brain. It has been thought by some that the thalamus and cortex here are playing a game of echo, a burst of electricity in the thalamus following a burst of electricity in the cortex, each writing its part of the record. If one part fails or is too feeble, the push for the other is lacking, the attack blocked.

An epilepsy draws some of its electrical character from the part of the brain that excites it. Attacks may start apparently from anywhere in the brain, mind appear close, but the relation between brain and mind as always elusive, and the mind as always slipping away from any but literary or philosophical definition. From the scalp of a person who never has had an attack it may be possible to tell by the electrical signs whether he belongs to the part of population that might expect to have an attack. The epilepsy is latent. But, also, there may be epilepsy and no evidence in an abnormal electro-encephalogram. Sleep may bring out an abnormality. The sleep may be induced with a sleeping pill.

Hippocrates two thousand years ago did object to epilepsy being called divine, saw no divinity in it, took it for granted that a damaged brain was responsible, taught that any convulsion must be regarded as serious until proven otherwise. He was a careful doctor. He believed that epilepsy like other sicknesses had an exact cause, which must be sought, and a treatment found. The seeking continues. The other seeking is less common but somewhere continues also, to discover in these sick minds something concerning the nature of all mind. Epilepsy, the word, means attack from without, but our literal age has never been in danger of imagining spirit or god or demon to steal from outside into a skull, as the age of Homer might have, or the Middle Ages. Our age seems sometimes to disport as cocksure in the thin film of life roofing the planet as a fly unimpressed by the wonder of its walking upside-down across the parlor ceiling. The fly notwithstanding is fragile and the swatter is lifted and the wonder is fact. Even the most down-to-earth modern may sometimes think it would be easier to get a grip on the fantastic in his own mind if he were permitted to imagine spirit or god or demon to steal into these epileptic minds, he therefore less apologetic for his feelings, partly affrighted feelings, when next he enters a hospital where there are housed together, say, four thousand epileptics.

# BRAIN STIMULATION

## Summoned

The surgeon stimulates a human brain. When he does, if he is reflective, he may consider that he simply has taken us farther along a line that began with those first stimulations last century of the dog brain by Fritsch and Hitzig, and that stimulation of the human brain of Mary Rafferty by Dr. Bartholow, in Cincinnati, Ohio, which seemingly sent her to her Irish saints ahead of schedule.

The boldness of brain surgery when it has added to it the boldness of calling forth with electricity occasional flashes of mind, as can be accomplished notably in the speech-associated regions of some disturbed brains, must stand among the medical successes of the century, eerie success, or, depending on how one looks at it, not eerie at all. Dog, cat, monkey, chimpanzee, dolphin, and the human being, the brains of all have been stimulated by the hand of the researcher or the surgeon.

If it is the brain of the human being, and if the reason is epilepsy, the prerequisite steps for the surgery are those for any brain surgery: the patient's story, the X rays of his skull, X rays of the arteries of the neck and brain, electro-encephalograms, echograms, anything else, all the tests. There follows the diagnosis, the decision to operate, the consent of the patient, the informing of the family, operating room and anesthesiologist scheduled, surgeon's and associates' hands scrubbed, and scrubbed, in order to dig from the deeper layers of the skin as many as possible of the bacteria that always dwell snugly there, then the powdered rubber gloves, the green gowns, and the play once more is on.

Augering and sawing through the plates of the skull is routine, as is the exposing of the brain. Which side, right or left, which lobe or lobes, would have been established beforehand, usually correctly and to everyone's satisfaction, yet the detailed localization, that may need to be most precise for this brink-of-disaster surgery, might be helped to be if words could be made to come out of that mouth when that brain is stimulated electrically. The anesthetic is a local.

As for the substance, the flesh, even to the experienced eye that may appear quite normal, also under the microscope, the abnormality something not visible, something chemical. The surgeon's knowledge loads on him the worry that the smallest bit of brain can have life-and-death importance. It is in fact paramount that this brain be stimulated directly, the epileptic focus located with the greatest possible precision. The voltage that he now uses, the shape of the electrical pulse, rate of stimulation, are the harvest of years of trial and error.

He stimulates. Point. Point. Point. The idea in what he is doing, the sight of him doing it, if that had not struck amazement into those two Germans and into Dr. Bartholow, then the total *mise en scène* surely would have, the drama, added to by the ultra-violet light that for long operations is expected to keep the circum-ambient as sterile as can be. His stimulating is a nudging by the vulgar outside of this most intimate inside. Nudge at one spot—a single muscle off somewhere contracts. Nudge at another—the patient speaks out, says that something tingled. At another—clear-cut mind. It shines out. Anyone there at the operating table recognizes it. He would probably have known beforehand what was going to happen, still it is unexpected.

For an instant mind shone out. For an instant one thought one saw the slopes of Fuji. It was not something inborn in this patient. It was something he experienced. It was this-life. He acquired it. The scrappiest scrap it was, but it was his, and it was mind, and it was forced out mechanically; an electrode did it.

Scars in the brains of soldiers struck by shrapnel in World War I became epileptic foci and were cut out in numbers by a pioneering German surgeon. Scars in the brains of soldiers of World War II, and in the brains of civilians abnormal from many causes, birth anomalies frequently, were cut out in even greater numbers by a Canadian surgeon. Both those surgeons were in their time known around the world. The Canadian's patients, since they were being operated on under local anesthesia, could hear, could answer questions. A patient might be befuddled, dizzy, and suddenly he would do or say what he had no intention of doing or saying—it was that electrode. In the twelfth century the population might have hanged the surgeon instead of letting him shortly step from the operating room paring his fingernails while the bandaged patient was being wheeled away.

Part of the Canadian's staff is stationed in a gallery behind glass. A microphone lets conversation go back-and-forth between him and them. Each point of brain stimulated is labeled with a ticket dropped onto the pia, that web of membrane that lies over the substance and has the nourishing small blood-vessels. The tickets are numbered #1, #2, #3. A mirror fitted diagonally over the open skull allows the operative field to be photographed through a window in one of the walls. The photographs could make one imagine one was looking at a planted garden, the tickets marking the plants, geranium, jonquil, marigold. Meanwhile on sterile paper with sterile pencil the surgeon every now and then is jotting a note, afterward will be left with two kinds of on-the-spot evidence.

A small girl's epilepsy, cured by surgery, is talked of many times and has been over years in interested circles. While an infant she was given an anesthetic and for some unknown reason, and not recognized at the time, had a small hemorrhage under the dura of her brain, suffered a passing paralysis. Later, when she was seven years old, she was walking in tall grass of a meadow, her brothers in front of her, and a man came from behind and said: "How would you like to get into this bag with the snakes?" Terrified, she started running. Her brothers saw the man and her mother remembered every detail of the ugly experience. Later still, the girl began to have nightmares, and in them relived the experience. At eleven she had an epileptic seizure and in it also relived that experience. More seizures. More reliving of the experience. The experience was apparently her dominant memory. Her mind kept returning to it. And in that manner it was borne into her brain. At fourteen they brought her to the Canadian. He studied her case, decided there was hope, exposed her right temporal lobe, found an old scar, stimulated at many points in the neighborhood of the scar. No pain. The brain itself is not pain-sensitive and the tissues around had been anesthetized. He stimulated, stimulated. Abruptly the girl cried out. "Oh, I can see something coming at me!" Realize that her skull was open, her brain exposed, she talking, and feeling that she was about to have a seizure, knew the aura, as every epileptic knows. Her mind had gone off to that long past experience, but she was also present in the operating room, and knew she was. "She seemed to be thinking with two minds," wrote the surgeon. She was in a way two persons. The one was continuing to relive that experience in the tall grass that somehow had latched itself to the flesh damaged at the time of the anesthesia in infancy, became more latched with each nightmare and seizure. Our brain probably does the same with any repeated memory, latches it deeper into the substance (not surprising that there is so much we cannot forget), only in the small girl there was the damaged flesh to somehow help, to somehow sensitize. The touch of the electrode was in other respects like a penny dropped into a picture machine. The picture began to reel off. Other pictures must have been nearby or right there, many pasts channeled through the same spot, the same neuron pools, and one wondered why there were no communication jams, then remembered again how everlastingly clumsy the finest electrode, and concluded it had most to do with this small girl's brain having summoned this particular ghost so often that it had the strength to push other ghosts out of the way. The surgeon believed his electrode built up a local electrical storm similar to that probably built up for any usual epileptic attack. Some electrical stimulations evoked merely a feeling. Some evoked a sense of familiarity—an old haunt revisited. So on. Some gave the impression of a record-player in which the record began where the needle was put down, and stopped when the needle was lifted off. How wrong that must be if we understood. But a voice did undeniably speak out. Surely the surgeon must often have been tempted to follow down in, dig down where that voice came from.

## TEMPORAL LOBE

### *Peers from Beneath Sideburns*

This part of the brain lies inward of the temple, lies inward of that flatness, on the side of the skull, hollowed somewhat, attractive in a gentleman if he has a distinguished cranium in a narrow tall head.

The Canadian surgeon won his most dramatic successes in the temporal lobe and its vicinity. He was seeking to cure epileptics, and those seizures that caused him and would cause anyone to reflect on the nature and the meaning of mind were related to disease in this vicinity. The Canadian called them psychical seizures. Commonly they have been called psychomotor seizures.

Hughlings Jackson earliest spoke of the dreamy state that these persons often experienced, and all who since have come into the field speak of the dreamy state, because it is characteristic, because it is mind-peculiar.

The small girl frightened by the man who offered her a place in the bag with the snakes suffered temporal-lobe epilepsy. After the surgeon excised the bad flesh the epilepsy was gone, but she still had the memory of the experience. So, that must have been cross-filed, deposited somewhere else in the brain, presumably the temporal lobe on the other side. This cross-filing (too crude a word) to the other side, whether it occurred, how it occurred, how extensively it occurred, has interested many scientists, more a few years ago, encouraged some to think confidently and even smugly about the relation of brain to mind. At least, something mental acquired within a lifetime of an individual seemed deposited in more than one place, for its security, one might imagine, and could be called forth by an epileptic seizure, or by an electrode, or by a mind's decision (whatever that means) to call it forth.

The Canadian was a scholar given to the formulation of hypotheses, and he conferred upon this region of the brain the term *interpretive*, which may have been assuming too much, but it shows how he felt about that called forth, the quality of it, and it was based upon a surgery done upon many human brains. Mephistophelean success he seemed to have had. His rubber-gloved fingers reached into breathing brains, his stimulator touched a point, the voltage was turned up, and something of this patient's past, distant or near, was also turned up. If the electrode was held long at a point the dream might expand, there be more of it, more of the story, the electrical field perhaps spreading. And this might happen too if the electrode was moved to different points.

What the surgeon was awakening with the electrode was a hallucination—his patient was seeing something or someone not in the operating room, hearing something or someone no one else was hearing. Sights, sounds, smells arose and there was nothing in the outside world to account for them. They were all inside the patient's brain-mind. Hallucination could be stimulated only from temporal lobes that had trouble—here it was an old scar—not from normal temporal lobes. Illusion could be stimulated also—there was something in the outside world but transformed into something that was not in the outside world.

The Japanese pilgrim—the actual one in Japan—on different vacations approaches Fuji from different sides, recognizes that he is taking in separate views, never a total view, and the best he can do is piece together his views in his own mind, which might be tantalizing, yet if he was a true oriental sage he probably was pleased that he would have views left for the rest of his life.

The temporal lobe is anatomically like a fist with the thumb pointing forward. It extends toward the middle of the brain, where is the diencephalon, and toward the base, where is what is called hippocampus. The neuro-surgeon when he operates is apt to turn back a larger flap than the nonprofessional would think he needed, but the neuro-surgeon is of course concerned with disease, and that extends where it does. Furthermore the temporal lobe, like other lobes, is more single in name than in fact, merges, though circumspectly you might say, with the neighborhood around. If a scientist tried to think of how it relates to mind he probably would merge it even more. Yet the neighborhood does have great definitenesses. To be remembered that only diseased temporal lobes have so far shown the capacity to hold and to give back the acquired, experiences, dreams, but no other part of brain has shown this. It does have some undeniable crucial capacity for binding past experiences to the present. When one reads the surgical reports one is not surprised that they should be dramatic, not surprised they should be elusive, surprised rather that they can be presented as clearly.

Epileptic attacks that originate in the temporal lobe and its environs, which is the commonest place, are apt to be accompanied by *déjà vu*, the patient feels that everything that is happening now happened before. Hughlings Jackson long ago pointed this out also. Someone suggested it might be owing to a partial shutting off of the blood supply, the nerve-cells here not getting enough oxygen, which if proven would of course alter nothing as to brain-mind. The attacks here may have other bizarre accompaniments, as bizarre odors. Indeed, the place in the overall is so bizarre that some clinicians think that insanity should be sought for here.

For a long time there has been evidence that the temporal lobes contained the highest levels for hearing. They should therefore be closest to the hearing mind. Some spots in some temporal lobes when touched with an electrode draw from the person a gross sound or block a sound. That is, if the person is counting out loud (under the drapes on the operating table) and the temporal lobe is stimulated at a point, the counting stops. The patient does not know why or what happened. He just stops counting. (Only somewhat similarly, but at another point of cortex, a person might know that his hand had moved and that he had not moved it, and that if he reached over with his other hand he could stop the movement.)

Since there are two temporal lobes, a left and a right, it is natural to ask not only what each does but also what it does for the other. The answers are not clear, and perhaps they are essentially impossible.

There have been many observations on damaged human temporal lobes, and some research on animals. Both lobes have been removed in a monkey. It produced a manifest frightful mental maiming, called temporal-lobe syndrome. All in the neural fields know about that syndrome. It was well described, nevertheless difficult here also to be sure of just what was being described. In general, it is difficult enough to climb into the mind of another man let alone a monkey, but at least the appearance was of a shattering of a world. Even so, the removing of one lobe in a human being might be not shattering at all. It depends on which. In a right-handed man the left lobe has been spoken of as noisy with mind life—many regions of the brain are spoken of as silent. In spite of what has just been said of the left temporal lobe in the right-handed, a surgeon can snip two and one-half inches off its forward tip, and no mind be lost. Apparently, no mind lost. However, let the surgeon not snip off more than that, because "if the scale do turn but in the estimation of a hair" there will result the desperately maimed, reminding us again how close the chasm of mind may come to the edge of brain, chilling any Sunday-morning philosophy, and advising that we still draw cautiously our conclusions as to mind, brain, life, death.

As to the noisy left temporal lobe, someone described it as playing its part whenever the mind was concerned with things seen and heard, sight and hearing, and the right, whenever the mind was concerned with space and was getting its data from touch and sight. That is trim. It may be absurd.

Telegraph lines, gross lines that have nothing really to do with the work of the temporal lobes, run low in the neighborhood of each lobe, and when these are ripped across by tumor there are vision defects, smell defects, taste defects, not much more learned from such tumors than might have been surmised from the neighborhoods damaged.

With some deficits the person seems to be receiving his sensations normally but is not able to give significance to them, have knowledge of them, this called agnosia, and there can be individual agnosias for the individual senses. For all of these reasons, and there are more, the temporal lobes and their environs have appeared the green countryside of our mind, rather leisurely, a place where things are happening, but not clear exactly what, and surely not how. It does appear in 1969 that the intent of evolution is downward toward the temporal lobes rather than forward or backward in the brain. Which also is probably absurd.

One might suppose, if one were supposing, that because the right and left lobes are so profoundly important to the creature they were set far apart on opposite sides of the head, with the brain-stem thumbed up between, and then if one lobe were destroyed by screaming savages come down from the hills in an automobile at 2 A.M., the other would have a maximum chance to escape; but one must immediately admit that evolution has never seemed much concerned with you or me, always only with the big us. In the monkey where both lobes were removed the researchers, who for years followed the behavior, summarized it by saying that similar destruction elsewhere in the brain changed the world around the animal, destruction of the temporal lobes changed the animal. One understood that the instant one read it, then the next instant was unsure. Bilateral removals have occurred in man, rarely, as one would know, and they have added little to our understanding.

# MEMORY TRACE

## *If Democritus Had Explored for It*

What is the impression left in the flesh of our brain by an event? What is the trace? What is altered? And what is the unit of such alteration? Dare one look for that? The ultimate mark? Democritus would probably have looked for it. One does shrink from the idea unit-of-mind and tries to get one's foot on structure.

What is the footprint, or rickshaw rut, left on the road as one reaches this stop and tries to see the mountain?

The unit could be a neuron. It could be a linking of neurons. It could even be the surrounding of neurons. In 1969 many surely would be tempted to think of something into which they could shove a micro-electrode. Many would be tempted to join the chemists and try to find some giant molecule able to alter an atom here or an atom there, then able to hold the alteration.

Whatever the unit it probably would be the same kind of unit in beetle, fish, frog, man.

There might be an electrical twist. There might follow, or there might just be an oscillation of sorts. The oscillation would lead to an etching into the flesh of the brain. Which is a sequence developed by a winner not of one but of two Nobel Prizes. He may not have had much evidence but he had plausibility and vitality. The etching would not be visible, not today, but someday might be electron-microscope visible, or some future-microscope visible, and it might be that one could not rightly use the visual word *etching*. But that would be the memory trace. That would be the mark. If it does not yet have an identity it has long had a name, engram.

For some persons with their heads over microscopes the attention has gone to those terminal knobs of the neuron. The knobs are visible with the ordinary light-microscope, were described by Cajal. Dendrites → cell-body → axon → terminal fibers of the axon-knobs. Evidence of changes in those knobs has been searched for. A few workers were convinced that with activity there was an increase in the number. It could not be proven, and it must be admitted too obvious that more knobs would be more memories, and more and more knobs would be a sonnet of Rossetti.

For a time it was thought the size of the knobs held the answer. They swelled with use and shrank with disuse. If a knob swelled it would bring that neuron nearer a neighboring neuron, touch it possibly, a region of touchings, a region sensitized, a region abler to recognize a signal that earlier had passed that way, hence a memory. Except for dubious changes with violent experimentation, not much was found. Implicit—yes. Inevitable—yes. Unless there is a retaining—this never was in doubt—the creature would have no way of carrying past to present, present to future. And he has. He has the briefly enduring, sustainedly enduring, short memory, long memory, and he has that long long long memory that would be species memory.

American students spoke for a time of reverberating circuits. Reverberate means literally to strike back. A nerve-impulse would go round-and-round in a chain of neurons, reverberate in that sense, a round-and-round that maintained itself, the chemistry of each link, each neuron, supplying the energy for that link. The round-and-round would be the memory, the mark, the trace, the engram. Everything about this is reasonable. But the search for even a single dependable engram has been like the search for the Holy Grail, no pilgrim pure enough to find it quite.

A circuit might lie still for a time, be nudged, begin again to circuit, dig deeper into the flesh, lie still, be nudged. That seems right for a memory. It is not to be trusted too easily because it seems right, but it is one of the ways a human being memorizes. A nerve-impulse would start at a spot, travel from the spot, return to the spot, keep that up, an hour, a day, till death, then after death in some ghostly ether would keep that up, till the next *Homo sapiens*, and so it would have been with all the circuits of all his ancestors.

Odd to think that in a philosopher's or a professor's cranium, up there back of his ready-to-wrinkle brow, are uncountable depolarizations, repolarizations, electrical link to electrical link to electrical link, till out of the last link bounces a major premise, a minor premise, a conclusion perfectly contrived, perfectly chaste.

How long might a trace last? By common sense, long. By experiment, what dependable experiment there was, short. In one nervous-system, a cat's, the trace lasted four one-thousandths of one second. For four one-thousandths of one second a trace could be detected, which memory might suffice to throw a glimmer back into the past of a sinful butterfly. By another technique, eighty seconds, suffice to span the undying passion of a light-o'-love canary. By another, six hours. For six hours a second message would find it easier to get through where a first had gotten through, hence a memory.

Those knobs, those blebs, are tiny bladders chock-full of chemistry, and they might be the place. Chemist and electron-microscopist have been and are busy there. This it will be remembered is at the synapse, at that junction between neuron and neuron, and at the near side. So refined twentieth-century man has become. He can speak of an effect on the near side of a synapse. Here would be the deposit in the fabric of the brain. Here would be the trace, or part of it. Here would be the fraction of memory, fraction of mind. Trace would add to trace, tremendous additions, until from the brain-end of one nervous-system, one born originally with tremendous ancestoral additions, there would burst forth onto our earth Newton's *Principia*, or Marie Curie's *Radioactivity*.

Electricity is everywhere in the brain. That cannot be gainsaid. Small voltages sum to large. Voltages shift. Battalions of electrical Lilliputians march north, east, south, west. What a commotion shy Galvani with his twitching frog shanks started in our world. (Lay another flower on his grave.) What a commotion does exist. This aspect of it is no obsession in a neurotic at MIT. This is one of Pavlov's "facts." The potentials are there. They can be measured. We trust what can be measured. Whether and where and how such a memory is embedded in the commotion, the future may, very likely will tell, but no need to wait on the future to say that if all the studies of all the laboratories were pieced together, every region of the brain could be proven dynamically connected with every other, as structurally connected. Anatomists from their side long have heaped up the evidence for our brain being one-world. Our mind to each of us always has been one-world. A passing thought might give a camel's-hair touch to that one-world, a touch that in terms of reverberating circuits might be a blow, start a spin that never stopped. The circuit would be a dynamo of its own kind. An infinitude of dynamos would bring back a whiff of garlic that twenty years ago diffused through old Caproni's restaurant, a small impressment of the spin would be the fuzz of a feeling that went with the garlic, her knee in contact under the tablecloth, a large impressment, the revulsions of the subsequent experiences that had nothing to do with garlic, except garlic started the spin.

On the Great Dean's Original Committee there was, surely, an engineer, a physicist, an electrician, a neuro-physiologist, a neurologist, a psychologist, a psychiatrist. The Committee decided not only that there should be Newton's *Principia,* Marie Curie's *Radioactivity,* the sonnets of Rossetti, but that so-and-so-many and these-and-these circuits, in this case operating flawlessly, should produce one flawless total from one mind, the "Ode to Autumn." A scientist of the early nineteenth century was notorious for having said that mind, therefore memory, was manufactured by the brain as bile by the liver, and we of the twentieth century smile patronizingly because we know something manufactured by reverberating circuits, or the like. If history proves it noon-sharp that our century discovered the gold, ought we be surprised? Socrates and Plato groped along the lanes of reasoning, whereas we tramp the straight road of laboratory research. Accordingly in tomorrow morning's *Enquirer* in solid print you will learn that more has been achieved in the last four years than in the last four thousand. Print settles such points. What would not need to have been settled is that our century possesses the faith (one of the faiths left us) to search in the peaceless electro-chemistry of the brain for the next great answer. A plain citizen needs to pinch himself to keep knowing that what the laboratories have up to now found, besides a complicated chemistry, is complicated electrical patterns, that could be related in more complicated but exceedingly flexible fashion. Memory running alongside them is somewhat but not an extreme assumption. Memory being the output of oscillation, reverberation, patterns worn among nerve-fibers, swellings of synaptic knobs, anything comparable, might be nonsense. The siren science might again have been feeding us opium. But probably not. A wilderness of electrical swayings the brain inescapably is, *Waldweben,* a wilderness in which the pioneering of man right today, after so many tries, might be delineating the ultimate order, ultimate precision, ultimate chill mathematics with which Nature has put together this her prime instrument, if the human brain is her prime instrument.

# WORM MEMORY
## *Little to Wash*

The planarian is a leaf-shaped flatworm, lives in fresh water, has been a convenient dab of material for the use of zoologists and others and recently taken up by the psychologists.

Planarians—worms—were conditioned. It was a modification of Pavlov's technique with his dogs. Pavlov would have been interested, would have paid those worms a visit if they had made the trip to Leningrad. He never cared to travel.

The worms were "trained" to "remember" accurately and to "respond" wisely, and if they did not they were "punished" and if they did they were "rewarded." It was found they did. Their nervous-systems could make engrams, whatever they are. The worms had memory. Odd that anyone but Jonathan Swift or Lewis Carroll should have thought of training a worm, but this is our half-century, and our half-century is our half-century.

Next, in some laboratories, the worms were sliced, and now it was claimed that the head-end remembered, but the tail-end remembered too.

That dropped experimentation to the molecular level where all would be stable forever. Furthermore, since the mid-fifties, and earlier, biologists have been diversifying their investments, have drawn them somewhat out of the proteins and deposited them freely in the nucleic acids. (Tell that to Aristotle if you get there first.) One nucleic-acid molecule, DNA, as we all know, has the blueprint, the code, and another nucleic acid, RNA, carries the messages from the DNA, that is in the nucleus, to the ribosomes, that are outside the nucleus, aided and abetted by still another RNA, and promptly the building-stones of the proteins, the amino acids, get restless, get marshaled, fall into place, form chains, and the chains because of their great number and the rearrangeability of their links can be as different as our lives. In fact, they are our lives. They are our memories.

It is toward the better understanding of such and other memories that much of the experimentation of our time has been directed, is directed.

Natural in our mid-century, and apparently factual, and unquestionably fashionable, for many scientists to conclude that memory is chemical, is not deposited in some pattern of neurons, as was thought for years, but in those molecules of DNA and RNA, that have that rearrangeability because of the almost infinite possibility of arranging the nucleotides that compose them, hence almost infinite possibility of the numbers of memories. Soon some researchers reported that it was specifically the RNA molecule in the worm that learned, because if the sliced tail-end in the throes of regenerating itself a new head was chemically treated so as to alter its RNA, the worm behaved random, erratic, addicted, hysteric, psychotic. It had lost its memories or they were thrown into disarray. The worm had gone off into the who-knows-where.

A rash of laboratory activity ensued. The RNA and the protein manufacture is basally determined by DNA, and DNA is basally inherited, but somewhere there is also the capacity to retain the acquired, which is this-life memory. Neurons have millions of RNA molecules. The neurons are stimulated, by electricity or a street scene, the RNA gives orders, there is some newness in the orders, some newness in the proteins, some newness in the memories. So the worms learn.

And some of this learning can be transmitted to others. Psychology can be passed on. (The researcher needs only to define the psychology.) A fragment of mind can be transmitted. (The researcher needs to define mind.) Memory can be passed on. (Memory defined.) Animals can be taught. Life can be taught. Animals are said to have been taught fear, the fear passed on, and it was not vague but specific fear.

Contrary opinions all along the line spoke out. Verification of the planarian experiments in some laboratories was declared impossible. The worms in those laboratories refused to cooperate. For those scientists the worms simply would not remember. Slicing them was therefore silly. But, the first group and their protégés stuck to their guns. They were more enthusiastic than ever. The chemistry was moving. Some found memory to depend either on RNA or on DNA or on both, though the dedicated continued convinced that it was RNA in spite of the fact that in other biological systems studied— United States, Europe, Asia, Patagonia—RNA almost always seemed directed by DNA.

In short, memory and mind at the molecular level were proving as troublesome to agree upon as at any other level, and that conceivably is the moral.

Researchers nevertheless explained farther. Researchers all still increasingly are chemists—physiologists all dissolving into chemists. The physiologist's neurons are less and less units in a complex telegraph system, a wired system, mind and memory rising out of the wires. They rise out of the chemistry. Some researchers, experienced with planarians, continued insisting that these soft-bodied flat creatures not only learned but what they learn they pass on to the next generation, to naïve planarians, make them sophisticated. Naïve planarians fed intellectualized planarians behave superior. Even feeding RNA or a precursor of RNA or a catalyzer of RNA improves memory.

That lifts experimentation back up out of the molecular level to the psychological level. And lifts it also from those lower creatures to the higher creatures, rats and men. In rats too it appears, to state it succinctly, that the injected juice of the taught improves the IQ of the untaught.

But—too bad—it is not true. Somewhere it seems there has been a mistake. The worms—too bad—misled us. It would be so socioeconomically satisfactory that chemistry and psychology should join hands, make themselves yet more useful than they already are, restore memory to the senile, wipe the wrinkles off their minds. It might require no more than a prescription for purified yeast RNA. The forgetful would be less forgetful. The slow-learner would become the fast-learner. A man might make over his world by merely conscientiously eating a few molecules with his saucer of breakfast-cereal, swallowing them at the same time he swallows his capsule of vitamin.

Today educated planarians seem illusory, but who can tell about tomorrow? Who can tell what may yet bob up?

# COMPUTER MEMORY

## *Wash It Off and Start It Clean*

The nature-built brain and the man-built both have memory. Everybody who wears out his eyes reading advertisements knows that the computer is the future, and that the computer can store something, give it out as needed.

Each of the two brains, nature-built, man-built, has a unit. For the nature-built it could be that reverberating circuit, could be an RNA molecule, could be somehow a nerve-cell. Cajal thought the last. Sherrington did. A nerve-cell plus its fibers would act like a battery, generate electricity, with a neighbor nerve-cell act like a switch, on-or-off, yes-or-no, dot-or-dash. Not a fast switch. The living is not fast.

For the man-built the unit—twenty-five years ago when all this began—was a vacuum tube. Later it was a transistor. Then it was any of many magnetic devices, the unit growing smaller, mechanically more efficient. Compared with a nerve-cell a vacuum tube is enormous, millions of times bigger, operates at the speed of light. Starlight is fast.

Vacuum tubes did their additions by the method of the speedometer, one plus one plus one. Addition, subtraction, all computer capacities used carry-over, or the equivalent, and a schoolboy could no more perform his additions without that moment of memory, carry-over, than a computer its arithmetic. A schoolboy holds many longer-lived fragments in his memory, like $2 + 2 = 4$, does not have to work them out every time he uses them. Might shoot up his hand in class and question whether $2 + 2$ always equals 4 but uses the fragment even while claiming it out-of-date. He may have large lumps of multiplication tables similarly in his memory.

The computer, like the schoolboy, can hold memory fragments, so-called bits, binary digits in words of memory, has them where it can grab them. A small computer may have four thousand so-called words in its so-called core memory.

That first huge IBM at 590 Madison had panels of vacuum tubes, each panel with ten rows of tubes side-to-side, ten bottom-to-top. There was a total of 12,500 such units. There was also a total of 21,400 electric relays. Science newswriters at the time and others who do try and often succeed in making things clearer for us might exemplify with the abacus, explain that a panel at 590 Madison could be thought of that way. A Chinese laundryman right today may use an abacus to compute his bills, a kindergarten teacher use it to teach arithmetic, medieval Europe used it, contemporary Russia and Japan do. A Japanese scientist in an American medical school had an abacus so constantly in his hand that it appeared an extension of his ten fingers, and the fingers an extension of his brain and mind. He had carried that abacus with him when he emigrated from Japan, a simple wood frame, wires strung side to side, beads slid along the wires, beads representing digits, wires place. Were a configuration of wires and beads held for a time, that is a memory, practical, holds a number long enough for it to be a directive to the next step in a computation. It binds. It time-binds. And living memory has sometimes been defined as time-binding, a lean definition, but a definition.

If traits in personality, trends in history, trends in language, trends in women's styles, shades in the colors of a sunset, points along the curve of a love affair, curve of a thought, curve of an ocean wave, curve of that electro-encephalographic wave, were assigned numbers, since a digital computer can track almost incalculable numbers at almost incalculable speeds, it could at every instant be thought tracking the past and using that to project the future, and this might begin to seem even to the plain citizen like living memory, if he were giving an ear to a computer-enthusiast. And it is no more illogical to accept as memory the dots and dashes that for decades have been transcribed as words onto yellow Western Union forms causing the ringing of telephones or doorbells, announcing birth, death, war, an occasional frayed yellow form then holding its memories in an album, old trunk, dusty drawer.

In the computers in use today the mechanisms spoken of as memory are recent, sophisticated, many kinds, even if in essence they be not different from that moment of memory in the abacus, or that store of types of machine memories, that have increasingly become fact, and that had been fact in Charles Babbage's imagination.

The electronic storage capacity at 590 Madison was eight 20-digit numbers. Whatever these represented could be registered, retained, retrieved, or, anthropomorphizing, could be experienced, deposited in memory, recalled. At an instant the current could be turned off, the flow of electrons cut off, all memories wiped off, neat as a hanging.

A magnetic switch could help to produce a memory. This is crude, but magnetism is fantastically employed in today's computer memory. In the magnetic switch the circuit is closed, creates a magnet, that pulls over a metal arm, that breaks the circuit, that releases the arm, on-or-off, yes-or-no. With such a switch all mechanical events instead of occurring in thousandths of a second occur in whole seconds. Billionths of a second are commonplace in today's computers, and machine-memory recall has the possibilities of employing that speed.

Our nature-built memory unit, whether it was operating between nerve-cells, or between atoms in molecules of the RNA, might also be on-or-off, yes-or-no. Experience would have left its traces among metabolizing switches, reduplicating switches, among switches that were always repairing and regrowing. In one's mind's eye one sees nerve-impulses unperturbedly moving over neurons; then one sees dots and dashes starting from living telegraph keys, moving over living wires, and at the end somehow transcribing onto what corresponds to yellow forms that would hold the affectionate thoughts of someone in Tokyo for someone in San Francisco, except that as soon as one says "affectionate thoughts" one stops, reflects, decides that the popular analogy of machine memory, machine mind, ought not to be allowed even to supply a figure of speech, clacking patterns repeating day-in-day-out for as long as the girl in her red shift remembers your telephone number. "Girl." "Red shift." "Telephone number." Those have the reality. Those have the possibility of the quick spreading of the kind of nuance that memory has, that makes any summing of reduplicating switches a painted death.

Computer-memory everywhere has employed twentieth-century engineering. Computer-memory can be written with beams of electrons. To an expert, writing with beams of electrons on a moving drum may be no more remarkable than writing with pen and ink on paper. Electron-beam memory could be rewritten and rewritten, would lead itself to modulations that, again, would impress the plain citizen, spin him off his common sense, especially if he were giving one ear to a computer enthusiast, say an electronics music-theoretician describing the mechanic travel from notes on a page of music to computer, to tape, to sound, to the 1969 Beethoven. Or giving an ear to the word-for-word translations from French to English. Or an ear to a physicist who foresaw as a reasonable expectation that orders might go to a computer in the spoken inflections of one's native language, the phonemes that the ordinary human being speaks, Zulu or Japanese or German, these converted to computer-English, the computer work on them, refer them to its memory stores, and answers and comments come forth in the Spanish that the ordinary Spaniard speaks. To an extent this is already happening, but, to keep our feet on the ground, the results are not brilliant. We are however assured that a computer will be built that will store more flexible memories, will accordingly make better translations, but one nevertheless suspects they may not incline anyone to think of Lafcadio Hearn's translations of Théophile Gautier, or the quality of the memory storage in Hearn's head that enabled him to attain both the accuracy and the curious fluidity.

Magnetic tape provides a memory. Magnetic devices have as said revolutionized the computer. Magnetic alloys, ferrites with magnetic properties, are in universal use. Adaptations too numerous to list have made possible the memorizing of volumes of information undreamed in the 590. Magnetic cores a few hundredths of an inch long may retain a memory, and a computer may have a million cores. One complicated core is said to have cost ten thousand dollars but worth it if the other thing said is true: it can hold more memories than the man-mind can accumulate in twenty years. Babbage's idea of store was so smart. It had in it such possibilities of enlargement. It is dull to repeat that reels of magnetic-tape with their memories are filed, are mechanically commandeered; when a tape is required, out of its file it jumps, and no longer required back into its file it jumps. Our own memories appear like ghosts, disappear like ghosts, and much of the time, and irritatingly on occasion, appear and disappear without summoning.

Babbage called those machine-parts where data are processed the mill, those where memories are deposited the store. And what we have today is a store with fabulous capacity, and a mill with fabulous speed.

Man's brain has an appropriate leisureliness, has a pliability that he sometimes recognizes as indescribable, his memories operating day and night. But so do the computer's. Old Thomas J. Watson can be in his grave and young Thomas J. Watson, Jr., home in his bed, and their computer-memories in the various metropolises, New York City, Poughkeepsie, Cambridge, steadily operating, steadily clacking. As well to keep realizing, though it is obvious, that man's head, his mind, his virtual mind, is also day and night in each of those places, New York City, Poughkeepsie. It is, in fact, in each of the machines, the 7090, the 7094. Also, tomorrow morning a well-slept 1969 mind, accompanied by his body, accompanied by his memories, can drop in to adjust, to repair, to introduce a new program, introduce a newly invented machine-part.

Punched cards have from the beginning provided a memory. They are stacked up all over the world. They were contrived by a Frenchman, a weaver of rugs to serve as patterns for his looms, and Babbage took the idea, fitted it into his own. Late last century punched cards were already used as machine memories in census calculations. Today they reach into everything. They have little changed. IBM cards remain nearly the only IBM item that has not changed. Myriads of memories have been stored on them, brought to the computer in stacks, dealt in, dealt out, restacked. The whole embroidered history, mankind's memory, might be on punched cards, the benedictions, the maledictions, the complete edition of *La Comédie Humaine,* no limit except cabinets to house the cards, buildings to house the cabinets, cities to house the buildings.

A man-constructed reverberating circuit has provided a machine memory. The original is obsolete but remains illustrative. An iron pipe was filled with mercury. At each end of the pipe a quartz disc touched the mercury. The one disc was electrically pulsed, changed shape, kicked the mercury, a ripple ran through it, reached the other disc, it changed shape, caused an electrical pulse to go out over a wire, and this, amplified, was fed back to the original disc, thus round-and-round. A mercury column eighteen inches long could hold thirteen hundred memories.

The machine-memory, the computer-brain, has had its effect on our lives, our countryside, our cities. The romanticism of the nineteenth and other centuries may be deader than ancient Thebes along the Nile, but the romanticism of Cambridge, Massachusetts, along the Charles is alive. That region of our country looks different, in some eyes beautifully different, to some hideously different, Harvard not as different as MIT, whose efficient buildings and suburban layouts appear drawn with a giant, no, oversize ruler. Possibly old Boston will escape. Possibly some wisdom in any revolutionary present saves something of the past for the future. "Yes, it might be too early to say that those computer-memories are really like ours." That was back on Madison Avenue. It was the nostalgic humility of the specialist of 1952. He shrugged. "Who knows where it all may lead!" He was correct. It has. Did his 590 reason? "Well, not as a father using the memories of his own youth to convince his son." he said. "And not as a pretty girl using her memories to convince her boy-friend." With such trivialities the specialist was attempting to descend to the level of comprehension of the crowds moving along Madison Avenue.

# HUMAN MEMORY

## *Never Washes*

How soft often is that which lies in a human mind. . . . A human memory . . . It may seem the whole mind. . . . He dead so many years . . . Such a gentle gentleman was Charley. . . . Just a black

hole for when he was not . . . Charley on that first June night . . .
Two standing straight and facing each other and not moving . . .
Silk . . . The silk unrolls. . . . Its patterns paled with time. . . .
Not all have paled. . . . Some were appliquéd with silver and stay
sharp. . . . Others paled quickly, then slowly. . . . Frailest of
brain-stuff . . . Difficult to believe it brain-stuff . . . Impossible
to understand it yet . . . In eight hundred years . . . In eight
thousand . . . Pearl Bryan's murder . . . Remember Pearl
Bryan? . . . They never found her head. . . . Hanged those two
dental students . . . The two went right up the steps of the scaffold
and never had had a chance to talk and neither ever told where the
head was. . . . Hanged them right across the river in Covington
. . . That a human memory should keep itself that sheer for half a
century . . . Incomprehensible . . . Except to the unimaginative
. . . That that edged edgelessness should be molecules reduplicating . . . Exactly reduplicating . . . Mathematically reduplicating
. . . One would like the peace of mind of the unimaginative. . . .

A memory always seems floating, quivering, like a soft mirage
over hot sand. Even the finding of the next number of a series when
counting, that common memory, that passing from one to two to
three, has a lubrication one would wish explained, because now it is
more like an exact machine operating in oil.

Whatever once is in our mind, is in. We are assured of that. We
think we forget, think it unhappily or nervously, but are assured,
and it is on the whole convincingly, that nothing is lost, may be
hidden but not lost, except what came in immediately before a
concussion, or before a sleeping pill. The last we all know except the
few who have the sturdiness and the nobility to pass up this one
night's sleep. Whatever truly took the trouble to get in cannot get
out unless it is cut out literally (can one believe it?) with the attached flesh by a surgeon, or by an apoplectic stroke, by a neat small
hemorrhage, by a neat small blood-vessel narrowed until the blood
no longer flows through, by an island of brain if not cut out, rubbed
flat by the erosion of old age. Nothing is merely erased, particularly
when printed in by feeling, or had specific reason for staying un-
erased, reason of guilt, shame, embarrassment, pride. There was a
dance way back that lamentably failed. There was a dance with a
redhead who let one think one was possessed of the rhythm of the

wind. There was a dance that stood still like a summer night. There it is—now. There it is as it was five times ten years ago. A track of rain on a window pane, a figure in the design of a lace curtain as twilight settled toward evening. The psychiatrist seeks his patient's earliest memory. He believes that it was the one that clung so long and could be brought back in such detail because it had good grounds for clinging, and it would be worth his knowing. The years may have reshaped it, but the reshaping at each point had good grounds too, and it would be worth knowing.

Sometimes a memory seems merely an emphasis in endlessly extending mind. A beam of light strikes there, and that emphasis looms out of the dark, at a supper, at an all-night wake. Why? Why was that light right? All was roughly the same at this point on other nights, so why this one night? Every human being is as perplexed by the trifles he remembers as by the importances he forgets—that influential citizen's name, the mayor's, which for a hundred civic advantages one ought to remember. "I know your name, Your Honor, as well as I know my own—what is it?"

Every human being sooner or later is perplexed also by the connection established between far time-separated memories. A man's whole body shivers—the humiliating experience in the telephone booth last May, 1968. His body quiets, straightens, his neck lengthens—that is May, 1911, the morning his father was the hero of Linn Street, which was three years and two months before Sarajevo and the outbreak of World War I. That war . . . That other war that followed that war . . . And the dead . . . The earth a rolling graveyard . . . But a rolling womb . . .

Any day any hour his mind may play such hide-and-go-seek with him. He tries to recall the theme of the second movement of the Ninth Symphony. Has known Beethoven's Ninth from when he was a child, yet all that will sing in his head is "White Christmas." In the restaurant last night they played it over and over, pretended the holiday season was gay, and this morning his brain plays it over and over, enough to drive him mad. The restaurant was full of other noises but his mind kept filtering them out, only not "White Christmas." Nevertheless, tomorrow morning, Tuesday, while he squeezes an orange, over the earth pounding and clear surely will gallop the theme of the second movement, then tiptoe. Must have been there yesterday else how could it be there tomorrow?

A poor unhinged young man sits in an outpatient clinic and a doctor plies him with questions and the answer to every question is the same bare phrase. Inadvertently, or so the doctor's skill makes it appear, a shot of color hits the young man's face, and the next instant it is as if a child's crockery savings bank had smashed, and memories spill and roll over the floor till the clinic-hardened doctor wearily shakes his head and the man looks haggard. Another doctor picks at a detail in a woman's story, picks and picks, suddenly tears fill her eyes, and he scoffs at the tears with that occasional studied brutality of his profession, but she remains as noncommittal as the front door of the U.S. Treasury. In desperation he put her under a light anesthetic and from her too memories spill and roll. They had all been locked behind that noncommittal front door. An eminent psychiatrist relates of a person who had lived on a ranch in Montana until she was three, years later in her dreams saw the old house, but always the outside. He put her too under a light anesthetic—into the house she went. That interior had been in her those many years, and it would be past believing except that each of us has had some experience to match it. One gasps at the weavings from inexhaustible spools, hum of inexhaustible looms, that must have been in the skulls of Dostoevsky, Dickens, Hugo, Zola, and in that stubborn Irish skull of James Joyce. We who are no geniuses cannot understand how there could be crammed into a single brain, where it apparently was, the stuff of *Bleak House, Crime and Punishment.* Ours are meagerer savings yet impressive in their way.

A man at a public dinner is fidgeting because he is the speaker, bites into an olive, grows absent-minded, finds himself in his grandmother's house, the pantry, and he turns and tells the toastmaster in the new tuxedo about his grandmother. He always remembers his grandmother pleasantly. The old lady was a spiritist. Nothing wrong in being a spiritist. He has himself never met a spirit but we have different talents. His tall uncle . . . He loved his tall uncle more than the other uncles and there were five and five aunts and that was only on his mother's side. . . . But, the toastmaster in the new tuxedo, from Detroit, now is absent-minded too, is looking down toward the table, possibly seeing his own uncle, possibly seeing his soup. Tall uncle, tall mostly because his nephew was so short, had reached his bony hand into a crock on the third shelf in the pantry and came out with lard up to his elbow. The old lady had switched crocks because someone was stealing cookies. On and on he went. Why was all that registered in the first place? Why does it speak out now, out of that olive? He has eaten ten thousand olives from that day to this, and it was not the olive, not the whole of it, but the seed. When his teeth bit the seed, that was the trigger, the kind the psychiatrist seeks, the memory that hangs to the seed, the earliest if possible, on the same principle that if it stayed that long it had good reason for staying, and what was the reason?

To Toscanini the story of Mozart's feat in the Sistine Chapel may have been comprehensible. Toscanini is dead and we cannot ask him, and maybe it was not comprehensible to him either. Mozart on the Wednesday of Holy Week went to the Sistine Chapel to hear the *Miserere* of Allegri, a carefully guarded composition, even the singers not allowed to take home their parts on pain of excommunication. Mozart listened. The *Miserere* over he hurried to his boarding house, wrote it down, all of it. It was in old church form, a double choir, no strong accents to help a mind remember. Two days later, Good Friday, he went a second time, heard the *Miserere* a second time, had put his score into the bottom of his high hat and while he listened made corrections. His family was frightened. The news did leak. He was summoned. Kapellmeister Christofori demanded the score, examined it, must have been as dazzled at the power of that memory as everybody since, found the score accurate, showered Mozart with praise.

Some memories step boldly to the front of the mind, as that with the olive. Some are timid. Some are methodic: a letter of the alphabet leads to a name, a name leads to last year, last year to eleventh-century Scotland, to Lady Macbeth, to Shakespeare. He said that memory was the warder of the brain.

If a man has an experience like that with the olive he tells it next day at lunch, and if the story goes well, tells it at dinner, and if that goes well . . . Each time he tells it, it enlarges. He is surprised how he plucks fresh details right out of the air. Were those part of it? One definitely was not. It occurred this morning at Third and

Walnut but it fit so perfectly and anyone would have included it. Thus does a memory become a luxuriant tree. Sends out twigs. Sends out branches. May produce mutants. May evolve an entirely new botanical species. Each twig has its bud. Each bud has its reason for being there. The brain is strange—stranger than any physiology. The mind is strange—stranger than any psychology. Memory may seem airily indifferent to what soil it is planted in, indifferent to what it helps itself to for the sake of its growth, and that is why memory underlies invention, imagination, tomfoolery, and the congenital liar.

# FRONTAL LOBE

## *Investigation with Gunpowder*

Phineas Gage was a competent laborer on the construction of the Rutland and Burlington Railroad, had an excellent mind, acceptable personality, good character.

That was one hundred and twenty-one years ago, 1848.

Phineas placed a quantity of explosive and a fuse in a drill hole, was starting work with a tamping iron 3 feet 7 inches long, 1¼ inches in diameter, and weighing 13¼ pounds, when something went wrong. The iron ploughed into Phineas's face, on through the orbit of his left eye, into his skull, through the frontal lobes of his brain, came out the middle of the top of his head, landed some distance away, was bloody and greasy. Phineas was stunned. He was driven in an oxcart three-quarters of a mile, walked a long flight of stairs to his hotel room, was lucid, pulse sixty, hoped he was "not hurt much," seen by a doctor, survived twelve years. The tamping iron amputated the forward part of the frontal lobes on both sides, achieved, as proven at post-mortem, what in our decades has come to be known the world over as a lobotomy.

It was a prefrontal lobotomy.

Phineas's trouble was in his two frontal lobes. The small girl's had been in her one temporal lobe. Memory is associated with the temporal lobes. What is associated with the frontal lobes is even harder to say. One feels uncomfortable when one begins to speak of any part of the brain in isolation, as one must, but one ought also to feel uncomfortable, and go on feeling uncomfortable, the parts and functions always proving interconnected, grossly or subtly.

In the history of medicine ever so often a solitary person has been struck down, in this case Phineas, the result some enduring advantage for the rest of us, for the health of our bodies, the richness of our minds.

Phineas's character changed. He grew profane. (Did not inhibit his self-expression.) Started enterprises that each time were abandoned because they failed to lead quickly where he expected. (Did not think ahead.) Was obstinate. Lost his adult point of view. Retained his passions. Something had happened, as stated at the time, in the sexual sphere. "The equilibrium of balance, so to speak, between his intellectual faculties and his animal propensities seemed to have been destroyed." That was a doctor speaking. "A child in his intellectual capacities but with the general passions of a strong man." For a time Phineas was a Barnum exhibit. For a time he planned a line of coaches in Valparaiso, stayed in Chile eight years, no coaches. Went to live in San Francisco. After other vicissitudes of a human life it got to be May 1861, when on one day Phineas was laboring in a field, the next was stricken with convulsions and dying. Tamping iron and skull can be examined by anyone who has the curiosity in the Warren Museum, Harvard.

Something was learned from Phineas about the frontal lobes. Something about the brain. Something about mind. Something about brain-mind. More might have been learned but the time was not yet ripe, would not be for another hundred years, when the surgical removal of great numbers of frontal lobes would provide great numbers of facts, not easy to interpret, differing with the person, none to be trusted too far, and seeming more to be distrusted as mind on all sides was studied apart from brain, by psychologist, psycho-analyst, psychiatrist. This studying-apart-from-brain might change now any day, might never.

# PSYCHO-SURGERY

## Investigation with Knives

Two chimpanzees continue the story. The two had appointments at Johns Hopkins University, were volunteers (so to speak), were to take part in a study of the common cold. The one's name was Lucy, the other's Becky. Having finished their obligations at Hopkins they went to Yale. This was a quarter of a century ago. In Yale they received intensive testing (Yale was early and always a place for psychological testing) to establish in numbers the power of their minds, given a battery of psychologicals, then their frontal lobes were investigated with knives, were surgically extirpated. This did considerable. One of the two before the operation had been "difficult," had a "disposition," let the world know it, flew into a fit of temper over any laboratory problem she could not solve, got worse from day to day, but after the surgery was a saint. The word used was saint. Later that year, 1935, an adventuresome neurologist in Lisbon, Egas Moniz, who knew about Lucy and Becky, and knew about some earlier attempts (frowned upon) to cure human insanity with operations on the brain, thought it worth risking the injection of alcohol into the frontal lobes of an asylum patient. The alcohol destroyed where it went. So that tissue was beyond function, good or bad. In later procedures the white fiber tracts joining frontal lobes to thalamus were cut through on both sides of the brain, the lobes in that way varyingly disconnected from the rest of the nervous-system.

The frontal lobes lie on the base of the skull above the eye sockets. They are front as anatomy. They are front as idea. And it is likely that we cannot anywhere in the brain know with any accuracy the ratio of anatomy to idea.

The operation of prefrontal lobotomy was comparatively simple yet might worry the surgeon because it was to disconnect what long had been thought the highest flesh. A special kind of courage there must have been in Egas Moniz. He later received a Nobel Prize and later still it may have been regretted that he received a Nobel Prize.

He had performed a planetary miracle, but miracles do not as we know remain miracles, this one for a few years, but men learn by them both during the period of faith, and afterward during the period of disillusion. Surely Egas Moniz had by leaps increased the experimental familiarity with the human brain. This pioneering began in November 1935, and by January it was possible for him to report on some twenty asylum patients whose mental illnesses he thought he had helped.

The original procedure was called leucotomy, after the instrument that did the work, a leucotome, a blunt hypodermic needle with a knife hidden in it. The leucotome was plunged into the brain, the knife released, rotated, and the brain-flesh thus cut through. Later the surgeons, styled now psycho-surgeons, largely gave up the leucotome, and there ensued a long series of modifications of technique till the operation got that name that for years was on the tongue of the world, lobotomy. It was less on the tongue after a time. However, the almost vulgar familiarity with the human brain continued. Aspects of traditional psychology were boldly re-examined by surgery, always in the course of operations that were attempting to cure human ills, and often did.

If the frontal lobes were lifted out of the skull the term was lobectomy. Lucy and Becky enjoyed lobectomies.

Following that first gamble of Egas Moniz, the number of disturbed persons operated upon rapidly mounted, so that by 1948 four to five thousand cases could be reported at a congress in Lisbon, a record that was afterward bettered, if bettered was the proper term. The modifications of technique indisputably improved the technique, which became more exact and controlled, the brain extirpations less extensive, even small, the original slashings rarely employed. Then the whole procedure began dying out, and is dead, by and large.

The clinical successes or failures or any moral meaning attaching to this surgery are not under discussion here. Drastic ills may require drastic measures, the patient's misery driving the physician to steps he otherwise might regard as barbarity. Life sometimes requires barbarity, as sometimes homicide, as often suicide. Whoever visits an asylum finds his judgment about treatment in a fluid state, unless he numbs himself by the frequency of his visits. The human mind is darker in an asylum, but, even while thinking that one should not numb oneself, should not get used to other people's dark, one does. A few who early began performing lobotomies remained routinely enthusiastic, others were reserved, others condemned the procedure, others were "intemperate doubters." The general field gradually got comfortable with its name, psychosurgery. Animal experimentation, as for all things in modern medicine, was brought in to test methods and, where that had any common sense, with results.

For the human being the operation was mostly performed through holes bored into the top or sides of the skull, one on the left, one right. With time, less brain was disconnected, all operative steps were more economical, instruments more precise. Only the bottom quadrant of one frontal lobe—that frontal bastion of us—might be separated off. Whatever the fiber tracks slashed, the cortex might be skillfully helped to retain its connections with the central system. Or slivers of different sizes of the outside surface might be merely undercut. That is, procedures were diversified, might in themselves be good, bad, better, worse. In the earliest days one occasionally had the impression that the surgeon was simply weighing off tissue, thirty grams for schizophrenia, twenty grams for involutional psychosis, fifteen grams for something else. It made one think of Shylock.

The probability is that brain should never have been imagined that hashable.

A thin electric wire, a cautery, was sometimes inserted an exact distance at an exact spot and an exact voltage of current delivered, the spot coagulated. Voltage, time, kind of current became standardized. The thin electric wire (insulated except at the tip) might instead be directed toward a spot in the thalamus, the spot coagulated, that procedure called thalamotomy. Claims were made for all the methods, always some balance struck between psychiatric relief and permanent loss of mind.

What may have been the most extraordinary of the operations was carried out not in an operating room but in a psychiatrist's office, the patient after a short rest on a couch returning home to enjoy his lunch, the whole no more bothersome, locally, than the extraction of a tooth. For anesthesia he was given an electric shock through the brain, then an instrument with a sharp point, that it is not nice to call an icepick, but called that, was driven by a mallet through the bone of the upper inner corner of the orbit of the eye, shoved in a measured distance, swept from side to side, an amount of frontal lobe separated off, the operation accurately named transorbital lobotomy. It has been stated (in print) that an enterprising psychiatrist could do fifteen transorbital lobotomies in a morning, which is not to belittle the operation but only to state what has been stated (in print), and no reason at all why a psychiatrist's office should be a less proper place than a hospital, though in those years one discovered oneself not immediately able to throw off one's outmoded prejudice that a psychiatrist's office was a place for talk with now and then a suggestion thrown in, now and then a prescription. Some surgeons, questioning the method, thought a surgeon ought always to be seeing where he was working, his eyes in every instant focused on the operative field, especially so because this flesh was agreed to have more than a casual relation to a man's inner life. But did it really matter? Hamlet striding after the ghost cried out: "And for my soul, what can it do to that, being a thing immortal as itself?" Which rhetoric would be directed not at lobotomist or psychiatrist but at life and death and salvation and whatever else. Let it be said, furthermore, that lobotomies were conscientiously debated by most doctors recommending them (like heart transplants today), conscientiously carried out, conscientiously examined a month or a year or even years afterward.

Lobotomy did, or could, or should, without question have taken us to one more station from which to view mind. One station cannot be expected to be as attractive as another. We are all of us huddled together, and each of us also huddled alone, young surgeon, Princess Margaret, crazy old woman, crazy young man, poor muttering human lump squatting among the pigeons on the curb at Trafalgar Square, each with his mind of a somewhat different titer, huddled somewhere on the road below the mountain. On most days we do not think of the huddling.

# THE LOBOTOMIZED

## *It Is Dimmed*

The result of frontal-lobe surgery? The result of lopping off this forward bastion of human brains?

To state it bluntly: it produced tens of thousands of Phineases.

They differed. Their illnesses had been different, the surgical proceedings different, the persons were different to begin with. But, taking this all into account, the result of the lobotomies upon the minds was rarely satisfyingly described, and not explained. The tens of thousands yielded comparatively little. One has a sense of the carnage. One must try to force oneself past that to the thoughts of mind and frontal lobes. Something common there was undoubtedly, but in the articles and books, and there were plenty, the common was not convincingly found. The surgery was neater than that with gunpowder and tamping iron; skilled toilers of the brain had taken up the quest, and they were adventuring not in dog or cat or chimpanzee but in the creature with frontal lobes par excellence, the creature best known to man, best able to explain what was going on inside himself. Nevertheless, interpretation was not easy. Interpretation of brain-mind never is easy, and this was the latest in evolution, and the mass of it, the sheer quantity of it, unique in man. No such simple formula as: this much frontal lobe cut away, this much mind cut away.

The lobotomized after the surgery had less usable brain substance, might or might not have less mental illness, might or might not have more definable mind. It was difficult to make one's way through the reports.

It might be enthusiastically stated that a patient after lobotomy is no longer a custodial problem, no longer requires hospital care, no longer is an expense to the state, can return to his family, get back his job. All of this is without question a medical result, and it should tell us of the work of the frontal lobes, were one not immediately stepping into quicksand. Even if the patient (or experimental subject, it seems) had had a normal mind willing to cooperate, one might have felt engulfed, but he had an abnormal mind carrying the handicap of a brain underneath that had been surgically cut through. It would take a most rare psychological acumen. That is itself rare. And here it is confronted with apparently so many items. The engineer-oriented thinks: why not drop them all into a computer? However, the items belong to human beings with human minds that a human-mind computer-programmer would probably never be able to convert to the bits and words of a computer. They do not have that kind of clarity. Furthermore, the computer in the time those tens of thousands of Phineases were being produced was not yet what it was going to become, if that could make any difference, as probably not.

It might (in a report) be enthusiastically stated that the lobotomized is less destructive. He has not the zest he once had to rip a fixture from a wall. This could mean that the presence of frontal lobes even in a sick person makes him less predictable, less inflexible, less socially acceptable, more interesting.

It might (in a report) be flatly stated that the lobotomized had a self no longer troubled about self. This could mean that the frontal lobes normally inspire an individual to trouble about himself, his private self, to dwell on that, possibly to seek to understand that. This is reaching rather far. Troubled he is, but hardly in any such clear-headed direction as to give himself the problem of trying to understand his private self. He is sick, usually seriously sick, mentally sick.

It might (in a report) be stated that the lobotomized has the aggressiveness to go out and meet a world in which previously he had been timid, and for this reason might even be earning good money. This could mean, freely interpreted, that the frontal lobes normally prevent a person earning good money, and that could mean, freely interpreted, that the frontal lobes lure the mind from preoccupation with the outside world to preoccupation with the inside world, which surely would make the frontal lobes too bad to lose. But that is of course more quicksand.

It might (in a report) be stated that the lobotomized is no longer worried about the future, does not concern himself with the future, and that could mean that frontal lobes deal with the future. Now they are gone and there is no future, nothing ahead to worry about, nothing ahead to expect.

Many things might be stated.

It should be pointed out, as it has been, that the gains from lobotomy might be as great as those surgeon-psychiatrists said, yet the person, imagining him equipped to stand off and look at himself, not agree that he has gained, rather lost. A psychotic, an insane man, in a lucid interval might very well swing around, curse the surgeon meditating lobotomy, or any comparable brain-damaging therapy, tell him straight: How can one human being dare decide for another that it is better for him to escape his tortured self in order that, for instance, with a dusty cap on his head, a dusty cigarette between his lips, a dusty cloth bag for an apron, with majestic dignity via a nail at the end of a stick he may pick up paper on the hospital grounds, or even all day have the privilege of being a clerk and sitting on the same side of a desk checking off something, then after supper in the evening make drawings with a lead pencil, and later walk three times around the block before going to bed.

The original massive amputations of the frontal lobes were meanwhile less and less frequently performed, only on the bitterly ill, schizophrenics with ceaseless hallucinations, voices that came from someone who was not in the room, sights that no one else saw, schizophrenics on whom all other therapy had been tried and none worked. Also, on the incurably depressed in whom psychiatric treatment plus medication had completely failed. Also, on those suffering the pain of cancer, death waiting, opium addiction established and no longer overcoming the pain.

When the amputations are reduced in size, as they were, there is reduced blunting of mind, reduced deterioration of personality, reduced scarring, therefore reduced chance of epileptic seizure. The last is a side effect in half the cases, one body-mind misery traded for another, a blunting, a gloom traded for convulsions, that bring their own gloom. Even when the surgery is no more than an undercutting of a small part of cortex, there may be scarring and epileptic seizure.

All of this may throw light on the world, on the mind of the lobotomized, but it is a confused light. One sees—but it is not clear what one sees. One looks—but it is not possible from this station to see far up the side and to the peak of Fuji.

Summarized—so far—and drawing from scattered reports, the frontal lobes normally might deal with self, future, imagination, no money in the pocket. None of that has anything like a single meaning, a single direction, does not filter out a frontal-lobe function. That inside a man's noble jutting forehead remains a part of his mystery. A peculiar fact, among hosts of peculiar facts in the mystery, is that the I.Q. of the lobotomized is unimpaired. That apparently has been a consistent finding. It is even improved in those patients where there had been continuous agitation.

Immediately after a massive amputation the patient lies in bed with eyes closed. He can hear, can answer, but is profoundly inert. He has lost sense of time. He thinks minutes to be hours, tomorrow to be yesterday. That seems something to hold on to. But other experts have other views, appear not to be observing the same situation, one insisting that the sense of time is a task divided between frontal and temporal lobes, one insisting that the frontal lobes are not concerned at all, but thalamus is. Sense of time for the psycho-surgeon gets a relativity (or even non-existence) psychologically, as for the physicist it has gotten a relativity (or even non-existence) physically.

So, the patient lies with eyes closed. Gradually he appears to return to the world. When he does return it is at once evident that something has been lifted off his mind, a Bostonian restraint. There may be an uncalled-for gaiety. Coarse joking. One patient swats a nurse on the rump. And this now might seem to mean that the frontal lobes normally instead of making a person less socially acceptable might make him more socially acceptable.

The frontal lobes in any reasonable understanding were evolved to work for the individual mind, even if that mind exists to perfect a movement (which is probably nonsense), because only by working for each individual mind can those lobes ever have been working for the vast community of minds that is species mind, or, what we of the West find it harder to imagine, working for or making themselves part of that East-Indian universal mind. Either way, one is not surprised to have it proven that the frontal lobes as revealed by lobotomy function somewhat differently person to person, mind to mind.

Summarized again—so far—it appears that the human stump left after lobotomy depends upon what the human being was before lobotomy, and upon his illness, and upon the surgery.

Confusion as to the work of the frontal lobes is natural, as must be obvious from any listing of the impressions of these honest-enough experts. Even the brain in this neighborhood, though it may as flesh look like brain-flesh elsewhere, may one day prove chemically or otherwise not to be the same. (There are definite chemical differences in parts of the brain.) The frontal lobes are a late neighborhood in the brain's evolution. They are possibly not as eon-frozen. They are possibly not as unvarying. They might accordingly not be as analyzable.

A deep depression sometimes settles permanently over the lobotomized, deeper than that which may have been there the years before, though observers usually emphasize reduced depression. It is manifestly reduced if it had been an agitated depression previous to the lobotomy. Self-absorption (from all the reports) appears reduced. Anxiety appears reduced. Reverberating circuits may be generating the anxiety at different levels of the nervous-system, as has sometimes been thought, and there may be a reverberating between frontal lobes and thalamus, a vicious circuiting over the white fiber tracts between the two, and there, also, is where the lobotomy cuts are (or were) made. Lobotomists agree about anxiety. It is reduced. A flattening of mind, a bleaching, is emphasized. The mind-flattened schizophrenic even in the heyday of lobotomy would accordingly not have been considered a good candidate. After lobotomy the pathologically groping grope no more. Nice not to grope, but not nice not to be able to grope. Most lobotomists claim, as said, that the I.Q. is unaltered, the intelligence unaltered, the capacity to pay attention unaltered, and the implication could be that the personality is unaltered, but that is not true. Even the patient may claim the contrary. Members of his family recognize losses.

It is easy to see in all of this that the surgeon-psychiatrist (psychosurgeon, lobotomist) is trying like the rest of us to get an impression of mind and is moving from point to point to see if he can ever once have a complete vision of the mountain. He cannot.

Naturally the surgeon-psychiatrist (psycho-surgeon, lobotomist) tries to see some good accomplished by his surgery, yet only a foolish surgeon could imagine that he creates mind by cutting away brain. He may cogently argue that the brain left behind after the surgery is thereafter less interfered with, as is argued (and proven) for the hemispherectomies where the removal of the bad half makes the good half work better. Those minds in any case are always altered, maimed, though they were of course maimed before the hemispherectomies.

The human mind to be a human mind requires some level of agitation. (No doubt, animal mind too.) After the massive operations the patient looks as if he might be content to play parchesi forever. In soldiers of World War II who had their frontal lobes shot away friends noticed tactlessness in persons who had been tactful, and if tact is defined as some sensitive restraint, that again would bring the highest toward the frontal lobes. (Self. Future. Imagination. Anxiety. Tension. Individuality. Tact.) Sometimes the lobotomized would be described as not only tactless but as acting equally grossly inappropriately for a gay situation as for a sad.

So, descriptions pile up, but not satisfying descriptions, and there are explanations but not satisfying either. Right today there are fewer descriptions, fewer explanations, and almost no lobotomies, not much psychiatric interest in them, nevertheless the reports of the past voluminous surgery has for purposes of throwing light on brain-mind continued to have interest. Some intellectual observers keep reiterating how in the lobotomized the power for abstraction did dwindle. (Self. Future. Imagination. Anxiety. Tension. Individuality. Tact. Abstraction.) The possible disposition to abstraction of the observer must always be kept in mind—so impossible to see anyone else and not to some degree see oneself.

With the practical, with the obstructing disadvantages that the lobotomized may seem sometimes to have lost, one does understand what was said earlier about the lobotomized being able to get a job. Some physicians say, too, that though he gets a job, he drops it, does not want it. His ambition is gone. Why work? Some who are not lobotomized have come to that conclusion too.

# LOBOTOMIZED PERSON

## *A Bicycle-Rider Observes It*

Single cases were described by a Swedish physician. As might be anticipated, to observe a single case imaginatively, to observe it before and after lobotomy, or even only after, might tell something dependable about the frontal lobes, and the observation need not necessarily be by a physiologist or psychiatrist or physician, though these would bring the advantages of their techniques and experience, but observed and described by a mind that had the cunning to enter another mind, say a combination of the minds of the brothers Henry and William James. To observe such a single case as against testing five hundred with batteries of tests by eager testers, or five thousand culled from the journals and presented to the computer, would of course not produce graphs, but might help us some as against not at all to understand how the frontal lobes relate to the rest of the brain, and relate to the mind, also to understand why lobotomy accomplished something for some disturbed minds, if it did.

This Swedish physician not only described single persons instead of populations but he used ordinary terms in describing them. Some others did that later. The Swedish physician seemed to do it intuitively. He had undoubted talent. He wrote his own histories. He did not neglect to perform the routine psychological tests but what was impressive was his common sense and his interest in these persons. Like a good doctor he wanted to keep in touch with them afterward when they were less ill, in some instances for years. He visited his patient's home, invited him to dinner in a public dining room, went bicycle-riding with him to watch how he behaved in traffic, talked to his family, to his friends. All in all the Swedish physician appeared as free of the compulsion to push a human mind into a textbook class as to report it in textbook fashion. He never forgot that these were single human beings with single human minds, approached each with the conviction that each was unique whether sick or well, and did not in his writing palaver about doing this. He reported a wife as saying she had been given a different husband by the lobotomy. This was not the same man. A mother said of a daughter whose frontal lobes had been disconnected that the deep feelings in her, the tendernesses, were gone; she had grown hard somehow. Another who had preferred classical music now preferred dance tunes. A wife said of her husband, "His soul appears to be destroyed." That last remark has been repeated many times by those who treat the mentally ill, expressing what these specialists had themselves noticed but for which they had found no such effective phrase.

The Swedish physician revealed his own mind while he was revealing those other minds.

One might conceivably come nearer an understanding of the work of the frontal lobes if one made a generality of that wife's remark: namely, the frontal lobes piece together those delicate shades that some have thought to be a man's soul, the headiest distillate of his mind. The distillate may be fairytale. But it could also be simply the unanalyzable. The Swedish physician was definitely not thinking of religion when he quoted that wife. It has been claimed that religious feelings are helped by lobotomy, the person less tense, less depressed, lighter in mood, though lighter may not be the character toward which religious feelings should move in our day lest they float off altogether. As for churchgoing, it has been pointed out that after lobotomy churchgoing may grow more compulsive than it ever was before.

An English neurologist thought that the frontal lobes were necessary during the period of life that mind was acquiring, and that after that one could more safely lose them. He drew his conclusions not from persons who had been surgical lobotomies but who lacked frontal lobes from birth, God preoccupied and dropped a nucleotide at a wrong place in the double helix. The English neurologist went on: he thought that the frontal lobes enabled our minds to acquire, the acquired then transferred to another part of the brain. If true, we would be needing our frontal lobes less after our mind had put away its savings and was living on them. The Englishman warned that many persons got old early, settled into that humdrum, but others stayed young late and for them lobotomy would be too bad. "She let herself be driven ten miles twice a month for a hairdo in icy winter until she was ninety-six." Had this old lady been noticeably melancholy in the heyday of lobotomy, and her hands tied, what a pity. Untied, she would have blasted the therapy-minded psychiatrist-surgeon out of his office.

The Swedish physician and the English neurologist do add some small clearness to our view.

Human frontal lobes are huge versus those of the ape, more so versus cat and dog, and if one looks lower than mammals one cannot be sure that there is even anything corresponding. This of itself says something about the frontal lobes. It has also put the burden of understanding them upon human studies.

# SIGNS OF EMOTION

## *Masks Its Feeling*

What a person is feeling, whether he has a sick or a healthy mind, may or may not be evident in the signs he gives out, usually is, and it depends again on the observer, one so often better at this than a hundred others.

A scowl appears to be gathering by the same stages in two persons. A grin appears to. A leer. A woman in the clinic each time her husband is mentioned curves upward her lip on the right side, then slowly the sign fades. That far, which is far, all is stereotyped, belongs to the species, is grossly evolutionarily selected, for survival. Sadness has its stereotyped signs. Gaiety has. One often thinks, masks. One thinks, ancient masks, those that covered the body head to toes. Eons fashioned and supplied each species with its kinds, sometimes a number of species with almost the same kinds. A dog like the woman in the clinic will curve upward his lip, then similarly slowly the sign fades. Man, monkey, cat, others may display that histrionic snarl.

Throughout his life, then, the individual remolds his species masks, but only somewhat, that somewhat, however, important to him, and mostly only the human individual remolds, but always to be remembered that it is the human eye that sees, and might not see perceptively the signs given out by a fruit fly.

Emotion is always somewhere behind the mask, feeling, strong feeling or weak feeling, subjective behind objective, and the scientist no less than the plain citizen may find it difficult to keep the one separate from the other, the *deus ex machina* from the *deus,* the painted laugh from the laugh inside.

As for the scientist's experiments, it is more feasible to perform them on the *machina.* Goltz successfully removed those two cerebral hemispheres from those three dogs. In his warm way of writing, and he could be accurate and still warm, it was sufficiently clear that signs that previous to the surgery had been held in check, after the surgery were released. He did not use the word *released.* That would come later. But signs seemed preternaturally ready to burst forth. It would have been more possible to make something of the signs than of the emotion, but he would have been in no position to speak of that either. Some believe that the two are one—perplexing to know what that means. Pavlov believed it—whatever it means. Pavlov had it in his nature, his intellect, his passion, to see no separation between the two.

Years later, there was a series of experiments. They are famous. They were by an American. Surgical techniques had improved. Ablations for a variety of purposes had multiplied, mostly for purposes of learning how the animal stands and runs, but what the American was seeking to do was locate accurately the site of control in the brain of the signs of one emotion, rage. He used cats. Many cats. He removed more and more brain in successive cats until he approached near the tailward end of the hypothalamus. The hypothalamus is just where it was—middle bottom of the brain on the brainpan. Here he found a few highly special cubic millimeters. These he must preserve if the cat was to display the signs. They must not have their connection with the nervous-system below interrupted. These millimeters were the rage center, that still today remains an established landmark.

Such a mutilated cat burst into the signs on slighter provocation than a normal cat. Something that was acting as a brake on the signs was lifted off by the surgery. With time the meaning of this display became clearer. The geography of the brain became clearer. If the experimenter's cut went farther toward the tail, even a trifle, there were no signs. They refused to be summoned forth. But if the cut went farther headward there would be some modification of the signs, something probably added; it might be difficult to see. Techniques continued to improve. It was possible now for an electrode to be implanted right within those millimeters, the animal allowed to recover. Then when the electricity was snapped on there was again a bursting forth. Signs of rage could be snapped on or off. Strange, of course. Or, just what you would expect. Depending on the position of the electrode the signs might seem more like fear, more like terror, more like aggression, more like defense. It was a grossly mutilated animal and yet every now and then there were the evidences of nuance of control. The implanting of electrodes grew technically by bounds, as did the surgical cunning, the shrewdness of the instruments. Everything suggested increasing delicacy in this experimentally induced behavior, but far from delicacy when contrasted with what we know, or think we know of our own signs. Everything warned that there must be care in interpretation. Everything also might remind one person of the computer character of his brain, and remind the next person of the exact reverse.

Interesting, or again just what you would expect, those millimeters (the rage center) were by evolution relegated to the same neighborhood in cat, in dog, in man.

When one sees a hissing cat—Willy, the tom, sending greetings to the hound next door—one recognizes that Willy's private life has indeed modified Willy's evolutionary life. One needs to look closely but then one cannot escape seeing that somewhere inside this respected male, despite any scientist determination to stay severely outside with the species signs, there are touches of complex individual signs.

By the old James-Lange theory signs would come first, and they cause the emotion. A man looks sad and that causes him to feel sad, and unquestionably looks and feeling, whatever the final truth about them, are intimate, which would not prove the James-Lange sequence correct. It does not feel correct, if one can say feel in this connection, but it still may be correct. The back-and-forth between body and mind is everywhere quick and it is not easy to speak for oneself let alone for this tom. That the signs and the emotion could

keep each other going requires no theory. Othello feels, shows what
he feels, feels the show of what he feels, feels more, this accelerating
until the Moor can do it, strangle Desdemona, she so lily-white and
frightened, he so dusky and unhappy, and in the audience not only
the excitable Italian, the mad Irishman, the hot Spaniard, but the
cool Englishman apprehends personally what is transpiring behind
the footlights. Apprehends, but with that anguish of the theatre that
is enjoyment. So much in our own personal signs of emotion is
layered and matted and not quickly unraveled, but is enjoyment,
too. A researcher would, as one says, identify more with Desdemona
than with a  mutilated cat and would be less alert to the delicacy
and detail of the signs of emotion in a cat. He might go so far as not
only to think a cat a machine but a simple machine, which ought to
make him freer to catalogue signs of rage, fear, et cetera, to write
them down. We prefer to catalogue. It is lazier. By a hop, skip, and
jump, then, we can draw in also motivation, shame, guilt, punishment, reward. This puts us, as one also says, in the swim.

The laboratory signs of rage were christened *sham-rage*. That was
long ago. Nevertheless, better to stay out of the way of that cat's
claws, for the claws in themselves are not sham. They unsheathe,
then scratch, they tear, and they appear always to have in them,
even the claws, something besides signs. The more brain left intact,
left attached to the rest of the nervous-system, if it is the appropriate brain, the more lifelike will be the signs. Or possibly it could be
stated this way: in the mutilated female cat, as against the unmutilated, there is that lack of nuance that the experimenter might
overlook, but that would be all and all to the beloved or the hated,
to another cat, to that tom. For perfection of signs of rage, or of any
other emotion, any experimental worker would concede that the
total nervous-system should be intact, all the autonomic, all the
somatic, all also of the direct chemical controls.

A cat in rage has its hair stand up, its pupils get large, blood leave
its skin, its pulse rate rise, blood pressure rise, spit accumulate—
those are autonomic signs. It hisses, lashes its enlarged tail, does its
nimble footwork, struggles, bites—somatic signs. The normal can be
thrilling. The mutilated is an interesting but a sorry sight. A human
being in a fit of rage also may be a sorry sight, the observer's own
emotions coming in, his disgust, his contempt. He thinks, if he is

given to this way of thinking, that what he is witnessing in this enraged human being is maximally the signs, minimally the emotions. The modulating parts of this human being's brain have by his rage been uncoupled from those critical signs-controlling millimeters. The signs are in control. They are free. They are giving a sheer display of themselves.

In other, later experiments it was discovered that when a brain-part called amygdala (deep to the temporal lobe) was destroyed, rage also freely flashed forth. Amygdala is small. It is complex. The word means almond-shaped. This is to say that apparently neither the tremendous destructions of two entire cerebral hemispheres nor the still considerable destructions in that later series of cats were necessary for the freeing of the signs of rage. It was enough to destroy amygdala. This should be on both sides of the brain. Surgical techniques had continued to improve. A thin electric wire—a cautery such as used in that one method of human lobotomy—was inserted into the brain an exact distance at an exact point. The current turned on. That destruction was complete. Occasionally the results were not what had been anticipated, might even be the reverse, and it was evident that diametrically opposite controls existed close together; also that there were these opposite controls for everything. Destroying amygdala on both sides with the rest of the brain intact freed sham-rage. Destroying the entire hemispheres freed sham-rage. Cutting through the brain-stem above the rage center freed sham-rage. In each case those critical signs-controlling millimeters were released. (It was Hughlings Jackson's release.) More than this, when that limbic lobe that wraps around the brain-stem was destroyed the result was the same but feebler. It all suggested that the controls on the masks, on the signs, were numerous, were interlocking, might be located in far-flung regions of the brain. Recent experimentation makes this still more probable. Everywhere there is check on check on check, and apparently, as device inside the machinery, the essence of the check was nothing beyond the old excitation and inhibition, the quantities varying, the places varying, the harmonizings fabulous.

In monkeys—this now was bewildering—with amygdala destroyed there was no lashing-out. Quite the contrary, a killer-monkey after the destruction was ready to be petted like a kitten.

He could be handled without gloves. He was sociable. He wanted plenty of sex, wanted to investigate everything with his mouth, lick everything, and if he set his teeth into friend or foe he did it with consideration. Freud's oral phase may have ready-made nervous-system patterns. The love of lovers may be a passing through the valley of suppressed amygdala. In trying to account for the difference between cat and monkey it is possibly worth recalling that the monkey dines largely on vegetables and goes with colleagues on picnics, whereas the cat eats meat and slinks along alone. Cat is carnivore. Cat possesses the bloody beautiful nuance of signs of carnivore. Monkey and cat are different in their stars, their behavior, their genes. One says that last—different in their genes—on so many occasions that one does not stop any more to think of what a fantastic thing it is. Monkey and cat are different probably in their proteins, their enzymes, their RNA, their DNA, and each monkey different from each monkey, cat from cat—but after one has lived on this planet awhile one is not surprised at anything. Monkey and cat and Holstein (a bull weighing a ton and a half was sold a few months ago for a quarter of a million dollars and one was not particularly shocked) are different, and every beast is different from every other beast, and every fish from every other fish though a thousand hoisted silver-shimmering in a net look so alike, and different almost surely out into the signs of rage or fear of every creature's psyche.

With amygdala destroyed a monkey behaved as if he adored the world, a cat as if she detested it, and the adoration and the detestation were both separate and intimate geographically in the brain. The vicious Norway rat followed roughly the behavior of the killer-monkey, the dog followed the cat. If the brain above amygdala was carefully removed, so that amygdala was free to play with unimpeded force upon the rage center, the cat was milder, less cat, less quick to defend her slight body. Future experimentation might show other brain-parts to be important but the general theme will very likely not be altered.

And the theme applies to us. We might in fact anticipate, simply because of our more-brain, more disarray of signs, especially among sick human beings, or among normal human beings when the times are sick. There is a voluminous literature on tumors and surgery of the brain, but only a few cases show anything resembling rat, cat, dog, monkey. What this means, partly, is that tumors are casual and surgeons cautious and the absolute number of the cases are truly few. Yet, every now and then, a patient is reported as displaying the periodic bouts of meaningless rage, or meaningless sexual passion. The chronicle in its total does read like a latter-day *Pilgrim's Progress* with Rage and Placidity and Fear and Defense and Amygdala and Limbic Lobe and Hypothalamus written as proper nouns. When the nervous-system localities are left as nature decreed them, the creature, if a human being, shows the signs of one who is watching with appropriate decorum interspersed with appropriate excitement a street fight under his window, a murder trial, a big-league baseball game. Destroy Amygdala, he becomes Placidity, being of the monkey family, and the fight and the trial and the game are as stimulating as a cup of tea to Hercules.

# FACE

## *Refines Its Mask*

Imagine a monologist impersonating Dylan Thomas, or Charles Dickens, exaggerating somewhat, when suddenly one sees the muscles under the skin of his face. If the light is just right one may see a single one stand out quite by itself.

Anatomists give us drawings and photographs, the weeklies have occasionally reproduced them, and we all know how our face may look when the skin is dissected away and the muscles cleaned, how distinct they may be. For purposes of printing, their edges may have been sharpened a bit.

In a man's face especially his thoughts and his feelings express themselves to anyone who concentrates on the face, or, better, does not quite concentrate. We may know considerable about that foreigner of whose language we know nothing. His face tells us. Muscle contraction, and the tone of those small muscles, and the state of the small blood-vessels that keep changing the color of the face, all together and joining may speak very clearly of what the

man is feeling. During a conversation in one's family one often does
not take the trouble to hear, because one is seeing. The story is
there. Who has not tried to read the dream in the face of a sleeping
child by some minimum of contraction around the eyes, around the
mouth, over the nose.

Animals who have no words to lay over their feelings reach us in
the same way.

Some of the muscle-fibers of the right side of a human face cross
anatomically over to the left, interweave, and that changeling, the
emotion-loaded human creature, wittingly and unwittingly is pulling his muscles all directions with his nervous-system strings.

They pull outward the angles of his mouth and a smile remolds
into a laugh, but then remolds the other way, into gloom (bad news
has come over the telephone), remolds into petulance (he has said
what he shouldn't). This is gross statement for what may be infinite
refinement of the signs of emotion. Darwin stated that in human
grief the angles of the mouth droop while the brow furrows in the
midline. That at the time Darwin wrote was not gross statement.
What Darwin was saying was then all new. It is still enough new in
this vicinity today. Darwin carefully observed facial expression and
set everything down in his *The Expression of the Emotions in Man
and Animals*. Painters observe facial expressions. Writers do. We all
do professionally and nonprofessionally. The thieving merchant
does—observes any slightest weakening in our expression.

In the face of a cat the expressions may be more lightly laid
down, may not be. In the face of a canary, the enraged male, expressions surely are more lightly laid down than in us, but in some
not exactly identifiable manner we see or sense what we are seeing.
In the female canary the placid thoughts, if we may be permitted to
call them thoughts, seem no more than a breeze over a wet surface,
versus the tempest from below that a moment before lifted the
features of the male and with his feathering darkened the color of
his face.

Your chronically happy colleague, whom you must day after day
work next to, has his silly expressions that reveal his silly emotion,
that bores you to death. Your TV tenor may be famous for the silly
expressions that are coupled with his silly ideas. So it is too with
your blustering drill-sergeant. Your football coach. The lamb of a

man made temporarily courageous because he has had four Martinis, or permanently courageous because he has syphilis of the brain. In all of these the signs of emotions are, and should be, not only more vivid but often more genuine than the emotions. The signs have been of more use, must be at least as ancient, and as rehearsed and exercised by the species. In an old tart the muscles of her face may appear stripped; she contracts them not with emotion but only with the signs of emotion, with her will, which makes it all the plainer that she is only a time-resisting corpse.

Animals have had electrodes implanted at critical points in their brains, and these animals, as newspaper readers know, press keys to stimulate happiness, press, press, press, happiness, happiness, happiness. The wide-spread as against the laboratory interest in those animals is that the mimicry of pleasure can so mechanically be given to flesh, just an electric key. Our day piles up one more instance of machine-cunning, and one more instance, and one more instance, never gets tired of doing it, and indeed it is the laziest thing in the world for a researcher to do.

A pleasure-center in the brain has been defined, verified, and even if it were some day for some reason proven an artifact, no harm in having called it a pleasure-center. The white rat will just the same go on pressing the key, press, press, press, happiness, happiness, happiness, and from the pleasure-center muscles will be directed to operate in unison, and the happy animal will press without stopping, forgo body needs, may become so fatigued with happiness that it collapses. Meanwhile so many microelectrodes have been and are being implanted into so many animal brains, also gingerly into so many human brains, that one could not escape the passing thought that they might be, mostly, not altogether, another of the day's fashions. Once a researcher gets a notion he can be as busy with it as a monkey let loose in a roomful of chairs. Of course, he also is a monkey, the highest, the highest primate, and that curiosity could be quite honest, monkey-honest.

We might want all the more consciously and carefully to watch the small muscles of his monkey face.

We have, then, invested with words a sleeping-center, a waking-center, an appetite-center, a satiety-center, and here it is a pleasure-center.

An actor manipulates the signs by what to an outsider watching the rehearsal of a play sometimes seems a pressing of keys. He up there on the stage sometimes easily reveals a separation of the signs of emotion from the emotion. The actor orders a grin. Orders a grimace. Orders dignity. Has trained himself to release signs on command. Some of the greatest actors have written on the subject, have claimed they were best when cold, their brain pulling the strings, the puppets performing. The trained brain knows that it can depend on its body-parts, the proper muscles will properly perform. Earlier those actors had signs and feelings united, but tonight signs are marching onto the stage alone, the mind having learned how mechanically to cut off or draw in parts of the nervous-system, the actor able confidently to reach into his cupboards of costumes and jewelry, pick the masquerade that fits the scene, we in the orchestra paying $8.80 to watch his skill. Other actors have claimed that they must feel six nights and two matinees. (How tired they must get.) One does not need to be an actor to recognize the two alternatives. A coed on a date recognizes the alternatives: ought she tonight let herself freely express the inside of her, or just again play her expression-machinery? The dictionary defines a date as a social appointment with a person of the opposite sex. Dogs have dates too.

# EMOTION

## *It Feels*

Emotion is old in the world and in each of us. A green fern has none, we think. A green snake has much, we think. We have more.

A list of the emotions might begin with the vivid one so much studied for the signs, rage. Far down the list might be a dull one, but an emotion, determination. Somewhere in between would fall joy, exaltation, love, adoration, heaviness, dejection, misery, gloom, then, to move upward again, fright. Odd how even each word starts something, smog to sunshine. May start a generosity, too, toward those who have the difficulty of stepping even any short distance beyond the words that are so permeating: psychologist, psychiatrist, physiologist.

Somewhere far up that list would have been loneliness, the missing of somebody, anybody, everybody, nobody, the missing, missing, missing, dead damned missing. A human being may be ashamed of loneliness. He may run away and hide it under something else, for instance deportment.

There are the complicated emotions, several layers, that sometimes are begged out or snared out by a physician or a dramatist or a novelist. There are for instance the chagrins. There is the passion of the stingy to save every penny, to eat every cookie, to let nothing be taken away from them, this posing as morality, no-one-will-cheat-me, I-never-cheat. Then there are the many aspects of to-have-and-to-hold, sexual excitement, sexual amusement, sexual maladroitness, sexual ludicrousness, sexual faking, sexual disgust. There are the more plebeian appetites, all tapering off with lessened body vitality, with surfeit, pampered by idleness, going on mechanically. Hunger then may be almost noble. Thirst may be. Thirst may also be torturing.

Envy.

Jealousy.

Humility.

Hostility. Not the hostility of the professional jargon of 1969, but of back-fence gossips and of everybody's relatives.

Yearning.

Enthusiasm.

Affectation. Can be an insufferable woven-in emotion.

Maternal feeling has affectation usually, has pride, has narcissism. A fine unused word that once was, nevermore. All of woman born know how genuine and how fierce maternal feeling sometimes can be, nurtured and swollen and modulated by the mother while she is carrying inside her the fetus that is getting more and more acclimated and comfortable while she is nervous about her own state, planning her future, planning its future, day-dreaming, night-dreaming. Then, finally, comes the Fifth Act, the pushing, the quaking, the stretching, the tearing, the cutting with the scissors, she wildly convinced that now surely she has earned the right of possession, but as the pains die down remembering again, bitterly, that no one can possess anyone.

A plot of untenanted brain undoubtedly can still be found where an electrode can still be planted, the current switched on, something scream out, something bleat out, conceivably only an emphasized silence. But, though for the signs of emotion there were many experiments, and a history of experiments, and a physiology for the emotions themselves, as for any experimental brain-mind, there is not much, the not-much also strewn with inference. There is behavioral phrasing. There are endocrine data. There are autonomic data. Alongside each emotion the physiologist might write autonomic accompaniments: blanched face, blushing face, dry fingers, thyroid signs, adrenal signs, gonadal signs.

Fear.

Fear seems a reasonably pure emotion. Purest? Alongside fear the physiologist might write: over-rapid heart, rising blood pressure, temporary breathlessness, pale skin, eyes wide open, muscles set to let the creature flee or freeze it stock-still. There may be trembling. There may be sweating. There may be many kinds of body imbalances.

In and around and at the back of everything, at the back also of happiness, only farther back, there might be, would be, all the anxieties, the normal ones of every Johnny, the abnormal ones of some Johnnys, life-enhancing, life-depressing, continually shifting. The word *anxiety* is overused. Everybody says that it is overused, then uses it. Its crescendo and decrescendo are a running accompaniment with each human being's story, and many a dog's, and many a cat's. We need it to keep us alert, to prepare us to attack or be attacked. Lacking it entirely we are ill. It may be a mood, a state of tension more pleasant, less pleasant. Possibly it is the commonest emotion in the world, yet someone has settled scores with it by defining it as "concern with future consequences." A few persons may experience it only rarely or slightly. They are the adjusted, the equilibrated, the damned, that is, who shall be gassed or suffocated or whatever other pleasantry there is waiting below, or the righteous, who are able every night to go to sleep with their self-approval, and every morning to rise with it, and with their shallowness. Someone else has settled scores with it too as "unresolved tension."

Music, even music, has been described as anxiety. A musical theme becomes a passing from the unresolved toward the resolved, but just before the resolved is reached there is a quick turn to another unresolved, and so on, the listener moving from blissful anxiety to blissful anxiety. Macbeth moves to the murder of Duncan by a succession of anxieties, and after the murder anxiety piles upon anxiety, keeps everybody in Scotland anxious, and everybody in the theatre, until at the last immediately before Macbeth is slain the long unresolved meaning of the witch's prophecy that "none of woman born shall harm Macbeth" resolves itself in the revelation that "Macduff was from his mother's womb untimely ripp'd."

The uses of emotion?

Besides its being synonymous with being alive (above and beyond the biochemical and the cellular levels), it whips up life, prods the daring for life, protects body and head for having dared, but distrusts, therefore keeps the head on the lookout.

Without emotion there is nothing that could be called mind. Clarification of thought depends on it. Who has not lowered his voice to get cool his thought, got it cool, tried to reason in that cool, failed to reason, put all off till tomorrow, yawned, grew dull, grew annoyed at being dull, blazed into anger, got awake, stayed awake, then toward morning reached an obvious conclusion? What a fool one can appear to oneself toward morning. On a historic occasion the conclusion has been more than obvious. Galileo, facing the Inquisition for a mathematical derivation, that our earth was a rolling ball rolling around the sun, did not make his additions and substractions without emotion, because the man who grasped that our earth was a rolling ball grasped that fact hard or would not have grasped it. A mere mathematician, no. A mere mathematician would not have faced the Inquisition. Galileo was willing to face prison or death for a fact. Socrates died. Christ died. Those are extreme examples but the feeblest scientist would admit that emotion, besides readying our muscles for action, intensifies our convictions concerning the action.

Put another way, emotion is the clarifying heat of the mind, or the ice of the mind, that mind that the physiologist said was for the ultimate perfecting of the body's movements.

Hate may marvelously clarify. Shouting out may. Psychiatrists advise us to shout out. A high love clarifies. Arm in arm for an evening and everything is so clear, for an evening. In Beethoven there was a furnace of fire, and clarity. In Shakespeare there was a furnace of many temperatures, and clarity. What those two brains were like! What a brain and a mind can be like! Yet a physiologist comes with an electrode and a psychologist comes with a black box, and both come with a white rat. In the depressed young couple who travel from the hills to the large city and sit on the stone steps, there is no furnace, no clarity, and they sit.

Emotion prolonged is mood. Emotion getting out of hand is mania. Often one thinks that emotion exists in packets that have compression values, low behind the enjoyment of the sight of a child waving shyly from the back of the automobile just ahead, or the sight of the calamine blue of a January sky, and high behind a murderous plan, behind an erotic anger. He bites at his wife, with or without his teeth, knows that to bite at his wife is uneconomical but who else is it legal to bite at, and he has such trouble remembering what the biting costs, the bad effect on the bank account, the bad effect on the children, the bad effect on his sleep, therefore on tomorrow's gain, therefore in that equivocal state he begins to work on next year's tax return until the fire in him is ash, not enough even for a tax return, so he slumps before his TV like any simpleton till he decides to go to bed and see what can be done to make up with his wife. Galileo was willing to live for a mathematical derivation, to rise to heaven, to find the derivation wrong, but he and his psyche would have risen together, we hope. To die must be a breath-taking emotion, and no pun meant.

# EXPERIMENTAL NEUROSIS

## *Svengali*

The scientist can imbalance the emotions of an animal. He can push farther and produce neurosis. Persons with experimental talent have been doing that, producing neurosis, for a third of a century, in dogs first, later in a variety of animals, goats, fish, spiders. If a spider is driven neurotic he weaves neurotic spider-webs and is claimed to reveal the type of neurosis in the flaw in the web, type of the flaw, place of the flaw.

Pavlov produced experimental neurosis originally by chance, then produced it in dog after dog, soon was doing nothing else, producing, studying, treating neurotic dogs. He cured them too, not entirely. While he was doing that he was steadily reconstructing his scheme for his higher nervous activities, and after a time had put into his scheme the thinker, the artist; and then these sick brains, as he saw it, of the neurotics.

Pavlov had known neurotic dogs in the streets of Leningrad, as we all know them in our own streets, and he found them among the strays lured to his laboratory for conditioned-reflex experimentation. But, by 1930 he was manufacturing them, became highly proficient at it, and on the basis of neurosis in dogs laid down the laws, as he saw it, for neurosis in man. Later he established a small psychiatric human clinic. He was old by then. He began reading psychiatry. He never had previously, in general was not a reader, only belles lettres, as he said, on his summer holiday.

He was utterly convinced that the wrecked human psyche that goes to the psychiatrist, and that came to his small clinic, was in mechanism not different from the wrecked dog psyche that he produced. He was upsetting the higher nervous activities of these dogs—he never would have said psyche and still never said minds. With the upset nervous activities he upset the bodies, and he knew how his experimentation and his thinking would extend outward, and he would not have been surprised that widely separated movements, like brainwashing, like hysterical snake cults, would later seek their nervous-system explanation in his dog findings.

The first neurotic dog was another of those laboratory accidents that writes history. He had presented the dog with the problem of distinguishing between an ellipse and a circle, no-meat for the ellipse, meat for the circle. Soon the dog had made that differentiation. It drooled saliva for the circle, none for the ellipse. Pavlov knew therefore that that dog had learned.

The ellipse was a long narrow one, a ratio of 9 to 2. Pavlov altered the ratio, 9 to 3, then 9 to 4, so on, waited each time until the dog had learned, had made the differentiation, and continued this until the ratio was 9 to 8. It may even give the human eye some trouble to differentiate that from a circle, and for the dog it was too great a problem. Pavlov persisted. The problem became an agony. Pavlov persisted. Abruptly the dog's behavior changed. It was peculiar behavior now. This was a mentally sick dog. Pavlov would not have said it so. He would have said no more than what he regarded as fact, that this dog had a brain, which was a delicate instrument, could be overexerted, or wrongly exerted, subjected to stimuli that were too confusing. There was a clash. In a human brain the same confusion would cause the same clash. Months of rest with sedation were required before that dog was able again to do a day's work. (*Work* was Pavlov's term for what a laboratory dog did on a laboratory day.) Even after recovery there might be relapse. Pavlov called what he had accomplished experimental neurosis. The phrase stuck. For sedation he used bromides.

A flood in Leningrad upset the minds (brains) of a large population of dogs. The River Neva rose in the kennels, some dogs were drowned, others escaped but the water had been up around their necks, and in consequence of that terror the behavior of these dogs was so altered that anyone would have admitted they were neurotic. Pavlov would also not have used an emotional term like *terror*. But this was life neurosis as the other had been experimental neurosis. Those River Neva dogs also required months of rest before returning to their work. Superficially they might seem cured but they were not. Pavlov would play Svengali. He would trickle water under the door where a dog was alone in its kennel. That was enough. That would be enough too for you and me, make a Poe or a Maupassant tale, water steadily advancing under our door locked from the outside.

In the course of his long study of neurotic dogs Pavlov became convinced that whether a dog did or did not easily slip into neurosis depended upon its temperament. Dogs were born with their temperaments. Dogs' higher nervous activities were genetically different. Pavlov decided that any extended study must recognize those differences, and accordingly began classifying dogs. No classification proved completely satisfactory. The one he settled on was that which the Greeks used for human beings: choleric, sanguine, phlegmatic, melancholic. He decided he would find those temperaments, and then breed them. He would breed standard dogs. He would have enough of them. Then upon standard dogs he would perform quantitative experiments and arrive at quantitative laws for the deviant mind. It was ambitious for this early point in this type of animal experimentation, and it was like him. He was well on the way with it when he died. More and more enthusiastically he waved his wand and altered minds, dog minds, but whoever saw those dogs found it easy to believe that he was seeing the characteristics of any neurotic mind. Pavlov's dog-hospital was an old-fashioned state-hospital.

In a dog that was by nature an inhibited animal he would intensify inhibition, which was to intensify shyness in a human being born shy. He would undermine a strong temperament by castration. He would present an intelligent animal with a problem that was simply too complex for it, like that of the ellipse and circle. Collision in the nervous-system was Pavlov's statement for what had happened there. He performed hypnosis like any psychiatrist, but animal hypnosis. He continued to believe the hypnosis explicable on the lines of his theory of sleep. For him hypnosis still was partial sleep. It was a brain with a patch of excitation and all around it inhibition. Pavlov's way of wording it was: one or more analyzers asleep.

Pavlov would make a dog cataleptic. He would say that he had produced a plastic condition of its nervous-system and on that account was able to mold its body into bizarre postures that the dog would maintain, like the cataleptics known for generations in lunatic asylums. He would crack a nervous-system by pounding it with stimuli too powerful for it. Would produce what he called transmarginal inhibition that would spread, that would irradiate. He excited what he called paradoxical and ultraparadoxical states. It was a new language for new mechanisms. No need precisely to understand. No need fully to believe. Pavlov believed. The states have before and since been turning up here and there on the human scene. If Pavlov ever thought of chemical mechanisms for either neurosis or psychosis, he never said so. Everything was environment versus nervous-system. He thought he had produced paranoia, and it looked like it, a dog with a fine intellect regarding most things but mad toward one. The inmates of his laboratory were undoubtedly inmates. They worried the observer. Pavlov's interpretation always remained conditioned-reflex interpretation. A psycho-analyst digs back into the history of a human patient to find the destructive experience, and to find it is to treat it; Pavlov supplied the destructive experience, as that problem of ellipse-and-circle poisoned with meat-powder meaning, made the dog sick, then treated the sick dog. Pavlov often said that if stimuli are strong enough or the nervous-system weak enough, neurosis results. He believed psychosis does too.

# A CASE
## *Method in Its Madness*

This is not the case of a dog.

This is the case not of a young man, but certainly not old, a blond, soft-spoken, and one was always having to prick up one's ears to hear what he was saying and one wanted to hear. He looked as if he might have a story in him. But everyone has a story in him. On his side, he wanted to be heard, nevertheless regularly slid off into lowering his voice until it was out-of-hearing. A feminine type but a father. He had sired three children, had gotten them as far as high school, supported them, and they were healthy enough. Three was the right number; families were having three; families used to have four; and long before that they might have eleven. It would mark you as outstanding in the community if yours was the family with eleven, and the newspapers would say in the death notice that you had been a good citizen. That is, everything argued this man's more than willingness to meet his social obligations. For a long time he had been having trouble with his foot. It was really bad, deformed. It had chronically gotten worse and looked as if it were going to go on getting worse. A spasm in the muscles of the leg had shortened the leg, had twisted the foot inward, and at the present time it was bound down with contractures. That word *contractures* explains itself. Not only was the leg drawn up on the thigh but the thigh on the abdomen, the whole limb withered. It was really absolutely useless. Painful? Of course. He had been bedridden since March.

His back ached like a toothache. Indeed, touch any of a dozen places on his body and he squirmed. An intern checked off gout, arthritis, joint tuberculosis, a list of textbook diseases. (Then the intern went off to his coffee-break.) As for living with his illness, as people say, this man did it. He was not in the least the complaining type. His life was in no danger, and he did not pretend that it was, but occasionally he did say that that was unfortunate. There was not one of his friends would not have defended him hotly had someone suggested he had done this to himself. But he had, in a way.

It began far back. There was this and that and the other thing, slight signs. They had not seemed worth mentioning. He recalled how he had begun to be uncomfortable when he was just sitting thinking. He had the habit of doing that, was philosophic, his friends said, went over matters in his mind. Anyone would be going over matters in his mind, he thought, the times being what they were. As to his condition, he remembered scrupulously everything he was ever told. Once he was told that he must have had an undiagnosed poliomyelitis. But this was not true; medical opinion was definite, the deformity could not possibly have been left behind from an old poliomyelitis. On and off he wavered about that. How could one be sure? The doctors ought to know but they do make mistakes. There had been a galaxy of doctors. He listened to them all. He had a genuine interest—objective, he thought. He thought he was the kind of person who could honestly have that kind of interest. Even so, the diversity of medical opinion was bound to do something toward sapping one's courage. Plenty of non-medical diagnosis had blown his way, of course. It always does. All he had been told, or half told, some of it entirely apocryphal, and all that he told himself, the complete account amounted to a conspiracy in which he took part. Everything presented itself to his mind in as unfavorable a light as possible, and he helped in that. Everybody helped. Tongues do wag too freely. Everybody whispers, and nobody thinks that anybody hears, but everybody hears, even hears when there are no whispers. Everybody knows everything, especially about another person's illnesses. Everybody thinks he has the right to know. So-and-so has had a colostomy. Now keep that a secret! So-and-so has had a second heart attack. What business are such facts to anybody? The beasts in the field do have one advantage over us: they do not gossip.

Rheumatic fever had been suggested by one doctor, but rheumatic fever leaves marks on the heart, and another doctor said there were none on his.

Of course, his pain was fact. That was sure. Pain was all day and all night stamping about his body. He said that, but he made light of it. The contractures themselves would have caused pain enough. The pain was utterly explicable. Some of the pain may not have been as explicable but it did not on that account trouble him any the less. When one of our limbs gets out of order, the other limbs get out of order too, it seems.

He had repeated the history of his case so many times that he was tired of doing it.

In one of those repetitions he ought to have noticed, and noticing might have waked him to something about his mind, that the time of the onset of each new set of symptoms corresponded to some new difficulty dropped on his life. The correspondences were almost too pat. The last was when his dear wife without announcement deserted him for a real man. That word *real* was a neighbor's, but he heard it. From the day of his marriage, the neighbor went on, his wife never had been able to stand having him around. She was quite frank about it. Once in a rage she blurted out: "He makes me puke." That was a strong thing to say. She probably did not say it twice. She probably did not mean it.

She undisguisedly used him as a male escort. This he recognized and accepted. "She suits me fine," he said with such regularity that a friend of his came to the conclusion that her neglect of him, her cruelty toward him, made up to him for the luck of his having a superior wife. Everybody saw that she was superior. The difference between them was so great. Everybody liked her. She was bright and she was contemptuous, interestingly. He liked her too. Even now he liked her. "You can't help liking her." He also said that with regularity. He would call her his "little wife" and say she was "pretty enough." In the last three years on a dozen occasions she did not come home all night, and when a neighbor remarked about that, he did blow up, and this was not at all characteristic of him, and when he cooled down he said over-quietly that every woman had a right to some private life, and marriages were just too sticky. It was before his latest relapse, which was the worst of his relapses, that she finally dumped everything on him, the three children, the household, fled. That did upset him. Desertion, humiliation, children, debts, all put on a person equipped to meet only the luckiest of lives.

The line from a disturbed life to a disturbed mind to a disturbed body is seldom as straight as this, and the line here can be trusted, as far as any such line can be.

So, he lies there. He likes a neat bed. Is a gentleman. Wants to express himself to his physician in good English. Wants to give the least possible trouble to his nurse. It would have surprised him, enraged him, then would have begun to worry him, if anyone had implied or said that he was hiding away in that snug bed, hiding from the world, a big-bodied man, letting a nurse take care of him.

Actually, the nurse was impressed by how he ignored his deformity. She was just as impressed as anyone else.

One thing was fortunate—his religion. At least it was fortunate at this point of his life. He was devout. He expressed his devoutness. He folded his hands in the old-fashioned way when he prayed. His piety helped him accept. He mashed his toe when he knocked the lamp off the table next his bed, insisted it was nothing, whereas everyone cringed at the weight of that lamp and the look of that toe. His mother sighed. She said: "All his life he has been that way, met everything that way, cheerfully." An excellent woman this mother, the neighbor said, was so considerate of him; when he was having a visitor she took it upon herself to answer every question directed to him, saved him the exertion. She was outspoken about his trouble and knew exactly where to put the blame—his wife. The neighbor agreed. The wife was the villain. Everything began with that marriage. He had been such an affectionate lad even as a baby.

With so much having been said about him by everybody, so much suggested, so much and such kinds of things having happened, anyone would have figured something, but it should be kept razor-sharp that no one who had seen him would have been apt to underestimate the severity of his deformity. It would have been prima facie ridiculous to get the idea that that part was in his mind. That was in his body. It was beyond repair. A surgeon might help in a palliative way. A wise and compassionate physician might clarify to the man's mind how these things came about, give him a better view of himself, and this might help or harm or probably do nothing. A psychiatrist might do the same more professionally, might also probe back farther into his past, or nudge him to do the probing, especially that far past that it is difficult or impossible for anyone to probe without assistance, the time before any of this started, before any imagined poliomyelitis, any rheumatic fever, any wife, back there where the hell of infancy is supposed to smolder. If the man survived the inquisition, and the psychiatrist, and the bank account, and he not ashamed to look into the face of his mother, or even if he was ashamed, some deep reason for the illness might have been discovered. This would have helped his thoughts. But it would not now have helped his deformity. That was past helping.

If the psychiatrist was skillful he might even have uncovered the special occurrence, because there must have been a special occurrence, that led to the foot. Why just the foot? How did the foot get into this? Most of the psychiatrist's findings would have been provocative, might not have been clinching. (Better that way.) Any such clarification might ward off some future added trouble, first to the mind, then once again to the body. (Who can tell?) So, he lies there. One hates to embarrass him or embarrass oneself. One finds oneself never looking into his face. One finds oneself in one's mind right there in his room reviewing his story, trying to fasten when and where and how he began deceiving himself, became so expert at it, so addicted to it. A stranger entering the room, even a friend who knows as much as anybody, would have his thoughts, at least his eyes, drawn toward the man's foot and away from the man's face and mind. That is what the mind wishes.

The foot is doing its job.

# PSYCHO-SOMATIC

## *The Head Has a Body*

Twenty-six centuries after Homer, women got the peptic ulcers, and the next century, ours, men got them.

If the ulcer figures are not just figures, since God Almighty fashioned Eve from Adam's rib rather a long time ago, and since anatomy and physiology could not have changed much in the interval, something else must have. It could have been society, which does change enough to nag at the human emotions, and the nagged emotions nag at the human body, and if the nagged body is damaged, that is the psycho-somatic.

A generation or so ago an authority writing *On Simple Ulceration of the Stomach* pointed out that after an ulcer had healed it was apt to recur "from strong mental impressions." Ulcer is frank disease, can be seen with the X ray or during surgery or at post-mortem, and something in the mind did gnaw at the flesh of the stomach; the gnawing was related directly or indirectly to the "strong mental impressions."

Freud (Austrian), Pavlov (Russian), Cannon (American), brought their different talents to work on aspects of the psycho-somatic. Freud defined the unconscious, defined conversion, gave them reality, left illustrations, cases where mind then body were disturbed.

Pavlov at about the same time was introducing experimental neurosis, producing animals with severe neurotic problems, mind-sick body-sick animals that, after he had made them sick, he found ways of treating and curing.

Cannon dwelt on the physiology of pain, hunger, fear, rage, leaned on Darwin's study of the signs of emotions, kept his laboratory experimentation partly in that direction, partly on the autonomic nervous-system.

The achievements of those three men, different in value and in force, were largely completed in the first third of the century. Since then practicing physicians have frequently made incontrovertible that body illness can result from emotional illness. Psycho-somatic medicine—the illnesses, the causes, the diagnoses, the treatments— will in the future either have become fatally defined or shown to mean something else.

Many outside of medicine, protesting campuses, protesting blacks and whites, protesting teen-agers and the police, protesting school-teachers and schoolboards, the for-and-against glue, the pill, abor-tion, are daily leading us away from some of our older thinking. Earlier in the century many besides Freud were doing the same with still older ways of thinking. Ibsen. Strindberg. Chekhov. Nietzsche. Literary men and anthropologists and evolutionists were aiming their fire against nineteenth-century prudery, against what today seems to have been gross promulgation of ignorance. The myth of the happy family circle, the myth of man's good will to man, the myths of sex today are discussed by parent and pedagogue and priest and by the children. The gestures and postures of fornica-tion are freely exposed in automobiles, on the stage, and in the streets. *Ghosts* is a dated play. O'Neill fails to shock. A group of high-school students reviews James Joyce's *Ulysses* and remains matter-of-fact. "All of us sat on the floor and asked each other straight-shooting questions." Despite the emancipation, neurosis ap-pears on the increase, and psychosis, and the psycho-somatic. That may be sharper recognition of the illnesses, partly.

The illnesses are believed to occur in one of two ways: as symbol, as altered physiology. As symbol: an example would be that hysteri-cal paralysis of an arm, the arm symbolizing the useless or the unwanted, person, thing, event. As altered physiology: an example would be the excessive battering of an autonomically controlled organ by an autonomically elevated blood-pressure.

The damaging blows across the border from mind to body excite hypotheses, observations, experiments, broaden our view of mind, and give us nothing too clear. In conditioning techniques, during the period of delay between the stimulus and the response it is simple to produce disturbing signs in the stomach, in the heart, elsewhere, these far outlasting the stimulus, the heart still pounding and racing after the experiment is over. That might be one way heart damage could occur. In hypertension, high blood-pressure, because of, say, an unrelieved silent rage, blood-vessel walls are battered, the vessels narrowed, notably those of the kidneys, the heart having thereafter to move blood through narrowed vessels, this costly both to heart and kidneys, the sick man possibly stricken by uremia, possibly dying after three days and nights of Cheyne-Stokes breathing that reverberates through the house. In another man or woman, instead, there may be a holding back of tears, a holding back of rage, a holding back of secrets, and that might go over, by one theory of the mechanism, into a holding back of breath with damage to the bronchial tubes, the diagnosis asthma, though the cause of asthma still often seems a mix of physiological factors and medical misunderstandings. Asthma has been called the "grand-daddy of psycho-somatic diseases."

How mind digs into the inside of molecules and changes the position of atoms, how that changes the inside of a cell, how that spreads to other cells, to whole organs, this all continues to inspire a guessing that keeps our thoughts on fascinating aspects of life, with mind always coming like light through cracks.

There are physicians who insist that even in a disease proven to be caused by bacterium or virus it is the side-effects on mind that are the reality. These physicians believe what they say. Their opponents—the two occasionally are near to enemies—insist that mind never causes disease. Disease to them requires the microscopic, or the otherwise visible, or the countable, or the chemically traceable. Amazing to have lived a lifetime on this mind-glorified and mind-dirtied planet and not be able to imagine mind responsible for disease.

In practice psycho-somatic medicine uses all the tools. It brings in all the disciplines, physiology, chemistry, pharmacology, surgery, internal medicine, psychiatry. Mind (emotion and intellect), and demos (individuality and society), and brain (limbic and other systems), and body (organs), run around in a circle of cause and effect. There are psycho-somatic books, psycho-somatic journals, psycho-somatic societies. Seminars. Conferences. Lectures. Clinical investigations. Laboratory experiments. Chairs in colleges of medicine. On some days our world seems only ambitious and adolescent.

Human beings have been typed in ways reminding one of Pavlov's typing of dogs. There is the hypertensive type, the ulcer type, the accident-prone type. Men of action belong in the hypertensive. Men who require the love they were deprived of as infants belong in the ulcer. Such irreverent oversimplification would offend professional opinion, but it is indicative. One investigator argued that an abnormal body-part resulted from an abnormal gene, an abnormal behavior from the abnormal body-part, an abnormal mentality from the abnormal behavior, a back-and-forth between them all, or any, the partners in the square-dance dancing each other around until there is such alteration in the flesh that even a hard-headed county coroner must admit there is pathology, damaged flesh.

The wish to connect mind precisely to the flesh is also a sign of the times. Sherrington's dualism has lost by lengths to Pavlov's monism. Pavlov's fusion of subjective and objective falls within the comprehension of any Simple Simon. If someone in the crowd objects, most in the crowd are baffled by what he could possibly be objecting about. Even the matter-and-energy-are-one of the physicist gives support to the mind-and-energy-are-one of the psychosomaticist. Yonder in the morgue lies a tramp. Yesterday he had a hemorrhage in his brain, collapsed in the street, could not hear, could not see, could not speak, today is dead. The brain-cutter looks into his brain. Plain as the man's bulbous nose is the bursted artery and the clotted blood at the precise point the intern in the ambulance said. Hypertensive type, said the intern, wrote it into the chart. Dead, said the intern, at 3:05 A.M., wrote it into the chart. The tramp's mind and body lived together, were broken down together, died together.

# NEUROSES

## *Works Its Way Out*

We are more concerned with the neuroses than with any other illnesses excepting that hostile triad, cancer, stroke, heart disease. What in generations past a person thought his private, his personal problem, or did not think a problem, did not think a neurosis, in fact did not know there was neurosis, now thinks, now classifies, now turns over to a specialist, as he turns over his tax affairs to another specialist.

Also, what were formerly severe mind disturbances the psychiatrist and chemist and other professionals have made less severe; at least, they are less severe; at least, they appear less severe. On some days, notwithstanding, bothered minds seem as numerous as the heads under them. The heads are individually bothered, but they are heads, and similar, and their types of trouble have long been classified. The classifications have never been quite satisfactory, hence new classifications, and though much has been learned, much has been changed, much also remains.

Psychopathic personality used to be a phrase on many tongues—a decade ago. Psychopath and constitutionally inferior—two decades ago. Character disorder—today.

Ladies and gentlemen then as now, 1909 as 1969, have thrown the changing language as sobriquets at each other, or at absent friends, in a cocktail bar. We choose to explain our friends, when they are not like us, as mental aberrations. With the specialist this is something else. With him the terms, when he tries to clarify them, even when he clarifies them wrong, as a good part of the time he must, help to bring some order, or some further order, into the eternal bewilderment of mind despite his often adding his own additional bewilderment.

One point more, before touching a few spots merely to indicate an area: changing terms with changing times are somewhat the changing knowledge, not change in the bothered minds. These change largely on the surface. These are affected by a society that changes largely on the surface. In the depths even the bothered minds apparently change most slowly, if at all. That, too, has usually been the conclusion of the philosophers of history concerning all minds, the bothered and the less bothered.

Psycho-dynamics is the mechanism of mental illness. Psychotherapy is its treatment. That essentially has not changed, only the terms.

The overall difference between the neuroses and the psychoses, until there is completer understanding, is most safely considered a difference at their roots. That essentially also has not changed, and not the terms. Both are different from feeble-mindedness, from mechanical or gross chemical damage to the brain, from addiction, cleptomania, criminality, possibly from states listed under character disorder or neurotic character.

In the neuroses the physician looks carefully at the society around the patient, in the psychoses looks at the mind.

Hippocrates knew hysteria. Hippocrates' remedy was marriage. Marriage may have been the best remedy in the Greek world, as often in ours. Hysteria in Greek medical opinion was attached to the womb, *hyster* meaning womb, the womb imagined to get loose from its moorings, skate about the body causing signs and symptoms. In that general neighborhood, the womb, but mostly in its psychological suburbs, we today, too, believe that signs and symptoms arise. The idea as accepted by Hippocrates was accepted for centuries. In the ancient world there would have been no hysteric men.

When late last century psychiatry entered its modern phase hysteria became for a while the overall title for the neuroses, today is one class, still the largest.

In the neuroses illness of mind frequently becomes illness of body, a turnover called conversion. It has been called that for half a century. A mind in trouble finds itself a body-illness. If not a body-illness, a behavior, a determination never again to dress like a mouse but wear a vermilion waist with a midthigh purple skirt and the most bouffant hairdo in town. It may be any quirk, anything that provides a pillow for the mind to lay its head. The damage may come another way. A man may have an excessive, a hopeless feeling of futility. He may be convinced that he can do no more with his life than he has, so drinks, keeps that up until he can in fact do no more, must rest his dead-drunk head on a pillow of drink. The neurosis usually spreads around. His wife—disappointed—flirts. One of their older children—disgusted—commits armed robbery, and all society is touched by that, and the police, and the psychiatrists. The social conversion is seldom that clear-cut or widespread. But the inclination to look at everything widespread has grown with the generations, and parallels the increased prying of society. But it began with the man, his need to get relief from the gnawing at his mind if he must stay alive. His neurotic behavior had been—weary word—a defense.

As a result of the growing tolerance, so far as it is tolerance, so far as it is not neurosis in the total society, the mind of the community blames single minds less, which sometimes is fortunate, sometimes unfortunate, both sides of which we all know. Murderers are blamed less. They blame themselves less. One eminent psychiatrist has brought himself so far as to be able to see the punishment as the crime. Anyone having perpetrated anything may cry expectantly from his prison cell: "Have pity on the poor neurotic." We are much products of our environment, but most of us, when alone, know that our genes, or whatever, include our way of reacting to our environment. A wilted carnation (maybe) knows that it is a carnation, and suspects (maybe) that it has some veneer of free will, however it defines free will, some responsibility, even regrets that it did not somewhere acquire discipline, make itself, if not happy, somewhat less subject to catastrophe, help its society in the same direction.

Hysteria, today's hysteria, that one class of today's neuroses, has many guises but probably the same mechanism. A woman suffers a sudden frightening reduction in power, cannot stand on her legs, sinks to the pavement, a man thinks she is having an apoplectic stroke, nervously asks everybody around whether he should go for help, while another man keeps slapping her face as she lies there on the wet cement. The reduction in power may be in a single limb, may be anesthetic and not paralytic, but, either way, not explainable by any anatomy of nerves. Anatomy is the same for everybody but the losses of the hysteric would require a new anatomy. A stocking-foot anesthesia has the sharp upper border of a stocking, and no anatomy for that, and a glove-hand anesthesia the sharp upper border of a glove. Most frequently the hysteric is a woman. She extracts satisfaction from her suffering. She endures in silence. She is apt to be young. Has had life's usual difficulties. May not be too intelligent. May have a good idea of what her mind is doing. "I know I am a neurotic." She says that while she puts lipstick on lips that have too much. She must not have too clear an idea of what her mind is doing or she might be discharged from the clinic as a malingerer, and that might make her worse, or better, and even a skilled physician might not be able to predict which. And there would be two points of view also about what malingering is.

Each of us sees the operating of some of the trickery in his own mind. He feigns a headache to escape the boss's esprit-de-corps dinner, after a while has the headache. The boss's dinner is the illness of mind; the turnover, the conversion, is the headache. That is a normal enough conversion. There may be most abnormal ones. There may be blindness—so as never to have to see the foul earth. May be mutism—so as never to have to speak to or of or on the foul earth. Persistent vomiting—to make plain what one thinks of this fourth smallest of the nine planets. Immobility—never again to have to move below the merciless firmament. These interpretations are too pat, but the actions may be also, keep the observers fooled for a lifetime, the person himself fooled even if he gets as far as suspecting he could be.

Hippocrates knew anxiety. We are all apt sooner or later to run the full scale of normal anxiety. Some run the scale every day. Anxiety can be a symptom of any neurosis. It can itself be an unmistakable neurosis, called then anxiety-neurosis, a severe mental illness. The sufferer frequently is a man. He says there is no reason for the way he feels. There is. There is (probably) some hidden conflict—another weary word. He lives in unrest. It is genuine. Contrary to the hysteric he feels not less body power but more. He too experiences that turnover called conversion. It may go to any organ, heart, stomach, a joint. (Pavlov would have said that that joint had been neurotically conditioned and that he could produce the joint experimentally.) The symptoms of the anxiety-neurotic may be diverse, a pressing in the belly, in the orbit of an eye, in the skull, anywhere. A mind let loose is inventive. When the conversion is to the nervous-system the man may think he is going insane—if he thinks he is he probably won't.

Depression, some level of it, also is normal. It varies person to person and situation to situation. A person who has had misfortune and is reacting to his misfortune is reasonably depressed, the depression is called reactive. Many depressions are not reasonable, are neurotic, may be slight, may be painful, but really not harmful, are spread through the populations of the world and our town, increase disproportionately around the holidays, seem then to come out of their corners. Everybody has experienced degrees of them. Then there is psychotic depression, profoundly different, probably not merely some unresolved conflict in a man's life, probably some difference in the molecules of his brain. Psychotic depression is not quickly over.

The sufferer may slash his wrists, attempt hanging, take an overdose of sleeping pills, some unoriginal method of suicide and usually bungled, but not always bungled. There is that in his brain-mind that would wreak the final havoc upon itself. This psychotic needs to be watched if we continue to think it is our business to keep everyone alive. On the occasion when this psychotic does not bungle, our obligation to save him who does not wish to be saved will have been frustrated, and there will be satisfaction in that too, relief at needing no longer to stand guard. Whoever attempts suicide is peremptorily classified as neurotic or psychotic, and (this is remarkable) criminal. He may be neurotic of course. "My Tony has committed suicide again." He may be psychotic of course. A supremely sane suicide never gets credit.

Hippocrates may not have known what we call traumatic psycho-neurosis. The person may have had an accident but he may also not have had an accident, a real one. Psychiatrists say that if he had the accident he would be less apt to have the neurosis, would have "cathected" the neurosis into the accident. Or it may have been a small accident, not much body-damage, nevertheless blamed for a prolonged illness. "Never was right since that day." The man's family and his doctor may keep the neurosis going, though likely that the person was a neurotic from far back, merely pushed his mind's burden onto his body. Blames the soulless factory. It may deserve the blame. If the money compensation promises to be sizable, that helps the neurosis. Accident may follow accident, the man have a talent for accidents, accident-prone, needs accidents for the importance they give him, breaks his bones, lacerates his flesh, and his friends say he is nice but awkward and his wife screams at him to keep his hands out of the china closet.

Any one of these neurotics at any point may feel guilt, then feel shame because he felt guilt, and so on. He may need to feel full of sin, feel full of penitence, therefore rid of responsibility. Henrik Ibsen's characters often thrive on guilt, as those of a dozen contemporary writers.

The wars showed neuroses latent in many soldiers. Combat or thought of combat was intolerable to their minds, and symptoms appeared. To be sure, bullet-lacerated flesh is real, as is death, but the equilibrated mind is presumed able to meet adverse reality either with successful illusion formation or unflinching intelligence.

Our streets and parlors and Sunday supplements buzz with psychiatric terms. This person or that is in the genital phase, anal phase, oral phase. People love such words, wallow in them. It is part of the neurosis of the times. The skinny boy next door assures everyone that old Mr. Ross is obsessed and compulsive.

Hippocrates probably knew compulsions and obsessions. There are elegant compulsions, ways of eating potato chips, snuffing cigarettes, peeling off nail polish, twirling a dinner-ring, counting without seeming to count one's change, little scenes that multiply before anyone's eyes. We all are compulsive, could not get done our normal life without being, or are convinced we could not. But, again, there are the body-destructive society-destructive compulsions of the abnormal, persons who must shoot persons whom they have not so much as met, persons who must wash their hands two hundred times a day. A nightwatchman in a college goes from room to room and before he opens each door, before he touches each knob, puts on a glove, refuses to turn that dirty knob without putting on a glove. The psychiatrist says believably that that glove covers something else in the man's past, the man's mind, that he shrinks from touching. Troubled human minds grope in dark rooms, in sunny streets, New York, Osaka, Rome. Troubled bodies bow, scrape, are full of politeness, full of palaver, all to balsam raw depths. Every one of us is obsessed some of the time, compulsive some of the time, listed as normal but shading off into the abnormal. And then there are those others who are abnormal most of the time.

The neurotic is living in the same reality that we live in, mostly. The psychotic is cut from that reality, mostly. It is a long-standing conclusion that he is living in the only world he can endure. He requires his delusions (his room electrically wired by an enemy), his profound responsibilities (he too large to have come freely from his mother's womb and therefore damaged her irretrievably and doomed to hell). He requires his voices. He lives with them.

This could all have been said long before Freud plunged in.

# FREUD

## *Helped to Comprehend Itself*

Often Freud has seemed a dark fisherman off in the morning with rod and notebook, stubble on lip and chin and cheek, darker as he got older, fishing for species never yet classified, fishing where others fished but they did not know how. Occasionally what he pulled in was a glimmer not a fish. Occasionally he seemed to have a fish but it slid off his hook. Occasionally a haul. At night that fisherman wrote into his notebook the catch of the day, not what he had hoped to bring back, what he brought. Peculiarly honest he was for a fisherman—was born fisherman.

Freud would have allowed the figure of speech, Sherrington been vexed over it, Pavlov vomited upon it, Cajal turned his back on it.

The one was Austrian, the one English, one Russian, one Spanish, four who in the first half of this century were the pillars under all study of nervous-system and mind, and, dead now, still are the pillars. Each could express himself, and each was allowed a span of life long enough to get himself expressed. Freud died at eighty-three. Sherrington at ninety-four. Pavlov at eighty-six. Cajal at eighty-two. Each was born in the short span between 1850 and 1857, the explosive *Origin of Species* with its smash-hit for determinism published in 1858 when they were children. Each was scientifically creative. Each had a tough mind and a tough body, had to have, worked as men work who change some patch, and change some patch each did, worked *comme un bête,* Claude Bernard said and Freud repeated.

Sex-begins-in-infancy was Freud's most revolutionary discovery, and the fact that he could make it sketches the first lines into that mind-lined face. Sex was what he meant, literally, infantile sexuality. In 1900 that was repulsive. In 1969 it is matter-of-fact, or shrugged off. When the idea sex-begins-in-infancy occurred to him, he thought, as everybody knows in this sophisticated day, that the parents or the uncles were the sex aggressors, attacked the infants, thought he was forced to this extraordinary conclusion by his own and his patients' self-revelations. But impossible—too many wicked parents. What to do then except suspect the infants. They were the incestuous; as they grew older, phantasied, pushed the incest over onto their parents, made them responsible. (Not that there was never an incestuous parent, because there was, of course.) That region of the mind where the incest was pushed became increasingly central in Freud's files on the mind, and as the years passed he filed there more and more, everything that had been or could be repressed. The term *repressed,* and what is barnacled to it, we of this generation have known all our lives, first had difficulty with it, and with the other allied terms, then became amateur analysts, as everybody in the western world, to the disgust of the analysts. Sometimes we were annoyed at the roadblocks we thought Freud placed with terms, though we granted he needed to introduce them because his subject was fresh, improbable, then probable, then improbable, so on, less improbable the second and third times around. Frequently we were open-mouthed at his insight. Frequently we wondered whether this was not all enlightened charade.

For the nonspecialist what Freud first in the world presented to the human mind was the human being with the dubious privilege of crossing the threshold of sexuality not once but twice in his life, as infant, as adult.

An idea like that required, besides observation, fancy, and Freud's was a fancy everlastingly checking itself, cruel to itself. Undoubtedly, Freud was the creative writer as well as the scientist who writes, should be read with that at the back so that his interpretations do not get a literalness they probably do not always have. Of course, he won no Nobel Prize for writing, for literature, won no Nobel Prize for anything, the committees never adjudging him deserving, which can be explained, yet when one thinks how long he lived, that is, how long he was around, how lately he died, how tremendous his influence in the world (regard that good or bad), the fact of no Nobel Prize is arresting.

To the Victorian, spoken-out sex was obscene and the thought that every move in life on the surface or underneath was saturated with it seemed evil; more evil, and preposterous, the thought that birth should have begun the long melee, that incest should have been part of our innocent infant mind. Indeed, the thought of this lifelong life of sex could weary anyone on some rainy day. Who before Freud would have conceived the sexual as from the start built into the upper end of the digestive tract, lips for play? The newborn mouths everything, a rubber nipple, a rag doll, slobbers over the dirty doll as twenty years later over the beloved. The daydreaming of the adult could therefore be construed as a playing with imaginary dolls long after those of Christmas-past had been stored with their broken legs in the attic. Freud said that every nursemaid knew the facts. Where is that remarkable nursemaid? Freud remembered his own nursemaid, dead, repeatedly fished her, one felt, from the dim back of his mind into the forward part. It was not surprising that Freud's mind, his utterly unique mind, should have gone further and found symbols for the sexual in the simplest matters, in the activities of the nursery, the activities of the garden, of the kitchen, of course in all church steeples whatsoever, and that his thinking should in 1969 overtly and covertly reach into the street life of the world, into the hippies, into the writers, their novels, their movies, as well as into the treatises of philosophers and churchmen, even into the writing of one pope.

At four or five the child forgot its sinful infancy, buried it, resurrected it when society and biology thought proper, in the glow of puberty, sex with nature's cool intent. From that long sojourn of it in us came so much that motivated our mind, gave it its shades of color, unexpectednessess appearing in unexpected places at unexpected moments in unexpected forms, making psychology even more scene-within-scene than it had been before, filling the stage with Hemingway beards, geisha eyebrows, bouffant hairdos (for a time), shorter miniskirts (for a time), belladonna eyes, the submerged powdering the nose of the manifest before the lady herself opened the door and with every gamin in the street staring went down the front steps into the moonlit plaza, the candlelit church, the chandelier-lit ballroom.

Freud would have allowed that figure of speech also.

In his interpretation of the mind, and every interpretation interprets the interpreter, Freud granted hypothesis, as how could he help, though he talked often as if he stood on bedrock, and often did. More persons realize this in 1969 than in 1937 when he died, and many more than in 1900 when *The Interpretation of Dreams* was published and did not seem an important book.

There at the turn of the century this stubbled fisherman fished day and night, with cryptic bait, with conviction that fish were just waiting to be caught, as it seemed they were. He lured the single neurotic mind, was even better at this than at theorizing, also lured the mind of the century (the mind of many professions beginning therefore immediately denying him) to look at facts like infant jealousy, infant love, infant hate, emotions that we today watch expressing themselves in our infants playing on the lawn, are able to watch with recognition and intelligence and often amusement because of the way Sigmund Freud and a few others watched mind in infant, in child, in the whole blessed human family. We sometimes say blessed to avoid saying cursed. Good child and bad child never in the future would be what good child and bad child had been in the past. Society may be watching too much, watching for more than is there, watching here as everywhere too factually, but that would be only society's over-pay for Freud's originality. It would be too late now if we foolishly wished it to escape an observer and an imagination like his. He is somewhere about us, seen and unseen, day and night, interprets our waking state, interprets our sleeping state, partly creates, no doubt, our present comprehension of both states.

Freud was willing to share credit for his crucial insight into the oedipal with Sophocles, that Greek, that first Freudian. It would have been a temptation to eavesdrop on Freud and Sophocles in the halls of Hell or on the other bank of the Styx, modern Vienna meeting ancient Athens, Freud assuring Sophocles that he need not have been thinking oedipal for the oedipal to have been the motivation of his great drama, *Oedipus Rex*. His depths had come up and had done the thinking for him, better thinking on that account. Freud would have gone on talking, was always the teacher, would earnestly have defined the oedipal to Sophocles, who would have been listening with some credulity, some incredulity, squinting his mind's eyes, his experience having been that poetry might lurk in the dawn of any new idea, and one must miss no spot of poetry.

Having discovered the oedipal in *Oedipus Rex,* and the kernel of

a new psychiatry, and having come to regard as universal the impulses of that infant boy who became king by killing his father and
seducing his mother, Freud sought to plumb literature everywhere,
urged others to. He tried to solve what was underneath that Scottish
monster and pathetic human being, Macbeth, Ibsen's Rebecca West,
other fictive characters, particularly of the theatre, was convinced
one could find why they thought as they thought, behaved as they
behaved, why the plays were written. Freud believed that the
greater the dramatist the more of his own life would have been
introduced into his drama, not obviously, concealed, and the dramatist not know it, was alerted as all of us to the surface but not to
what was below the surface, the persons of his play being fusions of
their minds with his mind, the writer's.

And what scientific justification that might seem to the literary
and psychiatric critics whose ambition anyway was to peek everywhere beneath the covers of the writing to discover what could be
of the writer, drag him out where the reader might see him naked,
and the critic have the satisfaction of being seen too.

The name William Shakespeare was to Freud a pseudonym for
Edward de Vere, Earl of Oxford, who wrote the plays, and Edward
de Vere loved his father, who died, and hated his mother, who too
quickly married someone else. Edward de Vere *must* therefore have
been Shakespeare, *must* have created Hamlet, who *must* not have
been able to kill his uncle, Claudius, because that would have been
to kill himself, who in his depths had been covetous of his father's
bed, not his throne. Imagine Queen Victoria let alone Shakespeare's
Elizabeth sitting comfortably in the theatre with that notion in
every head in the seats around. Times change. Five years ago, 1964,
the Shakespearean year, if the notion did not come through the
Hamlets as they were played its absence was called attention to,
even by the people in the audience, who are all Freudians, or not.

It is possible that Freud never believed that Shakespeare in life
was the Earl of Oxford, that he simply snatched at that chance
because it fitted his theory of the neuroses, and because it was startling. Writers snatch at the startling.

Curious to reflect that while scholars and pedants with more or
less literary competence, and latterly with more or less psychiatric
competence, have vied to explain Hamlet, the Prince has continued

to stand there timeless and still enigmatic. Of generations of explanations Freud's was the most unexpected. He thought Shakespeare had put at the heart of the play a conflict that had existed in readers and theatre-goers for the last three hundred and sixty years and all the two thousand years before. The oedipal of *Oedipus Rex* was direct, Oedipus did kill his father and did fecundate his mother, and the oedipal of *Hamlet* was hidden, Hamlet merely did not recognize that he wanted to kill his father and wanted to fecundate his mother. The two thousand years had driven the oedipal underground.

It was the fire from below. It was the harassment under the play, and there must be harassment, some kind, under all theatre. Who is not ready to pay Broadway $8.80, if he can be sure of two hours of harassment? Up to Harold Pinter's last play and beginning with the late great Eugene O'Neill the American theatre would have been thin diet had Freud not left it his psychological conceptions.

Freud was a neurotic, said so, noted that he had gotten more normal than he had been four or five years before he started on his psycho-analytic self-cure. A dark neurotic fisherman, who even looked so as with the years he weathered and grayed, said we were all neurotic. We are, no doubt, but surely most neurotics who reach the psychiatrist are suffering a different neurosis than that he was saying we all suffer. Freud was able to study his own neurosis, did with persistence and a tough objectivity; one always forgets he was a small-bodied man. His neurotic torments were the common ones but uncommon for what he extracted from them, how he wrote about them, with economy, vividness, without technicality, straight at the reader, simple, sometimes almost garrulous. He was a voluminous correspondent in spite of his pinched time, the letters enabling him to battle his way to conclusions, to clear his thoughts; most of us must achieve our clarifications by expressing ourselves to someone. Another mass of Freud's unpublished letters is shortly to appear. Someone competent estimates there are thousands. He was also a diarist. It is reported that he burned one batch, all of his papers, in his twenties, again in his fifties. To say that a man was a diarist, even if not dictionary true, then to say that he burned all of his papers, might cause anyone to nod his head, and it does sketch lines into the portrait. We see in another way that he was small-bodied. We think with sympathy of a troubled person.

A bad period followed his father's death, "an intellectual paralysis such as I have never imagined," and then it was that with determination he began to interpret his dreams. This—everybody also knows—was the womb of psycho-analysis. The first psycho-analysis in the world was by Freud upon Freud. He could trust himself to start with himself, expose his mind to his mind (what that came to was in a new way expose all of our minds to our minds), peer through the fog into the water below, fish for whatever swam or floated. Every scrap had meaning he was convinced and was to go far to convince the rest of us, every word from our mouths, choice of phrase, verbal slip. Freud's self-analysis remains singular. No one who reads the history of it could regard it as literary fancy, though it was that too. He kept at it, a half hour a day apparently, for the rest of his life.

In the course of it and of the analyses of many patients and of a ruminating that was different from Sherrington's or Pavlov's or Cajal's or anybody's he put together his idea of the structure of mind, gave it the three levels, not proving that three was necessarily rigidly the truth, because Freud's had also the human brain that goes on helplessly dividing, but, important to him, he could thus effectively hammer into us his theories that were fresh, difficult, more so at the time, especially at the time. But, it may have been something more innocent still. Woodrow Wilson who changed the world in another way and whom Freud is purported to have studied analytically, was innocent, or call it what you will. Strong men may be. Children may be. Also they may not be. Animals in some sense are—and they kill. Freud said that the higher animals could be assumed similar to man, their minds would have the three levels too. The three levels began increasingly to be spoken of well beyond the psycho-analytic world and its university rim. The levels are, everybody knows, id, ego, superego, words worth defining not for the definitions, there being definitions of them enough, and in matters of mind all definitions are too clear, Freud's also, but mostly to etch more lines into his portrait.

The id was the primal, wilder in one and not so wild in another, shaded from the daylight of our life. It was the at-birth, the before-birth, the ancient, the racial. It began far back and in places far off. It was the compressed source of the mind's energy. It was the spring of the wound clock that through our lives unwinds. Saint and sinner drew from the id. The closest translation from Freud's German, as translators simplify, was *instinct*. To this id year on year Freud added, and if the added was not always lucid it could mean that our depths never are. Freud never thought they were. Our depths would have, then, both what was shoved under by us and shoved under before us, family, history, species, much doubtless that the intellect cannot ever resolve though Freud believed everything down there finally resolvable, what appear warring forces, wrestling opposites that clash and thrash.

The word *id* like any word, like any fact, gathered layers. For Freud, these were especially the masquerades of the fecund, sexual rage, sexual glee, sexual deprivation, sexual anything, that whole dry firewood always ready to burn down the house. It, the id, seemed always the infinite roguery of the reproductive urge plus a fraction of self-defense. Over it Freud threw the dusk of his often romantic prose—that presumed realist. He separated the love in-stinct from the death instinct, the latter brought to him with force, it was said, by the death of his daughter. The death instinct clove apart, the love instinct pulled together, love and death to Freud and to most of us being both fact and symbol, introducing into life flower and weed, art and insanity. Freud spoke of the Kingdom of the Illogic. He grasped at phrases. He had talent and power for descriptions of the mists of the mind, descriptions that his followers often froze into laws. Sometimes Freud's descriptions leave over the reader a Wagnerian music-drama rinse that washes off, however, as soon as he chooses to tell of the analysis of some single disturbed mind. It is there that time will have deserted him least, where he recognized the id's part in a man's or woman's life, that drove the life, that was always a stirred past inside the present, and another past inside that past, and another, as the analyst finds when he pages back and tracks the subsequent in the previous. Because of Freud at every hour in every country except the Soviet Union and China and a few others some worried mind is encouraged to follow paths back in the direction of the id and forward to where those paths are leading, thus revealing to the mind why the present is what it is, this revelation clearing the emotions and leaving the person more than a hope that the future may hereafter be less destructively servile to the past.

As for the ego, it was sculptured from the stone-of-the-hills of the id by the chisel and hammer of reality, always under the watchful eye of the superego, the boss. The sculpturing proceeded wherever and whenever the id went forth, early and late, on its *Wanderjahre.* Freud's went, ours goes, off in the morning, on through the day, never stops, cannot, the restless thing, around the nursery, around the earth, around a city stupid or not so stupid, Vienna or some other place, a few streets, up the one and down the next, a parlor, the bedroom, the butcher shop, the church, the synagogue, the slushy rush of New Year's Day, sweltering July, an excursion to Schönbrunn on an August Sunday. In the course of this the ego acquires ego-strength or sinks into ego-weakness, but always remains tied to that infant ego that was shaped when everything was fresh, had its uneroded power, cannot therefore be reshaped easily. This we know because of Freud and those who followed him, though we have the illusion, perhaps not illusion, that we knew without them, that everybody always knew, which is one kind of historic illusion. The infant ego had under it the id's hottest fire, but with time the fire cooled, was ashed over, the heat lessened, but never lessened so far as not to have left the energy to squabble at home, to squabble abroad, squabble, squabble, to fight it out with oneself, and with anyone, to the last cackle.

Finally—the superego. No need to be exact about that either. Freud was not. He frequently dropped such a phrase as "more complicated than we have here described," or "my latest formulation." Freud was the best of the Freudians. The superego was the authoritative, could be people, could be objects, could be ideas, could be ancestral precepts, cultural precepts; it accumulated somewhat in a lifetime but mostly was the past. Superego and reality confronted id—as is said every day. Superego was the eternal parent—as is said every day. Stern parent or not so stern, and not the one or the two but the entire corseted crowd, tradition, history, the hierarchy that in our time seems usually to be leading toward the sociologist or the psychologist. But it could be the Secretary

General of the Longshoreman's Union, the Chairman of the Board of the Stock Exchange, the President of the United States or the University, Socrates or Billy Graham, all past or present who loom over us, advise us, get us grants, and if we do not keep one eye on them choke us. Superego differed culture to culture but principally the form, the tilt of the hat, length of the skirt, diamond in the tiara, whether or not on a holy day we went into the temple wearing straw sandals. It was the orthodox. To Ibsen it was the *ideal,* a trap. To Freud it included what was left of the conscience. Capitalize it and to the existentialist it became the embodiment of what to less ironic minds would have been Jehovah or Buddha or Christ.

To comprehend all or any of this we must in the opinion of some analysts wait till we have ourselves been analyzed, which would enrich us, surely, but most of us will have to get along with our poverty. Each of us has a mind and an idea of mind and must agree to be satisfied, must also insist on being allowed to do his thinking about mind with what mind he has, no matter how that annoys anyone. Sherrington gave succinct statement to this inevitability. "Science, nobly, declines as proof anything but complete proof, but common sense, pressed for time, accepts and acts on acceptance." It would be best if that common sense had no indoctrination, not religious, not psychiatric, not scientific.

The brain as brain, the 1,350 grams of near-to-fat, did not enter Freud's scheme, nor did Cajal's microscope, nor Pavlov's higher-nervous-activities, and not Sherrington's reflexes. Freud sidestepped the work of those other three giants of his day, may not have known much about any of their work, would have sidestepped it if he had, sidestepped also today's computer mind, macromolecule mind, information mind, anthropologist mind, though he did himself dabble in anthropology, and psychiatrists ever since have dabbled in it. Of course he would have sidestepped today's brain experimentation, that appears sometimes to expect to latch the small brains that compose our big brains to the small behaviors that compose our lives. Mind sans brain was the territory of Freud's explorations. He never doubted brain, his determinism including everything or trying to and often having the possible misconception that it did. Early in his career he attempted to fit mind squarely on brain, drafted a manuscript, postulated clever nerve-cells, not as Cajal had shown them to be but as it would have favored Freud's theory if they were. He never published the manuscript. It was a naïve performance on the part of a genius who was many things but rarely naïve, though naïveté is part of the portrait too. His conclusion? We were living in too early an age to fit mind on brain, so he let it be mind for mind's sake, psychology pure and simple, his own. He summarized: "I have to conduct myself as if I had only the psychological before me." It was the psychological detached from the flesh of the brain, but grossly or subtly attached to the sex flesh of every Adam and Eve. The riddle of mind's inner nature he did not pretend to solve. That naïve he was not. His words were "defies all explanation or description." This fisherman fished through layers of fog, through depths of water, made his hauls, then noted that the greatest haul of all eluded and might always elude the hook and net of the intellect. Near the end of his life he spoke of "the still shrouded secret of what is mental."

That last must be recognized about Freud's view of the mountain, view of the mind, insofar as he placed it or was able to place it where we could recognize it. This was closer to Sherrington's view than to Pavlov's, not as impoverished as Sherrington's, not as imperious as Pavlov's.

In Vienna, he occupied that famous house up the steep sidestreet, 19 *Bergasse,* where he had three small rooms, a waiting room, a room for interviews, a study. There he conducted his daily sessions, nine hours, thirteen hours, guided neurotic after neurotic to pick up layer after layer. His reading and writing, important to him, he put off till night, typically offered a lady, in whom he was interested as colleague, and as lady, an hour of discussion if it could be after ten o'clock at night. Vienna medicine lay below him. He said he hated Vienna, where he had spent most of his life, toyed with the idea of quitting it, going to England, Australia, never went till a crazy world drove him out. Possibly he loved Vienna. No one knew better than he how love and hate cohabit. America he hated; Rome, the ancient city, he loved. Vienna gave him isolation. He considered the isolation not given to him, but forced upon him, he left with it, but probably preferred it, certainly profited from it. To the people in that street his days must have had a monotonous order. His clan was small. He broke with one group of it, held to another. A group might desert him en masse, as did the Jung-Swiss group, and though this may have saddened him he was dry about it, anticipated it, understood it. He was outspoken, enjoyed his humor, coarse often, in controversy was logical, ruthless, truthful. Social opinion in the world was all of the time changing. He spurred the change. With bluntness, argument, hypothesis, fancy, fact, he did his part to wipe out the old moral order that he believed had produced the neurotics he treated. Unwittingly he may have done his part to wipe out all moral order, in the respect that, at the moment, this sometimes seems to have occurred, the ethical replaced by the psychological, the old world loosened before the pillars under the new are firm. The sick came to him from many places. Psycho-analysis spread through many countries. It sprouted branches that forgot their trunk. Sometimes it was as if a flame were being fanned in desert grass, curiosity about sex the wind. The righteous recognized that they too had the curiosity. Events preceding World War II drove him as a Jew to England where he died a loitering death of cancer that gnawed its way through his jaw and cheek and into the orbit of his eye. Some canny and evasive border, it is not possible yet to say how broad or solid, was by Sigmund Freud added to the edges of the human mind's comprehension of itself. His body was cremated, the ash put into one of his Grecian urns, his approach to the mind living on, not so much what he thought mind was, but his technique for thinking of it.

# DREAM I

## *Freudian Wish*

Freud's book *The Interpretation of Dreams (Die Traumdeutung)* was his greatest, most think, he thought. From the time he published it, nearly seventy years ago, up to the time of his closing illness, he revised it year after year. He had youth when he wrote it. He wrote rapidly. The book runs. It may be somewhat slow toward the end, may be too long, may be or feel unfinished, but Freud was expecting the subject to go on developing from where he left it, knew no doubt that it ought to feel unfinished, ought to be incomplete, allowed open ends to remain open ends, had the competence to do that, recognized that there was surmise, that there must be error. Besides textual revisions he made revisions also in the overall ideas. He knew this must go on. Of the foundations he felt confident, thought they were sound, and two-thirds of a century later they stand in a way that would make anyone think they might a long time, long after their superstructure has crumbled, and it will crumble, has crumbled, but of course is not the rubble on the ground that one might think if one listened to some, these often the ones who have dwelt most inside the edifice. That is somewhat as with Pavlov.

Freud's book is dream cover to cover. Psycho-analysis rose from it. Psychiatry and indeed the life of our time has been profoundly touched by it. People read it who read nothing else of Freud. Not a difficult book to read. Freud knew and said that with it he had struck his great blow, would not strike another such, no one having the power for two. "Insight such as this falls to one's lot but once in a lifetime."

The essential conclusion of the book was that if the meaning of a dream were searched for and found it would invariably reveal a severe logic running through its ostensible illogic. Its insanity was sanity. It was as determined as everything else in body and head. It was an extension of Bernard's constancy. Freud would hardly have said that. It was a piece of the homeostasis. He would not have said that either. Psychiatrists speaking of homeostasis would come later. But undoubtedly to him a dream was part of both. A dream was no random spot of awakedness in an otherwise sleeping brain, no Pavlovian analyzer half-inhibited, and it also was not the flashes of a pale night-lamp carried through a house darkened so as to get rest for the work of the next day. He might have admitted something like the last. But the idea of dream being random, being a moment of illumination in a mind that was for the time of sleep a widespread inhibition, would have been absurd. We may be sure that he never heard of it. He would not have been annoyed at it but contemptuous of it, that anyone could so brashly sweep a broom over the patterned meaningful tapestry of dream. That would have worried the responsible scientist in him.

The dream laboratory and dream research support purpose in dream. We require it. When we are deprived of it we are ill. There are many experiments on dream-deprivation. Nevertheless, the intent in the content of the dream the researcher does not much concern himself with. Freud did.

Freud was the first to see economy in the content of dream. In sleep everybody has always seen economy, for body and mind. And Freud thought that dream existed to guarantee sleep. Dream enabled us to go on sleeping. The body must get rest though Freud was not concerned with that, but the mind must get its kind of rest, and dream helped it to that. Again, today's dream-researchers have come to a similar conclusion. Dream is physiological. It is part of the normal machinery. For this the dream researcher brings laboratory evidence in an era when laboratory evidence is sacrosanct. It is safe to say that the researcher never has had Freud's insight, keeps off from where Freud spent his life, in the hidden, sometimes thinks that he would not care to have Freud's insight, sometimes seems unaware that dream has content. Freud not only brought attention to the intricacy of the content and to what he believed led from that to aspects of our human life that have and that have not been revealed. Freud's genius gave him an earned indifference to the anatomy of the brain, and he might have been indifferent too (for himself) to the apparatus that would be invented to study dream. His study needed no grants, needed three meals and himself.

He tells dryly how *The Interpretation of Dreams* was received by book-buyers. Two years after publication the sales were two hundred and twenty-eight copies, three hundred and fifty-one after six years. (Should be comfort to authors.) This did not make Freud doubt himself, but deepened his natural irony, probably depressed him on and off.

The book had its origin in his own dreams. He recounts a number, chills his prose, states that he left gaps wherever the substance would too much expose him, drew in a friend to advise him. He wanted to present the necessary facts for the interpretation, but was no Jean Jacques Rousseau, kept his private life as far as possible to himself, but autobiography like biography is nearly impossible, the biographer often getting himself in the more when he is trying to keep himself out. Freud did try to be frank with his reader; trying will not always manage it.

The first of his general dreams to be publicly analyzed was his famous Irma dream, and when one read it, so many years ago, one's astonishment was that all of that could be in a dream, or could be put into a dream. One was apt not really to believe any of it.

He continued analyzing dreams. He says that by the time he was writing his book, which may have been half through his career, he had analyzed more than a thousand. Analyzed the dreams reported by others. Repeatedly he referred to his famous Dora dream described elsewhere in his writing. Often he dissected a dream episode for episode, phrase for phrase, word for word. He explained that symbolic interpretation of the totality of a dream, saying what the whole of it means, which had been the common way previously, was not his way. He must break the dream into pieces, examine each piece, afterward fit them together, make a history.

Out of this type of study he reached his conclusion, not entirely original with him, that the dream was the fullfillment of a wish. It remained his theory. It was not all he thought of dream but it was overall. The dreamer was given something he wished, in the simplest case something he would have to wake to get, and so, because he had gotten it, in his dream, he could go on sleeping. The dream had had purpose. It defended his sleep. This Freudian interpretation became with time the classical psycho-analytic interpretation, spread through the psycho-analytic world, the psychiatric world, medical world, literary world, world. Dream-researchers would sometimes think it invalid, or think worse, would come up with a precise controverting example, and Freud himself in successive editions of his book, as he had foreseen, modified his interpretation, but the remarkable fact is how little he did modify it, and how it branched.

Every analyst recognizes *The Interpretation of Dreams* as the beginning of his field. So, that field began in dream, and not figuratively.

Early in the book Freud recounts other theories of dream. The ancient ones were not theories, not hypotheses to be tested, the dream simply a direct message from the divine. Freud put the theories in their historic order, commented on them, picked out what led to his, was meticulously careful about crediting, and one might say that he could be because his was by then so secure. He related what he learned. He developed his conception of the dream's structure, mechanism, purpose, revelatory power of the character of a mind, especially neurotic mind. Dreams of patients appear in his voluminous writing. He started a new language, made a new vocabulary, and psychiatrists increasingly used that vocabulary in their reconstruction of human stories even when at a later date they began to doubt Freud, as many did, thought he claimed too much.

Freud's original reason for studying dreams was to study the neurotic mind, help in its cure. Disturbed though these human beings were he nevertheless dragged from them what one might think must disturb them further, this night-life that had in it so much of their day-life. He scissored out fragments, sought to relate each fragment to something in the symptoms, brought together the fragments, laid the resulting picture, as completed as he could get it, before the neurotic, who, seeing his mind from earliest infancy illumined by his dreams, was able to contemplate his whole panorama, the whence, might get a grip on the whither, be helped.

Freud believed that via dream he was making each neurotic more intelligible, neurosis more intelligible, also psychosis. He believed the psychotic to live a life of dream, to be always or mostly in dream, that only by staying in dream could he maintain some balanced relation to the world, and it remain bearable, his suffering being unbearable as soon as he stepped into so-called reality. Here too dream, this life-dream, was serving as wish fulfillment, a long-sustained fulfillment instead of getting something to enable one to stay in bed for a night.

Not only the psychotic required his dream, the neurotic his dream, we all did. Dreaming helped, in its way, to maintain the level of the universe, a level demanded for galaxies and stars and planets and atoms, and of each of our infinitely small selves with this halo of mind around them.

Freud was concerned with those small selves. Many found their way to 19 *Bergasse*. Each must dream. Each when his dream was interpreted realized the wish and whatever else was in his dream.

Dogs must dream no doubt. Freud's cat. The dream-researcher has come to the same conclusion though it may seem more banal.

Somewhere along the line every dream-researcher and every psychiatrist is apt to pay lip service to Freud, the father. One well-known researcher into sleep regularly paid that lip service but also regularly added a touch of sarcasm about the father, possibly eased himself for regularly finding his own thoughts so much less significant. The same researcher added his touches of sarcasm about Pavlov too. Jealous world. We are all part of it. Even in our sleep, in our dreams, we do not escape it.

In spite of our familarity with Freud's wish-fulfillment conclusion it still remains startling that he could think we dream to preserve our sleep, that dream had that as its function. Formerly we might have said that it had no function, which now from all sides we know it has. Freud thought it satisfied our night-worries, night-wishes, let us sleep on oblivious of the clock. At moments that has seemed unashamed invention, made us wonder whether Freud might not have needed his ironic intelligence to save him being victim of his ironic imagination.

Herodotus would have devoured Freud's book, might have thought him too hesitant about keeping in the anecdotal, though Freud was really not hesitant, only the anecdotal must be meaningful for his ideas, or have the possibility of his thrusting the meaningful into it, into what he heard from his patient, drew from his patient. Homer too would have devoured Freud's book. Caesar not—Caesar would not give heed even to the ides of March. Cleopatra would have paged the book a while, let it fall, a slave pick it up, she prefer to float in her unanalyzed mystery of the Nile. (That was the Vivien-Leigh-vixen Cleopatra, not the Vivien-Leigh-queen Cleopatra.) Is it conceivable that the wily woman, Cleopatra of the Nile, believed that divinity through her dreams was telling her what to do? Psycho-analysts also devour Freud's book, not every one of them, but many chat with each other repeating chapter and verse, assure us that Freud's theory has been verified in thousands of analyses. Many regard it as fact not theory. Our dream, that bizarre mixture of pleasure and unpleasure, plays the fool with us for our good, serves us with what we wish, and we stay in bed. "Oh how I hate to get up in the morning."

Freud's plebeian illustrations are often quoted. A man is hungry between midnight and morning, so dreams himself a banquet, therefore does not need to wake, put on his shirt, walk to the corner, buy a sandwich, because of that agitation not be able to fall asleep the rest of the night. An intern knows he should be awake, visiting a patient in the woman's ward, therefore dreams himself into the ward, not as intern but as patient, double-locks his sleep, is in his bed where he wants to be, and compelled to stay there because patients are forbidden to get out of bed. Freud himself had a recurrent dream in which he drank a glass of water, induced the dream by eating anchovies, sated his thirst with dream-water, did not so much as have to lift his glass. (One of those informal experiments that would not even be remembered except that it came off the pen of an accepted master.) Most of the time wish-fulfillment was less revealed. It took on disguises. The interpreter became an expert at seeing behind disguises, even when they were superimposed, his adroitness at this sharpened by his daily work and by his knowledge of the symbols that dream uses, his conviction that every word in the crossword can be fitted into its proper place if one keeps at it, years.

# SUFFERING

## *It and Its Brother*

The pilgrim below Fuji stops at a shrine to pray, then at another shrine, and another, and one of all his prayers is sure to ask for some understanding of his suffering.

Suffering and pain are mind and body in a juxtaposition.

Suffering.

In animals suffering is less than in man, we think, but their pain may be the same. We cannot be sure even of that. In animals pain is less remembered than in man, less reflected on. At least, that seems likely. A man needs not to be too penetrating to draw some line between his suffering and his pain. If he is more penetrating the complexity of his suffering and the simplicity of his pain become apparent. That may not be mere size of brain. There are mammals with larger brains where pain would have larger space to echo and re-echo in, if suffering is produced that way. It could be. But man's intuition is that it is something else, something special. That may be only man's vanity.

A simple sensation can in man's mind gather to itself what makes it memorable for years, gather what is agreeable, what is horrible, the pain long over but what it gathered to itself is here now. Dog and man have similar pain-receptors, similar pain-paths, but that which happens along or at the termination of those paths, if that is where it happens, is not the same. The meanings are different. Man's mind attaches meanings, dwells on meanings, his naked sensations going into him, his mind clothing them, embroidering them, and then they come out with meanings. "We look before and after." Man's mind makes extraordinary the naked sensation that was ordinary. Out of pain it shapes suffering, as out of the cemented-together sand of sandstone it shapes a gargoyle, out of blackness of ink shapes a mystery.

Now and then a physician has claimed that suffering on occasion has wrought strange changes in the anatomy of a human body. A woman added to a sensation in her abdomen a swelling such that even her unfeeling husband began to think she was suffering. He had some pity for her. However the swelling was caused, it was of use to that woman's mind, of undoubted practical use in her daily affairs, won arguments about having supper in a restaurant instead of at home, provided a good excuse for going to bed or staying out of bed. Whether the swelling was truth or fiction, the entertaining of the possibility of its being truth by an excellent medical observer, or even the actual physiological producing of the swelling, is more imaginable since Freud. That same woman, an obvious hysteric, years earlier had successfully rejected a sex-loaded seaside honeymoon on the grounds that she might get hayfever, insisted that she had gotten hayfever at a seaside resort on a vacation when she was fourteen. Young husbands can be such fools. But physicians are fooled too.

Pain sometimes draws pleasure to it. A child had an irascible father who punished her with slaps, then immediately was sorry, lavished affection on her, so her mind concluded that slaps were a pleasure, welcomed pain. When that child became a woman, and broke a leg, it might be difficult or impossible for her physician to appraise her pain, to separate pain from suffering, to separate it from that mental mixture that included pleasure.

Idiot children have been known to mutilate themselves with a wild enthusiasm, pull out clumps of their own hair, pain and pleasure join in bacchanalian frenzy.

A small girl skinned her shin because everything since morning had been going wrong, and that very night her parents left her with the baby-sitter whom she hated, and to the pain in her shin she added the suffering in her heart, and she remembered that night when she was an old woman, would rouse from a doze and mumble of her long-dead father. She still spoke of him as "Pops." That faraway night with all that it drew to it might obviously have affected her life. She was a simple old woman. She would not have thought of trying to separate pain from suffering for the interest of doing so. A Calvinist might have dwelt on the fact that she had had plenty of recompense for everything, the pain, the suffering, the general bewilderment, that they had purified her soul. That is possible. But it is not especially pertinent to anything.

The martyr while he is waiting for the flames from the green fagots to reach him is probably suffering, probably several kinds of suffering, and no matter how successfully his mind is battling to deceive him, with his wish for nobility supporting him, when the flames arrive there is pain and the suffering is probably little or none. The two are being burnt apart.

We all have a fringe of varying width for any sensation. We call it affect. The intelligent woman in the black satin sheath with the double bourbon in her hand could explain to us what there is to know about affect. While she explains she sadistically squeezes the glow out of the tip of her cigarette, gives her fingers a twinge, adds a minuscule of pleasure to a minuscule of pain, explains that too, in her black satin sheath. An intelligent human mind can be so busy with meaning that it is quite stupid in a special way.

Pain is sure to accompany pleasure. "Our sincerest laughter with some pain is fraught." Hate is sure to accompany love. Nothing is but that it moves toward its opposite. Nothing mental long remains where it was. Wait, and the tide will be on its way toward the other shore.

Worry is suffering. Worry often does not know what it is worrying about. Tears can be streaming down a face. Why? We all recognize that no answer may be forthcoming. The human mind is the most worrying construction on this planet, forever picks in yesterday and today to scatter the seeds of worry for tomorrow. Distrust may quickly spread. Distrust is worry. From morning to night a man moves from trust to distrust. He distrusts (vague) that first woman, his mother, continues (less vague) with that second woman, his wife, and the third, whoever she is, then falls in love with and distrusts (most vague of all) his daughter, and through everything there is his background distrust of both futures, the one of this week, the one called eternity. And distrust is suffering, but in the midst of his suffering the man stubs his toe, and that is pain.

The tendency of pain or of any sensation is to bring to its acuteness the muffled sensitiveness of the rest of the mind, and bring in more and more nervous-system no doubt, and more and more universe, all beginning at once to spread in circles. Fully comprehended, pain would again detach itself from suffering, the man for this one experiment of his life become a philosopher, allow pain to remain pain. That ordinarily is bearable. The other sometimes is not. He screams at his pain, moans at his suffering.

Serious illness besides pain and whatever other uncomfortable sensations may be a mat of suffering. Fright, depression, disgust, contempt, bitterness, apathy, the grisly heads of thought rise and are present at the same time or in succession, add themselves to the pain, with thrusts of some apparently unrelated feelings, even affection, the whole knotting itself into a complex hard to define, hard to pick apart, the patient now so completely miserable that even while under one dose of morphine he begs his doctor to stay somewhere near in order to administer the next.

The dose dulls the pain that may then not further excite the suffering, but the suffering still begins to excite itself. A priest interrupts, pays his semiweekly visit, or a rabbi, or a gentle neighbor, stays an hour, speaks of anything, of how we human beings know nothing, and the patient for that hour misses his appointment with the syringe. Suffering may so stir itself into the pain, and into the other sensations, into memories near or far, into the revealed, into the hidden, that its full accounting if it could be made might be a description of the entire person. Suffering more than pain has upon it the tone of the temperament, the tone of the history, oscillates with mood, with luck, goes low, goes lower, is less, is gone. If a man can lay his head on the shoulder of a friend and pinch her arm at the cocktail hour he may suddenly have eased pain and suffering both, to know they have returned with force at 9 P.M.

A surgeon relieves a particular pain by cutting the pain-path. His results frequently have been embarrassing, differing so from his prediction.

The surgery in itself may be successful with a single operation or in itself unsuccessful after a series. The surgeon may first cut there where the nerves lead from the sick flesh. Then he may cut where the nerves wrap themselves around the arteries, strip the blood-vessels, as is said. Cut where the nerves enter the spinal cord. Cut inside the cord. Inside the brain. The localities are increasingly dangerous, require increasing skill, more imagination, more daring. If after the operations the patient still is twisting in his bedsheets, the wires down but the dispatches appearing somehow to get through, the surgeon may consider lobotomy. He formerly considered it oftener and sooner. The pain-path was not severed in lobotomy. The patient kept his pain, winced if pricked with a needle, but he no longer dwelt on it, on its possible hopelessness, on the falsehoods he believed he had been told. The brain-flesh somehow added a pain-mimicking unrest to his mind. Now that particular brain-flesh is severed. The suffering is gone. The pain is there but the meanings are stripped off. Even death may not seem important. The patient has stopped being afraid. After a while one may think that he was better off when he was afraid, when he had his old suffering. An ignoble emptiness surrounds him.

Phantom limbs have presented baffling mind-body problems. The phantom and the limb again juxtapose mind and body. A limb has been amputated but it has stayed in the mind. Someone might remember back to loss of body-image, loss of the fact of having a body, but phantom limb is different. When a person has lost body-image he denies the existence of the limb he still possesses, whereas with phantom limb his mind still possesses the limb he has lost. The phantom may be painless, or painful, and around the pain may be a heavy umbra of suffering. Patients tell similar tales. When one amputee's dead limb was shoved into the incinerator—the incinerator miles away—he experienced a frightful flare-up of pain. Surgeons have reopened healed wounds to reduce such pain, have trimmed stumps of nerves, followed the nerves back into the nervous-system, still not been able to exorcise the phantom. One theory was that reverberating circuits kept the alliance of pain and suffering reverberating, circuiting, and when a circuit low in the system was cut, a circuit higher up began circuiting. There was no proof. Freud might have considered treating the phantom with psycho-analysis and when one reflects on how pain gathers suffering to it psycho-analysis could seem logical.

# SLEEP

## *Off It Goes*

The following was written in 1606. Everybody remembers it because everybody reads *Macbeth,* and everybody sleeps or tries to.

> Sleep that knits up the ravell'd sleave of care,
> The death of each day's life, sore labour's bath,
> Balm of hurt minds, great nature's second course,
> Chief nourishers in life's feast. . . .

That could not have been written in 1969 because there is no one to write it, but that almost affection for sleep may come to anyone deprived of it, as Macbeth was.

A child leaps into the day, in a few minutes is as bright as dawn, is dawn, is the whole world, is shy, is asleep, is awake. "I had a nap." The adult does not do that well. The adult during his waking hours lets his mind wander, here, there, forces it to stop, forces it to concentrate, forces himself to explore, to act, is carefree one moment, is anxious the next, suddenly has grown so tired, the eyelids just close, the retinas have the curtains fall in front of them, some touch of light may still get through, some signal travel to parts of the brain, but not enough, the neuron pools stay quiet, and there is that balm of hurt minds, sleep.

The reasons for sleep could be simple. To the student of sleep they do not stay simple. To one student they did. He was satisfied that sleep was merely an ancient habit acquired for some forgotten reason—that was simple.

Sleep surely is one station on the Tokaido Road from Tokyo to Kyoto, the sacred mountain over there, up, yonder, in fog.

During sleep the living machine does not stop, would be dead if it did, but continues at some minimum, stays in one place.

A hedgehog, one species, sleeps twenty-two hours of the twenty-four. The human infant sleeps most of the day and night, and, despite some educators and some physiologists, sleep is probably not a conditioned reflex established anew for each arrival on the planet. Anyway, the infant sleeps and sleeps. Then it is a child and sleeps less. Then it is grown up and sleeps mostly at night and on Sunday afternoons, and even for that requires bourbon or coffee. Everybody knows somebody who can only sleep after coffee. There are the persons who half-sleep all day and solidly sleep at night.

Much data on sleep is written in books, usually descriptive. Increasingly of late, however, the writing has become explanatory, and the explanations increasingly bound to experiments.

The most natural would be to find that to be asleep was just not to be awake. Then the study would turn to the waking brain and the waking mind. What waked them? What keeps them awake? And that negative approach has been satisfactory but not entirely. Sleep does suggest positiveness, suggests it might have its own brain and mind characteristics, and at least it would seem sensible to listen to this or that theory, and be willing to lay it aside.

What are the body signs of sleep?

The neurologist's reflexes are diminished, may be absent. The knee jerk, that reflex, may be depressed beyond arousal. Those remarkable reflexes that hold the body upright falter. Our muscles rest. As we know, single fibers, or great masses of fibers, may not only rest but rhythmically rest. Blood-flow lessens and one could say that in this respect the whole circulation rests. Between each two beats one can think that the heart, the contracting organ, rests. Between each two nerve-impulses, thousandths of a second, the nerve-fiber, the transmitting machine, rests. A pleasant word, rest. When the boy cut the earthworm in half, each half took its siesta, but a worm is a worm, has no cerebral hemispheres like ours and to us sleep is our kind. Our dog's too. Our cat's. Everything in the house sleeps. Some old dogs sleep away their lives. Some old men. Some old women. Some young women. Most adults spend a third of the twenty-four hours in one or another sleep-posture, change posture, may think they do not but laboratory observation says they all do. One American celebrity was partly celebrated for saying that if he slept a small fraction of every hour, four hours in the total, he had had enough. According to the newspapers a student stayed awake for eleven days and recovered from the loss by merely sleeping seventeen consecutive hours. An Arab policeman spent his third in the vertical stance leaning on his gun at a busy intersection in Cairo directing traffic, asleep. At least, his eyes were shut and his body swayed.

Difficult to look accurately at one's own sleep or one's own waking. Many have tried, and wondered. Sleep, human sleep, has perplexed philosophers and plain citizens from long before Freud and Pavlov and Darwin and Homer. An alive brain lies under it. We know that. The bewildering electrical states of an alive brain lie under it. Sleep is far from fleeting death though writers have called it that.

The sleeping mind frequently has been considered the reality, the waking mind the unreality, by sentimentalists and mythologists, but by nervous-system scientists too. It continues easier to think of all mind as the reality, the only reality conceivably.

Some of us are miserably self-conscious toward our sleep. It does not help us sleep. One gentleman, on the contrary, said that he just told himself to think of nothing, and he fell asleep. The nothing may of course have been something, and exceedingly positive in its effect on brain and mind; the gentleman did not have the kind of detachable intellect that might have known. Another, a bad sleeper, stated it as his day-mind always waiting to step out from his night-mind. Another, hearing this, stated that his own night-mind stepped into clarity and went back into obscurity many times a night, and that he envied a neighbor whose didn't step into clarity until the alarm went off at 6 A.M., when there was just clarity enough to reset the clock, to step out into final clarity at 7 A.M. That wonderful extra hour! This man drank himself dead with vodka night after night. A man's sleep may be as light as a leaf or as leaden as lead and all depths are experienced sooner or later by every mind-observing mind. If we understood better the leaden condition we might understand better the sun-white waking condition, and what on this rolling globe would it be as personally worth understanding?

Gross body fluctuations occur with sleep. The pulse rate drops, it may be twenty beats. The blood-pressure drops, it may be twenty millimeters. The kidneys put out less urine, the morning urine then more concentrated. The sweat glands put out more sweat. Body temperature lowers though that of the skin may rise. Carbon dioxide accumulates. Breathing may change in character, become irregular, a gasp, a bump, a snore, when a man's wife wakes him and the breathing becomes regular again. Eyeballs roll upward and outward or any direction, may oscillate slowly once a second, then rapidly many times a second. Pupils constrict. There are numerous facts clear or unclear. One-half, three-quarters, ninety-nine-one-hundredths of us goes off somewhere, the night-mind coming into a glory that the day-mind recognizes or suspects but cannot ever quite apprehend. The day-mind meanwhile has been given surcease from the unutterable if we think as Eugene O'Neill, or surcease from all the pleasant little duties, if we think as Isaac Watts.

# EARLY THEORY

## *How It Goes Nobody Knows*

There is evidence that the brain never sleeps.

Aside from that, how do we get sleep? How can the top-end be thrown out of gear, or into another gear, the engine left idling, as has been said, for three or eight or twelve hours?

Eighty sleep theories have come and gone, most of the early ones obscure enough to be buried in a bibliography, but easily illustrated.

After supper more blood goes to grandmother's stomach and bowel and skin, some of it diverted from her brain, it deprived of fuel and oxygen, and grandmother loses consciousness, sleeps. By this old theory the brain during sleep is partially suffocated. It never was a probable theory, and blood-flow measurements during sleep apparently prove the brain to have not less blood but the same or more. We would not understand why the more-blood, except if the mind is intensely active during sleep the brain might need the more-blood. At least, there is no evidence of anemia. It is always possible that a small controlling area of brain is getting less blood, and quick shifts of blood from one part of the brain to another are suspected and recognized. Not anyone nowadays believes that insufficient blood or insufficient oxygen gives us this regular, this peculiar, this partial disappearance.

Another early theory had the work of the day pile up acid, lactic acid, and that brought on sleep. But, the brain uses lactic acid normally for energy. Also, we sleep when body and brain have not been hard-working, may sleep well after a lazy day. Furthermore, we do not confidently know what a hard-working brain is. As to a hard-working mind we think we know better, and probably do.

Another early theory had the brain pile up not a by-product but a poison. There was a "hypnotoxic" substance, hence sleep, and when sleep had reduced this, we returned to the waking state. To investigate the substance, spinal fluid was drawn from a sleeping dog and injected into a waking, who fell asleep. However, a researcher found that when spinal fluid was drawn from a waking and injected into a waking, that one fell asleep.

Undoubtedly, the phenomenon of sleep is one of the fifty-three stops along the Tokaido Road, one station from which to get a glimpse of the sacred mountain, sometimes nearly clear, most times lowering and distant.

The early ideas of sleep hardly indicate the ingenuity that went into them, or how much was learned from them. They did take emphasis off of the brain as a message-over-wires instrument. Other theories did not do that, were more nerve-transmitting, might therefore be considered more neuro-physiological.

All day our muscles with their nerves are tense and working, or not working but nevertheless tense, then evening comes, they are relaxed, fewer stimuli travel from muscles to brain, this therefore less beat upon, less roused, and we sleep. That is sleep. A light sleeper might insist that his body often was supine and collapsed, he not fidgeting, lying there, 3 A.M., wide awake, just did not take the dip into sleep. We all know the dip. It seems to occur unexpectedly, despite that it has occurred night after night after night. We are here and we are gone. That is the kind of occurrence that makes our mind always a stranger to us, the two of us looking at each other. We know our mind better than anything in the world, yet are shy toward it, vaguely uncomfortable with it. Anyway, a bad sleeper who has been a bad sleeper all his life and rather accepts it, cannot easily be convinced that he should look to his unrelaxed muscles and not to his unrelaxed thoughts for his trouble.

All eighty theories, discarded for one reason or another, have had their historic or their literary interest, have also thrown some light on sleep, some light on mind.

Even the way a person approaches his bed has an interest. It may be bizarre. A small girl regularly dives in headfirst, and nothing seems right about that, ought to wake her for a week, but her brain-switch is new and young and freshly oiled, and off she goes in half a minute. Her affairs are not as scrambled as they will be. Possibly too she does not go off as instantly as she seems to, possibly loiters somewhere in the stages, A,B,C,D, or 1,2,3,4. She babbles considerably. From what she says, she sneaked in one more ride on a picket gate, went skiing down a chasm, climbed the mountain on the other side, nothing upsetting enough to wake her but exhausting enough to assure the final blackout. The evidence seems to be that in the first part of sleep we plunge into the deepest depths, pass quickly through stages 1,2,3,4, stay much in the last stage, less as the night advances, generally go back and forth.

A prominent citizen approached his bed methodically, allowed himself time to unbend from the strains of the day, nagged at his wife, patted his dog. An executive, naturally. Sprawled in a chair. A second chair. A third chair. Gradually he got himself to the room he knows so well. Removed his pinching shoes, his squeezing collar, slipped into his gaudy color-blind rayon, slipped into his bed. Decidedly, he did not dive. Pulled the bedding up to his neck, expensive linen. Folded back the top sheet, neat as an envelope. Unkinked his joints to the accompaniment of self-conscious yawns. Had shut the windows against noise. Turned on the air-conditioner, not to cool the air but to produce the steady hum that would screen out other noises. Previously he had drawn shut the tailor-made curtains against the morning light. And now he drew shut his eyelids. (This hero never takes a sleeping pill.) Following two hours of deliberately shutting off his thinking he decided on a warm bath. Immersed his skin in the monotony of water not too hot not too cold, eased himself into it, stretched himself out in it; if only he dared enjoy it. (This hero never takes a half-pint of vodka.) He came to a resolve. He would make adjustments in his calendar. As it was, he was spending a month in Florida in winter, a month in Europe in spring, two months in the White Mountains in summer—he would take off more. Also, he would play more golf. Had he been born in the low categories, he would now have dried, sat in a chair, begun reading the fine print of the stock quotations, which would cause no agitation to his brain and would tire his eyes, those lights turned down, turned out.

# RECENT THEORY

## *Even Now Nobody Knows*

The inside of skulls has increasingly been the researcher's laboratory. (The inside of every part of the body.) Sleep has been investigated with gross surgery done upon the animal brain, and, in the human brain, fine electrodes applied to the surface, or pushed distances into it, even implanted, left for periods of time.

A cat's thalamus (that solid "chamber" lying below the canopy of the cortex) has had an electrode implanted, time allowed for healing, the thalamus shocked, and the cat has fallen asleep. Before she did she curled up as cats do and it was natural sleep because if the researcher approached with a package of sausage she sniffed and woke. The sausage has acquired fame. The experiment appeared to favor the existence of an active sleep center, a place that when shocked induced sleep, and sleep not just depressed wakefulness. Other experiments favored a wakefulness center. Experiments can be clear-cut. Sleep still is difficult to define, has relations to the brain that still are difficult to limit, this engine not under the hood of a Cadillac.

Sleep. Arousal. Wakefulness. Attention.

The four are linked in our everyday thinking, linked also in the psychiatrist's probing of the psyche, and the physiologist's study of the working of the brain.

Sleep inside us does often feel like an on-off event. Surgeons know an area of brain where they go carefully, that might be considered an on-off area, life and death uncomfortably close, sleep and waking close. Sleeping-sickness at post-mortem has shown nerve-cells destroyed there. Latterly the pharmacologist has seemed to prove the blessed-cursed sleeping tablets to act there.

Microelectrodes are pushed into single cells (cat's brain) or around single cells, and from the news that comes out, from the recordings, the researcher can separate excitatory from inhibitory cells. The study of sleep, that is, has been brought down to the dimensions of a microelectrode. The study could be construed as a search for unitary sleep-neurons and unitary wakefulness-neurons. One thinks for the hundredth time of old Democritus' search for the unitary. Some of the present-day ambition is that microelectrodes have become popular. Some is the temptation to repeat old studies with a new tool. Some is no doubt genuine curiosity about sleep, to dig and to find something further about the way the nervous-system works at this function. Single cells in numerous neuron pools at numerous depths have produced data, the computer brought in to help, and adding also no doubt its part to the idea of the brain operating probabilistically, an idea that is more and more attractive, and, each time one looks into one's own mind, seems likely. There is a perpetual balancing of inhibition and excitation in all directions at once, with always some kind of magnificent average struck.

Nevertheless the 1969 picture of sleep is fundamentally not different from, say, ten years ago. Sleep remains commonplace, bewildering, alluring, and through it Fuji half-disappears half-reappears in fog.

When the brain-stem is cut across high-up a profound sleep settles on the animal. That experimental surgery was performed long ago. At the time, the thought was already developing that the low stem kept the high stem aroused, that the high kept the cortex of the total brain aroused, and so we were awake. Numerous experiments made it almost unarguable that the stem forever intervened between a sense-stimulating world and the cortex, to rouse it.

When that cut brain-stem of the profoundly sleeping animal was, on the raw surface toward the top of the brain, electrically shocked, the animal waked. An electrode woke it. On the contrary, if the core of the brain-stem was surgically scooped out, a sleep descended on the animal from which it never woke. If there was no strict sleep-center, that neighborhood nevertheless plainly had a primordial concern with sleep and wakefulness. Loosely speaking, a switch could be thrown there and good-night till tomorrow. The body's thermostat was not far away, and the daily cycles of body temperature were sometimes described as paralleling and even somehow determining those other cycles, sleep and wakefulness.

It was fifteen or so years ago that sleep experimentation began to concentrate on the brain-stem core, on the reticular formation in there, that intricate weaving and criss-crossing of nerve-cells and nerve-fibers which runs the length of the core. Sense paths (pain, vision, hearing, smell), on the way to their specialist areas of the cortex, sent side branches into the core. That is, sense-messages on the way to the cortex telegraphed collaterally into the stem. Thus that sequence: core stirred, cortex stirred, creature stirred, in a state now to pay attention to the messages that were being delivered. Messages also traveled down into the spine where were the final controls for neck, arms, legs, the parts that perform anything the body is called on to perform. So, as has often been said, the mind is alerted above, the body alerted below, the creature able to think and to move.

To repeat: sense messages, pleasurable or dangerous, cold toddy or hot lady, stimulate the brain-stem, it stimulates the cortex, sleep departs, our mind is aroused, and we attend, act, think.

One recalls the physiologist who believed the mind existed for movement, for the next motor act, so that the act might have the greatest possible perfection, the psyche simply the latest evolutionary creation for muscularly meeting an attacking world with effectiveness. The mind during the time of sleep would of course have lost that effectiveness, but would be preparing for a greater effectiveness later.

The cortex also stimulates itself, though the how of this is not so neat, but it must, or it surely seems it must; at least, each of us seems all day to be finding that his nervous-system at its top is able to stir itself. Or, to speak with unphysiological abandon, the mind is able to wake itself, to know itself to be awake, though only God knows just how it knows.

The part played by fatigue? Where in this scheme does the fatigue of the day enter? That again is not as demonstrable as common sense suggests. Gradually vast numbers of the nerve-to-nerve linkages do probably in some way become depressed, are harder to reach, their thresholds raised, a block thus placed in the way of the stimuli of life, the creature deprived from underneath, getting drowsy, settling into a light sleep, a heavy sleep, passing through the stages, in-and-out, up-and-down, till tomorrow morning the thresholds are dropped to day levels again, the networks reachable again, the mind attentive again, enslaved again.

Nothing has been said of sleep's periodicity. Some of us prowl through the night and some prowl through the day, but most sleep at night. On a day-night planet resulting in a society with day-night habits learning could have set the periodicity, though we recognize that the nervous-system must have had the particular pliability for this learning built into it by a joining of hands of genetics and evolution. They left us a system both rigid enough and pliable enough. In Bergen, up near the Arctic Circle, the streets were awake and even crowded in what ought to have been the sleep-period of a twenty-four-hour June day. Citizens claimed they needed little or no sleep during that period, which may have been partly the traveler's too literal acceptance of whatever he was told. People anywhere who claim they never sleep usually lie or are mistaken. Each of us does however find that he can swing around his twenty-four-hour schedule, change the one-two ratio of sleep to wakefulness, also find that he is sleeping too much or sleeping too little. On the other side of this is the fact that in numerous arctic expeditions, where observations were made not casually but exactly, men did maintain the one-two ratio despite the all-dark or the all-light of the planet. An animal with its entire cortex surgically removed still shows some periodicity. That is, the brain-stem itself has some clockwork, though what the psyche of a brain-stem might be one must despair having any idea of.

Frédéric Bremer, eminent and imaginative physiologist, in his picture of sleep related cortex to brain-stem. This was years ago, but Bremer's discoveries do not lose their freshness and charm and they continue to be a restraining force in sleep theory. Bremer was much responsible for our knowing that the brain-stem wakes the cortex. In his picture he brought stem and cortex together. The cortex he conceived as made tired by the affairs of day, the stem as watchman on guard over it, its master. Fatigue spread in the cortex. The master nodded. The watchman watched. At a point the watchman decided, now. He threw the switch and that was good-night till tomorrow. Bremer did two unforgettable experiments. He used cats. In the first cat he cut between brain and spinal cord, produced a brain separated from trunk and extremities, a brain with no nerve-lines coming to it from below the head, but those from the parts of the head undamaged. This cat slept and waked as whole cats do, in rhythm. Its eyes during wakefulness followed persons as they moved around the laboratory. Would have interested Edgar Allan Poe. In the second experiment Bremer cut higher, between cortex and high brain-stem, produced a cortex separated not only from trunk and extremities but from the parts of the head and face, especially from the labyrinth, these parts of the unmutilated animal battering the cortex day and night. This cat slept a deep sleep and like any sleeping cat its eyes rolled downward and there they stayed. Its eyes did not follow persons as they moved around the laboratory.

# PAVLOV'S THEORY

## *He Thought He Knew*

Pavlov's favorite word all his laboratory life was *facts*. He wanted nothing but facts, probably did not always get what he wanted, and sometimes surely did not relish what he got, like all experimental workers, the greatest, the small. But he got much. His theory of sleep he did not consider a theory. It was one of the facts. Sleep was the spread in the cortex of *internal inhibition* as he called it. He had discovered it during his conditioned-reflex experiments. Internal inhibition spread and the creature slept. Hypnosis was partial spread. Sleep and hypnosis and internal inhibition he considered essentially the same process. Much that is known today about inhibition, the chemistry, was not known in Pavlov's day. Many physiologists not only doubt Pavlov's internal inhibition, but his theory makes some of them scathing toward him as a person, toward a man who died in 1936. His fame meanwhile persists and grows, in recent years appears to outgrow Freud's. Pavlov is one of the figures of the past most present to the present. He turns up everywhere. He was many-sided as an experimental worker, and complicated, not a simple and innocent mind functioning in a research laboratory.

Internal inhibition was to him as real as the sun and snow of the north, might conceivably be as real to the physiology of 1969 were he alive to repeat and correct and expand his experiments with the techniques and instruments of 1969, the growing knowledge of the night-busyness of our brain, and were he alive to argue for and to defend what he believed, and, of course, to change his mind.

The inhibition was, then, part of conditioned-reflex action, and with the conditioned reflex Pavlov was prepared to account for inanimate and animate, for Theodore Roosevelt and William Jennings Bryan, and the crowds that listened to both, and, later, Mussolini saluting, giving a conditioned response with his chin to the conditioned stimulus of the crowd at the Rome International Physiology Meeting, where Pavlov pointed it out; was also prepared to account for the behavior of kittens and cats and dogs and neurotics and psychotics, for everything that the sleeping and the awake creatures of earth do.

This special inhibition, sleep, as Pavlov understood it, was protective, prevented exhaustion of the chemistry of the delicate brain-cells, particularly of the cortex. A brain was active day and night. Inhibition got it its rest, moment's rest, prolonged rest.

Fail to give the dog its dab of meat when the bell rings or the buzzer sounds or the light flashes, inhibition begins. Repeat the fail-to-give and soon the dog sleeps. The man in the street might say that we dull when things we expect do not happen. So we do. Pavlov would have agreed. Give a difficult problem to a dog, it sleeps. Students in a classroom avoiding the frustration of a difficult problem or a difficult lecture, sleep, roll in their seats. If it is a girl in a classroom who risks sleeping she first carefully protects her hair by making a pillow of her purse. Lengthen the delay between stimulus and response, between bell and meat, the dog sleeps. In a hundred laboratory situations, corresponding to a hundred life situations, the dog sleeps. Inhibition begins at a point in the cortex, as Pavlov back in the first third of the century understood it, or the inhibition begins at many points, the points coalesce, inhibition spreads, moves down into the stem, where by many lines of evidence there is a final shut-off, almost mechanical (but much interpreted) for the waking mind. Soviet Union disciples today claim Pavlov said that. It is not in print. But the remainder, his internal inhibition, hypnosis, the spread of internal inhibition as sleep, he himself documented with abundant experiments published finally in an array of languages. Out the window into the night went the mind, something went, wakefulness had ebbed, and had left behind it the silent lake of sleep.

To the human mind, despite a lifetime of good or bad nights of sleep, not much in the universe seems more strange. Fuji looms, but in the night we can be only yet more vaguely sure that it looms.

Instead of inhibition having moved down in the brain it might have moved up or left or right or in or out or all directions, only not into the vital controls for circulation and respiration, that must never sleep. Ostensibly we would die if they did. Ostensibly parts of the brain were always accepted as always awake. Yet, in late years, what is suspected is that no part of the brain is ever fully not awake.

Pavlov studied sleep originally because his laboratory dogs fell asleep. This disturbed his experiments. It made him irritable. So he decided to study the situation, to find with conditioned-reflex experimentation what sleep meant. His dogs were to pay for their sleepiness by contributing facts about sleep. He respected his dogs.

In one well-known experiment he selected twenty notes of a musical scale, made inhibitory stimuli (no-meat) of nineteen of them, an excitatory stimulus (meat) of the twentieth. With the dog-brain thus prepared the experimenter related that he played the musical scale. First there was a succession of inhibitory notes and the dog fell asleep, stayed asleep until there was the one excitatory note, when it waked, ate, fell asleep again when the inhibitory notes of the rest of the scale once more depressed its waking mind. An astounding experiment even before one began to think of explanations.

Hypnosis supplied facts. Pavlov thought hypnosis was an inhibition in perhaps a single analyzer, as Pavlov conceived analyzer; this asleep, the hypnotist able therefore to perform Machiavellian maneuvers with the unresisting parts of it. A dog's limb could be lifted by the experimenter where no dog would lift it, the limb remain there, as a poor woman's arm may for hours in an insane asylum.

It should be illuminating to those who watch human behavior that an occasional scientist has seemed to scorn Pavlov exactly because of his imagination, his fancy, his vitality, his (what amounted to) playfulness, his lightness. He was light both in body and mind. All that he did, and he did many relevant experiments, and no doubt thought that he did many more, has with scrupulous avidity been combed by those who do not believe as he believed.

So, according to Pavlov's theory, there were always spots of internal inhibition in the brain, of partial sleep, and if these moved far enough and in the right direction the dog heaved a sigh, as dogs do, its body collapsed, its mind went out the window. Had anyone ever asked Pavlov whether dog and man slept by the same mechanisms he would have bristled. There could be no question here. Brain was brain. And what Pavlov knew Pavlov knew. He would have said peremptorily that he could produce sleep, vary sleep, measure sleep, in a dog; and, if the circumstances were right, in a man, who was an evolutionarily later dog.

# DREAM II

## *Has a Night-life*

When we were young we sometimes woke and wept and needed to be comforted because something had happened while we were asleep. We might occasionally instead shout with joy because fairies had stepped from a fairytale, a Charlie Brown from the funnies. Hungry baby Anna Freud called from her sleep for "stwawbewies and omelet and puddin'," and her father could not conceal his satisfaction at having found dream material in his own family.

Anybody's dream may make us listen, may even in 1969 mystify us, for a minute, help us to see the ancients as having been like us, see them sympathetically, also help us to take another hard look at sleep. There is in its fabric this embroidery of dreams, that can be heavy, can be light, can be eerie, can be dull.

Half-roused by the plumbing, we think it was the plumbing, in the next apartment at 4 A.M., or by some plumbing inside our head, we dream. We wake. We sleep again. Wake again—this second crash could only have been the plumbing. Tomorrow we will do something about that. At breakfast we tell our annoyance, and tell our dream, tell it well, too well for accuracy. Nevertheless, what we tell, what a man or a woman tells at breakfast, may interest us more than we ever were in him or in her, make us realize there is more story there than we thought. No story ever in that man in the daylight. What peeked out from him by night was not at all like him. He is a stiff sophisticate. If only he would dream more! The century has made all dreams suspect—yes, if he dreamt more we would know more.

Memory supplies the data for dreams. It must be memory. What else could it be? It is memory—it may be memory for a while unrecognizably rearranged. Perhaps it is something else. We feel it is.

It always seems memory under glass.

It can excite any of the emotions. There can be fright. Bismarck's dream had fright: that defile narrowing to where his horse could no longer turn around, could only go on, on. The fright of dream has the responses of day-life fright, thumping heart, bursting temples, sweating, drooling, autonomic signs, somatic signs.

If one has time, and if one has the memory that holds last night's memories, the details, one always is tempted to take the dream apart, see through that always distorting glass, that theatre illusion gauze, figure out what did really happen, why it happened. The analyst does that. He nudges us. He only appears to sit immobile in his chair. He will somehow have helped us to fit the pieces together, to relate today to yesterday to long ago. About the illusion gauze itself he usually does nothing. It has no items. He wants items. A face. An action. These are of use to him, to his reasoning.

Any dreamer might want to know about that illusion gauze. Each time it is gone so suddenly, evaporates, it seems, that color, that mood of *mise en scène*. A wave laps the sand, wets the sand, disappears into the sand. He wakes. The unreality of night is still for a while around the reality of day. It may make him think of a Blake drawing not quite seen, an eighteenth-century gavotte not quite heard, a Maupassant tale written toward the last of Maupassant's haunted life.

But does a dream have color? Does the illusion gauze have tints? We may ask that of ourselves every now and then. Does a dream have smell? There was Haydn—a dream does have sound. It takes some backward-looking to be certain. Yes, it was Haydn. Franz Joseph Haydn. Then—Haydn dissolved into the boss. Where did *he* come from? Haydn just dissolving into the boss might have waked anyone.

On a hundred occasions, surely sometime when he just wakes up, a man says to himself that everything about the mind is strange, the waking, the sleeping, the dreaming, stranger than anything any Japanese ever saw when he looked up from one of Hiroshige's fifty-three stations toward legend-capped Fuji.

What a night!

There was that street back of Carnegie Hall. It must be Fifty-sixth Street. Who could mistake that stage door? As realistic as a concert ticket with four edges. A relief to be seeing it from the street side, otherwise one would be in there playing the Beethoven Violin Concerto to five thousand people! Six thousand! Seven thousand! Eight thousand! Nine thousand! Ten thousand! Eleven thousand! Twelve . . .

One opened one's eyes. It was still dark. One was afraid to look at one's watch—hated to see the lost time written on the watch's face.

One closed one's eyes. There was a hymn. From childhood. Not thought of for ever so long. The words were remarkably exact. "Sleep, my child, and peace attend thee . . ." Lafcadio Hearn said he heard voices in his dreams. The congenitally deaf say they see vivid colors. Most opposite things are said. If what is said is printed, that makes it authentic, also removes what a dream cannot have removed: its quality.


# DREAM III

## *Laboratory Steps In*

In 1952 a young scientist observed that the eyeballs of sleeping infants moved under their eyelids. The movement was back-and-forth and rapid, fraction of a second, and the eyeballs were moving together. Later, this was labeled rapid-eye-movement-sleep, REM-sleep. It came in bursts, spurts. Still later, the time between the bursts of REMs, when there were no REMs but the person asleep, was labeled non-rapid-eye-movement, NREM-sleep. The 1952 observation by the young scientist was made in Chicago where problems of sleep had been studied for years, but no one there or anywhere had observed the eye-movement. Might seem merely an interesting observation. It turned out more, brought the study of dream into the laboratory, because older persons were found also to have the REMs, and when they were waked they said they had dreamed. The movement went with dreaming. Odd that eyes could have been doing that for ages unobserved. He in Chicago who saw first, saw first. One wonders whether some mother somewhere had not observed it in her baby in her lap, but that would not have been the same kind of observing.

In the ensuing years dream laboratories sprang up in place after place, United States, France, elsewhere, and before many more years the National Institute of Health was offering grants of more than a million dollars in support of dream research. There was a society of dream researchers. There were local branches as numerous as those of the coagulationists. There was an international meeting, a Frenchman, a German, a Russian, a South American, talking different languages around one subject (raising for the thousandth time the tenuous hope of one world). There were graduate students. There were graduate programs in colleges of medicine, investigators metamorphosing into administrators, dream articles piling up, computers retrieving the salient, and the rest.

A consequence is that we know with greater dependability when we dream, how many dreams per night, how long a dream lasts, it being apparently not the instantaneous affair it was thought, but lasting apparently the same length of time as the event enacted. Et cetera. We cannot be said always to have cared whether we did or did not have all of the new knowledge, but we had it. At the same time we might be losing the dream. The research scarcely was concerning itself with the hallucination that dream is, with its emotional or literary or analyst content, appeared mostly unaware or uninterested. No doubt the research wished to stay severely objective, severely mechanical, severely scientific. It was valuable research of its own kind.

The dream-laboratory?

It is practical to take the moment when the researcher enters for his all-night job. His subject, a man, is paid. He is lying on a couch in a cubicle. He will probably fall asleep. The researcher is not in the cubicle, has his own cubicle, or does not have one, just a place where he can without disturbing the sleeper watch him and make his recordings. The watching is apt to be through one-way glass. The researcher is throughout employing present-day physiological instruments. He is pegging the brain. At least, he is pegging it more than the mind, because, no matter how one thinks of brain-mind, the mind keeps largely aloof from most of this experimenting.

The subject has fallen asleep He has passed from the awake-stage through the various sleep-stages. These are well established though with some arbitrariness that is inescapable in 1969. An electro-encephalograph is recording the stages. Physiologists use letters or numbers for the stages, 1, 2, 3, 4, we being here too in the telephone-dialing-age where soon we should be relieved altogether of the luggage of words. How fast it has come!

The subject on the couch continues sleeping. Electrodes of the electro-encephalograph are attached to his skull. An electrode for recording eye-movement is taped near his eye. From under his chin runs a wire for recording body-muscle activity. From inside his nose for recording respiratory activity. From the left chest for the action of the heart. Nevertheless he sleeps. First the sleep was NREM-sleep. After a while it passed from NREM to REM, the dreaming-sleep. It is REM now. His eyeballs plainly are oscillating under the eyelids. Once one has seen that one could think one always had but just paid no attention to it.

A hypodermic needle has been stuck into his vein and taped down so as not to break if samples of blood are to be drawn—usual in surgery but not usual in dream research. A stomach tube is in his stomach to record what goes on there—still less usual. A special kind of plethysmograph is registering changes in the size of his penis—not usual at all. Psychiatrists have made everybody, and themselves, and researchers so aware of sex behind every door that the sex organ itself gets into some of the experiments, though one suspects that even a surviving early analyst might be surprised at how valiantly his sons and grandsons were carrying on. A rubber tube, hose, is belted around the man's chest to record in the old-fashioned way, also, respiratory activity. Other apparatus. Yet despite all apparatus the subject continues to sleep, is hired to sleep, so sleeps, perhaps helped by the virtue felt by the laborers for science (or for scientists). One researcher did admit surprise that with all of this going on around a human head it still could take its part in this active process that requires a minimum of self-consciousness at least to start it. We all have been told of the soldier who the night before a battle sleeps, and Pavlov over and over told us that the animal's brain was built tilted toward sleep, and that a spreading inhibition was sleep. What Pavlov did not know was the degree of excitement there could be in a brain while arms and legs and trunk are limp in sleep, which is one of the facts that dream researchers have found and taught. If the dream-subject has trouble falling asleep on the first night, that is called first-night-effect, and the subject gets over it, or if he does not, if he is a healthy insomniac, he is discarded, or given sleeping pills.

Cats frequently have been subjects. Cats too must sleep. At least they sleep. But do they dream? They have REM-sleep and NREM-sleep. All mammals examined have. If REM-sleep in them too is dreaming-sleep, then of course they dream, and what does a cat dream about? Who would not like to know? The Bible speaks of the dream of animals. The Bible is full of dreams that for the most part never occurred in the way related. Something occurred, something was added to what occurred, something meditated, something turned into meaning, and the meaning stayed through the centuries.

In the researcher's cubicle the paper of the electro-encephalo-graph keeps rolling.

It is ninety minutes since the first REM-sleep for this night, and according to previous research there should now be another spell of REMs, and there is. The electrode for eye-movement is recording. There has been a reduction of general body activity, also recorded. The wired subject (curlered, crimpered) has passed through the various sleep-stages, the electro-encephalograph recording them. There are the periodic spindles of sleep, the "bursts" of fourteen cycles per second, which is stage 2, then a sprinkling of so-called delta, slow and high voltage, which is stage 3, then all delta, stage 4, and now back to stage 1, which is light sleep, dreaming-sleep, with REMs. There will be three to four or five more spells of REMs before the night is over.

Thales, measuring Greek, had much to do with the beginnings of all of this, and from him it continued to spread outward, produced glorious mathematical periods occasionally, but ours surely must be the golden age of universal measuring.

When the sleeping subject passes from NREM-sleep to REM-sleep and the eyeballs report, that is the cue. The researcher rouses the subject, questions him. Eighty percent of persons so questioned say they had a dream, relate it, or part of it, or hesitate, are blocked. Such blocking of the content of dream, as other somewhat unusual characteristics of it, have always offered to the psychiatrist or the psychologist the possibility of uncovering something hidden, or half-hidden, something with meaning for that mind. But the dream-researcher is more concerned with the mechanics of the occurrence of the dream than in the dream, and more in the brain than in this presumed exhalation from it. Pavlov would have been. One some-times wonders whether, if there were enough successful avoiding and denying of the subjective, there would actually be less and less of it in the world until there was none, and the brain would be mind, as the computer expert thought that, when the computer was excellent enough, it would be mind.

Many small discoveries have come from the dream research, and many explainings of the discoveries, some alluring, none as alluring as the dream itself, a good dream. And a good dreamer. Researchers believe there is no such person as a no-dreamer. The REMs say so. The no-dreamer, or the apparent no-dreamer, proves harder to wake. Everybody REM-sleeps. Even the animal surgically deprived of its cerebral cortex REM-sleeps—that fragment of nervous-system REM-sleeps. It gives off rapid-eye-movements, REMs. The previous questions arise. Does it dream? What does that fragment dream about?

The paper of the electro-encephalograph keeps rolling and will till morning.

For the dream-researchers there would accordingly be three phases of brain-mind: REM-sleep, NREM-sleep, the wide-awake state. Why not? The brain's relation to those three phases has been carefully investigated, is better understood; the dream, *das-Ding-an-sich*, possibly not much better understood. The best-known of French workers has located the source in the brain of the REM activity. It is in the pons, the part of brain-stem that long was recognized as controlling the motor activity of the eyeballs.

One might say that everything fits.

As for the use to the researcher, the REMs let him break in on the sleeper when he dreams. The researcher can cut in on the dream. He can stop it. He can entirely prevent REM-sleep. He can deprive the creature of it. For the length of a long experiment, days, the creature might not be allowed to snatch a wink of REM-sleep. The consequence? Does he behave just as he behaved before? Is he normal? Is REM-sleep absolutely necessary to him? Does he require dream? Would a creature die if he never dreamt, never REM-slept? He would not—that might of course depend on the fortitude of the researcher. Would he go mad? No. Be irritable? Have an excited brain? Yes. Parts of the brain undoubtedly are busier during REM-sleep than at other times. (Odd always as here that mind should be thinking of its own brain.) So the brain can be highly active as we lie inactive. Researchers point to the relevant fact that ulcer crises, asthma attacks, heart failures often occur in the middle of sleep.

Everything fits.

The body's reflexes meanwhile are depressed, motor activity depressed. That fits too. It leans toward former ideas of the state of the nervous-system during sleep.

This rough waking of subjects, human and cat, not letting them stay in dreaming-sleep, forcing a cat to crouch twenty-one hours of twenty-four on a brick with water all around and no chance to go limp and doze, might seem a crude (even cruel) way to study the cobweb dream. As for the crudeness, that depends on how one looks at it. It really is no more crude than sawing open human skulls, stimulating human brains electrically, causing the minds to see in the present what they saw in the past. And both are fascinating crudenesses, if crude. What the human dreamers dream, the Queen Mab fabrications, are not crude.

One returns spontaneously to the dream of cats, the cat mind, the idea of the cat mind, and concludes, after five minutes or half an hour, that we do after all live in a world that has much besides wars and politics, and that one lifetime is too short, despite sore throats and hot-wet Sundays. One understands Pavlov—he would like to have been frozen and then thawed out in a thousand years to look around.

Dream-research is a comparatively new research. The first high enthusiasm may have dampened some, which would be the history of any new research, but undoubtedly it has excited speculation, has caused a re-examination of old beliefs, old superstitions. It also can make one wish to re-examine ordinary facts about the dream condition.

Hard muscular labor decreases dreams. Meaning? Hard mental labor increases them. The decrease and the increase can now be definitely established, or so it seems. Dreaming increases dreaming. Attention on dreaming increases it. Anxiety does. Illness. Sexual preoccupation. High REM-activity is reported to accompany the dreaming of the sexual act.

Thus can the researcher go on adding any and all kinds of findings. He can even get bored doing it. The night-brain as brain plainly is demonstrated busier than was formerly believed, though it should always have been believed busy—it was not categorically known how much it dreamt. Is that so much more to know? Yes? No? Immanuel Kant's theory was that we dreamt after heavy food so as to keep alert the reflexes that keep digestion going. To Kant that was the physiological use of a dream. Sounds almost ridiculous. Kant had a good mind.

# DREAM IV

## Dream of Infants

It was in infants that the rapid-eye-movement-sleep was first observed. It should mean, or could mean, that they already are dreaming. The newborn there in the crib is dreaming. Maybe he dreamed before he arrived. Anyone would be piqued to know what that dreaming was about. However, the chance is that the infant is not dreaming. A researcher is a researcher and as long as possible stays away from the psychologically odd. Confronted with REMs in the newborn he is apt to relax his original conception of "dreaming" sleep, look for some other or some additional explanation, some other purpose for the REMs.

The newborn sleeps much of the twenty-four hours, shows REMs and NREMs, and may be in the "dreaming" sleep eight hours of the twenty-four.

He also when he is being studied has the electrodes wired near his eye, to the top of his soft skull, other instruments wired, and thus wired he blessedly sleeps.

Production of sleep (not dream) is believed to occur as follows. It might be at night, might be a somnolent afternoon, and sense-organs are ceasing to be stimulated from outside the body, ceasing to be stimulated from inside, especially from the contracting of the tensing muscles, the stimulation of the brain-stem in turn ceasing, and of the cortex in turn ceasing, so the creature sleeps. The reverse for waking. This is for the matured nervous-system and has been recognized for two decades. Is it the same for the immature nervous-system? Probably. The infant is less active in the world, is stimulated less, sleeps more. And it does REM-sleep. Is something else accomplished by the REMs for the infant than giving it dreams? Is there dream of infants? Does the day-and-night hammering of the world in the newborn, as also somehow the REMs, stimulate the developing brain to develop? That has been a question. We do of course not know what mind in a dream essentially is any more than what mind in the waking state essentially is, but, slurring over that, does the dream of infants assist their nervous-systems to become completed nervous-systems, and their minds to become minds?

The stimulating world via the senses gets to the ancient brain-stem that has charge of the vital signs, breathing, heart, nutrition. Does dreaming help to connect, to make functional the connection of that ancient to the much less ancient forward brain, that relates to this individual going about his business in the world? Pavlov already was convinced that truly and fully developed conditioned reflexes, that belong to that individual as individual, depend on the cortex of that forward brain.

Why do we dream?

From the analyst's point of view there has not for half a century been much doubt. Different analysts would say it differently, emphasize differently, but the dream is some kind of mental clearing, mental riddance, allows for an escape into disorder of the tightly stricturing order of the day.

From the researcher's point of view there has also of late been less doubt. He suspects purpose not in the dreams of the newborn, which he hesitates to believe in, but in the REMs, which are fact. He thinks he has some beginnings of an understanding. He gets rid of his difficulty with the idea of dreaming newborns by looking for (and seeming to find) one reason for REMs in the adult, and another in the fetus and infant. In the adult they concern the mind. It dreams. By dreaming it rids itself of what would in some way clog it, the REMs necessary for the mind, not to dream being destructive for the mind, and there is now experimental evidence that that may be fact. In the fetus and newborn the REMs concern only the brain. Lacking them the brain fails to get something it needs for its maturing. Dream in some way stimulates the brain to become a completed, a connected, a working organ, whatever the consequence of that for the mind. Fetal and infant brain are not yet sufficiently nudged from the outside, so they must be from the inside, and by that nudging the brain and then the mind get to be what in our lives we seem to know them to be.

Infant and adult and life and brain and mind are finally rolled into a oneness, magnificent, and still mysterious, if you like.

Our thoughts naturally wander again to scattered facts we all know about dream, scan them possibly. Women long have been said to dream more than men. Is it true? Why would it be? Dogs are said to dream. "My dog runs—chases rabbits in his sleep." Et cetera. Dogs do have REMs. Pups? Do only domesticated animals dream? And in that case is it only as part of their association with man? Or do all animals dream? Does a canary dream? It may jerk suddenly in its sleep, shake itself, have vacant eyes for a moment. A pigeon too. The dream of animals, as well as the dream of infants, is a matter to fret the mind of any man. It spreads wider the biological terrain of dream. It spreads wider our philosophy.

Children, long before any of the late experimental investigation, were said to dream as early as their second year or earlier. The experimental investigation would support that if the REMs always and only mean dream. Newborns, as found, and young children have the rapid-eye-movement. If the rapid-eye-movement were, however, evidence not for some brain developing, but for the newborn dreaming in the adult sense, that it arrived on earth dreaming, and that for months afterward it dreamt during half its sleep, which would be two-thirds of every twenty-four hours, this would mean something on the human scale tremendous. It would have shaken Socrates, the philosopher, and Aristotle, the scientist, and Herodotus, who was, quaintly, both. Nothing is for nothing. If the neonate does need the dreaming-sleep for the physical development of his brain, it may well be that REM-sleep in him is without mental accompaniment. The mental would add day by day. The neonate there in the crib needs to laugh, probably, for his development, for his life's sake, at least he does laugh. He needs similarly to perform other movements, life-subserving movements, at least he performs them. It might be the same for the REMs. Is he happy when he laughs? We do not know. That there is some accompaniment of contentment every mother is convinced, but that might be *her* contentment.

The newborn may have needed to dream in the womb. He may have needed to dream in order to climb out of the womb into daylight. He may have dreamed his way from another life into this life. Reincarnation may be fact. Mind may go on from mind to mind to mind. A present life may be a literal continuation of previous lives. Ancestral memory may be dream. In that case the researcher may be making the incomprehensible comprehensible, may be carrying us back to Wordsworth, farther back, to the Middle Ages, farther back.

Some dream-researchers have speculated as freely as poets. "Life is a dream," one of the best known has said and often repeated that. Our world is no longer a place of meditation and confabulation for poets and philosophers only. Our poets often seem faint of heart, self-conscious, and possibly therefore so unhappily noisy. Our poets are afraid not to trust our researchers, and possibly therefore we could conclude that they are not sufficiently poets to see with that fierceness which might push researchers aside and tell them not to be so cocksure. When indeed the researcher confronts the poet with the "crucial" experiment the poet is bound to be embarrassed, feel stupid, not know what to say, what to do.

The times are out-of-joint for poets, are as never before in-joint for researchers. Accordingly some who would have been poets, who are poets, born poets, who were poets in their cribs, have found their way into science, and some of those have discovered that they could be poets there too, that imagination and fancy are useful everywhere. Dreams are a land between, reach toward science, reach toward poetry, play in the garden between the two houses.

# DREAM V

## *Creativity*

"Kubla Khan" is the most famous instance of a dream doing direct work.

Coleridge had taken laudanum, very likely. He had toothache, very likely. Everybody in England had toothache, but, to be remembered, not everybody in England wrote "Kubla Khan." "In consequence of a slight indisposition, an anodyne had been prescribed, from the effects of which he fell asleep in his chair." Those are Coleridge's words. He is speaking of himself. The slight indisposition does for some reason, one cannot put one's finger on it, start a feeling of invention, which may not have been intended and the words are ordinary, but, whether or not, out of the ensuing dream rose a vision that for its kind may not be surpassed in any language, except the visions of Dante.

> In Xanadu did Kubla Khan
> A stately pleasure-dome decree:
> Where Alph, the sacred river, ran
> Through caverns measureless to man
> Down to a sunless sea. . . .

Of Coleridge's taking anodyne we know bitterly. He took laudanum until it destroyed all of his capacity for literary production, alienated his family, his friends, including the prig Wordsworth, finally left him only Charles Lamb and some indefinite persons who seemed to roam around him and in whose house he lay year after year. The laudanum did however not destroy the strange Coleridge quality, dream quality, that still was there when the emaciated head was sinking dying between two pillows. For this we have the testimony of Charles Lamb, who visited him every day, was Charles-Lamb honest, a Charles-Lamb reporter, which is to say watchful and compassionate.

In the explanatory note printed with "Kubla Khan" Coleridge tells us that he was writing down the dream when a visitor interrupted, and the rest was lost. Literary analysts accept that account, though one cannot escape remembering that Coleridge's "Christabel," where there was no question of dream, at least none was raised, also for some reason, some block, some something, stopped before its end, and the second half of even *The Rime of the Ancient Mariner* may not be as finished as the first half.

The introducing of the "slight indisposition" may have had concealed in it the hope to cover somewhat the fact of his addiction. No one likes everyone to know he takes drugs. De Quincey at the beginning of *The Confessions of an English Opium-Eater* states carefully that his addiction was under the, to use his words, coercion of pain the severest. Coleridge does not particularize toothache or laudanum in the formal note, only the indisposition, the dream, the writing, the visitor, the unfinished "Kubla Khan." The poem is the most internally finished Coleridge ever wrote. It is short, swift, visual like a dream, breaks off as do dreams. The hook-and-eye connecting of images (hook-and-eye is his phrase) and the glossed language may both have been facilitated by the laudanum, but warp-and-woof the fabric was Coleridge fabric, as De Quincey's was De Quincey, as the genuine inside of any lesser creativity is, opium or no opium.

Essentially, whether Coleridge did or did not have the dream is irrelevant. He almost certainly did. We all dream so-and-so-many dreams per night, as the dream-researchers have documented for us, and so-and-so-many daydreams per day, meaning three billion dreaming heads and their dreams rolling with the planet every twenty-four hours; and one "Kubla Khan."

More satisfying for the literary analyst or any other analyzing analyst, one would think, to reflect on the instance of Edgar Allan Poe, that mind that tramped in nightmare beyond dream. Through his rapidly uphill fame Poe everywhere exposed his disorderly downhill life, but the nightmare in that mind always ordered, always clear.

He had his actress mother who died when he was three. He had his father who disappeared when he was two. And a father is a father, but also is the husband of a boy's mother. So, child-Poe is child-Oedipus, if anyone likes, would in his unconscious kill his father, wed his mother, and this child also exposed on a wild mountainside, the emotionally wild urban milieu of Poe's childhood. He grows to be a man, takes as wife not his mother, (she's dead), but his child-cousin, thirteen years old, and like many another artist, like Dante Gabriel Rossetti notably, merges the lines of those two faces that he knew best, the corpse face of his tuberculous mother, the always dying face of his tuberculous wife, different faces in the gross, and out of them creates the faces of the women of his tales, Morella, Berenice, Ligeia, Eleonora, also creates, copies off his mind, their deathly pallor, their deathly allure. Add that Poe remained childless. Add the tale "Loss of Breath," and give it the over-interpretation it has been given: namely, that loss of breath is Poe's own loss of potency. Poe accordingly becomes impotent. Maybe he was. Impotence itself, by a similar over-interpretation, becomes death and has the hues of death. Add a quick summary of several of his death-in-life stories. Add his predilection for employing as sculpturing material a ghastly cold lava, plus a sculpturing talent for lava, and what an irresistible invitation has been extended to a first-year resident in psychiatry to diagnose necrophilia, corpse-love. To dis-diagnose necrophilia would require at least a second-year resident. Add an unrelated fact—because it is called attention to though one does not see why and it could hardly have much to do with creativity—that he had trouble borrowing four dollars out in Fordham. (Four dollars in Fordham where McLuhan, as some of our obsolete Gutenberg-movable-type print tells us, would this year be drawing a salary of one hundred thousand.) Add that Poe like the late Brendan Behan appeared to drink himself into who-knows-what dreams, but highly creative dreams surely, drank himself blind, alternated brandy with opium, they say, the dazed unhappy man carrying his hemorrhaging child-wife from chair to bed, eventually placing her coffin on his writing-desk. Dream and reality seem now a frightful mix of soured white milk and shiny black coffee.

But—he did write "The Raven." One's own temperament may not incline one toward a poem like "The Raven," but that does not prevent one reflecting that the whole hosts of drunkards or near-drunkards and addicts or near-addicts of America and the whole hosts of its psychology-bitten awake-and-sing critics could in the genre neither produce its equal nor explain its peculiar force. "A limited genre." "Quite so." Poe was a dreamer and a poet, his kind, a singular kind, and let anyone account for the dream in any poet if anyone can. His life seems rifted, cloven by a too lucid creativity, while the person, the human body merely, was transported through Richmond, West Point, the Bronx, Philadelphia, Baltimore, in which last place he died at forty, presumably of delirium tremens, and, whether or not, his inscrutable restless genius died with him. The scrutable is his life, and, like much in anyone's life, is irrelevant. In the Foreword to Marie Bonaparte's overdone *Life and Works of Poe*, which Foreword she asked Freud, her teacher, to write, and he conceivably therefore not able to evade writing, he says: "Investigations such as this do not claim to explain creative genius, but they do reveal the factors which awaken it and the sort of subject matter it is destined to choose." Marie Bonaparte may not have been that modest. Freud was modest. Again and again he shows himself the best of the Freudians.

Now move off to the Orient and consider for a moment the way the creativity of one Oriental, a Japanese, can strike one Occidental. He is the greatest of Japanese artists probably, or one of two or three. Human faces in Hokusai's line-drawings often appear reality exaggerated, reality seen through that illusion gauze that every dreamer knows. Hokusai's sketches are of the hard Japanese daily life but always with that gauze dropped down in front. Even in his paintings real and unreal touch the lines, black lines frequently done with black Japanese ink, a bristle of a woman's hair, streak of eyebrow, hollow of cheek, jawbone, the violent thrust of a man's extremity. A face to terrify a child rips through a Japanese lantern, gets stuck there, as one could think, lantern and face immortalized, the lantern being the *mise en scène,* the face being the actor.

Hokusai ranks with Shakespeare and Beethoven. Shakespeare wrote, Beethoven composed, Hokusai painted.

Steeped in nation Hokusai was, helplessly Japanese, helplessly himself, helplessly creative, hundreds and hundreds of paintings and drawings. Of the man we know only a few facts, not one with the melodrama of the whole of Poe. Hokusai endured, if you like, this difficult planet until he was ninety. Was poor. Stayed in the same faubourg in Yedo, today's Tokyo, that is on the large island of Honshu; he was always vacating one chilly room and moving into another, sleeping on straw-matted floors inside rice-paper walls, the houses always somewhere close by the Yoshiwara; painted the women of the Yoshiwara but made no specialty of it as did Utamaro. Painted anything living, dog, bamboo, fish, a samurai, a bridge. A bridge can live of course. Worked always at lightning speed, never blotted out, it was said, apocryphal no doubt. Had trouble with a nephew, like Beethoven. Had trouble with the police, like Beethoven. All of us have troubles, a new trouble little or big every day, and all of us daydream and nightdream; and one Hokusai. Probably he accepted as did everyone around him that divinity spoke through dreams, the tutelary divinity, the ancestor who pushed his way across the generations and appeared in his mind, and he made a straight copy of what was in his mind, as Coleridge claimed he did, as that scientist did from that laboratory manual at 3 A.M. Hokusai bequeathed us his own face too, a landscape of wrinkles, harshly believable and unbelievable to Occidental eyes, no doubt to Oriental eyes too, at least what is there was seen through the illusion gauze. One fancies Hokusai in lantern-light squatting by the urori warming his not-very-warm rice over not-very-warm charcoal. With chopsticks he shoveled the rice into his mouth from a green bowl, dozed, dreamt, peered curiously strainedly at something, waked, worked, sketched. Under the ink-brush a robber swung his sword, slayed, was slain, two bloody heads lifted, died. What he peered at was something reflected and refracted in his special mind; born with that.

And now Shakespeare. "To sleep—perchance to dream: ay, there's the rub!"

From those words Shakespeare moves into his great soliloquy. How quiet immediately when one comes on Shakespeare. How far immediately from nightmare. It is not dream either, but nearer. We have had that soliloquy inside us so long that we think it is part of us, think: "I could do it." Try. Paraphrase. Let Hamlet speak your paraphrase. Listen. You will be saddened because you will have had demonstrated anew your deep lack of the depths of poetry.

This has this much to do with creativity: *One mind can.*

There are times when we wish that Shakespeare had left a journal, then recall what analysts did with Beethoven because he left a journal, or left so-and-so-many scribbled sheets of paper. It makes no difference to Beethoven how any analysts analyzes, nor to the idea, Beethoven. The *Eroica* is there where it was. But it could make some difference to us, to what our banal thinking has made of us. A sheaf of Shakespearean letters would be a temptation nevertheless, even one good-length letter. We might not learn anything about the nature of creativity, or about Shakespeare's dreams, whether he was an insomniac who batted about in his bed, whether he nightly as a preliminary doused himself with ale, but most definitely we would learn that he was not one to write himself out interminably in notes as did poor Coleridge, display himself naked as did poor Poe. Yet, a journal, even a short journal, must on some page give us a flash of that towering mind's view of itself.

He wrote the plays we do not know where, corrected them probably on the stage because he was an actor, played the ghost in

*Hamlet.* For this there are the extant playbills. There are the documented evidences of several kinds. There are the copies of the actors' parts, this and that printed play, a few versions exhumed by poor scriveners scribbling under the moon to buy themselves bread and cheese. There is the great fire that burned the Globe Theatre to the ground with the playbooks of the playwrights. Shakespeare is inconspicuous throughout but his reputation growing. Then somehow there are the Quartos and finally seven years after his death the great Folio of 1623. How subject to vicissitude all of it! Suddenly one realizes that Shakespeare might not have survived. No Shakespeare in the world! But God, let alone Shakespeare, is not dead, and Quartos and Folios were saved.

Shakespeare, that man, lived. Shakespeare was born: that mind was born.

Whatever being-born may one day mean either to chemistry or philosophy Shakespeare's mind arrived on earth mostly complete. It never in one lifetime could have accomplished anything beyond a polishing-off of some of its roughnesses. In 1969 we conclude not with satisfaction that mind is an arrangement of atoms. The sophisticated chemist-geneticist has however not in these matters gotten us too abysmally far. Fundamentally, we do not know what mind is, though any man's wife without sending for any analyst may know or think she knows the twists and turns his particular mind has taken under the pressure of birth, infancy, adolescence, mother-in-law, old age, and how that mind has made use of life to modify what it came with. Shakespeare's seems to have been able to make use of everything, and this, the able-to-make-use-of-everything, does not characterize creativity either, characterizes dreams only somewhat, but worth mentioning, though it directs our attention more to the substance of the created work than to that special energy, if it can be called energy, that causes substance to live. Substance is approachable. Imagination, the ease of the poetry in Shakespeare, the labor of the poetry in, for instance, Michelangelo, are hardly or not at all approachable, understandable.

Shakespeare may have napped in the afternoon as did the older Hamlet. If he napped he dreamed. We had to wait three and one-half centuries for dream-researcher and psycho-analyst to make us utterly sure he dreamed. He had to dream. We are scientifically convinced he had to dream. He probably always snatched a dream when he snatched a nap. And of a night he dreamed so-and-so-many dreams. He REM-slept, rapid-eye-movement-slept, and NREM, non-rapid-eye-movement-slept. Even cats do, if someone has not scooped out part of their brain. Then, Shakespeare's mind went the Freudian unconscious-preconscious-conscious back-and-forth and the id-ego-superego back-and-forth and whatever other backs-and-forths. Also, we can be sure he free-associated. Cats may do that. A cat squatting on a warm radiator with a slit of eyelids open and eyeballs moving is besides REM-sleeping doing something psychologically inviting, though the interpretation would require a cat-psychiatrist. For Shakespeare we might go scientifically on, no doubt, find a word-sequence that fingerprinted his creativity as Ernest Newman found three ascending notes to fingerprint Beethoven. It was an empty finding. Shakespeare and Beethoven when they slept staggered their dreams in eighty-to ninety-minute periods, unless the physiology of sleep was different in the seventeenth and nineteenth centuries than in ours. The eighty-to-ninety makes the planetary dreaming like the grains of sand, everybody, everywhere. Creativity is not like that.

The word *dream* comes into the plays, but it is always the poet using a word, a thin veil of the illusion gauze dropped even in front of the word, that is not at all the word the physiologist uses. Concocted dreams occur here and there in the plays. Then, Shakespeare questions whether we may not all always be dreaming, our life a dream, as he says, surrounded by a sleep, and this the present-day dream-researcher questions too, means it otherwise, states it otherwise, but something in his work has drawn him to such an idea. Hard-headed Pavlov said again and again and many years ago, as we know and also come back to again and again, that the brain was built tilted toward sleep, that to be awake was perpetual arousal, and he thought of dreams because, as he did also say, if he were given two hundred years he might fit dream into his scheme. The two hundred indicates that canny Pavlov was aware of difficulties.

Did Shakespeare daydream? Daydreaming unquestionably is not the same. Daydreaming would have mind-proofed him against theatre frustration, yet he could not have done much of it, not in the usual sense, not idly, since even with his genius he would have needed to keep disciplined thoughts one following the other if he was to achieve not only the unparalleled quality but unparalleled quantity of the masterpieces. He finished them first to last in something like sixteen years. Began late—his early thirties. Stopped early—late forties. Whatever dreaming is, whatever creativity is, his creativity did not loiter, his mind did not relent from that balanced-imbalance that every self-observing mind knows necessary if it is to remain itself, not to speak of producing play after play. As for the blank verse that buoyed the plays along, Shakespeare must have been able to produce it as his heart pumped blood, ordered his breakfast porridge in blank verse, blew out his midnight candle. One begins to think that blank verse was simply his prose, then one comes on a prose passage so limpid, so rare that—well, he could write prose too.

The daydreaming, the not-quite-awake, the half-somnambulistic state, Shakespeare's sharpness would dispose one against believing he used it directly or exactly or at all. That half-somnambulistic state has been considered the productive state. Tschaikovsky considered it so, Schubert did. We do not know how they induced it, brought it out of the who-knows-where, hard walk on a dark night, a plunge of their heads into cold water, vodka for Tschaikovsky, Rhein wine for Schubert. Nietzsche spoke of the productive state as *Rausch*, should have understood it, in some state like *Rausch* wrote the whole of *Also Sprach Zarathustra* in eleven days, or so he said. But Shakespeare? He appears to sit cold sober while his unconscious sprawls over the universe, a loose unconscious in the grip of a tight conscious. One always imagines him calm, inwardly, amidst the hurly-burly of Blackfriars, or any London theatre, or any brawling tavern. Marlowe, fellow-playwright, was killed in a brawl. Those were hot times—ours may be hot but not that way. John Aubrey judges that Ben Jonson was jealous of Shakespeare. How could he help? But Ben Jonson did stress Shakespeare's gentleness. That was stressed also by Hemminge and Condell, fellow-actors, in the introductions to the Folios. That is stressed also by us in the twentieth century. Gentleness flows to us from every play, sometimes from every page. We feel without too much examining the idea that a supreme and dreaming gentleness might be near to the highest creativity, assuredly part of it. Gentleness helps a mind to stand off from itself. Shakespeare's seems to do that, seems as aloof and nobly dignified as that face, those eyes, that tall forehead, even if each feature is considerably sentimentalized in the Droeshout portrait of him.

Finally comes *The Tempest.* Finally comes the moment when Ariel is dispatched to do his master's last biddings on the island, and, with those biddings done, to get his promised release, airy Ariel to work no more for Prospero, maybe for no master, can sail like any old-fashioned spirit or elf or even witch wherever it pleases him, couch in the cowslip and sing:

> Merrily, merrily shall I live now
> Under the blossom that hangs on the bough.

If Ariel was created by Shakespeare to represent creativity, his, it might tell us something of what he himself thought of that process of the mind, his experience with it, how separable from the rest of the mind, how in some respects a foreigner, a servant called in, sent off. "Where the bee sucks, there suck I."

*The Tempest* and whatever parts of *Henry the Eighth* completed, Shakespeare would have been bouncing in a stagecoach from London to Stratford, there to count his moneys, adjust his affairs, make over his will, loll in the soothing neighborhood of his daughter, Susanna, reflect on this and that, eat, drink, doze, sleep, dream, maintain an amused expectancy, wait for April 23, 1616.

The chance is that with *The Tempest* he was bidding us all goodbye, and how charming if he was. He was making his stage-bow before the theatre of the world, and by way of ornamenting the bow did once more toss into the stage-lights the colored balls of his grace, excite himself with words, Ariel not free at all, not able to be free, Shakespeare not troubling to understand dreams or creativity, just dreaming, just creating. Then, there would have been that other indifferent day, that day of amused, or musing expectancy, when with the familiar words he would have helped his body to stretch out under that flat stone.

> Good friend, for Jesus' sake forbeare
> To dig the dust enclosed heare;
> Bleste be the man that spares these stones,
> And curst be he that moves my bones.

Desecration of graves was common in the seventeenth century and Shakespeare had ordered his to be dug seventeen feet deep, presumably ordered.

# PSYCHIATRIST

## *Shrewd Scrutiny*

Something of everything that has gone before has contributed to the making of the psychiatrist.

The person should have flair. If he has talent he has it. The late Alexander Woollcott, brilliant broadcaster, used the word, relished it, the snobbery in it, old French, middle English, swirled it over the radio during World War II when it was not yet on everyone's tongue. Flair is sagacity for odors. In the psychiatrist it is a nose for private news. He pokes it into human affairs. A nose better than most of ours, partly because he continually exercises it; is his business, theatre business.

And all the men and women merely players. . . .

Think of Sartre, Camus, Ionesco as psychiatrists. Think of Jung, Ferenczi, Piaget as playwrights. It is reasonable. Ibsen was a psychiatrist. Peer Gynt, Hedda Gabler, Master Builder Solness, what neurotics! Ibsen treated them—and us—by clarifying them on the stage. He expanded. He condensed. He expanded. He did not let out much about how he worked, except that he worked in his head, only finally on paper. Expansion was part of his genius. He knew his personae. Expansion is the capacity of the psychiatrist too. There are psychiatrists will blow a phrase into an act, a sentence into five acts, the play often get so tortured and thinned that it bubble-bursts into the theatre's rafters. But sometimes the psychiatrist plumps a five-acter onto the boards and does not lose a nuance, sticks to a cruel imaginative realism. If he has the greatest talent he makes a perplexing psychological story reach quite beyond his small off-Broadway establishment to the whole human family. How that novelist-dramatist-psychiatrist Dostoevski did just that, understood, forgave, loved not only the single character but the masses.

However, flair in her psychiatrist is apt to be of most use to a neurotic lady. He smells her out. She likes it, may hate it first, then likes it, then likes him, then likes him more, then all but tells him, then he tells her why, spells it out, explains her to herself (has memorized his own role), and in the midst of his performance she turns her face toward him as would an illumined seductive but not quite comprehending older child, has picked up the theatre atmosphere, and is convinced she is the star. But he is the star.

In late years it often has been thought that what the psychiatrist smelled out too exclusively was sex. Freud did. He did it honestly and certainly earnestly. It glamorized him, added notoriety to his fame, a jealousy-arousing fame, so it was inevitable that psychiatrists spawned by him should one after another, some of the eminent first, draw away from him; ostensibly, some of them, draw away from his too exclusive sex interpretations, overlook that he had considerably drawn away himself.

Sex is enormous in the theatre. Without it what plays would there be? Yet when one thinks of it one suspects that the psychiatrist actually reduced its dimension, got it into proportion. By following Ibsen and Strindberg and other playwrights of that half-century, Freud in turn was followed by the psychiatrists (and other playwrights) of our half-century. Thence to the actors, to the Berkeley campus, to the long hair and the short skirts, and every psychedelic-pantried automobile, banners up for freedom, often freedom simply to be neurotic. And freedom-found is promptly a bore. To be fighting for freedom is 1776. So it is necessary to invent more freedoms, and that naturally gets thin, till there is not much left to do except scream.

Sex's past pale poetic mystery is banished, is relegated to dead poets and untutored adolescents, if there are such. Most often the past-pale-poetic is just cash-and-carry. It is still adequate to get done a murder or a marriage, or, in its milder moods, to titillate, to compensate somewhat for something lost. Probably human beings in all times have pretended to enjoy sex, and certainly fight for it, long after ennui has settled upon its repetitiveness, explaining why everyone has looked for the renewing agitation of a full moon, or a birthday, or an anniversary, or a New Year's Eve. The drab century's erosion of romance did of course not stop everybody talking about sex. They talked even more. For this talk the psychiatrist had his specialist conversation pieces, his dialogue, his technical language, said "penis" as lightly as "liver."

When the neurotic lady is in tête-à-tête with him, not holding anything back, as she imagines, suddenly she wonders whether he will tell the plot of her story to the client coming in at the one door while she is going out the other? He probably will not. He is forever using that word *confidentiality.* In some branches of the healing art the healer thinks nothing of telling, always tells his colleagues at lunch, and, should he ever have the big chance, tells the newspapers. If you are President of the United States or France, or the Pope, or flit about in Hollywood, and something is done to your prostate, everybody reads about it tomorrow morning in London and Berlin with his bacon and eggs. But the psychiatrist's *mise en scène* ought really to contribute something to keeping secrets, especially the analyst's, that room, two shut in, fifty minutes, curtain down before, curtain up afterward, the one actor prodding the other actor, and in that sometimes voltaged atmosphere she daydreaming out loud, relating her night experiences, relating her memories. In such a scene the pair of them, the psychiatrist and the lady, are lovers, restrained lovers, and it could be that it is the restraint that sometimes lets these plays have such long runs, fifty-two weeks, or four times fifty-two weeks, sex put to something besides its humdrum use. This was of course in Freud's plan, the old matchmaker.

It is obvious by now that the psychiatrist and the psychiatrist's office are bound to be another station below sacred Fuji, though Fuji for the moment may not seem especially sacred.

A performance over, the psychiatrist stands up, walks around his room, opens the window, gets a breath of fresh air, rubs his hands, figuratively, goes into the hall for a Coke, basks in the probability that he has put a few more stitches into the mind-wounds of one more character.

Whatever we complain or proclaim is our all-human comedy, the psychiatrist is a big member of the cast. His role is a sign of the times. He himself is more in the cartoons than anyone except the politician. He was in last week's *New Yorker*. Do not look. He was there. The jokes about him are among the stalest of our joke life. "Here's your prescription—here's your bill." That's a joke. "The bill is the cure." That's McLuhan. The bill of course was also in Freud's plan, the old keeper-of-the-books. "Motley's the only wear." Freud could never have overlooked the motley in a Park Avenue or Harley Street therapist for pay communing fifty minutes with a golden-haired golden-pursed hysteric before she went to her beautician and afterward home for a nap until cocktails and dinner. "My analyst!" If the hysteric is one of the luckier she has a secretary who sees to it that she keeps her appointments, and a chauffeur who drops her off and picks her up. She is a fashionable hysteric for a fashionable psychiatrist, and there are fashionable sanitariums, the physician having become a landowner, landscaped lawns, cottage-type build-ings, an inquisitive child nevertheless peering furtively through the wrought-iron fence and having to be pulled-on to get him away.

Farce is always in season in that small planet we all now have looked at from the moon. Farce is in season though we don't always laugh. Farce is a form of drama that regularly introduces misery for contrast. Misery can be quite apparent at the short distance up of an airplane, when one looks down, but the hostess up there is sure to trip and splash bourbon, or water, over a stiff-shirted superintendent of schools so that there is still a touch of slapstick.

The hysteric may of course be the hysteric's husband, or the aging bachelor of the family who afterward sells his adventures to the magazines or the cinema.

Whatever such falderol at the periphery, at the center the psy-chiatrist's day-to-day is to help mind. The background theme of his drama is always mind in some trouble. A mind started quaking from before-birth (a gene guilty), at-birth (an obstetrician guilty), sub-sequent-to-birth (an environment guilty). That is, a brain was misbegotten, therefore a damaged mind; a brain was mechanically or chemically bruised, therefore a damaged mind; or, the other way, a mind was environmentally suffocated, therefore a damaged brain. For the last there is no evidence. Probably there never will be. Probably in spite of everlasting micro-electrodes stuck into every division of the brain, the physiologist never will discover for the psychiatrist what he is looking for. And it is toward them, the environmentally damaged, that the psychiatrist much of the time is directing his techniques.

The most important of these is listening. The psychiatrist listens. Someone says quickly that the priest listens. Yes, he does. Someone says that the family doctor listens. Yes, he does too, if he lays aside stethoscope, reflex hammer, drugs, and takes the time to listen. In general, listening has gone out of our society—listening not shouting. The psychiatrist is a listener, may interrupt his listening with ah, um, if he is that kind of ass. But he listens with concentration, with over-concentration; anyway, he wants his audience to see his concentration, listens visibly, is visible, makes himself visible.

Freud was an actor. He was actor-playwright, like O'Neill. His numerically small but select audience sees in him all the dramatis personae, mother-figure, father-figure, child-figure, the whole cast of the *Everyman*. When one remembers Freud's fox-eyed face, the slight contempt, if that is what it was, the smile never allowed to break through into a laugh, except, rarely, a loud laugh, but the smile also never brushed off, one remembers suddenly the cigar. The cigar was always there. The cigar was stage business. The cigar may have killed him and that is too bad though most of us die of our lives. His beard also was stage business. Beards are all over the place now, of course, but nine out of ten in our streets have not yet gotten used to their roles, are as little attached to the minds as the bouffant hairdos or the high-rise stockings. Freud's mind was attached to his cigar and to his beard. He may have overplayed Freud but he never ceased working at his role. He recognized that in the presence of intelligent acting and in the glow of the theatre every human being, neurotic or not, is readier to accept life for what it is, a place to make love, to laugh, to weep, to reflect, to find oneself by acting parts. Some psychiatrist's role may not be well played (always some role not well played) or it may be so well played that the actor forgets he is acting. Freud never forgot. Freud is described in that Vienna group as of a night sitting to the side, away from the middle of the group, middle of the stage, where some non-professional would think how-modest, but every professional knows that that was the one character no one would take his eyes off.

The psychiatrist's most effective property is talk. He may withhold it, sit in statuesque silence, but that surely is talk. On the contrary, he may erect a wall of talk between himself and the species, talk, talk, talk, which can be unfortunate, though he needs his wall. Each of us needs his wall. It is a human ambiguity. Even those two so close in the parked automobile, each has his wall. In *A Midsummer Night's Dream* the facts are spoken straight-out. Prologue says:

> Wall, that vile Wall which did these lovers sunder;
> And through Wall's chink, poor souls, they are content
> To whisper.

The psychiatrist may have thought back at the beginning of his career that what he was treating was essentially routine minds. Soon he found there were no routine minds, he only had not yet learned to see them beyond the patterns of his field, minds and lives always bewildering past anything he dreamt ten years ago, two years ago. He discovered with some surprise that he was having a growth of humility toward mind. A quite ordinary psychiatrist might discover that. Might not. Anyway, he now never could fancy himself going his Monday-Tuesday-Wednesday with the insouciance of the proctologist. He goes like Don Quixote, only Don Quixote had windmills, and mind is more in flight than windmills, more like the wind itself, always changing its speed, sometimes slowing to nothing, or like cloud, always changing its relation to other clouds, and to itself.

The heaping-up of complexities whenever the nonprofessional stops to ponder it makes him wonder why anyone ever got himself in this difficult profession. He got himself in partly, like the rest of us, because he must do something. He must pass the day. Must have money. Have flattery. Drain off frustrations. Stave off loneliness. Escape the clown he wishes he were. How much of the world is arranged that the clown, who is all that really matters, may strum his instrument, kneel before a lady and sing a song, conceal his personal folly in a role of folly, which is many a human being's destiny, and an occasional dog's, and a good destiny. "O that I were a fool!" Jaques says that in *As You Like It*. He tells us that he would turn folly to medical use.

> Invest me in my motley. Give me leave
> To speak my mind, and I will through and through

Cleanse the foul body of th' infected world,
If they will patiently receive my medicine.

A psychiatrist may have had autobiographical reasons. We all have autobiographical reasons for everything. Medical students come into medicine for the obvious ones. A student was convinced he had a bad heart, his grandmother had, he saw his mother die in a heart attack, so elected cardiology. Another when a boy sat too many Saturdays in dental chairs, so dentistry. Another was uneasy about his mind, so psychiatry. Lucky in the long run if he had that intimate a reason, helped him to help himself, and helped him to help his patients, brought his life (that really concerns him) into theirs. Freud was uneasy about his mind. He needed his analysis. He undertook it because he needed it, not because he liked to undertake it. No doubt psycho-analytic institutes, besides pedagogic reasons, and besides obediently stepping in the tracks of Freud, dead, have from their practical experience learned that it is wise to anticipate trouble in the minds of their novitiates, therefore made it obligatory for each to be analyzed, this in the face of the current somewhat suspect state of analysis, probably only current, a phase, a passing toward something somewhat modified. Something was too popular so it became unpopular.

If a psychiatrist had genetic reasons he was lucky, if he had a psychiatrist-gene or even a whole chromosome. The gene raised its head and said: "Psychiatry." The embryo in the mother's womb repeated: "Psychiatry." The postnatal product in the schoolyard watched everything his classmates did, how they had "problems," how they could not come to "decisions." He nodded. He understood. Such a one was made in heaven and will return to heaven a still better psychiatrist.

While on earth he will not have wasted much time in asylums, will not have spent his Sunday mornings in a review-of-cases in an old-fashioned state hospital, will have known how to avoid, how to shut his eyes against ugliness, will be like the bird-mother who has flung the whole lot of her disabled brood out of the nest; if a well-meaning human being puts them back she flings them out again, lets those flung out wiggle their way to death, will have no abnormal creatures around her; yet for the sake of the completeness of her bird-mind the human-mind might fancy that she would have wished as she sat there comfortably in her nest to have before her inner vision a picture of total bird-life; would have wished, that is, to include the asylum.

The golden-haired golden-pursed neurotic will be pleased to hear what has been said about asylums, urges that it be printed in *Harper's Magazine,* wants her neighbors to understand that the neurotic is not ill but the asylum-patient is. The golden-haired has read about doctors who claim that neither neurotic nor the asylum-patient is, that psychiatry is or should be all sociology. An occasional psychiatrist does indeed manage to spend his entire professional life as if asylums did not exist, and if ever he visits one it will be to attend an 8 P.M. meeting in a recently renovated plush auditorium from where because of the crowd of the anthropologists, psychologists, citizen-minded citizens, one cannot so much as see the lighted windows of the other buildings. The auditorium is air-conditioned, and aired.

On the contrary, there are what one might risk calling asylum-psychiatrists who see those buildings as their castle. (This surely would be 1929 because surely everything is different in 1969.) Such an asylum-psychiatrist is a foreigner, used to be. Or he did not graduate in his class with high-enough grades. He could not hoist himself into Massachusetts General in Boston, then into Queen's Square in London, the customary round. Possibly also, but only possibly, he was genetically embued with a concern for those farthest estranged minds. He preferred them around him. Someone is sure to say he was dedicated. It could be. It could be, too, that chance simply dropped him into this insane place, and he fitted. It is a place where less than a generation ago an intransigent psychiatrist might suggest that ether be pumped under all the doors, a place where bodies supported heads that seemed eternally seeking, mis-dressed women frequently, young or old, methodically hurrying, methodically slowing, methodically rocking in rocking-chairs, methodically caressing some object a mother-in-law brought, a caress without purpose, then a sudden purpose all over the ward because someone started a line and a dozen fell in, convinced that they also must go to the toilet, a place where a face might come right at you with that flat look that if it had been a soldier's someone would have said euphemistically "combat fatigue."

It is not easy to be a physician if one wishes to cure. It is not easy to be anything if one has 1,350 grams of brain. It is not easy at all to be this specialist who spends days, nights, years, meeting minds that are helplessly non-constant, non-homeostatic, non-equipoised, better today, tomorrow worse, their gyroscopes never preventing them rolling, pitching, corkscrewing, or the other way, becalmed in an Ancient Mariner sea, and the specialist bound to go from ship to ship and not permitted to get seasick.

He may be pure Freudian, neo-Freudian, anti-Freudian, Adlerian (this is only to indicate), Jungian, Pavlovian, logotherapist, family therapist, group therapist, existentialist, hypnotist. He may be that psychiatrist-researcher turned surgeon, who does investigate surgery on a dog's brain instead of a dog's bladder, or turned chemist, who injects this or that in minute quantities here and there in a dog-brain, maybe a human-brain, or turned physicist, convinced that if he could build the right computer he could produce a computer-simulation of mind, then disturb the simulation, then study the disturbed simulation. One would hope he might not convince too many innocents (though immortally it would make no difference) that the laws of the mind, in 1969, can be found by appropriate programming, that they are fixed like the laws of astronomy, chemistry, physics, that they are fixed not in flux, a flux that cannot yet be understood, if ever.

Today the psychiatrist is most apt to be psycho-therapist, an all-part casting, employ psycho-dynamics, employ psycho-analysis, not to the last refinement, conceivably not to the last excellence, gather data on early memories, on current husband, current wife, current children, current adored, predecessors to that one, a wife who sexually rejects, a husband who rejects, divorces, abortions, pregnancies in and out of marriage, use and abuse of the pill, homosexual tendencies, shames, guilts, hostilities, suicidal tendencies, murderous episodes. He may be a psycho-pharmacologist, seldom calls himself that even when he writes prescriptions all afternoon five days a week. Sometimes—sometimes—he is a different psychiatrist for every patient. He has the energy for that. A logotherapist immediately wants to call him a logotherapist, one who seeks to give individual meaning to individual lives, but that would be a mistake, too, because this one is apt to be pure Freudian today, Adlerian, Jungian, Pavlovian, logotherapist tomorrow. Better not saddle him with a system. He produces and acts with a minimum of methods, minimum of properties, minimum of craft-language, minimum of rehearsals. He is a special specialist for each special person (regards each as special) who comes into the stage-lighting of his theatre, who sits on a chair, lies on a couch, stands on his head if that is how he stands most balanced, while in front of him the psychiatrist, the star actor, plays his repertoire of roles, adapts role to audience, sometimes gets his repertoire mixed as we other actors do. When off the boards he may be an exceeding sane man, relatively.

# DRUG

## *It Is Bored*

De Quincey took opium. Coleridge took laudanum—opium—and wrote "Kubla Khan." The Chinese smoke the opium in pipes, at least did before they became responsible communists. Japanese drink saki. Soviets drink vodka. Collegians drink Coca-Cola, or sniff glue, or swallow a cube of sugar with a drop or two of something colorless, odorless, tasteless. Brave-New-World Aldous Huxley swallowed four-tenths of a gram of mescaline, watched what happened to him, wrote what happened to him, wrote interestingly, wrote so that all levels of intellect would read, and we do not know just what he accomplished toward a fashion in our always-ready-to-be-fashionable world. Others swallowed, wrote. Most just swallowed. An estimated eight billion stimulating amphetamine tablets are produced in the United States per year. Women in the Tennessee mountains, instead of glue, sniff snuff. Hashish, marihuana, cocaine, curls of smoke from Hong Kong meet curls of smoke from Delhi meet curls of smoke idling up from Cuernavaca or idling down from a cheap high furnished room in an off-Broadway hotel. Much of the American twentieth century puts itself to sleep with sleeping pills, or bourbon, or bourbon plus pills, then next morning looks for its mind somewhere around the bed, cannot find it, then finds it with the help of Dexedrine or caffeine or just four cups of moral black coffee.

Which could be said another way. A mind pours a drug upon its brain, chases the drug with another drug, or heightens its effect, prolongs its effect. Which also could be said another way. A brain pours a drug upon itself. A chemistry pours a chemistry upon a chemistry to alter a chemistry. The brain pours in doses, fifty milligrams, two thousand milligrams. Why? Why would a brain want to do that?

Some of the reasons are hidden where no chemist or psychoanalyst or sociologist or Addiction Research Center will find them. Some are as plain as a prize fighter's ears. That brain is simply hoping for an hour to escape the pain of belonging to the human family. For an hour to escape itself. Escape its restlessness. Escape its fear of death. Fear of life. For an hour to avoid the minds that come via the telephone, via a letter, via a knock at the door. For an hour not to have to prop its skull and neck on its trunk and feet. For an hour indulge a satisfaction not trusted for a minute, a siren illusion, a last embrace, a success that already is not a success. For an hour to accelerate dullness into gaiety, gaiety into anxiety, anxiety back into dullness, that circle whose diameter gets shorter at each circling. For an hour to be in the swim, be of this world, or that glue world. For an hour—if an old man—to wander backward into his youth that they keep telling him was luminous, happy all day long. Read Plato as he thinks he read him when he was sixteen. Watch Socrates prepare to die. Watch Xanthippe watch Socrates. For an hour to dissolve his wife. Or himself. In vodka. In whisky. In sentimental nobility. (Does the body distill a drug that is sentimental nobility?) In blind hard work. (A drug?) In sixteenth- or twentieth-century Puritanism. (A drug?)

The human species is drugged. Always has been. Has thought it needed to be. Or did need to be. Or wanted to be. The cutting edge of life had gotten too sharp for the human type of mind.

The animal type of mind? We do not know. Often a dog appears to let itself fall asleep at the earliest good-manners opportunity. Has any animal but man ever been known to kill itself? Commit hara-kiri? How often, in man, that was just a performance. How often it came off better than intended. How often it demonstrated that he was as good as his word.

Hara-kiri. Fuji. The Mountain.

In an experimental mood a first human brain has administered to a second human brain what would throw the second into a temporary insanity, psycho-phrenia, the second having volunteered for this out of vanity, stupidity, cupidity, twenty-five dollars, the reasons we all volunteer. Lysergic acid explosively revealed its curious capacities years ago. Lysergic acid spread. Lysergic acid became besides a drug an idea in the mind. Lysergic acid became a symbol of the right of any human being to do with his life exactly what he pleased. Lysergic acid passed on its rights and privileges to glue. Lysergic acid claimed it produced more than a temporary insanity, produced revelation, shone a light into the truth beyond truths: final clarification, final beauty, final expanded consciousness, apparently an understood state on some campuses, by some psychologists, by some philosophers. *Psychedelic* became a newspaper word, meant consciousness-altering, applied to any drug that did it. Without a doubt lysergic acid changed John Henry. He changed while he was watched and while he watched himself, got uneasy, got suspicious, got a flush, felt nausea, talked too much, talked of hearing colors, seeing sounds, touching smells. (May have begun to wish that he had not let himself in for this damned adventure, experiment.) Wondered whether he would be the one who stayed changed forever—if this was New York, might rush to Bellevue. In a few hours, eight or ten, he might be all John Henry again. But—they say—the experience might of itself repeat itself, without the drug, diabolic uncertainty. Some psychiatrists definitively do not regard what occurred to be insanity. Some have administered these popular drugs in the course of therapy. Also, administered them to lift depression, to tone down elation, to achieve at one and the same time depression and elation, loosened someone who was not responding to Freudian interviewing, bring him where interviewing was possible, other bona fide and not so bona fide reasons.

Lysergic acid as every newspaper reader has known for years came to the attention of science because of a laboratory mistake. A chemist got it into himself, afterward took it deliberately, this was in Basel (Paracelsus' Basel), Switzerland, wrote in his notebook that he felt dizzy, lay down, had a delirium not all unpleasant, was both unclear and clear, a sense of standing outside himself, viewing himself, screamed, appeared mad.

He had poured something via his stomach onto his brain and somewhere in the neighborhood was his mind. He got on his bicycle, rode to his home which was close but that in that mind (where all reality is) was miles and miles, time-sense warped, his space-sense warped (time-space of the physicist warped, of the lobotomist warped), his report no doubt warped. Everybody seems to know more than one version of that famous occurrence. If Charles Baudelaire had heard of lysergic acid and could have brewed it in his kitchen, what flowers of evil would have been added to *Les Fleurs du Mal* and what artificialities to *Les Paradis Artificiels*.

Pavlov, the Russian, induced neurosis with two metronomes ticking at different rates, conflicting conditioned stimuli, named what he produced experimental-neurosis. The name sticks. What happened in the body, the face, the actions of the victims, dogs, did look neurosis, looked insanity. Were those conflicting stimuli producing an altered chemistry that left in the body a foreign chemistry that rationally might be called a drug, that acted on the brain (again somewhere in the neighborhood was the mind), that produced a behavior? Freud long and perhaps always believed that most of his neurotics, those victims that were not dogs but human beings, had resulted from some thwarting of their infantile sexuality. Did that produce a chemical? Today psychiatrists emphasize guilt, shame, aggression, defense, dependence, in-turned or out-turned hostility, in-turned or out-turned anything. Chemicals? Drugs? Life, the global experimentalist, produces its neuroses in known and unknown ways that sometimes appear all reducible to a mind examining itself too much, too destructively. Chemicals? Drugs? We owe it all to special actions on our special brain, a large brain, an instrument that the more we understand of it the more we suspect we do not understand. Our outrageous brain is the cause. We use it too much and we have too much to use. In Caesar's words we think too much—because we cannot avoid thinking too much.

For a time it seemed, still seems, that new chemicals to produce neurosis or psychosis were being discovered every afternoon. (Every afternoon is easily said.) Hallucinogens produce hallucinations, mis-stimulate a brain, so it mis-sees a world. Adrenalin is part of the jargon of our day. "He had a shot of adrenalin." Normally, adrenalin is broken down rapidly in the body, the break-down occurring in stages, its manufacture also in stages, and some of the stages being among the potent mind-twisters, so one could imagine—let us imagine—a chemistry stalled at a stage, and that sent the invitation that caused our dear departed grandmother to visit in our bedroom without turning the key. Chemists have delved into tyrosine, phenylalanine, tryptophane, amino acids with whose chemistry the body must normally work, and might work abnormally. Chemists have delved into the indoles, sterols, reserpine, serotonin, that body and brain employ in their operations, that may be present too much or not enough, an unnatural chemistry producing an unnatural mind, hereabouts one proven cause of inborn idiocy.

Tranquilizers for years were in the cartoons and on the musical comedy stage and in the gut and blood and brain of uncounted Americans. Tranquilizers still are. Tranquilizers, old ones, metamorphose into tranquilizers, new ones. Tranquilizers are part of the twentieth century's chemical faith. Molecules of a tranquilizer, with some peculiar arrangement of atoms, sidle up to molecules of a brain, with some peculiar arrangement of atoms, and a mind changes. Chlorpromazine. Meprobamate. Pentobarbitone. Rauwolfia. (All old.) Psycho-pharmacologicals there are of many kinds for many kinds of cures or hopes of cure, some discarded, some arriving, millions and billions of dollars in the turnover, drug houses printing and giving away magazines that one could think were *Life* or *Look*.

Mental hospitals discharge the sick en masse as a result of these types of medication (drug therapy) that may be praised as extravagantly as yesterday was lobotomy (surgical therapy) and day before yesterday shock (electrical therapy). Mental hospitals in consequence are quieter places. Society approves of quiet mental hospitals, quiet burial services, quiet graveyards. All seems to go on an assembly line whose fatal patience on some days does itself act like an hallucinogen.

Which all could be said still another way. A host goes to his cellar, accompanies a bottle to his dining room, agitates the bottle the slightest, proffers its contents in thin glasses to sophisticated guests, babbles about the year of the vintage. A family doctor proffers a capsule. A psychiatrist, fifty-minutes. A nurse, a suppository. A living body—the oozing of an endocrine or combination of endocrines or fractionation of an endocrine or failure of an endocrine or genetic absence of an endocrine. Result? A mind floats, or divides, or quits for the night, sees lush shapes in emptiness, alters shapes actually in the wallpaper, executes any of the thousand tricks that the head-end of this creature performs, to amuse itself, to numb itself, to treat itself, to inspire itself, to understand itself.

Which could be said an ultimate way. This beast, this accident or otherwise in an indifferent universe, this product of earth and air and water plays upon its highest regions with new configurations of earth and air and water. A lofty one, a beast, studies the configurations, changes a benzene ring, a hydrogen, a carbon, a methyl group, makes a crafty new compound. (That beast is the chemist.) Another dispenses the compound. (The pharmacist.) Another experiments with it. (The pharmacologist.) Another prescribes it. (Physician or dentist.) Another forces it to help in a mind diagnosis. (Psychiatrist.) Another informs the younger generation of its curse. (Narcotics administrator, congressman, President of the United States, schoolteacher, preacher, rabbi, priest.) Another informs that younger generation how it opens, how it ennobles the spirit. (Psychology professor, possibly Harvard, or Southern California.) Molecules insinuated into a brain built of molecules alter them, alter a mind, alter a person. But do they? Is he inside there altered? Is the observer of himself ever able to be honest enough, detached enough, shrewd enough to state accurately that he is altered, let alone how? Would De Quincey have been able? He was not in fact. Baudelaire? Poe? Poor Poe. What is altered could be façade. Is everything façade? In some persons on some evenings when the humidity is so-and-so it seems so. By his own alchemy the prime alchemist transmutes his alchemy, and with his energies so fortified, or mollified, or modified, invents a jet engine, a mathematical equation, a madrigal, and when night comes is proud with an earned pride and as he tramps to his apartment from the interdisciplinary seminar holds aloft his long head, casts hot glances toward the January stars that remain ice-cold.

# COMMUNICATION

## *It Is Lonely*

A tomato does not communicate with a tomato, we believe. We could be wrong. The rule possibly is that those of the living communicate who move over the earth on their own power: feet, wings, belly muscles of some crawling thing.

A female sparrow plunges into the air and a male sparrow flutters after. She has told him. One small mind (and brain) has communicated its indignation to another small mind, and with such clarity that it reaches across to our faraway species with its big mind, biggest presumably. A honeybee by a navy-signal technique communicates to a swarm of bees the direction and distance of the nectar, does it in the patterns of a dance, faster if the nectar is nearer, slower if farther, startling facts dulled only because we have had to listen to them so often. The facts have been questioned— facts always are—because the new generation questions, must, wisely or not, and the world moves on. With more reflection, the facts seem not as startling as does that other fact, that a human mind should have figured them out. A red ant zigzagging in a crack in the pavement nudges another red ant, as one American another at Forty-second and Madison, a hearty nudge, the American's, a slight nudge, the ant's, so slight that we looking down from so high above could not hope to detect the urbanity, if urbanity, or the communication, if communication; surely not detect, say, the dim beginnings of art or science should they be there, and it could be argued they are. Impossible for an ant with its small nervous-system to tell more than a short story, but those dolphins who click and whistle, male and female, while a researching mammal electronically listens or millions over TV look and listen, ought to be able to write a novel. Any of the huge-brained ought. A dolphin novel. An elephant novel. Either expressing himself with Oscar-Wilde glitter to a world of dolphins and elephants—that's attractive.

Tiny nervous-systems surprise us, insects'. Man is the noisiest of the insects. Noisier than the locust. Noisier than the cicada. Those, the authorized, hold forth mostly at night. Foreign forever, insects. Their taste for sugar—the dilution they can detect—is like ours. But what else? They wear their skeletons outside, soft parts inside, and this ought to have made them suspect from the start, that they would show themselves for what they are, in reverse through and through. Yet a cricket voice to the human ear and mind can have a warm quality, should be even warmer to a cricket ear and mind, if a cricket has a mind. It has of course. By the time evolution got around to manufacturing bodies of insects, crickets and fireflies, mind must have been well along on its evolution. (Mind *is* an evolution?) In some countries insect voices have inspired the poets. In Japan, Lafcadio Hearn collected a good-sized volume of Japanese poems where insects' rhythms, timbres, pitches are transcribed to human rhythms, timbres, pitches. Hokusai scattered insect drawings through his *Mangwa,* his sketch books, also painted them, and those insects communicate straight out of Japanese ink or paint.

Last Saturday night along Stony Brook—that's Princeton, New Jersey—insect voices filled the air. They got into Paddy's ear. She scratched her ear, then woke. Paddy is a spaghetti-eating Irish terrier. No, it was not the insects. She emitted a carefully architectured growl, her vote in the night town-meeting, loud enough to indicate how she voted, and having got rid of that was able to fall asleep again. What waked her was a voice evolved much later on on earth, a fisherman's. The fool was preparing for dawn somewhere down the river.

When the communication is by posture or movement, that's gesture. Animals use gesture even if the term is not in their dictionaries. A bird will feint with its wing to threaten a trespasser as a prize fighter with his arm to threaten the other fighter. The prize fighter will at the same time gesture with jaw, mouth, eyes, or an expressionless expression. Someone will think expressionless expression exaggerated, but will think it less so when reminded of Cassius Clay at the weighing-in glaring for a full minute straight into the face of his adversary, which the newspapers said he did. Many animals communicate with movements in their faces; monkeys, gorillas. A bird's tiny eye (part of its face) watches a man's huge eye to anticipate what that monster will do next. A macaw creates extraordinary drama with an eye, the pupil of it largely.

Pantomime goes farther. Pantomime is the carrying-out of a piece of dumb show, as when a cat cats up to any empty milk saucer and looks down, or a dog marches like Douglas MacArthur from parlor to kitchen and stands himself in front of the refrigerator and looks up, or a pigeon pigeontoes to a water bottle that has dried and if the owner of the coop is too stupid to understand sticks its bill several times into the bottle's trough, then waits with pigeon decorum while the bottle is refilled.

Vocalization goes farthest. Vocalization must have begun long ago, mind far on its way, the primeval forest full of sounds, and ears evolved to hear the sounds; with a miraculous nuance ears do. It would be instructive to anyone on a June evening at half-past-nine in a suburb of New York to shut his eyes, the better to concentrate his ears, and with help of tree toads and other such to recapture that forest. Somewhere in it there would have been a quorum of macaques, two hyenas expressing their frustration, three mynas who had gotten hold of some scandal, all those creatures blabbing. A frail wren would have been awake, humming to hurry along the night, as a maid in the kitchen finishing the dishes may. This sunlit moonlit starlit planet has been and is noisy with communication. Meaning? Explicit meaning part of the time. A warning. An invitation. A rejection. Less explicit part of the time—the wish with the raising of voice to shove away that age-old emotion, loneliness. Some state like loneliness does seem to reach up and down the scale. The creature has been too much alone, never liked it, wants to put something into the air even if only self-heard in order to make this hour less heavy, not knowing precisely what it is doing, as most of us do not know precisely what we are doing. Therefore, it communicates with just anybody, everybody, nobody. That maid humming in the kitchen was communicating with nobody that anybody could see.

At some place, at some time, in some manner, late in earth-life there was evolved a brain that let a mind communicate with symbols. Secondary signals Pavlov called them, secondary conditioned stimuli that excited primary conditioned stimuli that excited conditioned reflexes, the ambitious Russian thus thinking that he had brought communication too within his ken. Perhaps he did. Whether or not, symbolization must have appeared and disappeared through hundreds of thousands of years of nervous-system development, long before those colored paintings on the walls of the earliest caves. An understanding of symbols—not as Maeterlinck or Freud—has been proven possessed by the modern dog and chimpanzee. If the symbol was some juxtaposition of human sounds, say the two syllables "table," a dog mind (brain, 65 grams) has been known to attach the syllables not only to the one kitchen table it was good sense for this dog to recognize but to any and all objects with four legs and a top. If the symbol was a coin, one kind among other kinds, a chimpanzee mind (brain, 350 grams) has been known to instruct its fingers to collect coins indiscriminately from a coin machine, separate them into two groups, valuable, valueless, and, after a period of happy hoarding, to choose one of the valuable, insert it into a different machine, buy a banana. In studies of chimpanzees in the wild there has apparently been some evidence that the voicing and hearing of symbols could go over also into the beginnings of a capacity for abstraction.

Finally, the human mind (brain, 1,350 grams) collects, hoards, chooses, arranges colors, lines, shadows, employs the twenty-six letters of the alphabet or the fifty thousand ideographs if the country is China, and with the combinations of these, that may be colossal, conveys its thoughts, never quite satisfyingly, to the twenty or hundred heads propped up in front of it, or to the scheming dreaming heads with bipedal locomotor machinery under, hurrying toward Forty-second and Madison.

# PHONATION

## *It Is Noisy*

If the question is just put bluntly: "What produces a great voice?," anyone answers: "Genius." As part of the genius there is sure to be an instrument able to operate flawlessly, though not necessarily flawless.

In man this instrument includes windpipe, lungs, the breathing apparatus generally, walls of abdomen and chest, diaphragm, throat, mouth cavity, nasal cavity, so-called articulators, tongue, soft palate, lips, teeth, perhaps gums, then the various muscles that go with these and their blood-vessels and nerves. Using this total, man ejects the spoken and the sung, using much of his body as foundation. A voice may shake a foundation.

Phonation is the production of voice. Speech-sounds require voice. And speech-sounds are at the periphery of what at the center is the puzzle of language.

This instrument has fascinated singers, singing teachers, speech teachers, physiologists, physicists, otolaryngologists, engineers, also the architects of concert halls. High-speed cameras have made more understandable the top of the windpipe—the larynx. A man may invest everything he is in his larynx. If a hail-fellow-well-met temporarily loses the use of it because of laryngitis he might as well be dead. Often it has been described as a wind instrument. It has flexible reeds, vocal cords, two thin folds of muscle and membrane that meet at an acute angle, leave a chink between, and when air from the lungs blasts up at them they vibrate, which sets the air around that body vibrating, which sets eardrums vibrating, and a voice is heard. Fleshier false vocal cords assist. In a vibrato the singer lets the vibration be heard as vibration. The dimension of the chink, the force of the blast, the length of the cords, their tension, all contribute to produce the initial tone, and each can be changed by muscles inside and outside, cartilages used as levers. When the cords are further loosened or tightened, separated or drawn together by the professional, Anna Moffo or Joan Sutherland or Leontyne Price has started hurling a note into the new Met, and any person on a Wyoming mountain hugging his transistor in the dark knows where and when and who; and has added something about the mind of each singer.

The singer herself listens to her voice as it comes through the air but also as it comes through the bones of her skull. Two sources. Two listenings. These correct each other and that is the way timbre, volume, pitch are achieved, in some persons absolute pitch, and one isolated human being's timbre, volume, pitch. If either of those two sources is blocked as by a cold-in-the-head the voice may be so altered that we cannot make out what is being sung or, if recitative, said.

All of this is phonation, and is going in the direction of mind, will become mind, is mind, but the mind end of the definition is by no means as clear as the rest, is not clear at all.

A larynx can lengthen, can shorten, the basal voice accordingly high-pitched or low-pitched or anything between. That then is reshaped in pharynx, nose, mouth, these chambers having varyingly sensitive walls that by fractions of a millimeter are pushed out here, pulled in there. The mouth is subject to vast training. As the voice advances through these passages it gains in complexity. A voice can be delayed. Can be whispered, shouted, cuddled, coaxed. Can be made to pipe forth, explode forth, glide glissando through a tunnel of flesh. Can be articulated until it reveals perhaps better than any part of our body our capacity to individualize. Mind, not deliberate mind of course, is behind all of this. Resonance of voice is increased or decreased by the caverns of chest and head. Man's indescribable wish to express, and success at expression, requires this only somewhat describable machinery. A consequence of this wish and of this machinery is that on the portico of the Country Club at night we hear a coo, recognize whose, possibly why, or a shush, an ah, something bogus, a grunt, a snicker, a hiss, and who would not reflect on the vocal confusion in the Ark when the pairs were watching the rain.

A soprano sings a high tone in perfect pitch, seeks expression for the purest in her spirit, and while one listens to her, and when her mind and the instrument are obedient to that spirit, one can imagine oneself in some high tower of earth where one can glimpse the place from which last Sunday morning she thought she came while piously she folded her handkerchief and demurely in her choir-seat adjusted her buttocks.

How like some strained sculpture of Michelangelo besides how vocally knowledgeable may be the contortions of a great singer's mouth when one looks at her head-on. One sees lips and teeth, around them face and neck, sees the leaping shadows cast by machine-parts operating farther back. The grimness shocks one. Animal shows through. Species shows through. Race shows through. A larnygologist sent puffs of air through the larynx of a dead dog, produced the bark. That surprised him. Leonardo did the same for a goose, produced the honk. That did not surprise him. An intellect like Leonardo's would no doubt have distinguished quietly the qualities given to a tone from below the larynx, from above the larynx, and he very likely reflected on how the differences in voice-anatomy of man, wolf, lion contributed to their power to spread an emotion, a thing of mind, fright, through the world.

Having arrived at the vocal cords, that blast from the lungs was halted, much, little, infinitely little, infinitely infinitely little. The cords vibrated in their lengths for the fundamental, in fractions of their lengths for the overtones. Then, above the cords, eddyings were introduced among the quivering molecules, shades among the tones, till there emerged the voice of a German world, differentiated from an Arabic world, from a Balinese, five-month-old Ann from seventy-year-old Ann, a natural baritone from a fake. If the voice came mostly from the chest, singers said chest register, if from the head, head register, if from between, middle register.

Vowels are somewhat prolonged tones and have varying complication, more, less, mind making use of the nuances. Consonants are tones checked in throat or mouth. With "t" the checking is clean. A nasal tone like "ng" twangs because the soft palate has not altogether shut off the nose, sound leaking through it. In paralysis of the soft palate all tones are nasal. The rolling of an "r" occurs when the blast of air strikes the tongue as it rests against the palate, the tongue now the reed.

If mind keeps pouring into these products of phonation the molds of words, phrases, sentences—that is not all of it—there results the speech that marks a nation, a southerner of the deep South versus a graduate of Groton in New England, East Side New York versus West Side, the guttural, the simpering, the adorable, the affected, the genuine, the warm. Add dawn, or, should there be no dawn to hand, add a remembered dawn, years ago, the remembered voice-sounds, and one may well remember them, in those first streaks of daylight following the one sleepless night no man or woman ever regrets—a coolness now edges the warm. Explain all of this at a seminar whose announced subject is "The Evolution of the Speech Apparatus in Air-breathing Mammals."

Eons built these structures that produce the inherited, and besides the inherited are able to produce the results of individual learning, that began the first time the infant ear filtered his mother's voice from the voices around. A sensitive French observer states that meaningful attention has appeared on the second day of life. The infant would first have generalized the sounds, as we easily heard that it did, then more and more refinedly differentiated as it more refinedly imitated. One would expect this to have taken time. It did. Month after month an infant legato was carefully modulated. A year-and-a-half-old male will have traveled a great distance in this world of speech-sounds. Over the telephone he produced inflections, hurried tones, slowed tones, changed volume, plugged in a consonant, and while practicing his telephone gymnastics might have had the back of his hand on his hip as he saw his mother with the back of her hand on her hip when she gave orders to the grocer. Everything of his body, if we could see, was joining to express that mind. In due course consonants were not accidentally but deliberately planted. An audible question mark fell at the end of a sequence that asked a question. "Where da go?" At last, a great day, the family had the right to be sure that Aunt Susan had been named Hooi. Much of Hooi was stored in that male's brain-mind before this monotonous "Hooi, Hooi, Hooi." The name was commissioned to do things. The chief himself had a name, knew it. Names were splattered over objects, persons, actions. The incomprehensible was cracked by the comprehensible. We thought we heard that. Then we knew we heard it. Explicit articulations were attached to explicit situations. Life and voice were increasingly integrated into a total. (So easy, and too glib, to say it that way.) This continued, with diminishing freshness, through kindergarten, school, a trip to Cairo, to Jerusalem, to Istanbul, until from the winds sweeping the earth, from the taxis and the crunching buses, from TV blaring out of every summer window, from the fundamentals and overtones of men and their animals, there rose a mighty surge, reminiscent, bitter, vulgar, dull, a splash of brightness, to the sensitive ears of an aging lady of musical talent sitting in her room in the dark of the seventeenth floor of Tudor Tower in New York City.

# LINGUISTICS

## *It Is Analytic*

Sundry educators have suggested that any noise coming out of any human mouth was language. They seemed to believe it.

On reading that the first time one thought it must be some new insight, or a new absurdity, and when someone labeled that linguistics, one thought linguistics must be absurdity, which, though it sometimes had been, one knew it sometimes was not. Nor was it new. Nor was it stepping forth by itself, but with the times. It was a sign of the times though much older than the times, older far than the immediate surge. Its base was in years, centuries actually, of effort to construct a science of language, to relate that to other sciences, to find the origins and follow the development and discover what was common in languages, and eventually, but remotely, to relate language to the human brain. That last if accomplished would have been the work of anatomist, physiologist, neurologist, with help from mathematician, psychologist, psychiatrist, computer.

George Bernard Shaw listening at London street-corners for speech peculiarities, creating his play *Pygmalion,* to become *My Fair Lady,* was one kind of linguistics specialist. He was a descriptive linguistics specialist. He was inquiring into the superficies of language not the depths but made it plain that this was worth doing. Henry Higgins, the expert of the play, on that rainy night when the theatres were letting out, was jotting down how the people in the crowd spoke, the animal sounds they made, some of these not comprehensible even to their neighbors, and Higgins was localizing where these persons came from, what environ of the City, what corner of the Empire. Higgins' house in London was a not-complicated physics laboratory. There were phonograph records, a phonograph, a gas flame, various paraphernalia for the study of voice mechanics, the items dated, as 1969 by 1999, but the items were mere stage properties. The play was only partly linguistic. It was sociologic. Henry Higgins recognized how speaking one English kept a man or woman in a deprived place in society, and how the acquisition of another might lift him or her up into Buckingham Palace.

Insofar as Henry Higgins did bat Eliza Doolittle away from animal sounds and toward a literary English, Shaw might, particularly as the play metamorphosed into the musical movie listened to in cities and villages, have advanced spoken language, and the language professions, by a millimeter. Some linguistics textbook-writer coming home at midnight from the movie probably wrote a bit on his textbook, and a bit more carefully and less humorlessly than he had during the day. Shaw might more than he intended, with one light comedy, have contributed to lead descriptive linguistics in the direction of the honest teaching system that is part of its job. Shaw was a teacher. Great comic writers usually are. Aristophanes was. And the theatre is a powerful place to teach.

The descriptive with the human type of mind does however not long stay descriptive. It begins to dig in. Shaw mostly wanted the two-hour traffic of the stage, his *Pygmalion* nevertheless digging some depth into the philosophy of language, giving some urgency to our formal and informal study of it. Shaw was socialist, essayist, analyst, vegetarian, anti-vivisectionist, fraud, but also that scholar of language, and the same questions kept rising in his time as in ours, less polemically, less fired by our late awareness of a deprived world at our doorsteps, specifically a language-deprived world. Shaw forty years ago would, furthermore, hardly have been a member of a Linguistics Union or on the Advisory Board of the *English Journal*. He was a farce-loving Irishman, therefore avoided several instructional blocks: solemnity, over-interpretation, airtight logic, panicky forecasting of the wave of the future, as it might be by someone with a name in last-week's *Saturday Review* or next month's *Atlantic*.

One wishes one might have his humor, and his good humor, when someone orates frenetically about the new linguistics. One wishes one did not get annoyed at such a suggestion as that whatever is muttered or grunted in bars and bazaars deserves the designation language, that any squirt or sound should be included, any individual's tonal mind-wanderings, and, when the phonated has become the verbal, that one verbal communication is no better than another, one man's speech no better, and, pursuant of this imbecility, that *Romeo and Juliet* and *Julius Caesar* need not be patiently fostered in the high-school student who falters over an English composition, or, if the enthusiasm mounts to the rafters, that Shakespeare's language is in fact no longer language, his cadences no longer music. At the last one hoots, but quickly saves one's breath, reflecting that whoever ought to hear would not, would be quoting too loudly from his automobile-slogan stickers: LEGALIZE ABORTION; HE KEPT US OUT OF WAR; GOD IS DEAD.

It is possible to be ignorant of some shape of the field of linguistics, that has grown to a million-dollar field, and still pick up enough from the crammed air, sympathize with the wish for freedom of language, everybody's wish for every freedom, surely freedom from the blinders of the old grammar, also freedom for each child to discover its rules for himself and herself, like Adam and Eve. Or go the other direction, listen too agreeingly to the lazy who do not wish to take the trouble to let organized language become disorganized at some point in favor of vitality. Or the trouble to keep recognizing that ambitious language renovation is in part no more than a keeping in step with an organized-disorganized society that for good reasons but also for neurotic reasons must always be ripping up something. Or one can unscarringly chide oneself for lolling in one's armchair rereading *Pride and Prejudice* and *The Importance of Being Earnest,* assuring oneself that one's enjoyment of them is not only enjoyment, not only a gain for one's personal style in this world, but that some overflow from oneself will reach others, they choose then to loll in armchairs, on curbstones, and that this may be a true way, possibly in essence the accepted way of progressive teaching. Or one can reverse the direction again, feel apology at not having proclaimed that one always has been and always will be humble and obedient in the presence of the gigantic achievements come of the severity, of the knowledge, especially of the agonizing effort inherent in the perfecting of any human communication. One does not fear this ostensibly fresh field. One moves toward it. However, one does reject any easy disregard of the long past, any easy acceptance of the dogma that the latest theories of pedagogy, or any theories, stand on bedrock just because they are latest. Finally, one shrinks from slapping on that badge *linguistics* where one sees just anybody slapping it on, there being usually something wrong about badges except in a national election, and one or more wrongs could be pointed out there.

Americans—this is not on another subject—have frequently been assured that their contribution to the earth's noises is uncultivated, their phonation, not to get as far as their linguistics, puerile and unfortunate. A British actress made this clear. She explained them to themselves. "Americans today yell because in the early days of their country they had to yell to their cattle." Yelling had hung on as their mode of communication. She said that that had been unfair to the cattle, their ears. However, the raw country, she repeated sympathetically, did require the yelling and Americans never lost it, not in their loud theories either, not in their total minds. Each time the British actress explained, and she loved to, she made a horn of her two hands and bellowed through the horn, but what she bellowed was marvelously British.


# APHASIA

## *It Is Maimed*

Disaster in the brain may wipe out speech, all or some.

The victim will lie there.

Many do the whole day and half the night, stare at TV, a blessed use for TV.

One proud woman had a stroke, lay five years, then was shipped to Florida to lie more years in that peninsula of cardiacs and apoplectics. Her paralysis was on the right side. She was able to get herself from room to room. Wept too easily, but had an imaginative doctor and an affectionate husband who always found another way to help her over a spell. Her speech loss was near to complete. (Strange what such loss does to the human being, but, at least as strange, what even complete loss may not succeed in doing.) Abruptly a tortured effort showed in her face, not a disabled face. She was wishing to expel a word. The wish was like a light switched on, but the light dimmed, the barrenness returned, then she laughed, could hold back laughter no better than weeping. Began again searching for the word. The word was *strawberry* but what at last she burst forth with was *peach*. Her mouth, her face, her neck, her chest, like an opera singer needing to fill an opera house, came down on that one-syllabled *peach*.

Aphasia, loss of speech, has taught us much, possibly most, about that no-man's land between brain-flesh and mind.

A brisk eighty-two-year-old gentleman paralyzed on his right side sat month after month in a hospital bed, passed his time relishing irony, read his newspaper as often upside down as upside up, and with the military bearing of a reader in a public library who is there to get warm in winter. Once he broke the quiet (a quiet will suddenly fall over a hospital ward) with the single letter *u*, the next moment blurted *universe,* and when his newspaper was taken from him muttered monotonously *university, university, university.* It was as if he were trying to latch that word tightly among his few possessions—latch life is what one thought. One thought too of how there must be some border of feeling around speech, how without that border speech is not speech. Another gentleman in another hospital in another town repeated till it paled out, *"Don't remember, don't remember. . . ."* He was like someone in a nightmare. A single naked phrase was his bank balance after the crash. Another made a ward daft by echoing whatever anyone in any bed spoke, and one realized how correct but literal the neurologist's term for this, *echolalia.* Another was asked to touch his left ear with his right thumb, appeared to take that request duly under consideration, then duly touched his left ear with his *left* thumb, and when he had achieved complete agility with this, elaborately offered it as an answer to any question addressed to him.

The intonation that goes with speech may be lost, bleached out, the skeleton syllables remain. It is reported that a musician lost music, no other sounds. A psychiatrist might be suspicious of that— the musician's underneath could be manipulating for him, so that he never again would have to give a music lesson. A mathematician lost numbers. A psychiatrist might be suspicious of that too. Psychiatrists are suspicious.

Wrecks like these have been studied for generations by the neurologist, later by the brain-cutter, wrecks of war, automobile, tumor, hemorrhage, clot, narrowed blood-vessels. The details, flesh and speech, difficult to disentangle often, have added to the slow but accumulating analysis of brain-to-language. The overall intent of the neurologist has been to assemble a pieced-together brain-related speech, of the brain-cutter a pieced-together speech-related brain, both always working somewhere in the area of general mind. The results often have not been satisfying, though now and then they have been somewhat, and dramatic, the explanations nevertheless arousing caustic differences of professional opinion. Where there is dissatisfaction, this has been partly that any attempt to pin language to flesh like a waist to a dressmaker's model revolts us, the assumption that the soft words of two human beings with the lamps turned low could be thought that accountable. Mind seems demoted. Brain seems demoted. We ourselves seem demonstrated only more fatally earth, as we probably are, though unrealistic for such a reason to allow a depressed mood to settle on us, there being evidence enough, or ignorance enough, to let flicker the hope that we may in the end not be proven entirely masterpieces of dung.

About ninety years ago the French anthropologist and surgeon, Paul Broca, having reviewed published cases of speech loss and his own cases, and the reports on the post-mortems of the brains, thought he had staked out the territory of spoken speech as the posterior portion of the third convolution of the frontal lobe of the left cerebral hemisphere in the right-handed. (No special merit in knowing this geography; enough persons know it.) Broca concluded that this territory—Broca's area—must not be destroyed if the patient was ever again to talk. By hindsight we today think it not surprising that this locale should have been in the neighborhood of brain that controls tongue and lips and whatever body-parts are necessary to chip from raw sounds:

> . . . the morn in russet mantle clad
> Walks o'er the dew of yon high eastern hill.

Similarly for written speech, the territory was that of the muscles of the hand. Disaster there might not cause a paralysis of the hand yet its power to write words, the mind's power to imagine the written, was maimed or wiped out. Similarly for reading. The spoken, the written, the read are mind-soaked and interwoven in the brain, and, again, not surprising that this late evolutionary capacity of speech should have had intimately geared into it the machinery of vision, hearing, movement, therefore a crippling of it in some major or minor degree possible at numerous points. In minor degree the crippling might go unnoticed except by an alert neurologist. When the damage was in the association area for all three senses—parietal, temporal, occipital lobes—as happens, the crippling might be most unhappy.

A by-product of the study of aphasia has over the years been speech maps, speech atlases, and there have been speech cartographers, the last sometimes derisively called diagram-makers.

Early cartographers were succeeded or given support by late cartographers, by psychologists, by psychiatrists, speech students, speech therapists, linguists, philosophers. An erudite member of any of these groups might say that the pioneers were zealots, then (this has in fact occurred) march straight to a blackboard in front of his undergraduate university class, draw an outline of a brain, on the brain draw the so-called speech quadrilateral, four neat sides, mark the spots where speech could be damaged, and five minutes later say again derisively that the pioneers were diagram-makers. Obviously this does not mean that the erudite expert may not have added knowledge. He well may have. Yet it is plain that he is playing with ideas of the brain and mind as a child with a globe of the world. Anatomists have added. Anthropologists have added. Engineers have added analogies from telephone, telegraph, television. Even that geologist or paleontologist may directly or indirectly have sought among fossils for the earliest origins of language in the earliest origins of man. A chemist may look affectionately again at his DNA. This is a chemist bitten by the chemistry of evolution. Meanwhile, sitting behind his table, a secure physiologist might have decided to amass these different kinds of findings, rig them with nervous-system physiology and terms, and in the end believe that he is accounting for the spinnings of the poet, the duels of the court, the badinage of the drawing room.

Years ago a famous neurologist-physiologist set up a famous classification. Verbal aphasia: loss of the capacity to understand or use words. Nominal aphasia: loss of the capacity to understand or use nouns. Syntactic: loss of the capacity to keep a sentence in order. Semantic: loss of the power to extract intelligence from words. The famous neurologist had a tidy mind, an excellent mind, but one does nevertheless see a professor walking among the stars with a grammar in his hand.

Everyone suffers transient aphasia, at the close of a hard day, between midnight and morning if the telephone has waked him, after cocktails enough, drug enough, when he gets older. Pavlov complained as though a trick had been played personally on him— he was forgetting names! Everyone (it may be too early in life) forgets names, later proper nouns, later nouns, later just words, finally has left only baa, baa, baa. On the other hand, a person's rememberings, as we know, may be so bizarre that he is as fascinated by them as by what slipped away. A distinguished American woman-scientist returned after years to a distant country and was astonished at how she could snatch back names of members of families, attach them correctly to the wrinkled faces, but in a rice country, which this was, she could not recall the word for rice.

Electrical stimulation of a brain during surgery may light up a scene for the patient, a scene that was dark till then, and he speak out words appropriate to that scene. Everybody in the operating room hears the words. The whole maneuver remains more macabre than any whodunit, and it is a generation since it was news. Never did electrical stimulation call up words. Never did it start a patient counting, but with the patient under local anesthesia, as he would have to be, and counting, 1-2-3-4-5, at 5 his brain stimulated, his counting stopped. That is, numbers could be blocked electrically but not called up electrically. Anyone is free to use any such fact in any way he chooses to explain anything he chooses, and there are many facts.

The fitting of an aphasic's history to his post-mortem runs into difficulties. A man will die quickly after he has had a stroke, no opportunity in the minutes or hours or days he groped in his fog to study the detail of his speech, so no use to study the detail of his brain. Another apoplectic will live for years after a slight stroke, he and everybody forget just what did happen at the time, so again not much use to study his brain. Another suffers a cerebral accident (beloved phrase), is not able immediately to speak, as apparently with President Eisenhower, but in a week he has recovered sentences, in a year everything. He does not suffer a second stroke.

Survives twenty years. Moves to another city. Dies there. Is a member of a family with an idiosyncrasy against post-mortems. Back there twenty years ago he had told his story to a neurologist, and the neurologist is still alive, but he was always an absent-minded fellow. As for the man, the patient, he naturally felt no obligation to contribute to the next generation's understanding of brain and language, so he left no diary. After his death, in spite of his widow's idiosyncrasy and his civic prominence, his brain did get into the hands of a brain-cutter but—the brain-cutter was incompetent. Remarkable, when the pitfalls are reckoned in, that as much has been learned about brain-to-speech as has been. Remarkable, too, the disillusioned modern investigator with neither religion nor philosophy, the pathetic cold-in-the-head toiler with the ambiguities of his own mind and those of earth and sky and sea, that he should unravel as much as he has.

An occasional psychiatrist has found it possible to turn his back on the ruins of nervous-systems. He has found himself able to ignore the sometimes clear-cut damage to flesh associated with suggestive damage to mind, able to detach the intricacies of aphasia from intricacies of brain, call loss of speech loss of intelligence, then make intelligence global, not localized, in the brain. Why not? It avoids the geographic haggling of generations of neurologists, thrusts language into general mind, and disregards brain. Freud would not have gone that far. Sherrington might have. Pavlov went the whole other way, like a carpenter nailed mind to brain, nailed in language with conditioned-reflex hypotheses, and did this so efficiently that it might be difficult to pull out the nails if one thought it worth the trouble.

A colored woman sat in a chair in the clinic, had had a stroke, had made an excellent recovery, as doctors may phrase it, but her speech was gone. Her face on the right side did not move, had the smoothness of a basalt sculpture, the paralysis itself contributing some odd loveliness. She was touching. The corner of her mouth on that side drooped, moved with strong feeling though she could not move it of her own will, which is common. No fretfulness was in her. No confusion. "I know, but I can't talk." She was able to say that, repeated it, but it did not sound monotonous. This colored woman's problem was not that she could not see the shapes of words, or that her speech machinery could not produce the shapes, but her hearing could not make meaning of what was fed well enough into her auditory machinery. Testing bore this out. The comprehension of the heard was impossible. The function was gone. Never for the rest of her life would this woman be annoyed by the petulance of verbal wrangling. Spoken speech was henceforth mere tone. She was shown a photograph intended to ask a question. Vivid understanding sprang into her body. She looked as if she was also going to spring into an answer. Something blocked the road. "I know, but I can't talk." She repeated the six words. That was not the arresting, however, which was still that she could maintain such poise. She could sit in the midst of this trouble in her brain and in the midst of these doctors, and be calm. Far inside her there was a clear mind also. One might be wrong but one felt this. Conceivably she was even clearer—in there somewhere—than before the stroke. She always had been clear, exceedingly, her friends said. Except for her speech, they said, she was as she always was. Speech in this woman did seem an overlay on mind. Speech could be rubbed off and mind remain. A human mind was not a collection of language items—from time to time one has been tempted to think that. It was more enduring. It was more resistant to disaster. It hovered there beneath that basalt surface. She sat in the center of a clinic with thirty doctors and her loss and remained *that* woman.

# HAND

## *It Is Sly*

There was a party of great apes. (This was once upon a time.) They were talking. Suddenly one of them realized they were doing that remarkable thing, talking. Immediately he got suspicious and looked at his right hand. A fine party that was, those apes celebrating that they had taken an eon-long lunge in a new direction. But they did not know it—and wasn't that a pity?

When the first prehumans rose to their hind feet, wobbled for millennia, finally stood upright, achieved the erect posture, it must have been a sorry day but a touch of grandeur was on it. The top-end was free. The forefeet could help in the development of what was now to be a human body, especially in the development of that body's human hand, which would prove the most flexible tool on earth, invent lever and wheel, subsequently wrist watch and stethoscope and jet plane and missile to the moon and futuristic and surrealistic painting, and while getting that work done would take part also in the production of this most intricate of all types of communication, speech. Henceforth there would be a creature quicker than any to understand, and misunderstand, and—what is paramount—he would spread mind over the planet.

To what extent and in what manner the hand contributed to this we do not know, only that it did. The rest of the body was not idle, but we have more than an intuition that the hand led the way. Aristotle had said, "The opinion of Anaxagoras is that the possession of hands is the cause of man being of all animals the most intelligent." Aristotle did not think so. He thought hands were effect not cause.

The five-fingered reptile has a plan of hand in general like ours, anatomists tell us; therefore the original sketch was made long ago. Muscles, fascias, tendons, blood-vessels, bones, and the nerves that run to the centers and back from the centers, none is drastically different in us. There are differences in what because of its gross mechanics our hand can do, but the enormous difference surely is our enormous brain, its special character, the emphasis given to its parts, even the sheer space allotted to the hand in its cortex, as much as to the face, which includes tongue, cheeks, lips, all that is employed in speaking. Over the great span from reptile to us the fingers gained in the capacity to fold over an object, to grasp it, and the thumb in its capacity to touch, to oppose, any one of the four fingers; meantime the hand-brain was expanding into the huge communication center we know it to be.

A good hand plays many roles, strikes one sometimes as having mind tucked right in it, and no one would be surprised to discover how easily it is made self-conscious, as if it were suffering shame or guilt. In an Easter parade, in photographs of brides and grooms, on the platform at a commencement, abruptly one sees nothing but the hands. That flushed graduate would prefer we did not stare so at his hands, and we are not intending to, but we do, and we may suddenly, and apologetically, feel that we are intruding on some secret we have no right to know. Two hands stay stubbornly in a stubborn man's lap, or he drops them stiffly from his shoulders, or, the other way, gesticulates overmuch, discloses his nervousness. Pockets were designed, partly, to hide hands, as were royal blue suède gloves, but where hands all day hide themselves is in occupation.

A London shop takes advantage of that biological situation, wraps a package in a gay paper, attaches a red string to the package, a polished wooden ring to the free end of the string, and a lady slips her index finger through the ring, walks along the Strand with her package on ahead of her, one self-conscious hand given something to do, forgetting itself, its visibility diminished, her visibility increased, she decorated by a package. Along the Strand also there is always the lady whose package runs and pulls at a leash, again one self-conscious hand given something to do. A cigarette is a package neat and small. A smart-looking bearded young man lights one, puffs at it twice, rests it, taps off the ash, for a quarter of an hour keeps that up. (Today that smart-looking young man puffs at that coward, the cigar.) Canes and umbrellas communicate, in London. A gentleman in black cravat and striped trousers, skipping down the steps of St. Paul's at three o'clock in the afternoon, swinging his cane to the horizontal, then twirling it through 360 degrees, is bragging to the city that he has just kissed the bride and has no responsibility for her.

Every freshman medical student is taught that the left side of the brain in the majority of human beings is language dominant, more particularly one part of that left side, the person spoken of as left-brained. He is apt to be right-handed, right-eyed, right-footed, may lean to the right. Not all of these characteristics are in one person usually. Three facts: first, the language dominance of the right-handed may not be left-brained; second, the left-handed may be even less often right-brained; third, preference for left or right hand, called handedness, seems inherited but not fatally. The majority of cats and parrots have been reported left-pawed and left-clawed, the majority of rats and monkeys the reverse, so, were those reports to prevail, human dominances would not be unique. Cats in a more recent report were stated to have either left paw-preference or right paw-preference, and they stuck to the side chosen; difficult however to be sure that any of this can be established.

Formerly there was the conviction that if a child was left-handed do not try to make it right-handed because bad things would happen to it, stammering and neurosis. There was the other conviction, if a left-handed child was forced to fight its way through a right-handed world, that would produce neurosis. And still the other conviction, that the left-handed were all neurotics to begin with, so send them to the psychiatrist. Some neurologists at the same time were offering their opinion, that right-handedness was a matter of education, emphasizing that we live in a right-handed world, but not always emphasizing what is equally fact, that we live in an education-ridden world where a visitor from outer space might conclude that on this planet education invented mind.

Man's brain does seem cocked to one side, right or left, most frequently left, and the cocked character of his thoughts, their instability, their inventiveness, may be related to this. His brain is readier to be tipped, and he readier than earlier hominids to meet the topsy-turvy of urban and suburban life, able also to perceive and to reflect upon the restless vividness of life.

Hands in countless ways speak directly. There are the hands of the Japanese wood carver saying things to the wood. There are the hands of the French gardener saying things to the soil and the sprouts of green beans, advising them whether they should or should not come out in this changeable weather. There are the hands of the American jockeys, Arcaro, Shoemaker, and the long line before them and the line to follow, the hands of the jockey reaching to the mind of the horse, and the horse reaching back to the jockey, each correcting the other, the race won. Actors know the force of hands, understate with them, being professionals. Portrait painters sometimes cut them harshly off the canvas, or else paint them with conspicuous care, make them divide with the face the responsibility of communicating what this person has to communicate. For his $15,000 portraits Salvador Dali asked $25,000 if he included the hands, knew they were troublesome to paint. In photographs, weak hands (Churchill's) may go with melodramatic faces. Then there are the strong hands of the master painters. The left hand of Mary in Leonardo's *Virgin of the Rocks* stretches straight toward us, canopies that holy group, Jesus, John, the Angel, is canopying the world, an unearthly light shining from above, edging the fingers, throwing the palm into shadow, making us know that this is not only the most believable hand a painter ever painted but as gentle a thought as a human being ever thought. Blake in his drawings often used hands to convey his visions. He wrote:

> Tools were made, and born were hands,
> Every farmer understands.

In the infant the hand already shows the cunning of the species, soon starts on the acquisition of its personal cunning, life never letting it alone, it never letting life alone, becoming more and more the communicating instrument until one day in the heat of an argument a woman lifts her thumb, merely that *Homo-sapiens* opposable thumb, and her husband stabs her to death. Make a jury understand that. The years were not wasted on that hand. It said something exceedingly nasty.

# THE PERSON

## *It Is Flaming*

Under the stars, cold as cold ladies, dwells that speaking speck, the single human being. His story, or hers, began long before that night he became a member of the human race. That night we of the family thought he was the spit and image of his mother though the shape of his chin was his father's and he worked his mouth like his toothless Aunt Julia, though the truth was, if we had had the common sense to face it, he did not resemble any of us, not anyone anywhere, not even physically. He was unique. He was born unique. Two legs he had—that sort of thing. After a while he would resemble us more. He would begin to imitate us more, lose himself. Too bad. But inescapable. Only, why this fuss about this inexperienced performer who forgot his lines and was pushed without his make-up onto the stage? He is too vague. He is like the great vague out of which half an hour ago he came wiggling. It could be—perhaps—the fuss was that we thought we already saw there in the crib the single person. We would care about that. We would want that to survive, would want the single person proven again and again, might be concerned about saving even this infant's singularity, this piece of our flesh and blood and vanity, to keep up the precedent. How cheap if that were the secret: we wished to save everybody's to save ours, save ours from sinking into the sea of anonymity, save this bit of the veil of mind that got detached, floated off, would hang around the planet for so-and-so-many years, days, hours.

In the highest biological echelon the single person, that personality, is king. Christ, King of Kings, was a person, which is his poignance and his reality. Moses was. Buddha was. Herod. In the highest biological echelon the distinguishing of one from one is a chief concern, chief annoyance, chief delight, this distinguishing of mind from mind. Not tallness, not smallness, not a wart on a nose, not classic Athenian handsomeness, but mind—timid, bold, agonized, confused, assertive, assured, double-dealing mind—is the core. Body-parts make a difference, a simpering face, a short neck, a thyroid throat, a sway-back back, a clubfoot, but these too hold us sustainedly only because of what happens to the mind inside them.

Picture a congregation of Jews coming from an Israeli synagogue, Italians jammed in the square before St. Peter's on Christmas Eve when in his white the Pope steps forth at the lighted window, Chinese blocking the narrow streets of Shanghai, New Yorkers seventy-five stories down swarming from buildings at the end of a workday, even in those piazzas or ghettos the imagination reaches toward the solitary one, toward what one thinks is special in the mind of the tramp, the mayor, the pimp, the merchant, and especially special in the poet. Our human eye is that kind of eye. It goes moist over the personal. Goethe spoke of this. "*Grösstest glück der menschenkinder ist immer noch die persönlichkeit.*" (Greatest joy of the man-children is ever still the personality.) Notwithstanding, in some silly social-minded hour one wonders whether one ought to pay the attention to it that one pays (as if there were anything else to pay attention to), in the beauty parlor, the surgical amphitheater, at the races, at the PTA, in the art gallery, the concert hall, the theatre, frankly or furtively in church, in the analytic interview, psycho-analyst versus psycho-analysand, everybody or just one staring at everybody or just one, always trying to stare under the surface. Might we not see more if we merely glanced for an instant out of the side of our eyes? Glanced, and not analytically? And if that were not true, ought we then not, for our mind's sake, and all minds' sake, be willing to see less? Might not the intellect stare the single mind off the planet? Stare it into the unidentifiable? Stare it to death? Should we always be thinking: What is he thinking? What is he doing? Why is he doing it? How is he? What is his background for doing it? Should we be so earnestly trying to take apart what God or the Devil or just homespun evolution, or the three joined, have put together? Should we be so humorless now that we know there are the astronomer's and the physicist's light-years? For example, what would happen to a patch in the fabric of a person, say that woman's charm as communicated to that gentleman in black tie, if his intellect began to divide the charm into its component parts, did this historically, physiologically, chemically? Worked out the chemical structure of her charm, found it to be a three-dimensional, 3D, aggregate of polymeric macromolecules of gigantic molecular weight, but of course with this person's personal atomic configuration? Would she ever again be able to use hers on him? Communicate to him what he is hoping to communicate to her, communicate, not pass along on telegraph tape or as arithmetic from a computer? Should there not remain something not shrunken by calculation, by penetration, by interpretation? Something unexplained? Something inexplicable? Something that is? Something that pops up? Something that is innocent? Is not some innocence to remain innocent? Be fresh like a morning green-market? Born? Grown? Growing? Not measured and weighed? Not quantified? Must everything be flattened into carbohydrate, fat, protein, nucleic acid, and their descendants? Broken up into the billions of particles that build the millions of cells of those cold ladies' brains? Must the sociologist dig into the city directory to find the precinct of the Bronx we came from and derive us from that, from our deprivation? Or our possession? Must the priest or rabbi flatten everything into earthly and mystic? Must the psychiatrist flatten everything into hostilities, insights, aggressions, the phallic, the oedipal, the rest? Must even short-lived man of the streets choke what might otherwise burst through from his long-lived past? The personal appears so often to wish to stay with a man or woman or boy or girl and these appear so ready to destroy it by denying it the free impulsive character with which it arrived on the earth.

How does this uniqueness come about?

To the mathematician it is only mathematics. Considering the number of units in a human creature, any kind of units, the mammoth number of ways they can combine, and have been from the time of the parent of the parent of the parent, each simply must be different, by the mathematics, no further explanation necessary. Considering the number of genes on the chromosomes of that original nucleus, the fabulous numbers of possible arrangements of the atoms, and the still more fabulous numbers when the two nuclei relate after that night a male talked tenderly to his female on a motorcycle, each must be different. Modify that resistant inherited core in whatever ways are possible by those dialects in the Tower of Babel, or the Ginza in Tokyo, the phrases, idioms, inflections, uses, abuses, each to that slight degree also different. Considering the number of environmental influences that hang around each, the influences each in 1969 has been taught, overtaught, he must give heed to, the persons each has to meet and somewhat flatter, the detested female chairman of the Woman's Club, the male chairman of the Board of Directors, the mother-in-law with hemorrhoids who would like to move to southern California to live, the father-in-law who would like not to move to southern California but play golf right here in this county that is better than any in the whole United States, plus the clock in Tristram Shandy's kitchen, plus the classmates in shorts with all their sizes of hips and sizes of hopes in the dormitory at Vassar, each must be different. Each is a mosaic, ancient in most of the inlaid pieces of its colored stone, modern in only a few of them, the pieces ten million, or ten million times ten times times ten million, or any other number like the stars mapped and not-mapped by the Palomar telescope. Naked number could explain the multiplicity of that daydreaming nightdreaming single human head.

To that endocrinologist thirty-five years ago the person was all endocrines. To many a public-spirited citizen today the person is all interpersonal relations—the bad in the bad boy is blown up so big by the bad around him that the boy in the bad boy is contracted to nothing at all. But, start anywhere. Construct a model—to keep up with the times. List the parameters—to keep up with the times. Rights parameters. Economics parameters. Education parameters. Birth parameters. Infancy parameters. Immunological parameters. Metabolic parameters. Acute lack of oxygen does one thing. Chronic lack another. Lack of sugar. Lack of vitamin. So the removal of the daily threat of a cold-in-the-head by a winter in the sun of Arizona. So good news. So the eating of too much bread.

What stands without argument meanwhile, what asks no accounting, what in 1969 still can have no accounting, the person down in, the person large or small, is to each his faithful flickering light, so long as it flickers, so long as he lets it flicker. It is he assured how different from all others he is, assured most when he has had an exalted evening, given a solo performance, but assured still next morning when he drags his body from his bed, when he sits by himself on New Year's day, or sat the midnight before in a glum gloom studying the wine stain at the bottom of his glass and thought his green-black mood was because Cynthia in that dying year had returned him his ring.

What stands without argument, furthermore, that person, however he may eventually be explained, is streaked with genius rarely, his little is smothered into conformity usually, altered up-or-down only slowly in a lifetime, a happy fact when we admire it, a blooming bore when we don't. That little may have been the littlest little to begin with, and this farther worn down in a Spartan society that never suspects it is Spartan, but that little still is headstrong. In its skin it waits. On what does it wait? A shell of fat insulates it, a shell of adrenal cortex, a shell of progressively failing heart, failing speech, failing locomotion, failing eyesight, failing friends, failing mind, failing dreams. Dreams get thinner after childhood and in an old man may be thin indeed or absent, like his hair. A face bulges so one scarcely sees the eyes, but there at the middle puffs Harry Senior who fifty years ago had the desk next to the window in the third grade at school. We come, we go, we are the same, nine-tenths the same, or ninety-nine one-hundredths, or nine hundred and ninety-nine one-thousandths. We come cast for our parts, if born professional enough to have recognizable parts, Puck or Othello or

Cleopatra, born in our costumes, touched mostly in ostensible ways by religion, epoch, nation, family, puberty, a fatty liver, neurosis, success, bankruptcy, the days of the years of a life. The essence— that miracle in one opinion and militantly not miracle in most opinion—wanders not far from where it was when it arrived through that archway of the womb. We chisel and chisel but the stone is hard. The separateness of each—and separate it is for anyone who takes the trouble to look past the obvious—is so old-as-the-hills that what happens in one lifetime, one birth-to-death, one incarnation, one Socratic life-to-afterlife passage, one whatever current anthropological-sociological-pedagogical equivalent, could be irrelevant. Yet the sometimes-seeming great change that on closer inspection turns out almost no change at all, might be a reason, if we need a reason, for our coming, and a reason, if we need a reason, for our going. A pounding in his deaf ears may have modified a composition of Beethoven's, as even his rage at his hearing-loss, as even the silence that was always extending around him, as even the two-times-two of his various publishers, as surely the remembered noises of his beloved place in the country, of the stream running through, but Beethoven was Beethoven and not the stream, obviously. El Greco may have made use of his inborn visual obliquity, but El Greco was El Greco, conceived in Hell or Heaven or routinely evolved over the long road from masers to us. Coleridge was Coleridge and those words of his were Coleridge not opium. Lucrezia Borgia was Lucrezia and she was, besides a hysteric, a speaking, dreaming, forever self-creating but largely self-repeating, largely inexplicable, glittering person, and she was dead at thirty-nine; all that which would rise like a lavender cloud in and around the human mind for centuries was dead and gone at thirty-nine.

# THE WORD

## *It Is Noble*

Rarely, but it does happen, man states his case to God or Jehovah or, for anyone who feels those terms inappropriate in a universe of galaxies and masers, states his case into the hollow of time. Is that satisfactory?

> The voice of my beloved! behold, he cometh leaping
> upon the mountains, skipping upon the hills.
>
> My beloved is like a roe or a young hart: behold, he
> standeth behind our wall, he looketh forth at the windows. . . .
>
> My beloved spake, and said unto me, Rise up, my love,
> my fair one, and come away.
>
> For, lo, the winter is past, the rain is over and gone;
>
> The flowers appear on the earth; the time of the singing
> of birds is come, and the voice of the turtle is
> heard in our land.

Words may be the highest that the mind of man has achieved. In spite of the late flurry against words the Song of Solomon is still a good song, and is both grossly and subtly communicating. No other animal has the same possibility of perplexity in its reaching out to other animals. Yet man's words are poor things. They are always getting him into trouble. They never fall off his lips as he meant them. They excite the linguist, excite the traditionalist, infuriate both, are used by both as levers for their argument, notwithstanding the words say starkly what they are. That can be weak, can be powerful. The voice of my beloved, the feelings that rose in him he never was able to express, lost them in words. An entire mind may be lost in words. The winter is past, we shed our overcoats and tall boots but the words, those tricky traitors, still are not free. Or too free. What liars! What a delight! Through them the mind looketh forth even more than at the windows of the eyes, showing itself through the lattice. Words reveal us. Conceal us. Teach us. Bewilder us. "Words, words, words," says Hamlet of what he is reading but he is not indifferent to them as everything else he says proves. They are vulgar. They are wise. They are nothing. Everything. Statesmen screaming across frontiers egging each other on to human slaughter, administrators in parley with administrators to keep up the prestige of administrators, scholars caviling with scholars to keep exclusive the club of scholars, fishwives shrilling at their husbands, rough voices at the ballpark, a drunkard howling into the ear of his bar-sister, or into the ear of his satin-bodiced debutante of the year 1910. Words are the bubble and they are the soap. Within our brains they give the widest association for often the slightest stimulus. Give lightning sharpness to a cloudy recollection. Evoke tears. Wipe them away. Damp laughter. Cover an ugly act with a pretty excuse. Rip open flesh, then lay balm on the wound. Make more truthful. Make more false. Make more tolerable. Touch with reality. With unreality. Occasionally, occasionally, they ennoble. Occasionally, they reach beyond man. Occasionally, man speaks with self-respect to that God.

In the face of a dog we sometimes see the straining to understand. If only we could explain to the dog—in words—that we are only going around the corner to the movie and will be back in two hours. Does He up there glancing down (we are still so gravitation-minded) and seeing into our hearts, as people used to say, note our own straining to understand? We try with everything besides words, with music, painting, sculpture, haberdashery, test tubes, trigonometric formulae, new translations from the Greek or Hebrew. This body with the head on top seeking to find the meaning of itself, is that the cause of the frown moving across the brow of Jove, or is that brow only the brink of our old hill and the frown only the windy leaf-wrinkled October sky without any communication at all, without rage, without pain, without intent? A mind peering at a mind to find what a mind is, standing before its own glass anxiously to fathom its destiny, should one weep? And if not seeking to fathom, should one still more pityingly weep?

In a migraine attack, whereas two minutes earlier it was subject, verb, object, all in syntactic order, now a noun hangs in the air and will not come down to join its verb, and a sentence already on the way cannot remember where it began or foresee where it will end. Nevertheless, behind that blurred outside a mind sits (as that colored woman in the clinic behind her blocked speech), knows how things are, knows that the purpose of words now temporarily negated is to help a mind maintain its orderly relations to this earth, to those other heads that stray around it. Words help us foster the illusion that we know what we are about, bolster our effrontery to push forward on this journey, to discover if we can that it is not all illusion.

Off to the side bolstering us are Shakespeare and Aeschylus and Pavlov and Sherrington and the wordy St. Thomas Aquinas and Sean O'Casey and Eugene O'Neill. We see their façades, mountainous or neat, and something of what lay behind them.

Another time the words are merely continuing the antics, bells and tassels. The clown goes limp with drink or dreams, opens his mouth, the wind blows in, marvelous sounds come back out, words. Dear clown, when they are best he may not recognize them as his own. May think they are making someone else's music. Whose? The wind's? His own words amaze his own ears.

There arrives that morning, or evening, when each of us lets fall his final words, those his friends repeat to show they remember him. Those, too, can have a rightness that makes them the realest fact about that event. The man may speak them quietly, as Keats, or with ironic gaiety, as Voltaire, or with dignity, as Henry James, may be pretending to himself that he feels no fright, though previously he did always avoid looking straight at this day, his last on the funny earth, questioning how it would be, feared but also hoped that he might have a clear mind, so as to be present at his own death. Now he watches, to the degree he is able, squanders nothing. What he speaks is pure self, however trivial, however reeking of his lifelong greed, lifelong sentimentality, lifelong perplexity, or lifelong tender foresight. Some years ago in a picture magazine there were three pages of death-masks so trenchant that one woke in the night seeing them again. Yet even those masks, it was a satisfaction to know, could not rival the words that had issued from those heads. Under one, Jonathan Swift's, was written Jonathan Swift's own epitaph. Here it is. Here is what Jonathan Swift wanted said. *Ubi Saeva Indignatio Ulterius Cor Lacerare Nequit.* He wanted that cut into the gravestone. *Where savage indignation will gnaw at my heart no more.* His mask did have an inadequacy. It could not speak.

# INDEX

# ABOUT THE AUTHOR

Looking back, it seems all indirection. The earliest memories are scanter than other people's, only occasionally a clear one. Cheese—ridiculous. There was an inborn love of cheese, but what I mostly was fed on was bread; then on a half-dark wintery morning, because it was Christmas, I was permitted to walk what for my legs was miles and when the grocer had weighed the cheese he sliced off a slice from what was my mother's, gave it to me, because it was Christmas, and I could nibble it all the slow way home, and there was the tree.

My first day at school I do not remember, but a ravine through snow higher than my head, I do. The teacher, an old lady of twenty-two, had stopped at our house to give me her hand for the one-and-one-half blocks. A year then or so, and another teacher, a man, with a paunch, each recess leaned with his rattan from a low window-sill and indiscriminately beat boys, and from him I learned how to keep out of touch.

After that I traveled rapidly through high school and *Hamlet*, almost blocked by a father who believed in neither, assisted by a mother who believed in both. Then a professor of physiology with a glass eye and a long finger pointed to me after his lecture, I to come down to where he had been standing in the amphitheatre, and when I did he said severely I should be in the medical school, not in this dental school. So, I was in both. From the latter I earned the money that took me to Harvard briefly, to Europe extensively. Now I was a doctor and a teacher; taught bacteriology, then physiology, then neurophysiology, and after years had the privilege of sitting in seminars of psychiatry. Somewhere in the course of this I had found a pen, and that clarified everything.

*Design by Sidney Feinberg*
*Set in Linotype Caledonia*
*Composed by American Book–Stratford Press, Inc.*
*Printed and bound by Haddon Craftsmen, Inc.*
HARPER & ROW, PUBLISHERS, INC.

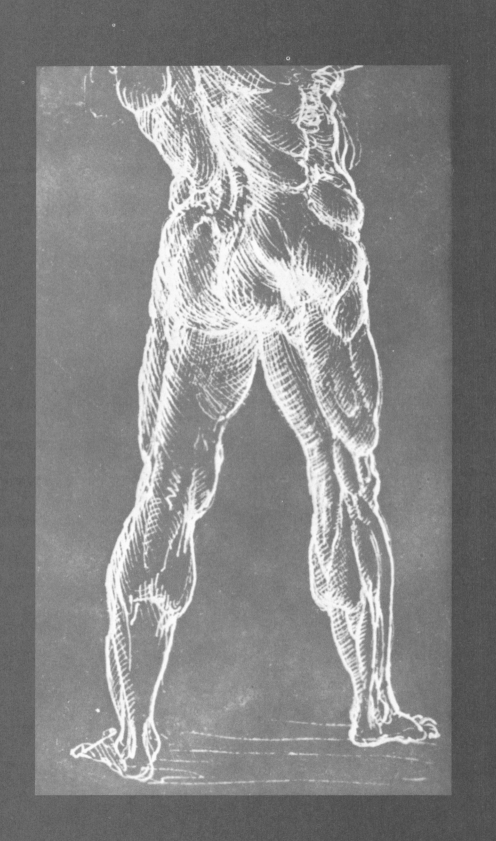